Graduate Texts in Physics

Graduate Texts in Physics publishes core learning/teaching material for graduate-and advanced-level undergraduate courses on topics of current and emerging fields within physics, both pure and applied. These textbooks serve students at the MS- or PhD-level and their instructors as comprehensive sources of principles, definitions, derivations, experiments and applications (as relevant) for their mastery and teaching, respectively. International in scope and relevance, the textbooks correspond to course syllabi sufficiently to serve as required reading. Their didactic style, comprehensiveness and coverage of fundamental material also make them suitable as introductions or references for scientists entering, or requiring timely knowledge of, a research field.

Pierre Michel

Introduction to Laser-Plasma Interactions

Pierre Michel
Lawrence Livermore National Laboratory
Livermore, CA, USA

ISSN 1868-4513   ISSN 1868-4521   (electronic)
Graduate Texts in Physics
ISBN 978-3-031-23426-2   ISBN 978-3-031-23424-8   (eBook)
https://doi.org/10.1007/978-3-031-23424-8

This Springer imprint is published by the registered company Springer Nature Switzerland AG
The registered company address is: Gewerbestrasse 11, 6330 Cham, Switzerland

Dedicated to Ila.

Preface

This book is aimed at students and researchers new to the field of laser-plasma interactions. It is motivated by the general lack of recent textbooks on the subject and by the renewed excitement about the field of inertial confinement fusion (ICF) following the achievement of nuclear fusion ignition in December 2022 at the National Ignition Facility at Lawrence Livermore National Laboratory (LLNL). Excellent and recent textbooks already exist on short-pulse, high-intensity laser-plasma physics; this book focuses on the interaction of long-pulse (nanosecond-scale), high-energy lasers with plasmas in the general context of ICF or high-energy-density science.

Most of the foundations of long-pulse laser-plasma interaction (LPI) physics were established in the 1970s and 1980s; parsing the massive body of literature accumulated on the subject over the past fifty years, and extracting and grasping the key concepts that are actually relevant in current experiments, can be a daunting task. One of the goals of this book is to provide an accessible introduction to some of the most important aspects of LPI physics in view of the current state of ICF research.

I would like to thank the many people who helped bring this project to fruition, starting with Frank Graziani and Félicie Albert with the high-energy-density center at LLNL who provided financial support for me to write this book. I also thank Richard Town, Nino Landen, John Edwards, Mark Herrmann, and the whole ICF team for their patience while I was pursuing this project and for their continued support throughout my sixteen years at LLNL. I gratefully acknowledge all the people I have harassed with my many physics questions while preparing this book, in particular Dick Berger, Tom Chapman, Laurent Divol, Josselin Garnier, William Kruer, Stefan Hüller, Eugene Kur, Denis Pesme, Wojciech Rozmus, Mark Sherlock, Yuan Shi, and David Strozzi.

I was fortunate to have some of the world's leading experts in the field review parts of this book and would like to thank them for their invaluable comments, corrections, and feedback, in particular Ido Barth, Farhat Beg, Jean-Michel Di Nicola, Russ Follett, Matthias Hohenberger, Pascal Loiseau, Paul-Edouard Masson-Laborde, David Montgomery, John Moody, Jason Myatt, Steven Obenschain, Mark

Sherlock, David Strozzi, and David Turnbull. I would also like to thank Andrea Macchi and Stefan Hau-Riege for their valuable advice about the editorial process.

Finally, I would like to thank Sam Harrison at Springer, for his continued support and for his patience (I believe we first discussed this book project at the APS/DPP meeting back in 2018). And above all, I thank my family, particularly Ramya and Ila, for tolerating my absence (mentally if not physically) for so many months.

Livermore, CA, USA   Pierre Michel
February 2023

Contents

Chapter 1
Fundamentals of Optics and Plasma Physics

This chapter provides an introduction to the optics and plasma physics concepts that will be used throughout the rest of the book. The optics section focuses on basic concepts of light wave propagation and Fourier optics, which will be useful when discussing optical smoothing techniques in Chap. 9. After a general introduction of plasma concepts, the plasma physics section emphasizes the description of plasma waves: it presents their fluid and kinetic descriptions, the wave energy, and action concepts and discusses the properties of acoustic waves in multi-species plasmas which can play a major role in laser–plasma instabilities and their mitigation, as discussed in Chap. 7. Electron-ion collisions, which will later play a central role in laser absorption (Chap. 4) and in the saturation of nonlinear kinetic effects (Chap. 10), are introduced next. Finally, the last section introduces the isothermal expansion of plasma in vacuum, which represents a type of plasma profiles often encountered in laser–plasma experiments.

1.1 Basic Principles of Optics and Description of Light Waves

1.1.1 Vacuum Propagation of Light; The Paraxial Wave Equation

In this section we are going to summarize a few basic optics concepts and methods that will be used throughout the rest of this book. The light propagation will be assumed to take place in vacuum for simplicity; propagation in plasmas will be the subject of Chap. 3.

A light wave consists in an oscillating electric and magnetic fields satisfying Maxwell's equations in vacuum:

$$\nabla \cdot \mathbf{E} = 0, \tag{1.1}$$

$$\nabla \cdot \mathbf{B} = 0, \tag{1.2}$$

© The Author(s), under exclusive license to Springer Nature Switzerland AG 2023
P. Michel, *Introduction to Laser-Plasma Interactions*, Graduate Texts in Physics,
https://doi.org/10.1007/978-3-031-23424-8_1

$$\nabla \times \mathbf{E} = -\frac{\partial \mathbf{B}}{\partial t}, \tag{1.3}$$

$$\nabla \times \mathbf{B} = \frac{1}{c^2}\frac{\partial \mathbf{E}}{\partial t}. \tag{1.4}$$

Taking the curl of the curl equations above and using the vector identity $\nabla \times (\nabla \times \mathbf{E}) = \nabla(\nabla \cdot \mathbf{E}) - \nabla^2\mathbf{E}$ leads to the vacuum propagation equation for the light's electric and magnetic fields:

$$(\partial_t^2 - c^2\nabla^2)\mathbf{E} = 0, \tag{1.5}$$

$$(\partial_t^2 - c^2\nabla^2)\mathbf{B} = 0. \tag{1.6}$$

The simplest solution to the wave equation in vacuum is the plane wave, $\mathbf{E} = E_0\mathbf{e}_x\sin(\psi)$, $\mathbf{B} = B_0\mathbf{e}_y\sin(\psi)$, with the oscillating phase term $\psi = k_0 z - \omega_0 t$. Inserting these expressions in Eq. (1.3) leads to $E_0 = (\omega_0/k_0)B_0$, and Eq. (1.4) leads to $\omega_0/k_0 = c$, i.e., $E_0 = cB_0$. The light propagates along $\mathbf{k}_0 = k_0\mathbf{e}_z$, with its electric and magnetic fields orthogonal to each other and to $\mathbf{k}_0$.

The light emitted by lasers is usually quasi-monochromatic and propagates along a well-defined direction $\mathbf{k}_0$. It will be convenient to describe the light wave as a quasi-plane wave,

$$\mathbf{E} = \frac{1}{2}\mathbf{e}_x E_0(\mathbf{r}, t)e^{i\psi} + c.c., \tag{1.7}$$

where $c.c.$ stands for complex conjugate. E_0 represents the envelope of the field and is allowed to vary in space and time—although these variations will typically be on spatial and temporal scales much larger than the "fast" oscillation wavelength $\lambda_0 = 2\pi/k_0$ and period $T_0 = 2\pi/\omega_0$.

The complex notation will facilitate the algebra later on and also allows to implicitly carry a phase term φ in the oscillation: indeed, writing $E_0 = |E_0|e^{i\varphi}$ leads to $\mathbf{E} = |E_0|\mathbf{e}_x\cos(k_0 z - \omega_0 t + \varphi)$. In other words: a wave always includes a constant phase term related to its initial conditions (i.e., such that $E = |E_0|\cos(\varphi)$ at $t = 0$, $z = 0$), but that term usually will not matter and can be "hidden" as the phase of E_0 when using the complex notation above.

An extremely useful approximation for dealing with laser propagation models is the paraxial wave equation. Inserting the expression for the field above into the vacuum wave equation leads to:

$$\left[\partial_t^2 - 2i\omega_0\partial_t - \frac{1}{c^2}\left(\nabla^2 + 2ik_0\partial_z\right)\right]E_0 = 0, \tag{1.8}$$

where we used $\omega_0^2 = k_0^2 c^2$. We will often assume that the light is purely monochromatic and that the envelope E_0 is independent of time (which can often be satisfied at least approximately if one considers temporal variations on time scales much slower than ω_0). This eliminates the time derivatives from the equation. Furthermore, if we assume that the envelope variations are mostly along the transverse directions (rather than along z), corresponding to beams with small divergence, then the $\nabla^2 = \partial_x^2 + \partial_y^2 + \partial_z^2$ term can be approximated as $\nabla^2 \approx \nabla_\perp^2 = \partial_x^2 + \partial_y^2$. This leads to the paraxial wave equation for the field envelope:

$$\boxed{(2ik_0\partial_z + \nabla_\perp^2)E_0 = 0}. \tag{1.9}$$

This equation can easily be solved by taking the 2D Fourier transform with respect to the transverse variables x, y: the equation then becomes

$$\partial_z \hat{E}_0(k_x, k_y, z) = -\frac{ik_\perp^2}{2k_0}\hat{E}_0(k_x, k_y, z) \tag{1.10}$$

(with $k_\perp^2 = k_x^2 + k_y^2$) and admits the following solution for the electric field at a position z as a function of the field at some initial location $z = 0$, expressed back in real coordinates x, y:

$$E_0(x, y, z) = \mathcal{F}^{-1}\left\{\mathcal{F}[E_0(x, y, 0)]\exp\left[-\frac{ik_\perp^2 z}{2k_0}\right]\right\}. \tag{1.11}$$

Here $\mathcal{F}$ denotes a 2D Fourier transform with respect to x and y,

$$\hat{f}(k_x, k_y) = \mathcal{F}\{f(x, y)\} = \frac{1}{2\pi}\int_{-\infty}^{\infty}\int_{-\infty}^{\infty}f(x, y)e^{-i(k_x x + k_y y)}dxdy. \tag{1.12}$$

The solution can also be expressed as a convolution; defining the convolution of two functions f and g as

$$(f * g)(x) = \int_{-\infty}^{\infty}f(\tau)g(x - \tau)d\tau, \tag{1.13}$$

the convolution theorem in 2D is[1]

$$\mathcal{F}\{f(x, y) * g(x, y)\} = 2\pi\,\hat{f}(k_x, k_y)\hat{g}(k_x, k_y). \tag{1.14}$$

Noting that

$$\mathcal{F}^{-1}\left\{\exp\left[-\frac{ik_\perp^2 z}{2k_0}\right]\right\} = -i\frac{k_0}{z}\exp\left[i\frac{k_0 r_\perp^2}{2z}\right], \tag{1.15}$$

---

[1] The factor 2π comes from our choice of a unitary Fourier transform definition for angular frequencies, verifying $\mathcal{F}^{-1}\{\mathcal{F}[f]\} = f$.

Fig. 1.1 Fresnel diffraction geometry, where a light wave's electric field in the plane (x, y, z) is expressed as a function of the field in $(x_0, y_0, z = 0)$. Fresnel diffraction is equivalent to the paraxial wave propagation model

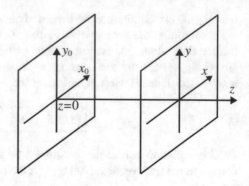

with $r_\perp^2 = x^2 + y^2$, we can thus express the solution for the paraxial wave equation, Eq. (1.11), as:

$$E_0(x, y, z) = E_0(x, y, 0) * h(x, y, z),\tag{1.16}$$

where

$$h(x, y, z) = \frac{-i}{\lambda_0 z} \exp\left[\frac{i k_0 r_\perp^2}{2z}\right].\tag{1.17}$$

Expanding the expression of the solution from the convolution, Eq. (1.16), we immediately recognize the Fresnel diffraction formula,

$$E_0(x, y, z) = -\frac{i}{\lambda_0 z} \iint E_0(x_0, y_0, 0) \exp\left[i\frac{k_0}{2z}\left((x - x_0)^2 + (y - y_0)^2\right)\right] dx_0 dy_0$$

$$\tag{1.18}$$

(cf. Fig. 1.1). In other words: the paraxial wave equation is exactly equivalent to the Fresnel diffraction model for the propagation of light waves.

1.1.2 Fourier Optics

Let us now consider the effect of a lens on a light wave. The lens introduces a dephasing for an incoming light wave; the dephasing depends on the distance from the axis $r_{0\perp} = \sqrt{x_0^2 + y_0^2}$, where (x_0, y_0) are the transverse coordinates near the lens' z coordinate, cf. Fig. 1.2, and can be expressed in the paraxial limit as (the derivation is deferred to Problem 1.1)

Fig. 1.2 Fresnel diffraction geometry, where a light wave's electric field in the plane (x, y, z) is expressed as a function of the field in $(x_0, y_0, z = 0)$. Fresnel diffraction is equivalent to the paraxial wave propagation model

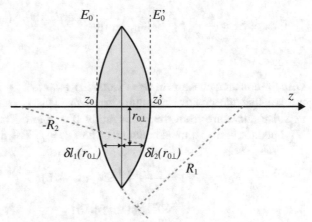

$$\delta\phi(r_{0\perp}) = k_0 n \delta l(0) - k_0(n-1)\frac{r_{0\perp}^2}{2}\left(\frac{1}{R_1} - \frac{1}{R_2}\right). \tag{1.19}$$

Here R_1, R_2 are the radii of curvature of the two surfaces with $R_1 > 0$ and $R_2 < 0$ for a biconvex lens, in accordance with the optics literature; $\delta l(r_{0\perp}) = \delta l_1(r_{0\perp}) + \delta l_2(r_{0\perp})$ is the lens thickness at radius $r_{0\perp}$ (as the sum from the two surfaces, cf. Fig. 1.2), n is the refractive index from the lens and k_0 is the light wave's wave-number in vacuum.

Using the definition of the lens' focal length f from the lensmaker's equation in the thin lens limit,

$$\frac{1}{f} = (n-1)\left(\frac{1}{R_1} - \frac{1}{R_2}\right), \tag{1.20}$$

the field at the output plane of the lens $E_0'(x_0, y_0, z_0')$ is then expressed as a function of the field at the input plane E_0 following

$$E_0'(x_0, y_0, z_0') = E_0(x_0, y_0, z_0)e^{ik_0 n\delta l(0)}\exp\left[-i\frac{k_0 r_{0\perp}^2}{2f}\right]. \tag{1.21}$$

The first exponential term in that expression is just a constant dephasing term; the second exponential corresponds to the curvature of the phase front of the incoming light by the lens, resulting in the focusing of the light.

The propagation of the field E_0' at the lens output up to some distance z is modeled by the paraxial (or Fresnel) propagation model. Inserting the expression for E_0' above into Eq. (1.18) allows us to connect the field after propagation to z to the field at the input plane of the lens E_0 (with $z_0 = 0$):

$$E_0(x, y, z) = \frac{e^{i\phi_0}}{\lambda_0 z} \iint E_0(x_0, y_0, 0) e^{-i\frac{k_0 r_{0\perp}^2}{2f} + i\frac{k_0}{2z}(r_\perp^2 + r_{0\perp}^2 - 2xx_0 - 2yy_0)} dx_0 dy_0 ,$$

$$(1.22)$$

with the constant phase term $\phi_0 = k_0 n \delta l(0) + k_0 r_\perp^2 / 2f - \pi/2$. From this expression we see that the positive curvature from the lens is exactly compensated by the negative curvature from the propagation (i.e., the diffraction associated with the $\nabla_\perp^2$ operator in the paraxial equation) when $z = f$. The field then becomes

$$E_0(x, y, f) = \frac{e^{i\phi_0}}{\lambda_0 z} \iint E_0(x_0, y_0, 0) e^{-i\frac{k_0}{f}(xx_0 + yy_0)} dx_0 dy_0 \qquad (1.23)$$

$$\propto \mathcal{F}\{E_0(x_0, y_0, 0)\} . \qquad (1.24)$$

This is the well-known result from Fourier optics, which shows that the lens performs a Fourier transform from the near-field (defined here as the location of the lens) to the far-field (defined as the best focus location, $z = f$). The 2D transverse image of the light wave amplitude at best focus thus corresponds to the power spectrum (for the transverse spatial dimensions) of the wave incident on the lens.

In our case, since we are most concerned with what happens in the plasma, i.e., at (or near) the best focus location of the lens, it will be more convenient to think of the near-field as the spatial frequency content of the light wave's electric field at best focus (i.e., in the plasma). This is easily done by virtue of the duality of the Fourier transform, which states that if $\mathcal{F}[f(x)] = \hat{f}(k)$ then $\mathcal{F}[\hat{f}(x)] = f(-k)$: thus by doing a change of variable $x_0 \to -x_0, y_0 \to -y_0$ in Eq. (1.23), we obtain:

$$E_0(x, y, f) = \frac{e^{i\phi_0}}{\lambda_0 z} \iint E_0(-x_0, -y_0, 0) e^{i\frac{k_0}{f}(xx_0 + yy_0)} dx_0 dy_0 \qquad (1.25)$$

$$\propto \mathcal{F}^{-1}\{E_0(-x_0, -y_0, 0)\} , \qquad (1.26)$$

with the spatial frequencies in the Fourier transform:

$$k_x = \frac{k_0}{f} x_0 , \qquad (1.27)$$

$$k_y = \frac{k_0}{f} y_0 . \qquad (1.28)$$

In other words: the light's electric field in the near-field (lens input plane) at a transverse location $(-x_0, -y_0)$ gives the spectral component $k_x = k_0 x_0 / f, k_y = k_0 y_0 / f$ of the light at best focus. This is also easily visualized in terms of wave-vector components emerging from the lens, as shown in Fig. 1.3: we see that a ray of light incident on the lens at a transverse location x_0 is redirected toward best focus with a wave-vector $\mathbf{k}$ with a transverse component $k_x \approx -\theta k_0$ in the paraxial

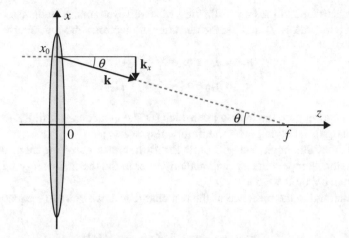

Fig. 1.3 Near-field (in the plane of the lens) vs. far-field (best focus) geometry: a ray of light parallel to z incident on the lens at a transverse location x_0 has an associated wave-vector with an x-component $k_x \approx -\theta k_0 \approx -k_0 x_0 / f$

limit ($\theta \ll 1$). Since we also have $\theta \approx x_0/f$, we get $k_x = -k_0 x_0/\lambda_0$: the field at x_0 at the lens thus corresponds to the spatial frequency component $k_x = -k_0 x_0/f$ for the field at best focus.

1.2 Basic Principles of Plasma Physics

1.2.1 The Plasma State; Debye Length and Screening

A plasma is essentially an ionized gas, i.e., a collection of "free" charged particles not bound to each other but interacting via their Coulomb electric field [1–3]. Plasmas come in many forms: they can be quasi-neutral (i.e., with equal amounts of positive and negative charges on average), or non-neutral, made of multiple species (electrons, positrons, multiple ions), and span a very wide range of temperatures and densities. The plasma state is sometimes called the "fourth state of matter," after the solid, liquid, and gas states. Like for gases, in plasmas the typical kinetic energy of a particle $(1/2)m \langle v^2 \rangle = (3/2)k_B T$ (k_B is the Boltzmann constant and T the temperature, and the brackets denote an ensemble-average) is much greater than its electrostatic potential energy $q\Phi$, where Φ is the electrostatic potential from all the neighboring particles.

A key parameter describing plasmas is the Debye length, which is a physical range within which plasma particles interact via the electrostatic potentials, but beyond which the interactions are "screened." To show this, let us consider a test charge q_t located at the origin of an arbitrary coordinate system $\mathbf{r} = 0$. We assume that the electrons and ions are following Maxwell-Boltzmann statistics, each with their own temperature T_e and T_i; the densities of electrons and ions are then

$$n_e = n_{e0} \exp[e\Phi/k_B T_e] \tag{1.29}$$

$$n_i = n_{i0} \exp[-Ze\Phi/k_B T_i] \,, \tag{1.30}$$

where n_{e0} and n_{i0} are the average densities (or the densities at infinity, away from the test charge), satisfying quasi-neutrality, $n_{e0} = Z n_{i0}$.

Throughout this book, **we will omit the Boltzmann constant from our notations**, i.e., the "temperature" T will actually refer to the thermal energy $k_B T$ and be expressed in eV or keV.[2]

The potential in the presence of the test charge follows Poisson's equation:

$$\nabla^2 \Phi = -\frac{1}{\varepsilon_0} [-e n_e + Z e n_i + q_t \delta(\mathbf{r})] \,, \tag{1.31}$$

where $\delta(\mathbf{r}) = \delta(x)\delta(y)\delta(z)$ is the Dirac delta function. Substituting the expressions for the densities from the Maxwell-Boltzmann distributions above that we Taylor-expand following $n_e \approx n_{e0}[1 + e\Phi/T_e]$, $n_i \approx n_{i0}[1 - Ze\Phi/T_i]$ per our definition of a plasma ($e\Phi \ll T_e$, $Ze\Phi \ll T_i$) yields:

$$\left[\nabla^2 - \frac{1}{\lambda_D^2} \right] \Phi = -\frac{q_t}{\varepsilon_0} \delta(\mathbf{r}) \,, \tag{1.32}$$

where we introduced the "total Debye length" λ_D, defined as

$$\frac{1}{\lambda_D^2} = \frac{1}{\lambda_{De}^2} + \frac{1}{\lambda_{Di}^2} \,, \tag{1.33}$$

and the electron and ion Debye lengths as

$$\lambda_{De}^2 = \frac{\varepsilon_0 T_e}{n_{e0} e^2} \,, \tag{1.34}$$

$$\lambda_{Di}^2 = \frac{\varepsilon_0 T_i}{n_{i0} Z^2 e^2} \,. \tag{1.35}$$

We recognize the left-hand side of Eq. (1.32) as the Helmholtz operator, $\nabla^2 + k^2$ with $k = \pm i/\lambda_D$ in our case. This operator is known to admit the Green's function $G(\mathbf{r}) = -\exp[-ikr]/(4\pi r)$, where G is solution of $(\nabla^2 + k^2)G(\mathbf{r}) = \delta(\mathbf{r})$; this

---

[2] To deal with temperature units in formulas, it will often be easy to introduce the ratio T/mc^2 with m the electron mass: since $mc^2 \approx 511$ keV, the ratio is simply expressed in practical units as T [keV]/511.

leads to the following solution for the potential (chosen to be zero at infinity, i.e., $k = -i/\lambda_D$):

$$\Phi(r) = \frac{q_t}{4\pi\varepsilon_0 r} \exp\left[-\frac{r}{\lambda_D}\right]. \tag{1.36}$$

This shows that for distances from the test charge shorter than the Debye length, $r \ll \lambda_D$, we have $\Phi(r) \approx q_t/(4\pi\varepsilon_0 r)$: we recover the Coulomb potential from the test charge. However, for distances beyond λ_D the potential decays following $\exp[-r/\lambda_D]$ as opposed to $\sim 1/r$: this is the plasma "screening" or "shielding" phenomenon, which means that an electrostatic charge in a plasma will be canceled (i.e., "shielded" from the rest of the plasma) for distances larger than the Debye length. Physically, given the expressions for the electron and ion densities in the Maxwell-Boltzmann distributions above, we see that this corresponds to the formation of a "cloud" of charge of opposite sign surrounding the test charge and cancelling it within a radius of approximately λ_D, as illustrated in Fig. 1.4.

We can verify that the test charge is exactly cancelled by going back to the expression of $n_e \approx n_{e0} + n_{e0}e\Phi/T_e$: the electron density perturbation surrounding the test charge is $\delta n_e = n_{e0}e\Phi/T_e$, showing that for $q_t > 0$, we have $\Phi > 0$ and indeed an accumulation of electrons around the test charge ($\delta n_e > 0$). The charge density perturbation (accounting for electrons only for simplicity) can be integrated over the entire space:

$$\int e\delta n_e(\mathbf{r})d^3r = \int_0^{2\pi} d\varphi \int_0^{\pi} \sin(\theta)d\theta \int_0^{\infty} e\delta n_e(r)r^2 dr$$

$$= \frac{e^2 n_{e0}q_t}{\varepsilon_0 T_e} \int_0^{\infty} re^{-r/\lambda_D} dr = q_t, \tag{1.37}$$

where $d^3r = dxdydz = r^2\sin(\theta)d\theta d\varphi$ with $r = |\mathbf{r}| = |x\mathbf{e}_x + y\mathbf{e}_y + z\mathbf{e}_z|$, meaning that the accumulation of electrons exactly cancels the test charge.

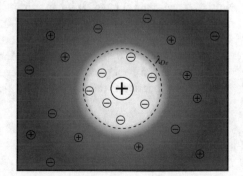

Fig. 1.4 Illustration of Debye shielding: a test charge (shown here as the "+" at the center of the figure) in a plasma tends to attract a "cloud" of opposite charges that cancels its potential and "screens" it from the rest of the plasma. The cloud radius is on the order of λ_{De}

Keep in mind that the picture of a surrounding cloud of charge is somewhat misleading, as in reality the particles in the cloud do not "stick around": rather, we have particles flying by the test charge at high speed that will be slightly deflected toward the charge (for opposite charges) or away from it (for same-sign charges), resulting in a higher average concentration of opposite charge near the test charge.

Now let us go back to our earlier criteria for the plasma state, which was that the kinetic energy of the particles shall remain much larger than the electrostatic interaction force between them, i.e., taking the case of the electrons, $e|\Phi| \ll T_e$ (which we used to Taylor-expand the expression of the densities above). Therefore, a typical electrostatic energy $e\Phi$ will be at most the value for the potential taken at $r \leq \lambda_{De}$, i.e., $\approx e^2/(4\pi\varepsilon_0\lambda_{De})$. The condition $e|\Phi| \ll T_e$ then becomes, after eliminating T_e via the expression for λ_{De} above,

$$\boxed{N_{De} = n_{e0}\lambda_{De}^3 \gg 1} \tag{1.38}$$

(up to a factor 4π).

The plasma state is thus defined by the simple criteria that the number of electrons in the "Debye cube" must be much greater than one. The number of particles in the Debye cube N_{De} is often called the *plasma parameter*.

1.2.2   The Plasma Frequency

When a charged particle (or a collection of charges from the same species) gets displaced from its quasi-neutral position in a plasma, the resulting electrostatic field will pull the charge back to its original position. The charge will then pass its initial position at some finite velocity, overshoot, and return—setting up a harmonic oscillation. The detailed study of plasma oscillations will be carried out in the section on plasma waves, Sect. 1.3, but in the meantime we can give a heuristic derivation of the fundamental oscillation frequency.

Consider an infinitesimally thin slab of plasma electrons displaced from their initial, quasi-neutral location by an infinitesimal amount δz: we then have a slab of positive charge at the initial location of the electron slab (Fig. 1.5) which creates an electric field E that acts to pull the electrons back. The field is estimated via Poisson's equation, $\partial_z E = \rho/\varepsilon_0$ with ρ the charge density, giving $E \approx en_{e0}\delta z/\varepsilon_0$ where n_{e0} is the background electron density. The force equation on the electron slab, whose location is denoted by $\delta z(t)$, then becomes

$$\frac{d^2\delta z}{dt^2} = -\frac{e}{m}E = -\omega_{pe}^2\delta z\,, \tag{1.39}$$

Fig. 1.5 Illustration of the plasma frequency: a slab of electrons displaced by an infinitesimal amount δz will be pulled back toward the region of positive charge left behind, setting up oscillations at ω_{pe}

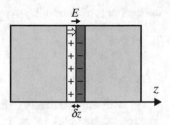

with m the electron mass and where we introduced the so-called plasma frequency ω_{pe} defined as

$$\omega_{pe} = \frac{n_{e0}e^2}{m\varepsilon_0}.$$

(1.40)

Note that if we define a thermal velocity for the electrons and ions as (cf. next section)

$$v_{Te,i} = \sqrt{\frac{T_{e,i}}{m_{e,i}}},$$

(1.41)

the (electron) Debye length takes the simpler form:

$$\lambda_{De} = \frac{v_{Te}}{\omega_{pe}}.$$

(1.42)

In other words: the Debye length is the distance over which an electron at v_{Te} travels during a time interval $1/\omega_{pe}$.

1.2.3 Equilibrium (Maxwellian) Velocity Distributions in Plasmas

In this section we summarize a few definitions related to the velocity distribution of particles in plasmas. The velocity distribution at thermal equilibrium (i.e., after a long enough time that collisions between particles have equilibrated the distribution) for a particle species $s \in \{e, i\}$ for electrons and ions is described by the Maxwellian distribution:

$$f_M(\mathbf{v}) = (2\pi v_{Ts}^2)^{-3/2} \exp\left[-\frac{v^2}{2v_{Ts}^2}\right],$$

(1.43)

with $v^2 = |\mathbf{v}|^2 = v_x^2 + v_y^2 + v_z^2$ and where we defined the thermal velocity as

$$v_{Ts} = \sqrt{\frac{T_s}{m_s}}. \tag{1.44}$$

As with any probability distribution, the meaning of f is that $f(\mathbf{v})d^3v$ is the fraction of particles in an infinitesimal volume element $d^3v = dv_x dv_y dv_z$ in 3D velocity space centered at $f(\mathbf{v})$. The volume element may be expressed more conveniently in spherical coordinates as $d^3v = v^2 \sin(\theta)dv d\theta d\varphi = v^2 dv d\Omega$, where $d\Omega = \sin(\theta)d\theta d\varphi$ is the infinitesimal solid angle element and $v = |\mathbf{v}|$. The distribution is normalized to 1, $\int f_M(\mathbf{v})d^3v = 1$; when we describe the velocity distribution of plasma particles the distribution will be normalized to the background density n_{s0}, in which case the expression for f_M has an extra factor n_{s0} in front.

We can also define a distribution for the *norm* of the velocity vector v, $g(v)$, defined as $\int f_M(\mathbf{v})d^3v = \int_0^\infty g(v)dv$ after integrating over the solid angle; we get:

$$g(v) = \frac{4\pi v^2}{(2\pi v_{Ts}^2)^{3/2}} \exp\left[-\frac{v^2}{2v_{Ts}^2}\right] \tag{1.45}$$

(the details of all the integral calculations in this section are reported as exercises in Problem 1.2).

Likewise, the distribution along one particular direction z can be obtained after integrating over the other two dimensions v_x, v_y, leading to

$$f_M(v_z) = (2\pi v_{Ts}^2)^{-1/2} \exp\left[-\frac{v_z^2}{2v_{Ts}^2}\right]. \tag{1.46}$$

The 1D distribution along one particular direction $f_M(v_z)$ and the distribution of velocity amplitudes $g(v)$ are represented in Fig. 1.6. We shall try not to confuse the distribution of velocity amplitudes $g(v)$ from Eq. (1.45) with the initial distribution $f_M(\mathbf{v})$ in Eq. (1.43) (which also depends only on v). An important feature of $g(v)$ is that the probability of finding a particle in the distribution with zero absolute velocity goes to zero. This is in contrast to the distribution along one particular direction in velocity space, for which the most probable velocity is zero.

Let us establish a few other useful characteristics of the distribution:

- Average velocity $\langle v \rangle$:

The average velocity is defined as $\langle v \rangle = \int v f_M(\mathbf{v})d^3v$, i.e.,

$$\langle v \rangle = \frac{4}{\sqrt{2\pi}} v_{Ts}. \tag{1.47}$$

Fig. 1.6 Maxwellian distribution of particles along one particular direction v_z or as a function of $v = |\mathbf{v}|$

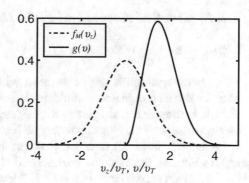

- Most probable velocity v_{mp}:

This is the velocity that maximizes the probability distribution, i.e., such that $\partial_v f_M(\mathbf{v})|_{v_{mp}} = 0$:

$$v_{mp} = \sqrt{2} v_{Ts}. \tag{1.48}$$

- Average kinetic energy $\frac{1}{2} m_s \langle v^2 \rangle$:

We have $\frac{1}{2} m_s \langle v^2 \rangle = \frac{1}{2} m_s \int v^2 f_M(\mathbf{v}) d^3 v$, hence,

$$\frac{1}{2} m_s \langle v^2 \rangle = \frac{3}{2} m_s v_{Ts}^2 = \frac{3}{2} T_s. \tag{1.49}$$

This is a well-known formula, since just like for an ideal monoatomic ideal gas, $(3/2)T$ represents the total internal energy. Since the distribution is isotropic, we also have $\langle v_x^2 \rangle = \langle v_y^2 \rangle = \langle v_z^2 \rangle = v_{Ts}^2 = T_s/m_s$.

To conclude this section we look at the situation of a Maxwellian distribution with a finite drift velocity $\mathbf{V}$,

$$f = (2\pi v_{Ts}^2)^{-3/2} \exp\left[-\frac{(\mathbf{v} - \mathbf{V})^2}{2 v_{Ts}^2}\right]. \tag{1.50}$$

In this case we can easily calculate the average kinetic energy (cf. Problem 1.2) and show that it consists of a thermal component as defined above plus the kinetic energy associated with the average flow velocity:

$$\boxed{\frac{1}{2} m_s \langle v^2 \rangle = \frac{3}{2} m_s v_{Ts}^2 + \frac{1}{2} m_s V^2}. \tag{1.51}$$

1.2.4 Kinetic (Vlasov) and Fluid Descriptions of Plasmas

1.2.4.1 The Vlasov Equation

The Vlasov equation is one of the most widely used descriptions of plasmas. An exact description of plasmas would involve solving the equation of motion for each particle in the plasma, accounting for the interactions between all particles via their electrostatic potentials—and possibly with externally imposed electric or magnetic fields as well. Since this is obviously intractable (except numerically over extremely small volumes), the particles are described via their distribution functions.

For a given particle species $s \in \{e, i\}$ (electrons or ions), the distribution function $f_s(\mathbf{r}, \mathbf{v}, t)$ is a function of seven variables (three-dimensional position $\mathbf{r} = [x, y, z]$, velocity $\mathbf{v} = [v_x, v_y, v_z]$ and time) whose meaning is that $f_s(\mathbf{r}, \mathbf{v}, t)d^3r d^3v$ is the number of particles in the six-dimensional phase-space volume $d^3r d^3v = dx dy dz dv_x dv_y dv_z$ at time t. It is usually normalized to the density of particles in the plasma,

$$n_s(\mathbf{r}, t) = \int f_s(\mathbf{r}, \mathbf{v}, t)d^3v \,. \tag{1.52}$$

A typical distribution function for plasmas at equilibrium is the Maxwellian distribution we described in the previous section, $f_{Ms}(\mathbf{v}) = n_{0s}(2\pi v_{Ts}^2)^{-3/2} \exp[-v^2/2v_{Ts}^2]$, where n_{0s} is the average density for the species s, corresponding to a "final" state of equilibrium due to particles collisions.

The Vlasov equation is the collisionless version of the Boltzmann equation applied to plasmas. The basic idea is that over an infinitesimal time interval δt, the particles initially located at $(\mathbf{r}, \mathbf{v}, t)$ will move to $(\mathbf{r} + \delta\mathbf{r}, \mathbf{v} + \delta\mathbf{v}, t + \delta t)$ with $\delta\mathbf{r} = \mathbf{v}\delta t$ and $\delta\mathbf{v} = \mathbf{F}_s\delta t/m$ where $\mathbf{F}_s$ is the force acting on the particles. Since these are the same particles subjected to the force, their number remains constant,[3] so we must have $f_s(\mathbf{r}, \mathbf{v}, t)d^3r d^3v = f_s(\mathbf{r} + \delta\mathbf{r}, \mathbf{v} + \delta\mathbf{v}, t + \delta t)d^3r d^3v$. This means that the total differential of f_s must be zero, or:

$$df = \frac{\partial f_s}{\partial t}dt + \frac{\partial f_s}{\partial \mathbf{r}} \cdot d\mathbf{r} + \frac{\partial f_s}{\partial \mathbf{v}} \cdot d\mathbf{v} = 0, \tag{1.53}$$

where $\partial_{\mathbf{r}} = (\partial_x, \partial_y, \partial_z)$ and $\partial_{\mathbf{v}} = (\partial_{v_x}, \partial_{v_y}, \partial_{v_z})$. This can then be re-written as

$$\boxed{\left(\partial_t + \mathbf{v} \cdot \nabla + \frac{\mathbf{F}_s}{m_s} \cdot \partial_{\mathbf{v}}\right) f_s(\mathbf{r}, \mathbf{v}, t) = 0} \,, \tag{1.54}$$

where $\mathbf{F}_s = q_s(\mathbf{E} + \mathbf{v}_s \times \mathbf{B})$ is the Lorentz force acting on the particle species.

---

[3] This is only true in the absence of collisions, since collisions are random and will apply differently for each particle, leading to diffusion in phase-space.

This is the Vlasov equation, describing the evolution of the distribution function of a given particle species (electrons or ions) in phase-space. It does not account for collisions: collisions intervene via additional terms to this equation, which then becomes the Fokker-Plank equation. It has the appearance of a total derivative ($Df/Dt = 0$) even though $\mathbf{r}$, $\mathbf{v}$ and t are independent: this illustrates the fundamental concept of the Vlasov equation which is that the phase-space density $f_s(\mathbf{r}, \mathbf{v}, t)d^3rd^3v$ stays constant along the phase-space trajectory of each individual particle "j" in the plasma, i.e.,

$$d_t f\left[\mathbf{r}_j(t), \mathbf{v}_j(t), t\right] = 0,\tag{1.55}$$

where $\mathbf{r}_j(t)$ and $\mathbf{v}_j(t)$ are the (Lagrangian) position and velocity of the particle j at time t.

1.2.4.2   The Fluid Equations

The hydrodynamic quantities describing a plasma, which depend on time and position, are obtained by averaging the quantity of interest weighted by the distribution function of the particles over the velocities. The first quantity is obviously the particle density n_s, which we already defined above as the integral of the distribution function:

$$n_s(\mathbf{r}, t) - \int f_s(\mathbf{r}, \mathbf{v}, t)d^3v.\tag{1.56}$$

More generally, we can express the ensemble-average of any function of velocity $g(\mathbf{r}, \mathbf{v}, t)$, taken over many realizations of the particle velocities, as

$$\langle g\rangle(\mathbf{r}, t) = \frac{1}{n_s(\mathbf{r}, t)}\int g(\mathbf{r}, \mathbf{v}, t)f_s(\mathbf{r}, \mathbf{v}, t)d^3v.\tag{1.57}$$

Next we have the fluid velocity $\mathbf{v}_s$ (not to be confused with the velocity variable $\mathbf{v}$ in phase-space):

$$\mathbf{v}_s(\mathbf{r}, t) = \langle\mathbf{v}\rangle = \frac{1}{n_s}\int \mathbf{v}f_s(\mathbf{r}, \mathbf{v}, t)d^3v.\tag{1.58}$$

And finally, the last quantity of interest to us in the context of this book is the pressure $\bar{\bar{p}}_s$:

$$\bar{\bar{p}}_s(\mathbf{r}, t) = m_s n_s \langle(\mathbf{v} - \mathbf{v}_s)(\mathbf{v} - \mathbf{v}_s)\rangle\tag{1.59}$$

$$= m_s \int (\mathbf{v} - \mathbf{v}_s)(\mathbf{v} - \mathbf{v}_s)f_s(\mathbf{r}, \mathbf{v}, t)d^3v,\tag{1.60}$$

where $(\mathbf{v} - \mathbf{v}_s)(\mathbf{v} - \mathbf{v}_s)$ is a tensor (dyadic). If the distribution function is isotropic then the pressure is a scalar quantity, proportional to the trace of the tensor:

$$p_s(\mathbf{r}, t) = \frac{m_s}{3} n_s \left\langle (\mathbf{v} - \mathbf{v}_s)^2 \right\rangle \tag{1.61}$$

$$= \frac{1}{3} m_s n_s \left[\left\langle \mathbf{v}^2 \right\rangle - v_s^2 \right]. \tag{1.62}$$

Note that if the distribution is Maxwellian, $p_s = n_s T_s$ leads to $\langle \mathbf{v}^2 \rangle = 3T_s/m_s + v_s^2$, consistent with Eq. (1.51).

The fluid equations are obtained by taking the "moments" of the distribution function, which means integrating the Vlasov equation multiplied by increasing powers of $\mathbf{v}$ (i.e., $\mathbf{v}^0 = 1$, $\mathbf{v}$, $\mathbf{v}^2$, etc.) over the velocity.

The first equation is thus obtained by integrating Eq. (1.54) over $\mathbf{v}$; we obtain:

$$\partial_t n_s + \nabla \cdot (n_s \mathbf{v}_s) = 0. \tag{1.63}$$

This is the continuity equation; it can also easily be obtained from first principles without using the Vlasov equation (e.g., cf. [2]). Note that the integration of the third (force) term in the Vlasov equation over velocity is zero due to the boundary conditions at infinity (f_s must vanish at infinite velocities in any direction) and involves an integration by parts for the $\mathbf{v} \times \mathbf{B}$ term.

Next, the second fluid equation is the fluid force equation, obtained by integrating the Vlasov equation multiplied by $\mathbf{v}$ over the velocity:

$$\int \left[\mathbf{v} \partial_t + \mathbf{v}(\mathbf{v} \cdot \nabla) + \mathbf{v} \left(\frac{\mathbf{F}_s}{m_s} \cdot \partial_{\mathbf{v}} \right) \right] f_s d^3 v = 0. \tag{1.64}$$

Integrating the first term in the bracket gives $\partial_t (n_s \mathbf{v}_s)$. The next two terms require more manipulations:

- For the second term, we rewrite $\mathbf{v}(\mathbf{v} \cdot \nabla f_s) = (\mathbf{v}\mathbf{v}) \cdot \nabla f_s = \nabla \cdot (\mathbf{v}\mathbf{v} f_s)$; the first equality comes from the dyadic dot product rule $(\mathbf{a}\mathbf{b}) \cdot \mathbf{c} = \mathbf{a}(\mathbf{b} \cdot \mathbf{c})$ for three vectors $\mathbf{a}$, $\mathbf{b}$, $\mathbf{c}$; the second equality is detailed in a footnote.[4] We obtain:

---

[4] Expressing the dyadic in matrix form, we have

$$(\mathbf{v}\mathbf{v}) \cdot \nabla f_s = \begin{pmatrix} v_x v_x & v_x v_y & v_x v_z \\ v_y v_x & v_y v_y & v_y v_z \\ v_z v_x & v_z v_y & v_z v_z \end{pmatrix} \cdot \nabla f_s = \begin{pmatrix} v_x v_x \partial_x f_s + v_x v_y \partial_y f_s + v_x v_z \partial_z f_s \\ v_y v_x \partial_x f_s + v_y v_y \partial_y f_s + v_y v_z \partial_z f_s \\ v_z v_x \partial_x f_s + v_z v_y \partial_y f_s + v_z v_z \partial_z f_s \end{pmatrix};$$

the j-th vector element can be expressed as follows, with an implicit summation over $i \in \{x, y, z\}$: $[(\mathbf{v}\mathbf{v}) \cdot \nabla f_s]_j = v_j v_i \partial_i f_s = \partial_i v_j v_i f_s = [\nabla \cdot (\mathbf{v}\mathbf{v} f_s)]_j$, hence the result (keep in mind that $\mathbf{v}$ is a variable independent of $\mathbf{r}$, i.e., ∇ does not act on $\mathbf{v}$).

$$\int \mathbf{v}(\mathbf{v} \cdot \nabla f_s) d^3 v = \nabla \cdot (n_s \langle \mathbf{vv} \rangle) \tag{1.65}$$

$$= \nabla \cdot [n_s \langle (\mathbf{v} - \mathbf{v}_s + \mathbf{v}_s)(\mathbf{v} - \mathbf{v}_s + \mathbf{v}_s) \rangle] \tag{1.66}$$

$$= \nabla \cdot [n_s \langle (\mathbf{v} - \mathbf{v}_s)(\mathbf{v} - \mathbf{v}_s) \rangle + n_s \mathbf{v}_s \mathbf{v}_s] \tag{1.67}$$

$$= \frac{\nabla \cdot \bar{\bar{p}}_s}{m_s} + n_s \mathbf{v}_s (\nabla \cdot \mathbf{v}_s) + (\mathbf{v}_s \cdot \nabla)(n_s \mathbf{v}_s), \tag{1.68}$$

where we used $\langle \mathbf{v} - \mathbf{v}_s \rangle = 0$, $\mathbf{v}(\mathbf{v} \cdot \nabla n_s) = \nabla \cdot (\mathbf{vv} n_s)$ (dot product rule) and
$\nabla \cdot (n_s \mathbf{v}_s \mathbf{v}_s) = n_s \mathbf{v}_s (\nabla \cdot \mathbf{v}_s) + (\mathbf{v}_s \cdot \nabla)(n_s \mathbf{v}_s)$ (cf. footnote[5] for details).

- For the third term, using $\mathbf{F}_s = q_s(\mathbf{E} + \mathbf{v} \times \mathbf{B})$ we can show after integrating by parts that

$$\int \mathbf{v} \left(\frac{\mathbf{F}_s}{m_s} \cdot \partial_{\mathbf{v}} \right) f_s d^3 v = -n_s q_s (\mathbf{E} + \mathbf{v}_s \times \mathbf{B}). \tag{1.69}$$

Putting the three terms back together, expanding $\partial_t(n_s \mathbf{v}_s) = n_s \partial_t \mathbf{v}_s - \mathbf{v}_s [\nabla \cdot (n_s \mathbf{v}_s)]$ (where we used the continuity equation) finally leads to the force equation:

$$[\partial_t + (\mathbf{v}_s \cdot \nabla)]\mathbf{v}_s = \frac{q_s}{m_s}(\mathbf{E} + \mathbf{v}_s \times \mathbf{B}) - \frac{\nabla \cdot \bar{\bar{p}}_s}{m_s n_s}. \tag{1.70}$$

For most laser–plasma interaction processes in unmagnetized plasmas, the distribution function will typically be isotropic and pressure a scalar quantity, $\bar{\bar{p}}_s \to p_s$ as defined above.

In principle one can continue to derive more fluid equations from the moments of the Vlasov equation; the next one, integrating the Vlasov equation multiplied by $\frac{1}{2} m_s \mathbf{v}^2$, provides the energy equation (connecting the time evolution of the temperature to the heat flux), etc.

We can see that taking the moment of the Vlasov equation introduces a fluid quantity that belongs to the next order moment equation, thus forming an infinite set of coupled equations. At some point this requires a truncation, which in our case will simply take the form of an equation of state for the electron or ion fluid. The simplest case is the isothermal fluid, where $\nabla p_s = \nabla(n_s T_s) = T_s \nabla n_s$. For a non-isothermal fluid, by analogy with ideal gases, we define the adiabatic index γ_s such that $p_s \propto n_s^{\gamma_s}$. The adiabatic index is also related to the number of degrees of freedom of the fluid N_s via

$$\gamma_s = 1 + \frac{2}{N_s}. \tag{1.71}$$

---

[5] The j-th vector component $[\nabla \cdot (n_s \mathbf{v}_s \mathbf{v}_s)]_j = \partial_i (n_s v_{sj} v_{si}) = (n_s v_{sj}) \partial_i v_{si} + v_{si} \partial_i (n_s v_{sj})$, from which we easily get by identification $\nabla \cdot (n_s \mathbf{v}_s \mathbf{v}_s) = n_s \mathbf{v}_s (\nabla \cdot \mathbf{v}_s) + (\mathbf{v}_s \cdot \nabla)(n_s \mathbf{v}_s)$.

For example, we will describe 1D fluid motion from a plasma wave in the next section, corresponding to oscillations of infinite "sheets" of plasma in the normal direction: in that case we have $N_s = 1$ and $\gamma_s = 3$. The expression for the pressure gradient then simplifies to $\nabla p_s = \nabla(Cn_s^{\gamma_s}) = C\gamma_s n_s^{\gamma_s-1}\nabla n_s = \gamma_s T_s \nabla n_s$ (where C is a constant and $p_s = n_s T_s = Cn_s^{\gamma_s}$).

The expression is similar when the fluid is assumed isothermal if we take $\gamma_s = 1$ (since in this case we simply have $\nabla p_s = T_s \nabla n_s$), so that in summary, the pressure gradient can simply be written as

$$\nabla p_s = \gamma_s T_s \nabla n_s \qquad (1.72)$$

with $\gamma_s = 1$ corresponding to the isothermal fluid.

Most of the processes discussed in this book will rely on the first two hydrodynamic equations only, continuity and force, which take the following form:

$$\partial_t n_s + \nabla \cdot (n_s \mathbf{v}_s) = 0, \qquad (1.73)$$

$$[\partial_t + (\mathbf{v}_s \cdot \nabla)]\mathbf{v}_s = \frac{q_s}{m_s}(\mathbf{E} + \mathbf{v}_s \times \mathbf{B}) - \gamma_s v_{Ts}^2 \frac{\nabla n_s}{n_s}, \qquad (1.74)$$

where $v_{Ts}^2 = T_s/m_s$.

1.3 Waves in Plasmas

1.3.1 Fluid Description: Dielectric Framework

As we shall see below, three types of waves can exist in a plasma (in the absence of external magnetic fields): (i) light waves or EMWs (for electromagnetic waves), with an oscillating electric field perpendicular to the wave propagation; (ii) electron-plasma waves or EPWs, with an electric field oscillating parallel to the propagation direction on fast time scales corresponding to electron motion; and (iii) ion acoustic waves (IAWs), also with an oscillating electric field parallel to the wave propagation direction but oscillating on the slow time scale of ion motion. Each wave is described by a dispersion relation, i.e., a relation that connects the wave frequency ω to its wave-vector $\mathbf{k}$, such that the wave is associated with an oscillation $\propto \cos(\omega t - \mathbf{k} \cdot \mathbf{r})$. We will first give a derivation of the dispersion relations in terms of the plasma dielectric quantities (similar to Refs. [1, 4]). In a following section we will also show a more direct derivation, similar to Ref. [5]; the assumptions used in this latter derivation will be better understood after being familiar with the more comprehensive dielectric treatment, which we present now.

1.3.1.1 Plasma as a Dielectric Medium

The motion of particles in plasma is connected to the fields via Maxwell's equations:

$$\nabla \cdot \mathbf{E} = \sum_s \frac{q_s n_s}{\varepsilon_0} \tag{1.75}$$

$$\nabla \cdot \mathbf{B} = 0 \tag{1.76}$$

$$\nabla \times \mathbf{E} = -\partial_t \mathbf{B} \tag{1.77}$$

$$\nabla \times \mathbf{B} = \sum_s \frac{\mathbf{j}_s}{\varepsilon_0 c^2} + \frac{1}{c^2} \partial_t \mathbf{E} . \tag{1.78}$$

By taking the curl of Eq. (1.77) and eliminating $\mathbf{B}$ from Eq. (1.78), we obtain after using a vector identity the following equation connecting the electric field to the current:

$$\left[\partial_t^2 - c^2 \nabla^2 + c^2 \nabla(\nabla \cdot) \right] \mathbf{E}(\mathbf{r}, t) = -\frac{1}{\varepsilon_0} \sum_s \partial_t \mathbf{j}_s(\mathbf{r}, t) . \tag{1.79}$$

Or, in Fourier space, with the hat denoting a Fourier transform:

$$\left[-\omega^2 + k^2 c^2 - c^2 \mathbf{k}(\mathbf{k} \cdot) \right] \hat{\mathbf{E}}(\mathbf{k}, \omega) = \frac{i\omega}{\varepsilon_0} \sum_s \hat{\mathbf{j}}_s(\mathbf{k}, \omega) . \tag{1.80}$$

By analogy with the macroscopic formulation of Maxwell's equations in matter, we can define the relative electric permittivity (more frequently referred to as the dielectric constant in the plasma physics literature[6]) ε such that

$$\nabla \times \mathbf{B} = \mu_0 \partial_t \mathbf{D} , \quad \mathbf{D} = \varepsilon_0 \varepsilon \mathbf{E} , \tag{1.81}$$

with $\mathbf{D}$ the electric displacement vector, which satisfies the following relation:

$$\nabla \cdot \mathbf{D} = 0 . \tag{1.82}$$

In general ε is a tensor, but for a non-magnetized, isotropic plasma it is simply a scalar. Using the Ampère-Maxwell law (the fourth of Maxwell's equations above) gives

$$\mathbf{D} = \varepsilon_0 \mathbf{E} + \int \mathbf{j}(t) dt = \varepsilon_0 \varepsilon \mathbf{E} . \tag{1.83}$$

---

[6] The permittivity is usually defined as $\varepsilon = \varepsilon_0 \varepsilon_r$, with ε_r the *relative* permittivity. However, for consistency with the rest of the plasma physics literature, we will refer to the relative permittivity as ε through this book, and call it the dielectric constant to avoid confusion.

Introducing the electric conductivity σ such that $\mathbf{j} = \sigma\mathbf{E}$ with $\mathbf{j} = \sum_s \mathbf{j}_s$ the total current from all the particle species, we obtain the following expression for the dielectric constant after a Fourier transform (we drop the hat from our notations for simplicity—it should be obvious from context whether a variable is in real or Fourier space):

$$\varepsilon = 1 + i\frac{\sigma}{\varepsilon_0\omega}. \tag{1.84}$$

Finally, we also define the dielectric susceptibility χ as

$$\varepsilon = 1 + \chi, \tag{1.85}$$

which like the dielectric constant represents the plasma response to an applied electric field. As we shall see below, the susceptibility is the sum of the contributions from the different particle species in the plasma (electrons and different types of ions), $\chi = \sum_s \chi_s$.

To describe plasma waves we look for the linear response of the plasma (via the perturbation in its fluid quantities n_s or $\mathbf{v}_s$) to the fields. We use a perturbative expansion: the zero-order quantities correspond to the equilibrium state, with a background density and flow velocity assumed uniform and constant (in the following we will take the background flow velocity to be zero for simplicity; we will derive similar equations with a finite background flow in Sect. 1.3.2). The first-order quantities are the response to the fields, i.e.,

$$n_s = n_{s0} + \delta n_s, \tag{1.86}$$

$$\mathbf{v}_s = 0 + \delta\mathbf{v}_s, \tag{1.87}$$

$$\mathbf{E} = 0 + \mathbf{E}, \tag{1.88}$$

$$\mathbf{B} = 0 + \mathbf{B}. \tag{1.89}$$

For each particle species s ($s = e, i$, also including the different types of ions in case of multi-species plasmas), the first two fluid equations (continuity and force) derived in the previous section are:

$$\partial_t n_s + \nabla \cdot (n_s\mathbf{v}_s) = 0, \tag{1.90}$$

$$[\partial_t + (\mathbf{v}_s \cdot \nabla)]\mathbf{v}_s = -\gamma_s v_{Ts}^2 \frac{\nabla n_s}{n_s} + \frac{q_s}{m_s}(\mathbf{E} + \mathbf{v}_s \times \mathbf{B}). \tag{1.91}$$

Linearizing these equations to first order gives for each particle species s, after performing a Fourier transform in space and a Laplace transform in time (with $\omega = -is$, where s is the usual complex Laplace variable):

$$\frac{\delta n_s}{n_{s0}} = \frac{\mathbf{k} \cdot \delta \mathbf{v}_s}{\omega}, \tag{1.92}$$

$$\omega \delta \mathbf{v}_s = \gamma_s v_{Ts}^2 \frac{\delta n_s}{n_{s0}} \mathbf{k} + i \frac{q_s}{m_s} \mathbf{E}. \tag{1.93}$$

Combined together, these two equations relate either the density perturbation δn_s or the velocity $\delta \mathbf{v}_s$ to the electric field:

$$\left[\omega^2 - \gamma_s v_{Ts}^2 k^2 \right] \frac{\delta n_s}{n_{s0}} = i \frac{q_s}{m_s} \mathbf{k} \cdot \mathbf{E}, \tag{1.94}$$

$$\left[\omega^2 - \gamma_s v_{Ts}^2 \mathbf{k}(\mathbf{k} \cdot) \right] \delta \mathbf{v}_s = i \frac{q_s \omega}{m_s} \mathbf{E}. \tag{1.95}$$

One can distinguish two types of waves in a plasma (or in general): *transverse* waves, where the oscillation associated with the wave is directed perpendicular to the propagation of the wave, and *longitudinal* waves, whose oscillations are along the direction of propagation. Assuming that the waves are described by a simple harmonic oscillation of the electric field, $\mathbf{E} \propto \exp[i(\mathbf{k} \cdot \mathbf{r} - \omega t)]$, a transverse wave is thus described by $\mathbf{k} \cdot \mathbf{E} = 0$, whereas for a longitudinal wave we have $\mathbf{k} \times \mathbf{E} = 0$.

1.3.1.2 Transverse (Electromagnetic) Waves

Let us first consider the transverse waves: these are the light waves propagating in plasmas. Since $\mathbf{k} \cdot \mathbf{E} = 0$, Poisson's equation tells us that there is no charge separation associated with transverse waves. The fluid velocity associated with the wave is also transverse, $\mathbf{k} \cdot \delta \mathbf{v}_s = 0$, and therefore Eq. (1.92) implies that $\delta n_s = 0$: the density remains constant in the presence of a light wave. Note that this is only true in the linear limit considered here (nonlinear effects will be introduced in later chapters), and if the background density is spatially uniform (an EMW propagating in a density gradient at a finite incidence angle can acquire a longitudinal component, which is at the origin of the resonance absorption phenomenon which will be discussed in Sect. 4.8).

Equation (1.93) then leads to the expression for the velocity of the particle species s:

$$\delta \mathbf{v}_{s\perp} = i \frac{q_s}{m_s \omega} \mathbf{E}. \tag{1.96}$$

This velocity corresponds to a quiver oscillation of the plasma particles due to the Lorentz force from the electric field (we will study the single particle dynamics in Sect. 2.1). From this it also results that $|\delta v_i / \delta v_e| = Zm/M_i \ll 1$, with M_i the ion mass: the quiver velocity of the electrons is largely superior to the ions' due to their smaller mass. As a result, the current is primarily dominated by the electron velocity, $\delta \mathbf{j}_s \approx \delta \mathbf{j}_e$, which implies $\delta \mathbf{j}_e = -en_{e0}\delta \mathbf{v}_e = \sigma_\perp \mathbf{E}$ (the $\perp$ subscript denotes quantities associated with transverse waves, as opposed to the $\parallel$ for longitudinal waves discussed later). Inserting the expression for the velocity above gives the electric conductivity:

$$\sigma_\perp = i\frac{\varepsilon_0}{\omega}\omega_{pe0}^2, \tag{1.97}$$

with $\omega_{pe0}^2 = n_{e0}e^2/(m\varepsilon_0)$. Inserting this conductivity in the expression for the current in Eq. (1.80), $\mathbf{j} = i(\varepsilon_0/\omega)\omega_{pe0}^2\mathbf{E}$, gives:

$$\left(-\omega^2 + \omega_{pe0}^2 + k^2c^2\right)\mathbf{E} = 0, \tag{1.98}$$

from which we obtain both the dispersion relation of light waves and their wave equation:

$$\omega^2 = \omega_{pe0}^2 + k^2c^2, \tag{1.99}$$

$$\left(\partial_t^2 + \omega_{pe0}^2 - c^2\nabla^2\right)\mathbf{E}(\mathbf{r}, t) = 0. \tag{1.100}$$

From Eqs. (1.84), (1.85) we also get the dielectric constant and the susceptibility for light waves in plasmas:

$$\boxed{\varepsilon_\perp = 1 - \frac{\omega_{pe0}^2}{\omega^2}}, \tag{1.101}$$

and

$$\boxed{\chi_\perp \approx \chi_{\perp e} = -\frac{\omega_{pe0}^2}{\omega^2}}. \tag{1.102}$$

Note that technically the dielectric constant should also include the ion susceptibility, $\varepsilon_\perp = 1 + \chi_{\perp e} + \chi_{\perp i}$, with $\chi_{\perp i} = -\omega_{pi0}^2/\omega^2$, however, as we discussed above the electron quiver in the light wave's electric field is much faster than the ions' and therefore $|\chi_{\perp i}/\chi_{\perp e}| = Zm/M_i \ll 1$.

An EMW can be described by several quantities: its electric field $\mathbf{E}$, the first-order fluid velocity $\delta \mathbf{v}_e$ related to $\mathbf{E}$ via Eq. (1.96), or the vector potential $\mathbf{A}$ defined as $\mathbf{B} = \nabla \times \mathbf{A}$ and $\mathbf{E} = -\partial_t\mathbf{A} - \nabla\Phi$. The vector potential is particularly convenient for laser–plasma physics when using the Coulomb gauge, since we then have $\nabla \cdot \mathbf{A} = 0$, meaning that $\mathbf{A}$ is exclusively a transverse quantity, which eliminates any ambiguity on the nature of the waves described by it (as opposed to $\mathbf{E}$, which can describe either a transverse or a longitudinal wave). It is also convenient to introduce the normalized vector potential **a** defined as

$$\mathbf{a} = \frac{e}{mc}\mathbf{A}. \tag{1.103}$$

With this normalization, the amplitude of the quiver velocity of the electron fluid δv_e in the oscillating field of the light wave is simply, from Eq. (1.96), $|\delta v_e| = c|a|$. In other words, $|a|$ is the ratio of the quiver velocity to the speed of light. In practical units, for a plane wave of the form $a = a_0 \cos(k_0 z - \omega_0 t)$, we have

$$\boxed{a_0 \approx 0.855 \times 10^{-9} \sqrt{I[\text{W/cm}^2]\lambda_0^2[\mu\text{m}]}}. \tag{1.104}$$

Relativistic effects start to occur when $a_0 \approx 1$, as will be discussed in Sect. 2.1.

The dispersion relation Eq. (1.99) indicates that a light wave can only propagate in a plasma if its frequency ω is larger than ω_{pe}: otherwise, $k = (\omega/c)\sqrt{1 - \omega_{pe}^2/\omega^2}$ becomes imaginary, $k = i|k|$, which for a monochromatic light wave of the form $A = \frac{1}{2}A_0 \exp[i(kz - \omega t)] + c.c.$ means that $A \propto \exp[-|k|z]$: the wave becomes evanescent and cannot propagate.[7]

Since the plasma frequency $\omega_{pe} = n_e e^2/(m\varepsilon_0)$ is directly connected to the electron density, a light wave with a fixed frequency ω will only be able to propagate in a plasma at densities up to a *critical density* n_c for which the plasma frequency equals the light wave frequency,

$$\boxed{n_c = \frac{m\varepsilon_0\omega^2}{e^2} \quad, \quad \frac{n_e}{n_c} = \frac{\omega_{pe}^2}{\omega^2}}. \tag{1.105}$$

The critical density is given in practical units by $n_c[\text{cm}^{-3}] = 1.115 \times 10^{21}/\lambda_\mu^2$, with λ_μ the wavelength in microns.

The dispersion relation also provides the phase and group velocities of light waves in plasma, $v_\phi = \omega/k$ and $v_g = \partial\omega/\partial k$:

$$v_\phi = \frac{c}{n}, \tag{1.106}$$

$$v_g = cn, \tag{1.107}$$

where n is the plasma refractive index,

$$n = \sqrt{1 - \frac{n_e}{n_c}}. \tag{1.108}$$

---

[7] There is, however, a finite residual electric field over a "skin depth" region, as will be discussed in Sect. 3.2.

1.3.1.3   Longitudinal (Plasma) Waves: General Description

We can proceed in a similar way for longitudinal waves, for which $\mathbf{k} \times \mathbf{E} = 0$. Applying $\mathbf{k}\times$ to Eq. (1.93) gives $\mathbf{k} \times \delta\mathbf{v}_s = 0$, which after using the vector identity $\mathbf{a}(\mathbf{b} \cdot \mathbf{c}) = \mathbf{c}(\mathbf{a} \cdot \mathbf{b}) + \mathbf{b} \times (\mathbf{a} \times \mathbf{c})$ gives $(\mathbf{k} \cdot \delta\mathbf{v}_s)\mathbf{k} = k^2\delta\mathbf{v}_s$. Inserting in Eq. (1.95) leads to

$$\delta\mathbf{v}_s = i\frac{q_s\omega}{m_s}\frac{\mathbf{E}}{\omega^2 - \gamma_s v_{Ts}^2 k^2}\,, \tag{1.109}$$

leading to the electric conductivity of longitudinal waves:

$$\sigma_{/\!/} = i\omega\varepsilon_0 \sum_s \frac{\omega_{ps0}^2}{\omega^2 - \gamma_s v_{Ts}^2 k^2}\,. \tag{1.110}$$

This finally gives the following expressions for the dielectric constant and susceptibility for longitudinal waves in plasmas:

$$\boxed{\varepsilon_{/\!/} = 1 + \chi_{/\!/}\,, \quad \chi_{/\!/} = \sum_s \chi_{/\!/s}\,, \tag{1.111}}$$

$$\boxed{\chi_{/\!/s} = -\frac{\omega_{ps0}^2}{\omega^2 - \gamma_s v_{Ts}^2 k^2}\,. \tag{1.112}}$$

Inserting back into the wave equation (1.80) leads to the general dispersion relation for longitudinal waves in plasmas:

$$\varepsilon_{/\!/}\mathbf{E} = 0\,, \tag{1.113}$$

where we used $\mathbf{k}(\mathbf{k} \cdot \mathbf{E}) = k^2\mathbf{E}$ since $\mathbf{E}$ is longitudinal.[8]

As we will see below, longitudinal waves in plasma (which we will simply refer to as "plasma waves") exist in two different forms: electron plasma waves (EPW), which are associated with fast oscillations of electrons on time scales faster than the ion motion, and ion acoustic waves (IAW), corresponding to ion motion on longer time scales.

Note that unlike for (transverse) light waves, (longitudinal) plasma waves do lead to a density perturbation ($\delta n_s \neq 0$) and create charge separation in the plasma. The oscillation of particles in a transverse vs. longitudinal wave is illustrated in Fig. 1.7; obviously in a real plasma particles would not be regularly spaced like here,

---

[8] Separating a variable $\mathbf{X}$ into a longitudinal and transverse part, $\mathbf{X} = \mathbf{X}_{/\!/} + \mathbf{X}_\perp$, we can write $\mathbf{k}(\mathbf{k} \cdot \mathbf{X}) = \mathbf{k}(\mathbf{k} \cdot \mathbf{X}_{/\!/}) = k^2\mathbf{X}_{/\!/} + \mathbf{k} \times (\mathbf{k} \times \mathbf{X}_{/\!/}) = k^2\mathbf{X}_{/\!/}$, where the first equality used the fact that $\mathbf{k} \cdot \mathbf{X}_\perp = 0$, the second used the vector identity $\mathbf{a}(\mathbf{b} \cdot \mathbf{c}) = \mathbf{c}(\mathbf{a} \cdot \mathbf{b}) + \mathbf{b} \times (\mathbf{a} \times \mathbf{c})$ and the third used $\mathbf{k} \times \mathbf{X}_{/\!/} = 0$. The equivalent relation in real space is simply $\nabla(\nabla \cdot \mathbf{X}) = \nabla^2\mathbf{X}_{/\!/}$.

Fig. 1.7 Illustration of particle motion in a (**a**) transverse (i.e., EMW) vs. (**b**) longitudinal (i.e., EPW or IAW) wave

however, this illustrates the general behavior where particles oscillate transverse to the propagation of a transverse wave, whereas they oscillate as "sheets" like an accordion in a longitudinal wave, creating regions of high vs. low density and generating space charge fields.

Like for light waves, plasma waves (EPWs or IAWs) can be described by several quantities: their electric field $\mathbf{E}$, the fluid velocity $\delta\mathbf{v}_s$ given by Eq. (1.109), the potential Φ such that $\mathbf{E} = -\nabla\Phi$ (again using the Coulomb gauge, such that Φ unambiguously describes a longitudinal wave whereas $\mathbf{A}$ represents a light wave), or the density perturbation δn_s connected to the other quantities by Eqs. (1.92) or (1.94) (unlike for EMWs, the density perturbation for a plasma wave is non-zero). Next we briefly present the two types of longitudinal waves existing in plasmas: EPWs and IAWs.

1.3.1.4 Electron Plasma Waves (EPWs)

Electron plasma waves correspond to oscillations of electrons on time scales that are too fast for the ions to respond, so that we can ignore their contribution and treat them as a static neutralizing background. We look for plane waves of the form $\mathbf{E} \propto \exp[i(\mathbf{k} \cdot \mathbf{r} - \omega t)]$; furthermore, we assume that the phase velocity of the EPWs is much faster than the thermal velocity of the electrons, $\omega/k \gg v_{Te} = \sqrt{T_e/m}$: this assumption ensures that the wave can freely propagate without being subject to strong Landau damping, which we will come back to later in this chapter. It also implies that electrons moving at velocities near v_{Te} or less (i.e., most of the electrons in the distribution) will only travel a small fraction of a wavelength over an oscillation period, preventing temperature equilibration between the high and low density regions (i.e., high- and low-compression) associated with the wave (illustrated in Fig. 1.7b): the electron fluid is therefore adiabatic, and we can use $\gamma_e = 3$, corresponding to a single degree of freedom (along the direction of the wave propagation). This also means that the wave carries a temperature modulation as well as a density modulation: indeed, under our simple closure of the hydrodynamics equations using the ideal gas equation of state $p_e = n_e T_e \propto n_e^3$ (cf. Sect. 1.2.4.2) we have

$$\frac{\delta T_e}{T_e} = 2\frac{\delta n_e}{n_e}. \tag{1.114}$$

This means that the compression from the wave in the high-density regions leads to adiabatic heating of these regions—and likewise, decompression in the low density regions leads to adiabatic cooling of the electron fluid. Note that this model ignores the more complicated effects associated with heat transport, which have been omitted here due to our truncation of the hydrodynamics equations by the equation of state.

The adiabatic assumption $\gamma_e = 3$ will be confirmed more rigorously when we derive the dispersion relation for EPWs using a kinetic model in the next section. We will also see that it is equivalent to the condition

$$k\lambda_{De} \ll 1, \tag{1.115}$$

where $\lambda_{De} = v_{Te}/\omega_{pe}$ is the Debye length. Assuming furthermore that $v_{Ti} \ll v_{Te}$, which is generally true due to the small electron to ion mass ratio, we have the ordering $\omega/k \gg v_{Te} \gg v_{Ti}$, and the susceptibility becomes (dropping the $\parallel$ sign for clarity)

$$\chi_e = -\frac{\omega_{pe0}^2}{\omega^2 - 3k^2 v_{Te}^2}, \quad |\chi_e| \gg |\chi_i|, \tag{1.116}$$

and the dielectric constant for EPWs is $\varepsilon = 1 + \chi_e$.

Inserting into the dispersion relation of longitudinal plasma waves Eq. (1.113) then directly leads to the EPW dispersion relation and wave equation:

$$\omega^2 = \omega_{pe0}^2 + 3k^2 v_{Te}^2, \tag{1.117}$$

$$\left(\partial_t^2 + \omega_{pe0}^2 - 3v_{Te}^2\nabla^2\right)\mathbf{E}(\mathbf{r}, t) = 0. \tag{1.118}$$

Just like for light waves, the dispersion relation indicates that an EPW can only propagate in a plasma if its frequency is larger than the plasma frequency. We will see in Sect. 3.4 that for a given wave frequency ω, a light wave or an EPW at ω follow the same trajectory in a plasma of varying density.

The phase and group velocities of EPWs are given by:

$$v_\phi = v_{Te}\frac{\sqrt{1 + 3k^2\lambda_{De}^2}}{k\lambda_{De}}, \tag{1.119}$$

$$v_g = 3v_{Te}\frac{k\lambda_{De}}{\sqrt{1 + 3k^2\lambda_{De}^2}}. \tag{1.120}$$

Note that the phase and group velocities can also be expressed in a form similar to the velocities of EMWs, $v_\phi = \sqrt{3}v_{Te}/\eta$ and $v_g = \sqrt{3}v_{Te}\eta$, where $\eta = \sqrt{1 - \omega_{pe}^2/\omega^2} = \sqrt{3}k\lambda_{De}/\sqrt{1 + 3k^2\lambda_{De}^2}$ is the analogue of the refractive index for EPWs.

Since an EPW can be described by several quantities, and the literature uses any of these quantities depending on the authors' preferences, it will be useful to summarize the relationships between them. These quantities are the electric field **E**, the electrostatic potential Φ, the fluid velocity perturbation δv_e and the density modulation $\delta n_e/n_{e0}$. We have:

$$\Phi = -\frac{m}{e}\frac{\omega_{pe}^2}{k^2}\frac{\delta n_e}{n_{e0}} = -\frac{m}{e}\frac{v_\phi^2}{1 + 3k^2\lambda_{De}^2}\frac{\delta n_e}{n_{e0}}, \tag{1.121}$$

$$\delta\mathbf{v}_e = -i\frac{e\omega\mathbf{E}}{m\omega_{pe}^2} = (1 + 3k^2\lambda_{De}^2)\mathbf{v}_{os,e}, \tag{1.122}$$

$$\delta\mathbf{v}_e = -\frac{e\mathbf{v}_\phi}{mv_\phi^2}(1 + 3k^2\lambda_{De}^2)\Phi, \tag{1.123}$$

$$\delta\mathbf{v}_e = \mathbf{v}_\phi\frac{\delta n_e}{n_{e0}}, \tag{1.124}$$

where $\mathbf{v}_\phi = v_\phi\mathbf{k}/k$, $\mathbf{v}_{os,e} = -ie\mathbf{E}/m\omega$ is the electron quiver velocity in the EPW (not to be confused with the quiver in a light wave), and where the electric field and potential are simply related via $\mathbf{E} = -i\mathbf{k}\Phi$.

1.3.1.5 Ion Acoustic Waves (IAWs)

Now for IAWs, we assume that the phase velocity of the wave is much larger than the *ion* thermal velocity (ensuring that the ion Landau damping remains small as we will see later in this chapter) but much slower than the electron thermal velocity, i.e., $v_{Ti} \ll \omega/k \ll v_{Te}$. The ion fluid can be assumed adiabatic under the same arguments as for the electron fluid in an EPW, leading to $\gamma_i = 3$ and

$$\chi_i = -\frac{\omega_{pi0}^2}{\omega^2 - 3k^2v_{Ti}^2}, \tag{1.125}$$

per Eq. (1.112).

On the other hand, the electrons (typically traveling at or near v_{Te}) have plenty of time to travel many oscillation wavelengths during one period, so they can equilibrate and be assumed isothermal, with $\gamma_e = 1$. Because $\omega/k \ll v_{Te}$ we have, also from Eq. (1.112):

$$\chi_e \approx \frac{-\omega_{pe0}^2}{-\gamma_e k^2 v_{Te}^2} = \frac{1}{k^2 \lambda_{De}^2}, \tag{1.126}$$

with $\lambda_{De} = v_{Te}/\omega_{pe0}$. Our choices for the equations of state for the electrons and ions will be confirmed in the next section when we derive the dispersion relations of EPWs and IAWs using a kinetic model.

Inserting the two expressions for the electron and ion susceptibilities into the general dispersion relation for longitudinal waves Eq. (1.113) gives the following dispersion relation for the IAWs:

$$\omega^2 = \frac{k^2}{1 + k^2 \lambda_{De}^2} \frac{Z T_e}{M_i} + 3 k^2 v_{Ti}^2. \tag{1.127}$$

If the oscillation wavelength is much larger than the Debye length, or $k\lambda_{De} \ll 1$, then we obtain the following dispersion relation, and its related wave equation obtained from Eq. (1.79):

$$\omega \approx k c_s, \tag{1.128}$$

$$\left(\partial_t^2 - c_s^2 \nabla^2\right) \mathbf{E}(\mathbf{r}, t) = 0, \tag{1.129}$$

where c_s is the sound speed defined as

$$\boxed{c_s = \sqrt{\frac{Z T_e + 3 T_i}{M_i}}}, \tag{1.130}$$

or, in practical units:

$$c_s \approx 3.1 \times 10^5 \sqrt{\frac{Z T_e + 3 T_i}{A}} \, [\text{keV}] \, [\text{m/s}], \tag{1.131}$$

where A is the atomic mass number and one must use the ionization level for Z.

A common simplification occurs when $Z T_e \gg T_i$; going back to arbitrary values of $k\lambda_{De}$ from Eq. (1.127), the sound speed for $Z T_e \gg T_i$ reduces to

$$c_s \approx \frac{Z T_e}{M_i} \tag{1.132}$$

and the dispersion relation becomes

$$\omega^2 = \frac{k^2 c_s^2}{1 + k^2 \lambda_{De}^2}. \tag{1.133}$$

Note that the $k\lambda_{De}$ parameter is also a measure of the deviation from quasi-neutrality associated with IAWs. Using Poisson's equation, Eq. (1.75), and substituting $\mathbf{k} \cdot \mathbf{E}$ from Eq. (1.94) (with $s=e$) gives

$$i\mathbf{k} \cdot \mathbf{E} = \frac{e}{\varepsilon_0}(Z\delta n_i - \delta n_e) = -\frac{m}{e}\frac{\delta n_e}{n_{e0}}(\omega^2 - k^2 v_{Te}^2), \qquad (1.134)$$

which after assuming that $\omega \ll \omega_{pe0}$ (as the IAWs' frequency is very low), simplifies to

$$Z\delta n_i = (1 + k^2\lambda_{De}^2)\delta n_e, \quad \text{or:} \quad \frac{\delta n_i}{n_{i0}} = \frac{\delta n_e}{n_{e0}}(1 + k^2\lambda_{De}^2). \qquad (1.135)$$

Likewise, since $-\chi_e/\chi_i = (\delta n_e/n_{e0})/(\delta n_i/n_{i0}) = \delta v_e/\delta v_i$ (cf. Problem 1.3), we also have $\delta \mathbf{v}_i = (1 + k^2\lambda_{De}^2)\delta \mathbf{v}_e$.

This means that for IAWs, the electrons never quite neutralize the potential created by the ion motion, resulting in some residual charge separation. The effect is more pronounced as $k\lambda_{De}$ increases, i.e., when the Debye length stops being negligible compared to the wavelength of the IAW: in this case, the electron shielding through the wave drops, because the electrons can (in the limit of $k\lambda_{De} \geq 1$) interact with ions several wavelengths away from them. In the limit of $k\lambda_{De} \gg 1$, the IAW dispersion relation Eq. (1.127) shows that the wave becomes non-dispersive, with $\omega \approx \omega_{pi}$, behaving like an EPW in the cold plasma limit (with zero group velocity).

Finally, the phase and group velocities of IAWs, taken in the limit of $ZT_e \ll T_i$, are given via Eq. (1.133) by:

$$v_\phi = \frac{c_s}{\sqrt{1 + k^2\lambda_{De}^2}}, \qquad (1.136)$$

$$v_g = \frac{c_s}{(1 + k^2\lambda_{De}^2)^{3/2}}, \qquad (1.137)$$

where we used $c_s = ZT_e/M_i$ in the limit of $ZT_e \ll T_i$.

It is also possible to include a finite background "zero-order" flow velocity in the IAW equation, i.e., assume that $\mathbf{v}_i = \mathbf{v}_{i0} + \delta\mathbf{v}_i$, where the flow is the result of macroscopic motion of the plasma. This will be described in the next section.

We now write a couple of useful relations between the potential, density, and velocity perturbations for an IAW, like we did for EPWs. Inserting Eq. (1.135) back into Poisson's equation, Eq. (1.134), leads to

$$\Phi = \frac{m}{e}v_{Te}^2\frac{\delta n_e}{n_{e0}} = \frac{T_e}{e}\frac{\delta n_e}{n_{e0}} = \frac{mv_{Te}^2}{e}\frac{\delta v_e}{v_\phi}, \qquad (1.138)$$

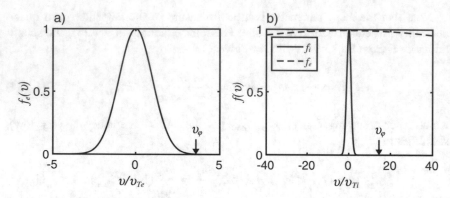

Fig. 1.8 Velocity distribution function and example of phase velocity location for: (**a**) an EPW, here with $v_\varphi \approx 3v_{Te}$; and (**b**) an IAW, with $(v_{Te}/v_{Ti})^2 = 20{,}000$ (corresponding, for example, to $T_e/T_i = 2.7$ in a helium plasma, $A = 4$), and $v_\varphi \approx 15v_{Ti} \approx 0.1v_{Te}$

and, using the continuity equation:

$$\frac{\delta n_e}{n_{e0}} = \frac{\delta v_e}{v_\phi} = \frac{\delta v_e}{c_s}\sqrt{1 + k^2\lambda_{De}^2}. \tag{1.139}$$

Other useful relations pertaining to IAWs are presented in Problem 1.3.

Figure 1.8 shows typical velocity distributions and phase velocities for EPWs and IAWs. For EPWs, the phase velocity $v_\varphi = \omega/k$ must be greater than v_{Te} in order to avoid strong Landau damping of the wave; whereas for IAWs, we must have $v_{Ti} \ll v_\varphi \ll v_{Te}$.

1.3.2 Rapid Derivation of the Wave Equations; Acoustic Waves in the Presence of Background Flow

In the previous section we discussed the different types of waves in plasma, and how to derive their wave equations by introducing the dielectric susceptibilities. Here we are going to show a more direct way to get to the wave equations for EPWs and IAWs; however, we are going to use approximations that have been described in the previous section. While this derivation is faster, it requires a good understanding of the more complete description from the previous section. We will also introduce a background flow for the ions (which represents an overall macroscopic motion of the entire plasma, as the ions will drag the electrons with them), and will derive the IAW wave equation in the presence of such a background flow.

- EMWs:

As we already mentioned in Sect. 1.3.1, EMWs in plasmas are commonly represented via their vector potential $\mathbf{A}$, since it is transverse by definition if we work in the Coulomb gauge ($\nabla \cdot \mathbf{A} = 0$). Taking the curl of Eq. (1.78) and combining with $\mathbf{B} = \nabla \times \mathbf{A}$ and $\mathbf{E}_\perp = -\partial_t \mathbf{A}$ leads to

$$(\partial_t^2 - c^2 \nabla^2)\mathbf{A} = \frac{1}{\varepsilon_0}\mathbf{j}_\perp, \tag{1.140}$$

where we used the vector identity $\nabla \times (\nabla \times \mathbf{A}) = \nabla(\nabla \cdot \mathbf{A}) - \nabla^2 \mathbf{A}$ with $\nabla \cdot \mathbf{A} = 0$.

The transverse current is dominated by the electrons because the ions move slowly in the light wave's electric field, $\mathbf{j}_\perp \approx \mathbf{j}_{e\perp} = -en_e \mathbf{v}_{e\perp}$. To first order, $\mathbf{j}_{e\perp} = -en_{e0}\delta\mathbf{v}_{e\perp}$ with $\delta\mathbf{v}_{e\perp} = (e/m)\mathbf{A}$ from the linearized electron momentum equation Eq. (1.93) leads to the EMW wave equation,

$$\boxed{(\partial_t^2 + \omega_{pe0}^2 - c^2\nabla^2)\mathbf{A} = 0}. \tag{1.141}$$

- EPWs:

The electron fluid equations linearized to first order like in the previous section are:

$$\partial_t \delta n_e + n_{e0}\nabla \cdot \delta\mathbf{v}_e = 0, \tag{1.142}$$

$$\partial_t \delta\mathbf{v}_e = -3v_{Te}^2 \frac{\nabla\delta n_e}{n_{e0}} - \frac{e}{m}\mathbf{E}, \tag{1.143}$$

where we used $\gamma_e = 3$ for the electron fluid as explained in the previous section.

Taking the time derivative of Eq. (1.142), the divergence of Eq. (1.143), combining the two via $\partial_t \nabla \cdot \delta\mathbf{v}_e$ and using Poisson's equation $\nabla \cdot \mathbf{E} = -e\delta n_e/\varepsilon_0$, leads back to the EPW wave equation Eq. (1.118),

$$\boxed{\left(\partial_t^2 + \omega_{pe0}^2 - 3v_{Te}^2\nabla^2\right)\mathbf{E}(\mathbf{r}, t) = 0}. \tag{1.144}$$

- IAWs:

For the IAWs we use the ion fluid equations linearized to first order—but now also include a uniform background flow velocity, i.e., $\mathbf{v}_i = \mathbf{v}_{i0} + \delta\mathbf{v}_i$ (we will see in subsequent chapters that background flow velocity is an important parameter for a number of laser–plasma interaction processes); the equations now read:

$$D_t \delta n_i + n_{i0}\nabla \cdot \delta\mathbf{v}_i = 0, \tag{1.145}$$

$$D_t \delta\mathbf{v}_i = -3v_{Ti}^2 \frac{\nabla\delta n_i}{n_{i0}} + \frac{Ze}{M_i}\mathbf{E}, \tag{1.146}$$

where we used $\gamma_i = 3$ for the ions equation of state and defined $D_t = \partial_t + \mathbf{v}_{i0} \cdot \nabla$ as the time derivative in the frame moving at $\mathbf{v}_{i0}$ (to avoid confusion we reserve the d_t notation for Lagrangian coordinates for the description of single particle motion, cf. Chap. 2). Combining again the total time derivative D_t of Eq. (1.145) with the divergence of Eq. (1.146) gives:

$$D_t^2 \frac{\delta n_i}{n_{i0}} - 3 v_{Ti}^2 \nabla^2 \frac{\delta n_i}{n_{i0}} + \frac{Ze}{M_i} \nabla \cdot \mathbf{E} = 0 . \tag{1.147}$$

The electric field is due to the charge separation from the electrons; the electron fluid momentum equation is now

$$\partial_t \delta \mathbf{v}_e = -v_{Te}^2 \nabla \frac{\delta n_e}{n_{e0}} - \frac{e}{m} \mathbf{E} , \tag{1.148}$$

where we used an isothermal equation of state with $\gamma_e = 1$ (cf. previous section). Neglecting the electron inertia, $\partial_t \delta \mathbf{v}_e \approx 0$ (which is valid as long as the IAW phase velocity is much smaller than the electron thermal velocity, $c_s \ll v_{Te}$: cf. previous section and Problem 1.3) gives $\mathbf{E} = -(m v_{Te}^2/e) \nabla \delta n_e/n_{e0}$; inserting in Eq. (1.147) and assuming quasi-neutrality, i.e., $\delta n_e/n_{e0} = \delta n_i/n_{i0}$ (which is valid when $k\lambda_{De} \ll 1$, cf. previous section), finally gives:

$$\boxed{\left((\partial_t + \mathbf{v}_{i0} \cdot \nabla)^2 - c_s^2 \nabla^2 \right) \frac{\delta n_e}{n_{e0}} = 0} , \tag{1.149}$$

where $c_s^2 = (ZT_e + 3T_i)/M_i$.

1.3.3 Kinetic Description of Plasma Waves

A kinetic model based on the Vlasov equation can also be used to describe longitudinal plasma waves. It involves a special function (the plasma dispersion function Z [6]) to describe the dispersion relation but is valid for any phase velocity and can describe arbitrary distribution functions for the plasma particles besides the usual Maxwellian.

The description starts from the Vlasov equation describing the evolution of the distribution function $f_s(\mathbf{r}, \mathbf{v}, t)$ for each particle species s (electron and any type of ion present in the plasma):

$$\partial_t f_s + (\mathbf{v} \cdot \nabla) f_s + \frac{\mathbf{F}_s}{m_s} \cdot \partial_{\mathbf{v}} f_s = 0 , \tag{1.150}$$

where $\mathbf{F}_s$ is the force acting on the particle species s. The linear kinetic description consists in developing the distribution function into a background component that depends only on $\mathbf{v}$ and a perturbation that depends on $\mathbf{r}, \mathbf{v}, t$. Physically, this means that the background distribution (typically a Maxwellian distribution of velocities, but not necessarily) is the same everywhere in space and at any time. The physical process we are considering (in our case the wave oscillations) must therefore occur on spatial and temporal scales short enough to assume that the background is uniform and constant.[9] The distribution takes the form:

$$f_s(\mathbf{r}, \mathbf{v}, t) = f_{s0}(\mathbf{v}) + \delta f_s(\mathbf{r}, \mathbf{v}, t). \tag{1.151}$$

We look for quantities that follow wave oscillations at the frequency and wave-vector $\omega, \mathbf{k}$; the only forces present are the internal forces from the electric field resulting from charge separation, which oscillate at the same frequency and wavenumber as the distribution perturbation:

$$\delta f_s = \frac{1}{2}\delta\tilde{f}_s e^{i\psi} + c.c., \tag{1.152}$$

$$\mathbf{E} = \frac{1}{2}\tilde{\mathbf{E}}e^{i\psi} + c.c., \tag{1.153}$$

where $\psi = \mathbf{k} \cdot \mathbf{r} - \omega t$ and $\mathbf{F}_s = q_s \mathbf{E}$. Similarly to the fluid derivation, we assume that $\mathbf{E}$ and δf_s are first-order quantities in a perturbation series. The background distribution $f_{s0}(\mathbf{v})$ is zeroth-order; f_{s0} and δf_s are connected to the background density n_{s0} and the perturbation from the wave δn_s (such that $n_s = n_{s0} + \delta n_s$) via

$$n_{s0} = \int f_{s0}(\mathbf{v})d^3v, \tag{1.154}$$

$$\delta n_s = \int \delta f_s(\mathbf{v})d^3v. \tag{1.155}$$

Collecting the first-order terms in the Vlasov equation along the oscillation phase $\propto e^{i\psi}$ leads to:

$$(-i\omega + i\mathbf{k} \cdot \mathbf{v})\delta\tilde{f}_s = -\frac{q_s}{m_s}\tilde{\mathbf{E}} \cdot \partial_{\mathbf{v}} f_{s0}. \tag{1.156}$$

Taking z as the direction of the wave and integrating over velocity leads to the expression for $\delta\tilde{n}_s$:

---

[9] It is also possible to allow the background to evolve slowly in time: this is the "quasi-linear theory," which is a useful tool to describe diffusion processes and the temporal evolution of the background distribution on longer time scales [7, 8].

$$\delta\tilde{n}_s = -i\frac{q_s}{m_s}\tilde{E}\int\frac{\partial_{v_z}f_{s0}(\mathbf{v})}{\omega - kv_z}d^3v\,. \tag{1.157}$$

Now that we have an expression for the density perturbation, we can look for the expression of the dielectric constant, which will in turn provide the dispersion relation of the plasma waves. Like for the fluid derivation, we are looking for $\varepsilon(k, \omega)$ such that $\mathbf{D} = \varepsilon_0\varepsilon\mathbf{E}$ with $\nabla \cdot \mathbf{D} = 0$. Poisson's equation, $\nabla \cdot \mathbf{E} = \sum_s q_s\delta n_s/\varepsilon_0$, can be recast into

$$ik\left(\varepsilon_0\tilde{E} + \frac{i}{k}\sum_s q_s\delta\tilde{n}_s\right) = 0\,. \tag{1.158}$$

To obtain ε we have to express the term in parenthesis as $\tilde{D} = \varepsilon_0\varepsilon\tilde{E}$. Substituting with the expression for $\delta\tilde{n}_s$ from Eq. (1.157) immediately leads to

$$\varepsilon = 1 + \chi = 1 + \sum_s \chi_s\,, \tag{1.159}$$

where the susceptibility for each particle species χ_s is expressed as

$$\boxed{\chi_s = \frac{q_s^2}{k\varepsilon_0 m_s}\int\frac{\partial_{v_z}f_{s0}(\mathbf{v})}{\omega - kv_z}d^3v}\,. \tag{1.160}$$

The kinetic dispersion relation of plasma waves is then:

$$\boxed{\varepsilon = 1 + \chi = 0}\,. \tag{1.161}$$

This expression is rather opaque in comparison with the simple expressions we obtained earlier with the fluid model (Sects. 1.3.1, 1.3.2). On the other hand, it is also much more general, since it is valid for any wave velocity (not necessarily satisfying the conditions $v_\varphi \ll v_{Te}$ of an EPW or $v_{Ti} \ll v_\varphi \ll v_{Te}$ for an IAW), as well as for multi-species plasmas (where the fluid approximations can break, as we will see in Sect. 1.3.6), and unlike for fluid quantities it does not make any assumption on the background particle distribution function or the equation of state of the fluids.

If, however, we assume that the background distribution $f_0(\mathbf{v})$ is Maxwellian, the expression for the susceptibilities can be expressed using a well-known special function, as

$$\boxed{\chi_s = -\frac{1}{2k^2\lambda_{Ds}^2}Z'\left(\frac{v_\varphi}{\sqrt{2}v_{Ts}}\right)}\,, \tag{1.162}$$

where $v_\varphi = \omega/k$ is the wave's phase velocity. The Z function is called the plasma dispersion function and was thoroughly tabulated and investigated in Ref. [6] (Z' denotes its derivative); its general expression is[10]

$$Z(\zeta) = \frac{1}{\sqrt{\pi}} \int_{-\infty}^{\infty} \frac{e^{-t^2}}{t - \zeta} dt . \tag{1.164}$$

The Z function has the following asymptotic limits:

• $\zeta \ll 1$:

$$Z(\zeta) \approx i\sqrt{\pi}e^{-\zeta^2} - 2\zeta + O(\zeta^3), \tag{1.165}$$
$$Z'(\zeta) \approx -2 - 2i\sqrt{\pi}\zeta + 4\zeta^2 + O(\zeta^3); \tag{1.166}$$

• $\zeta \gg 1$ ($\zeta \in \mathbb{R}$):

$$Z(\zeta) \approx i\sqrt{\pi}e^{-\zeta^2} - \frac{1}{\zeta} - \frac{1}{2\zeta^3} + O\left(\frac{1}{\zeta^3}\right), \tag{1.167}$$

$$Z'(\zeta) \approx \frac{1}{\zeta^2} + \frac{3}{2}\frac{1}{\zeta^4} + O\left(\frac{1}{\zeta^6}\right). \tag{1.168}$$

We can now verify that the kinetic dispersion relation Eq. (1.159) recovers the fluid dispersion relations for EPWs and IAWs by using the asymptotic expansions of the Z function. Note that the result can also be obtained directly by evaluating the velocity integral in the susceptibility analytically: this will be discussed in Sect. 1.3.4. For now we just show the "easy way" using the asymptotic limits of Z above.

For EPWs, no ion motion is involved ($\chi_i = 0$); assuming a phase velocity higher than the thermal velocity, the large argument expansion of Z' above in $\varepsilon = 1 + \chi_e = 0$ gives

---

[10] While the Z function is rarely included in scientific programing languages, the error function erf(z) usually is, and the two are connected by the relation

$$Z(\zeta) = i\sqrt{\pi}e^{-\zeta^2}[1 + \mathrm{erf}(i\zeta)]. \tag{1.163}$$

$$\omega^2 = \omega_{pe}^2 \left[1 + 3 \frac{k^2 v_{Te}^2}{\omega^2} \right] .$$ (1.169)

Solving the polynomial in ω^2 with the assumption $k\lambda_{De} \ll 1$ recovers the fluid EPW dispersion relation,

$$\omega^2 = \omega_{pe}^2 + 3k^2 v_{Te}^2 .$$ (1.170)

This also confirms the adiabatic coefficient $\gamma_e = 3$ for EPWs from Sect. 1.3.1.

Likewise for IAWs, we now use the small argument expansion of Z' for the electron susceptibility (since $v_\varphi \ll v_{Te}$), leading to $\chi_e \approx 1/(k\lambda_{De})^2$, and the large argument expansion for the ion susceptibility (assuming a single ion species), since $v_\varphi \gg v_{Ti}$, and obtain after inserting in $\varepsilon = 1 + \chi_e + \chi_i = 0$:

$$\omega^4 \left(1 + \frac{1}{k^2 \lambda_{De}^2} \right) = \omega_{pi}^2 \omega^2 \left(1 + 3 v_{Ti}^2 \frac{k^2}{\omega^2} \right) .$$ (1.171)

Solving the polynomial in ω^2 with the assumption $T_i \ll ZT_e$ leads to the same IAW dispersion relation obtained earlier in the fluid limit:

$$\omega^2 = \frac{ZT_e}{M_i} \frac{k^2}{1 + k^2 \lambda_{De}^2} + 3k^2 v_{Ti}^2 .$$ (1.172)

Again, this confirms our earlier choice of $\gamma_e = 1$ (isothermal) and $\gamma_i = 3$ (adiabatic) for the electron and ion fluids.

To make all this a little more concrete and tangible, we can plot the inverse of the dielectric constant $1/|\varepsilon|$ as a function of the phase velocity $v_\varphi = \omega/k$. Since the dielectric constant is in general complex, we take its norm: the natural modes of the plasma correspond to (real) frequencies for which $\varepsilon = 0$, so $1/|\varepsilon|$ will show peaks at the frequencies of the modes (i.e., EPW and IAW). The peaks' amplitudes remain finite due to the Landau damping of the waves, which comes into play via the imaginary parts of the susceptibilities as we will see in Sect. 1.3.4.

The plot of $1/|\varepsilon|$ vs. v_φ is represented in Fig. 1.9, for a pure hydrogen plasma with $T_e/T_i = 10$ and $k\lambda_{De} = 0.4$. We clearly see the two peaks corresponding to the IAW (at low velocity) and the EPW (at high velocity). Their velocities are in good agreement with the fluid model (cf. Problem 1.4).

1.3.4 Landau Damping

The kinetic model of plasma waves presented above is based on the assumption of Fourier modes $\propto \exp[i(kz - \omega t)]$ that exist for all times. What is missing from this

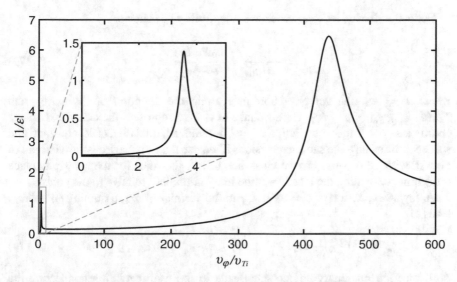

Fig. 1.9 Inverse of the norm of the dielectric constant, $1/|\varepsilon|$, as a function of the phase velocity $v_\varphi = \omega/k$ normalized to the ion thermal velocity $v_{Ti} = \sqrt{T_i/M_i}$ for a pure hydrogen plasma with $T_e/T_i = 10$ and $k\lambda_{De} = 0.4$. The peak at $v_\varphi \approx 3.5 v_{Ti}$ (inset) is the ion acoustic mode, and the one at $v_\varphi \approx 440 v_{Ti} \approx 3.2 v_{Te}$ (since $v_{Te}/v_{Ti} = \sqrt{M_i T_e/m T_i} \approx 135.5$) is the EPW

Fig. 1.10 Landau contour for weakly damped waves in the complex velocity plane

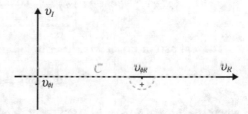

analysis is the issue of the pole of the susceptibility at $\omega = k v_z$, cf. Eq. (1.160). The correct treatment of the Vlasov equation, first derived by Landau [9], does not impose a solution that is a pure Fourier mode but treats the problem as an initial-value problem, where a periodic perturbation in space $\propto \exp[ikz]$ is imposed at $t = 0$ and the equation is then solved using a Laplace transform (instead of Fourier) for the time dependence.

The key result (cf. Ref. [1]) is that the integration must be performed along a contour that passes below the poles in the complex v plane (such that $v_\varphi = v_{\varphi R} + i v_{\varphi I} = \omega/k = (\omega_R + i\omega_I)/k$, where the subscripts R and I denote the real and imaginary parts). When the imaginary part of ω is negative and very small, corresponding to weakly damped waves, an appropriate contour can be found in the form illustrated in Fig. 1.10, with a small half-circle enclosing the pole.

For better clarity, let us rewrite the susceptibility as follows:

$$\chi_s = -\frac{\omega_{ps0}^2}{k^2 n_{0s}} \int \frac{\partial_u F_{s0}(u)}{u - v_\phi} du \,, \tag{1.173}$$

where $u = v_z$ (the velocity coordinate along the direction of the wave), and $F_{s0}(u) = \int f_{s0}(\mathbf{v}) dv_x dv_y$. Here we take $\omega \in \mathbb{C}$ to represent the complex Laplace coordinate s via the substitution $s = i\omega$: indeed, the dielectric constant and susceptibilities take the same expressions if we use the Laplace transform instead of Fourier with $\omega \to -is$ (but we must then follow the rules for the inverse Laplace transform concerning the location of the integration path relative to the poles).

A Taylor expansion of $\varepsilon = \varepsilon_R + i\varepsilon_I$ in the vicinity of ω_R, assuming $\omega_I \ll \omega_R$, leads to

$$\varepsilon \approx \varepsilon_R(\omega_R) + i\varepsilon_I(\omega_R) + i\omega_I \partial_\omega \varepsilon_R|_{\omega_R} - \omega_I \partial_\omega \varepsilon_I|_{\omega_R} = 0 \,. \tag{1.174}$$

Taking the imaginary components leads to the useful relation connecting the imaginary part of ω (i.e., the damping, or growth rate if $\omega_I > 0$) to the real components of the dielectric constant:

$$\boxed{\omega_I = -\frac{\varepsilon_I(\omega_R)}{\partial_\omega \varepsilon_R|_{\omega_R}}} \,. \tag{1.175}$$

The real part of Eq. (1.174) leads to

$$\varepsilon_R(\omega_R) = 0 \,, \tag{1.176}$$

where the term $\propto \omega_I \partial_\omega \varepsilon_I|_{\omega_R}$ is of order $\sim \omega_I^2$ and thus negligible.

To evaluate ε we use the integration contour C shown in Fig. 1.10 to calculate χ_s from Eq. (1.173); in the limit of $v_\phi \gg v_{Te}$, we can approximate the integral as

$$\chi_s \approx -\frac{\omega_{ps0}^2}{k^2 n_{s0}} \left[P \int \frac{\partial_u F_{s0}}{u - v_{\phi R}} du + i\pi \partial_u F_{s0}|_{v_{\phi R}} \right] \,. \tag{1.177}$$

Here P means the principal value and corresponds to the integration along the straight line in Fig. 1.10 on either side of the small half-circle; the second term in the square bracket corresponds to half of the residue at the pole since we integrate over a half-circle only. If the phase velocity is far in the tail of the distribution as is assumed here, we can reasonably assume that both F_{s0} and its derivative at $v_{\phi R}$ are very small (for example, cf. Fig. 1.8a), and therefore the missing segment near $v_{\phi R}$ does not significantly contribute to the integral. This first term in the bracket can then be estimated as follows:

$$P \int \frac{\partial_u F_{s0}}{u - v_{\phi R}} du \approx \left[\frac{F_{s0}}{u - v_{\phi R}} \right]_{-\infty}^{\infty} + \int \frac{F_{s0}}{(u - v_{\phi R})^{-2}} du \qquad (1.178)$$

$$= \frac{1}{v_{\phi R}^2} \int F_{s0} \left(1 - \frac{u}{v_{\phi R}} \right)^{-2} du \qquad (1.179)$$

$$= \frac{n_{s0}}{v_{\phi R}^2} \left\langle 1 + 2\frac{u}{v_{\phi R}} + 3\frac{u^2}{v_{\phi R}^2} + \ldots \right\rangle, \qquad (1.180)$$

where the first line was derived using an integration by parts, and the last made the assumption of small velocities compared to the phase velocity of the wave (the angular bracket is an average over the velocity distribution).

For the case of a Maxwellian distribution we already saw in Sect. 1.2.3 that $\langle v_x^2 \rangle = \langle v_y^2 \rangle = \langle v_z^2 \rangle = v_{Te}^2$ (and $\langle v_z \rangle = 0$), so with our definition of $u = v_z$ the integral and thus the real part of the susceptibility simplifies to

$$\chi_{sR} \approx -\frac{\omega_{ps0}^2}{\omega_R^2} \left(1 + 3\frac{k^2 v_{Ts}^2}{\omega_R^2} \right). \qquad (1.181)$$

For EPWs, since $\varepsilon_R = 1 + \chi_{eR} = 0$ this leads to the same expression for the fluid dispersion relation obtained earlier in Sect. 1.3.3, Eq. (1.169), where we had used the asymptotic expansion of the Z function.

The damping is given by the imaginary contribution of the pole in Eq. (1.177). For EPWs, approximating $\omega_R \approx \omega_{pe}$ to evaluate $\partial_\omega \varepsilon_R$ leads to $\partial_\omega \varepsilon_R \approx 2/\omega_{pe}$. Since $\chi_{eI} = \varepsilon_I$, with $\chi_{eI} = -\pi \omega_{pe}^2/(k^2 n_{e0}) \partial_u F_{s0}|_{\omega_R/k}$ we finally obtain

$$\omega_I = \frac{\pi}{2} \frac{\omega_{pe}^3}{k^2 n_{e0}} \partial_{v_z} F_{e0}|_{\omega_R/k}. \qquad (1.182)$$

If the slope of the distribution function is negative at the phase velocity of the wave, $\partial_{v_z} F_{e0}|_{\omega_R/k} < 0$, then this imaginary part corresponds to a damping: this is the Landau damping for EPWs. It means, in particular, that for a Maxwellian distribution function, a plasma wave will be damped and eventually vanish even in the absence of collisions. The physical meaning of Landau damping is linked to the dynamics of plasma particles whose velocity along z is resonant with the wave, i.e., for whom $v_z \approx v_\phi$: as we will see in Sect. 2.2.3, Landau damping corresponds to an energy transfer from the wave to the particles.

For a Maxwellian distribution of electrons, the Landau damping expression for EPWs becomes (following the literature we will often write $\nu_L = -\omega_I > 0$ for the damping):

$$\frac{\nu_{eL}}{\omega_{pe}} = \sqrt{\frac{\pi}{8}} (k\lambda_{De})^{-3} \exp\left[-\frac{1}{2k^2 \lambda_{De}^2} - \frac{3}{2} \right]. \qquad (1.183)$$

Fig. 1.11 Landau damping
of EPWs with a Maxwellian
background distribution
function

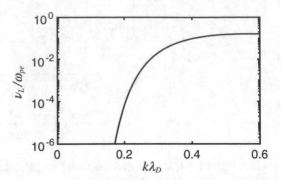

The Landau damping of EPWs for a Maxwellian background is represented in Fig. 1.11. The damping becomes significant for values of $k\lambda_{De} \gtrsim 0.3$: for values of 0.3 and above, the phase velocity of the wave is not much greater than the thermal velocity anymore (since $v_\phi/v_{Te} \approx \omega_{pe}/kv_{Te} = 1/k\lambda_{De}$), and there is a significant number of particles in the tail of the distribution function that can interact with the wave. This means that EPWs with values of $k\lambda_{De} \gtrsim 0.3$ are unlikely to be observed in plasmas as they will rapidly damp (unless they are forced by an external driver, like in the case of instabilities).

The ion Landau damping in IAWs can be derived in a similar manner; assuming $v_{Ti} \ll v_\phi \ll v_{Te}$, Eq. (1.181) leads to

$$\chi_{iR} \approx -\frac{\omega_{pi0}^2}{\omega_R^2} \tag{1.184}$$

and $\partial_\omega \chi_{iR} \approx 2\omega_{pi}^2/\omega_R^3$, assuming $v_\phi \gg v_{Ti}$. The electron contribution is obtained from Eq. (1.177) by taking the opposite limit as before, i.e., assuming very large (electron) velocities compared to the phase velocity of the IAW, $v_\phi \ll v_{Te}$; the principal value is then estimated as

$$P\int \frac{\partial_u F_{e0}}{u - v_{\phi R}} du \approx \int \frac{\partial_u F_{e0}}{u} du = -\frac{n_{e0}}{v_{Te}^2}, \tag{1.185}$$

where the last equality was obtained assuming a Maxwellian distribution for the electrons. Thus,

$$\chi_e \approx \frac{1}{k^2\lambda_{De}^2}, \tag{1.186}$$

consistent with our earlier results from the fluid analysis of IAWs (or the kinetic analysis using the asymptotic limit of the Z function) and leading to the same fluid dispersion relation for IAWs. The damping is again obtained from Eq. (1.175), leading to

$$\omega_I = \frac{\pi}{2} \frac{\omega_R^3}{k^2 n_{i0}} \partial_u F_{0i}|_{v_\phi R} , \qquad (1.187)$$

or for a Maxwellian distribution of ions,

$$\boxed{\frac{v_{iL}}{\omega_R} = \sqrt{\frac{\pi}{8}} \frac{\omega_R^3}{k^3 v_{Ti}^3} \exp\left[-\frac{\omega_R^2}{2k^2 v_{Ti}^2}\right]} , \qquad (1.188)$$

where we defined $v_{iL} = -\omega_I$.

In the limit of $ZT_e \gg T_i$, which has to be at least partly satisfied for the approximations used here to hold (since $\sqrt{ZT_e/T_i} \approx v_\phi/v_{Ti}$), the expression can be further simplified by replacing $\omega_R/k v_{Ti}$ by $\sqrt{ZT_e/T_i}$, i.e.,

$$\frac{v_{iL}}{\omega_R} \approx \sqrt{\frac{\pi}{8}} \left(\frac{ZT_e}{T_i}\right)^{3/2} \exp\left[-\frac{ZT_e}{2T_i}\right] . \qquad (1.189)$$

For both EPWs and IAWs, when the ratio of phase to thermal velocity (i.e., $v_{\phi EPW}/v_{Te} \approx 1/k\lambda_{De}$ and $v_{\phi IAW}/v_{Ti} \approx \sqrt{ZT_e/T_i}$ for EPWs and IAWs, respectively) is not very large ($\gg 1$) anymore, the expressions above for the Landau damping become incorrect and one has to use the Z function and its tabulation [6] to get an accurate response.

1.3.5 Wave Energy and Action

The energy density of a plasma wave is an important quantity which will later help us quantify the energy exchange between a laser and a plasma via wave-coupling instabilities. In this section we follow the description of Nicholson [1].

We us consider a plasma wave (EPW or IAW) with electric field

$$E(\mathbf{r}, t) = \frac{1}{2} \tilde{E}(\mathbf{r}, t) e^{-i\omega_0 t} + c.c. \qquad (1.190)$$

(the time-envelope $\tilde{E}$ also contains a rapid spatial modulation $\propto \exp[ik_0 z]$, which we do not need to extract). The energy density of the field at a given location in space is given by

$$\mathcal{E}_E = \frac{\varepsilon_0 \langle E^2 \rangle_T}{2} = \frac{\varepsilon_0 |\tilde{E}|^2}{4} , \qquad (1.191)$$

where the time-average is taken over an oscillation cycle. The field drives plasma oscillations, which have an associated energy density via the particle energy; the plasma component of the wave energy manifests itself via the dielectric constant ε.

To extract the contribution of the plasma particles in the wave energy, we connect the electric field to the current via Ampère-Maxwell's law, Eq. (1.78), which in the absence of magnetic field (plasma waves are longitudinal and do not have a B-field) leads to:

$$\nabla \times \mathbf{B} = 0 = \frac{1}{\varepsilon_0 c^2}\mathbf{j} + \frac{1}{c^2}\partial_t \mathbf{E}. \tag{1.192}$$

Since we are only considering longitudinal waves, all the relevant vector quantities (wave-number, electric field, current) are parallel so we can drop the vector notation, assuming that we are describing the scalar quantities along the wave direction.

Expressing $j = \sigma E$ where σ is the electric conductivity already introduced in Sect. 1.3.1.1, we have $\partial_t \tilde{E} = -\sigma \tilde{E}/\varepsilon_0$, which after multiplying by $\tilde{E}^*$ leads to

$$\partial_t |\tilde{E}|^2 = -\frac{2}{\varepsilon_0}|\tilde{E}|^2 \sigma_R, \tag{1.193}$$

where $\sigma_R = \mathrm{Re}(\sigma)$.

We now Taylor-expand σ near the wave frequency:

$$\sigma(\omega) \approx \sigma(\omega_0) + (\omega - \omega_0)\left.\frac{\partial \sigma}{\partial \omega}\right|_{\omega_0}. \tag{1.194}$$

We assume that the wave can have a slowly varying amplitude described via an imaginary component to the frequency ω_I (usually a damping, $\omega_I < 0$), i.e., $\omega = \omega_0 + i\omega_I$ with $|\tilde{E}(t)| \propto e^{\omega_I t}$, and with $|\omega_I| \ll \omega_0$. In this case we also have $\sigma_R \ll \sigma_I$,[11] and the expansion takes the form:

$$\sigma(\omega) \approx \sigma_R(\omega_0) + i\sigma_I(\omega_0) - \omega_I \left.\frac{\partial \sigma_I}{\partial \omega}\right|_{\omega_0}, \tag{1.195}$$

so the real part of the conductivity can be approximated as

$$\sigma_R(\omega) \approx \sigma_R(\omega_0) - \omega_I \left.\frac{\partial \sigma_I}{\partial \omega}\right|_{\omega_0}$$

$$= \sigma_R(\omega_0) - \frac{1}{|\tilde{E}|}\frac{\partial |\tilde{E}|}{\partial t}\left.\frac{\partial \sigma_I}{\partial \omega}\right|_{\omega_0}, \tag{1.196}$$

---

[11] Recall that $\varepsilon = 1 + i\sigma/(\varepsilon_0 \omega)$, so the real part of the conductivity corresponds to the wave damping, i.e., the imaginary part of the dielectric constant.

where we used $\partial_t |\tilde{E}| = \omega_I |\tilde{E}|$. Inserting back into Eq. (1.193) and multiplying by $\varepsilon_0/4$ leads to

$$\partial_t \mathcal{E}_E - \frac{1}{\varepsilon_0} \frac{\partial \sigma_I}{\partial \omega}\bigg|_{\omega_0} \partial_t \mathcal{E}_E = -2 \frac{\sigma_R(\omega_0)}{\varepsilon_0} \mathcal{E}_E . \tag{1.197}$$

This relation expresses the energy balance for a plasma wave: the first term on the left-hand side is the field energy density, the second is the energy associated with the plasma particles, and the right-hand side is the dissipation (wave damping). The two terms on the left-hand side can thus be combined to define the total energy density of the wave:

$$\mathcal{E}_{wave} = \mathcal{E}_E \left(1 - \frac{1}{\varepsilon_0} \frac{\partial \sigma_I}{\partial \omega}\bigg|_{\omega_0} \right) . \tag{1.198}$$

We now express the previous relations back in terms of the dielectric constant;[12] from the relation between the conductivity and dielectric constant, $\varepsilon = 1 + i\sigma/(\varepsilon_0 \omega)$, we get $\sigma_I/\varepsilon_0 = \omega(1 - \varepsilon_R)$, which after a derivation with respect to ω leads to

$$1 - \frac{1}{\varepsilon_0} \frac{\partial \sigma_I}{\partial \omega} = \varepsilon_R + \omega \frac{\partial \varepsilon_R}{\partial \omega} = \partial_\omega(\omega \varepsilon_R) . \tag{1.199}$$

Likewise, we also have $\sigma_R = \varepsilon_0 \omega \varepsilon_I$, and therefore the energy conservation relation above, Eq. (1.197), can be recast as

$$\partial_t \mathcal{E}_{wave} = -\frac{1}{2} \varepsilon_0 \omega_0 \varepsilon_I(\omega_0) |\tilde{E}|^2 , \tag{1.200}$$

where the plasma wave energy density is expressed as

$$\boxed{\mathcal{E}_{wave} = \mathcal{E}_E \omega_0 \frac{\partial \varepsilon_R}{\partial \omega}\bigg|_{\omega_0}} , \tag{1.201}$$

with $\mathcal{E}_E = \varepsilon_0 |\tilde{E}|^2/4$ as we saw above.

Finally, expressing $\varepsilon_I(\omega_0) \approx -\omega_I \partial_\omega \varepsilon_R|_{\omega_0}$ (cf. Sect. 1.3.4, Eq. 1.175), we have the expected relation:

$$\boxed{\partial_t \mathcal{E}_{wave} = -2\omega_I \mathcal{E}_{wave}} , \tag{1.202}$$

---

[12] We merely used the conductivity for convenience, in order to avoid expanding a term $\propto \omega \varepsilon(\omega)$ after Eq. (1.193).

meaning that the wave energy damps exponentially at the rate $2\omega_I$. In the absence of damping ($\omega_I = 0$), the energy of the wave is conserved.

Let us now give more explicit expressions for the energy density of EPWs and IAWs:

• EPWs:

For EPWs in the fluid limit the dielectric constant is (we only consider the real part, since we are only interested in the energy density, not the damping) $\varepsilon_R(\omega) = 1 - \chi_e = 1 - \omega_{pe}^2/(\omega^2 - 3k^2 v_{Te}^2)$. The real frequency of the EPW is $\omega_0^2 = \omega_{pe}^2 + 3k^2 v_{Te}^2$, so we easily find that

$$\omega_0 \left. \frac{\partial \varepsilon_R}{\partial \omega} \right|_{\omega_0} = 2(1 + 3k^2 \lambda_{De}^2), \tag{1.203}$$

and so the energy density of an EPW is

$$\mathcal{E}_{EPW} = \frac{\varepsilon_0 |\tilde{E}|^2}{2} (1 + 3k^2 \lambda_{De}^2). \tag{1.204}$$

One can derive another alternate expression using the density perturbation $\delta \tilde{n}_e/n_{e0}$ (the tilde denotes the slowly varying envelope): using the relation from Eq. (1.121), substituting $|\tilde{E}| = k|\tilde{\Phi}|$ and expressing the relation above as a function of the peak amplitude of the density modulation $|\delta \tilde{n}_e/n_{e0}|$, we obtain:

$$\boxed{\mathcal{E}_{EPW} = \frac{1}{2} m v_\phi^2 n_{e0} \left| \frac{\delta \tilde{n}_e}{n_{e0}} \right|^2.} \tag{1.205}$$

Alternatively, we can use yet another fluid quantity, the fluid velocity perturbation $\delta \tilde{v}_e$; since $\delta \tilde{v}_e = v_\phi \delta \tilde{n}_e/n_{e0}$ per Eq. (1.124), the EPW energy density can be expressed as:

$$\mathcal{E}_{EPW} = \frac{1}{2} m \delta \tilde{v}_e^2 n_{e0}. \tag{1.206}$$

• IAWs:

For IAWs (again in the fluid limit, single-species) we have $\varepsilon_R = 1 + (k\lambda_{De})^{-2} - \omega_{pi}^2/\omega^2$, where we assumed $3T_i \ll ZT_e$. Thus, we have

$$\omega_0 \left. \frac{\partial \varepsilon_R}{\partial \omega} \right|_{\omega_0} = 2 \frac{\omega_{pi}^2}{\omega_0^2} = 2 \frac{1 + k^2 \lambda_{De}^2}{k^2 \lambda_{De}^2}, \tag{1.207}$$

and the expression for the IAW energy density becomes:

$$\mathcal{E}_{IAW} = \frac{\varepsilon_0 |\tilde{E}|^2}{2} \frac{1 + k^2 \lambda_{De}^2}{k^2 \lambda_{De}^2}. \tag{1.208}$$

Like for the EPWs, we can use the relation connecting the electric potential to the electron density modulation in IAWs, Eq. (1.138), and substitute in the expression above to obtain, in the limit of $k\lambda_{De} \ll 1$:

$$\boxed{\mathcal{E}_{IAW} = \frac{1}{2} m v_{Te}^2 n_{e0} \left| \frac{\delta \tilde{n}_e}{n_{e0}} \right|^2.} \tag{1.209}$$

Alternatively, we can again use the fluid velocity perturbation as a variable, with $\delta \tilde{v}_i = c_s \delta \tilde{n}_i / n_{i0}$ per Eq. (1.139). Since $v_{Te}^2 / c_s^2 = M_i / Zm$, we obtain

$$\mathcal{E}_{IAW} = \frac{1}{2} M_i \delta \tilde{v}_i^2 n_{i0}. \tag{1.210}$$

Notice the symmetry with the expression for the EPW energy density above, Eq. (1.206).

Next we introduce the quantity of wave action density. The action density is simply defined as

$$\mathcal{A} = \frac{\mathcal{E}_{wave}}{\omega}, \tag{1.211}$$

where ω is the wave frequency. The action can be a useful quantity for laser–plasma physics, since it is a conserved quantity in three-wave instabilities, as we will see in our discussion of the Manley–Rowe relations in Sect. 6.4.6.

Physically, the action density represents the density of quanta from a wave: for example, in the case of a light wave, the energy density $\mathcal{E}_{EMW} = \varepsilon_0 |\tilde{E}|^2 / 2$ can also be expressed as $\mathcal{E}_{EMW} = N_0 \hbar \omega$, where N_0 is the density of photons (number of photons per unit volume) and $\hbar \omega$ is the energy of each photon. Hence,

$$\mathcal{A}_{EMW} = \hbar N_0, \tag{1.212}$$

i.e., $\mathcal{A}_{EMW}$ simply represents the density of photons from the wave. Likewise, we can conceive energy quanta for EPWs ("plasmons") or IAWs ("phonons"), with energy $\hbar \omega$, where ω is the frequency of the EPW or IAW, and interpret $\mathcal{A}_{EPW,IAW}$ as the density of plasmons or phonons (multiplied by the Planck constant $\hbar$).

It will also be convenient to introduce the complex *action amplitude* α, defined as

$$|\tilde{\alpha}|^2 = \mathcal{A}, \tag{1.213}$$

where the tilde denotes again a slowly varying envelope (or the peak amplitude if the wave amplitude is constant), $\alpha = \frac{1}{2}\tilde{\alpha}e^{i\psi} + c.c..$

From its definition, the action amplitude is proportional to any of the fluid quantities used to describe any type of wave in a plasma (EMW, EPW, or IAW), such as the electric field, potential, velocity perturbation and (EPW and IAW only) the density modulation amplitude. We obtain the following expressions:

- EMW: we have $\mathcal{E}_{EMW} = \varepsilon_0|\tilde{E}|^2/2$, so making use of the definition of the normalized vector potential $\tilde{a} = (e/mc\omega)\tilde{E}$ we obtain

$$\tilde{\alpha}_{EMW} = \sqrt{\frac{mc^2 n_c}{2\omega}}\tilde{a}. \tag{1.214}$$

- EPW:

$$\tilde{\alpha}_{EPW} = \sqrt{\frac{mv_\phi^2 n_{e0}}{2\omega}}\frac{\delta\tilde{n}_e}{n_{e0}}. \tag{1.215}$$

- IAW:

$$\tilde{\alpha}_{IAW} = \sqrt{\frac{mv_{Te}^2 n_{e0}}{2\omega}}\frac{\delta\tilde{n}_e}{n_{e0}}. \tag{1.216}$$

Notice the symmetry in these definitions. The dimensionless quantities $\tilde{a}$ and $\delta\tilde{n}_e/n_{e0}$ for EMWs and plasma waves (EPW or IAW), respectively, will be the most practical to work with when we investigate wave-coupling instabilities in later chapters.

1.3.6 Ion Acoustic Waves in Multi-Species Plasma

An important aspect of laser–plasma physics, especially in the area of ICF or HED science, is the fact that plasmas are often composed of multiple ion species. The presence of multiple species, especially when they have very different atomic numbers, can drastically change the properties of ion acoustic waves in the plasma by introducing multiple IAW modes, with different phase velocities and damping. In turn, this can impact the growth of laser–plasma instabilities—and as we will see in Sect. 7.1.4, introducing a new ion species in a plasma can be used as a tool to mitigate the development of instabilities.

In the following we simply assume a plasma with two species of ions, identified by the indices h and l for the heavier and lighter element, respectively. Heavy and light ions have atomic fractions f_h, f_l, such that their respective densities are $n_h = f_h n_i$ and $n_l = f_l n_i$ with $f_h + f_l = 1$. The average atomic number is

$$\langle Z \rangle = f_h Z_h + f_l Z_l \,, \tag{1.217}$$

connecting the electron and ion densities via $n_e = \langle Z \rangle n_i$ (the densities in this section all refer to the "background" values). The analysis is easily generalizable to more than two species (cf. Ref. [10]).

Each of the three particle species in the plasma (electrons, heavy ions, and light ions) is represented by its susceptibility χ_s with $s = e, h, l$. We start our analysis in the fluid framework, i.e.,

$$\chi_s = -\frac{\omega_{ps0}^2}{\omega^2 - \gamma_s k^2 v_{Ts}^2} \tag{1.218}$$

(cf. Eq. 1.112). We also assume that the two species of ions are at thermal equilibrium, i.e., there is a unique ion temperature T_i.

The first and most common situation is that of a "fast mode", i.e., an IAW whose phase velocity $v_{fast} = \omega/k$ is greater than the thermal velocities of both ion species—but of course, still much slower than the electron thermal velocity:

$$v_{Th} < v_{Tl} \ll v_{fast} \ll v_{Te} \,, \tag{1.219}$$

where $v_{Th} = \sqrt{T_i/M_h}$ and $v_{Tl} = \sqrt{T_i/M_l}$ are the thermal velocities of the heavy and light ion species, respectively. By "most common" what we actually mean is that the fast IAW mode is usually (but not always, as we shall see below) the least damped IAW mode and the only one likely to be present and measurable in the plasma, compared to another "slow mode" which we will talk about in a moment.

Assuming $T_i \ll T_e$ leads (to lowest order) to the simplified expressions for the susceptibilities:

$$\chi_e = \frac{1}{k^2 \lambda_{De}^2} \,, \tag{1.220}$$

$$\chi_h = -\frac{\omega_{ph}^2}{\omega^2} \,, \tag{1.221}$$

$$\chi_l = -\frac{\omega_{pl}^2}{\omega^2} \,. \tag{1.222}$$

When inserting in the general dispersion relation $\varepsilon = 1 + \chi_e + \chi_h + \chi_l = 0$ we obtain the dispersion relation for the fast ion acoustic mode:

$$\boxed{v_{fast}^2 = \frac{\omega^2}{k^2} = \left\langle \frac{Z^2}{M_i} \right\rangle \frac{T_e}{\langle Z \rangle \,(1 + k^2 \lambda_{De}^2)} = \frac{\langle Z^2/A \rangle T_e}{\langle Z \rangle \, M_p (1 + k^2 \lambda_{De}^2)}} \,, \tag{1.223}$$

where the averages in the angular brackets are taken over the two species of ions and M_p is the proton mass.

In many instances, this will be the expression to use when estimating the sound speed in a multi-ion species plasma. We can already notice some important properties, like the fact that the acoustic speed for the fast mode is generally dominated by the heavy ion species due to the dependence on Z. This means that introducing light ions in a plasma will not necessarily change the fast mode velocity enough to prevent significant Landau damping of the light ions in these waves. For example, take the example of CH, with $f_H = f_C = 0.5$: assuming $k\lambda_{De} \ll 1$, we have $v_{fast}^2/v_{TC}^2 \approx 48T_e/7T_i$, meaning that the fast mode velocity will be greater than the thermal velocity of the carbon ions as long as $T_e/T_i \gg 7/48$, which is almost always satisfied in most plasmas where the electron temperature is typically higher than the ion's. However, we also have $v_{fast}^2/v_{TH}^2 \approx 4T_e/7T_i$, meaning that v_{fast} will only be faster than the hydrogen thermal velocity if $T_e/T_i \gg 7/4$, which is usually marginally true at best in ICF plasmas, and sometimes not at all. The presence of hydrogen will typically lead to significant damping of the fast acoustic mode because there are enough hydrogen ions in the distribution that are fast enough to catch up with the wave and resonantly interact with it (this was measured in experiments as early as 1967 [11] and is the key idea behind the mitigation of IAW-driven instabilities discussed in Sect. 7.1.4). In fact, as was noted by Williams [10], when the electron temperature is only marginally higher than the ion temperature, the fast mode is not going to be significant due to its strong damping, and the mode with the least damping is the so-called slow mode, which we describe now.

The slow mode of a two-ion-species plasma is such that its phase velocity is at the same time faster than the heavy ion thermal velocity and slower than the light ion thermal velocity:

$$v_{Th} \ll v_{slow} \ll v_{Tl} \ll v_{Te} . \tag{1.224}$$

Fast and slow modes are illustrated in Fig. 1.12.

For the slow mode, we have the following approximations for the susceptibilities:[13]

$$\chi_e = \frac{1}{k^2\lambda_{De}^2} , \tag{1.225}$$

$$\chi_h = -\frac{\omega_{ph}^2}{\omega^2 - 3k^2v_{Th}^2} , \tag{1.226}$$

$$\chi_l = \frac{1}{k^2\lambda_{Dl}^2} , \tag{1.227}$$

---

[13] Here keeping the full expression for the susceptibility of the heavy ion does not complicate the algebra, unlike for the fast mode.

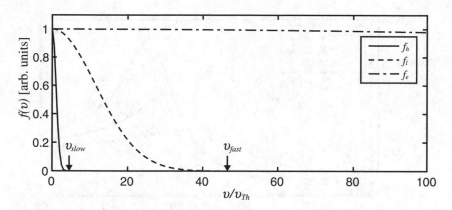

Fig. 1.12 Fast and slow IAW modes in a two-ion-species plasma (here H and Xe). The heavy ion, light ion and electron distribution functions are f_h, f_l and f_e, respectively; the velocity on the plot is normalized to the thermal velocity of the heavy ion v_{Th}. The slow IAW mode should satisfy $v_{Th} \ll v_{slow} \ll v_{Tl} \ll v_{Te}$, whereas the fast IAW mode satisfies $v_{Th} \leq v_{Tl} \ll v_{fast} \ll v_{Te}$

where $\lambda_{Dl} = v_{Tl}/\omega_{pl}$ is the Debye length of the light ions.

Inserting these expressions into $\varepsilon = 1 + \chi_e + \chi_l + \chi_h = 0$ gives the following dispersion relation for the slow mode:

$$v_{slow}^2 = \frac{\omega^2}{k^2} = 3v_{Th}^2 + f_h \frac{Z_h^2}{M_h \langle Z \rangle} \frac{T_e}{1 + k^2\lambda_{De}^2 + f_l Z_l^2 T_e/(\langle Z \rangle T_i)}. \tag{1.228}$$

We see that for the slow mode, the wave dispersion is primarily dependent on the heavy ions; the light ions play the same role as the electrons in shielding the electric field, since they travel much faster (on average) than the wave's phase velocity. Slow modes in multi-species plasmas were observed using Langmuir probes [12] and later using Thomson scattering [13].

An important note is that in most practical cases, it will be difficult for the slow mode to satisfy the condition from Eq. (1.224). In our example from Fig. 1.12, we used a mixture of H ($A = 1$) and Xe ($A = 131$), which has $v_{Tl}/v_{Th} = \sqrt{A_{Xe}/A_H} \approx 11$, which is pretty much as large a ratio as one can get with realistic plasma compositions. At best, we have $v_{slow} \approx 3.4v_{Th} \approx v_{Tl}/3.4$. In more typical ICF or HED plasmas, for example, CH, we have $v_{Tl}/v_{Th} \approx 3.5$, and Eq. (1.224) can never be quite satisfied.

What this means is that the fluid expansion used here to obtain the dispersion relation and sound speed is often inappropriate, and the general rule is that when working with multi-species plasmas, one should use a kinetic model instead of fluid, i.e., use Eq. (1.162) to express the susceptibilities of the different species. The full kinetic theory of ion acoustic modes in multi-species plasmas was first presented by Fried [14], and compared to fluid theory for typical laser–plasma conditions by Williams [10].

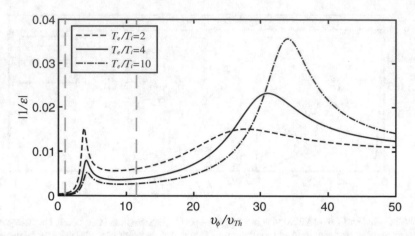

Fig. 1.13 $1/|\varepsilon|$ as a function of $v_\phi = \omega/k$ for Xe_1H_{99}, normalized to the heavy ion (here Xe) thermal velocity (and for $k\lambda_{De} = 0.1$). The two thermal velocities are indicated by the gray dashed lines (with $v_{Tl}/v_{Th} = \sqrt{A_{Xe}/A_H} \approx 11.4$)

Let us show a few examples to illustrate what we just talked about. First we consider the case of a hydrogen-xenon mixture with 99%H/1%Xe atomic fractions, following Ref. [10], since in this case the fast and slow modes are clearly separated. A high fraction of hydrogen needs to be present for both modes to be visible, otherwise xenon dominates. As we showed in Sect. 1.3.3 the modes correspond to the zeros of the dielectric constant ε for real ω (the susceptibilities are in general complex, with the imaginary parts corresponding to the waves' Landau damping). Therefore, to illustrate the different acoustic modes we plot $1/|\varepsilon|$ as a function of the phase velocity ω/k, and look for peaks which correspond to the different modes.

The plot of $1/|\varepsilon|$ for Xe_1H_{99} is shown in Fig. 1.13, for three different values of T_e/T_i. We can clearly see the slow mode (near $v_\phi/v_{Th} \approx 3.5-4$, in-between the thermal velocities of H and Xe which are indicated by the gray dashed lines), as well as the fast mode (near $v_\phi/v_{Th} \approx 30-40$). The fast mode is more prominent at high T_e/T_i, whereas the slow mode can become dominant when T_e approaches T_i.

Next we show the example of hydrogen-carbon mixtures, where the situation is more complicated, as we mentioned already. In our example we chose $f_H = 0.7$ (i.e., C_3H_7), for which the transition between the slow and fast mode is more clearly visible. A plot of $1/|\varepsilon|$ is shown in Fig. 1.14.

As we can see from the figure, the two modes are barely distinguishable, because the thermal velocities of carbon and hydrogen are not separated enough (compared to XeH above). The slow mode is mostly visible at smaller values of T_e/T_i, and coincides with the thermal velocity of hydrogen. As a result, both modes are strongly damped, and the sound speed in C-H plasma mixtures is very sensitive to T_e/T_i. Another important point is that the dominant mode, whether it is the fast or slow mode, will typically be strongly damped in the range of parameters relevant to ICF

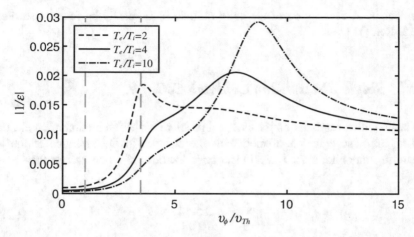

Fig. 1.14 $1/|\varepsilon|$ as a function of v_ϕ/v_{Th} for C_3H_7 and $k\lambda_{De} = 0.1$. The two thermal velocities are again indicated by the gray dashed lines (with $v_{Tl}/v_{Th} = \sqrt{A_C/A_H} \approx 3.5$)

conditions—as illustrated by the broad aspect (vs. narrow peaks) of the curves in Fig. 1.14 (Landau damping usually corresponds to the width of the resonance peak).

To conclude: because the damping of IAWs plays an important role for laser–plasma instabilities, as we will see in later chapters, engineering the composition of mixed-species plasma can be a powerful tool to control the development of instabilities. This will be discussed in more detail in Sect. 7.1.4.

1.4   Electron-Ion Collisions

Collisions play a central role in laser–plasma physics, especially for HED conditions with relatively long laser pulses and hot plasmas. They are at the origin of the collisional absorption process (also frequently called inverse bremsstrahlung), which is the primary absorption mechanism for laser–plasma experiments in HED conditions (cf. Chap. 4); they also play an important role in saturating kinetic effects, as we will show in Sect. 10.3.

In this section we restrict our analysis to electron-ion (e-i) collisions, which are significantly easier to describe analytically than e-e or i-i collisions as the large electron/ion mass ratio reduces the problem to elastic collisions between light particles (electrons) and immobile targets (ions). Electron-ion collisions are the process through which collisional absorption of laser energy occurs, and is one of the most important collisional effects responsible for de-trapping of trapped electrons in plasma waves, which sets the threshold for the "kinetic inflation" phenomenon as described in Sect. 10.3. Understanding e-i collisions will thus be sufficient to describe the most important collisional processes in laser–plasma interactions. A description of e-e or i-i collisions can easily be found elsewhere, like in Ref. [15].

1.4.1 Single Electron-Ion Coulomb Collision

We begin with a reminder on the classical problem of the interaction of an electron and an ion. The equation of motion of the electron in the presence of the ion (assumed immobile, since typical electron velocities are much greater than ions') is:

$$m\ddot{\mathbf{r}} = -\frac{Ze^2}{4\pi\varepsilon_0}\frac{\mathbf{r}}{r^3}\,, \tag{1.229}$$

where the ion location is chosen as the origin of the coordinate system ($\mathbf{r} = 0$), m is the electron mass and $\mathbf{r}(t)$ its trajectory. The dot denotes the time derivative d_t (and the double dot, d_t^2).

To proceed, we switch to a polar coordinate system (r, φ) in the interaction plane (x, y) of the two particles, with $\mathbf{e}_r = \cos(\varphi)\mathbf{e}_x + \sin(\varphi)\mathbf{e}_y$, $\mathbf{e}_\varphi = -\sin(\varphi)\mathbf{e}_x + \cos(\varphi)\mathbf{e}_y$: using $\dot{\mathbf{e}}_r = \dot{\varphi}\mathbf{e}_\varphi$ and $\dot{\mathbf{e}}_\varphi = -\dot{\varphi}\mathbf{e}_r$, the general expressions for the position, velocity, and acceleration are

$$\mathbf{r} = r\mathbf{e}_r\,, \tag{1.230}$$

$$\dot{\mathbf{r}} = \dot{r}\mathbf{e}_r + r\dot{\varphi}\mathbf{e}_\varphi\,, \tag{1.231}$$

$$\ddot{\mathbf{r}} = (\ddot{r} - r\dot{\varphi}^2)\mathbf{e}_r + (2\dot{r}\dot{\varphi} + r\ddot{\varphi})\mathbf{e}_\varphi\,, \tag{1.232}$$

and the angular momentum $\mathbf{L} = m\mathbf{r} \times \mathbf{v}$ and its time derivative $\dot{\mathbf{L}}$ can be expressed as

$$\mathbf{L} = mr^2\dot{\varphi}\mathbf{e}_z\,, \tag{1.233}$$

$$\dot{\mathbf{L}} = mr(2\dot{r}\dot{\varphi} + r\ddot{\varphi})\mathbf{e}_z\,. \tag{1.234}$$

Since the acceleration in Eq. (1.229) is purely radial, the component of $\ddot{\mathbf{r}}$ along $\mathbf{e}_\varphi$ must be zero in Eq. (1.232), which from the expression above for $\dot{\mathbf{L}}$ implies that $\dot{\mathbf{L}} = 0$, i.e., the angular momentum is conserved. Taking b as the impact parameter, as illustrated in Fig. 1.15, the angular momentum can be found by taking its initial value at $t \to -\infty$, $\mathbf{L} = m\mathbf{r}(-\infty) \times \mathbf{v}(-\infty)$, leading to

$$L = mbv_0\,, \tag{1.235}$$

where v_0 is the initial electron velocity.

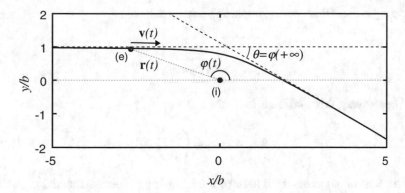

Fig. 1.15 Illustration of an electron-ion collision. Here we used $b/b_\perp \approx 3.7$, giving a scattering angle $\theta = 30°$ (with b the impact parameter and $b_\perp$ the 90° impact parameter)

The scattering angle θ can be derived by estimating the limits of the polar angle φ at $t \to \infty$ (cf. Fig. 1.15; we assume that $t = 0$ is the time of closest approach). To proceed, we introduce the new variable $u = 1/r$. From Eqs. (1.233), (1.235), we obtain $\dot{\varphi} = -bv_0 u^2$, and after using the transformations $\dot{r} = bv_0 du/d\varphi$, $\ddot{r} = -(bv_0)^2 u^2 d^2u/d\varphi^2$, we can transform the equation of motion in polar coordinates,

$$m(\ddot{r} - r\dot{\varphi}^2) = -\frac{Ze^2}{4\pi\varepsilon_0 r^2}, \tag{1.236}$$

into:

$$\frac{d^2u}{d\varphi^2} + u = \frac{Ze^2}{4\pi\varepsilon_0 b^2 v_0^2 m}. \tag{1.237}$$

A second change of variable to $w = u - Ze^2/(4\pi\varepsilon_0 mb^2 v_0^2)$ yields $d^2w/d\varphi^2 + w = 0$, with solution $w = c_1\cos(\varphi) + c_2\sin(\varphi)$. Taking the limit of $t \to -\infty$, for which $\varphi(-\infty) = \pi$ and $u(-\infty) = 0$, yields $c_1 = Ze^2/(4\pi\varepsilon_0 mb^2 v_0^2)$. The asymptotic limit for the velocity, $\dot{r}(-\infty) = v_0 = bv_0(du/d\varphi)_{-\infty}$ gives $(du/d\varphi)_{-\infty} = (dw/d\varphi)_{-\infty} = 1/b$, which in turn gives $c_2 = -1/b$. Then, taking $t \to +\infty$ with $\varphi(+\infty) = \theta$ and $w(+\infty) = -c_1$, we get from the expression of $w(\varphi)$:

$$-c_1 = c_1\cos(\theta) + c_2\sin(\theta). \tag{1.238}$$

Using the trigonometric identity $\cotan(\theta/2) = [1 + \cos(\theta)]/\sin(\theta)$, we finally obtain

$$\cotan(\theta/2) = -\frac{b}{b_\perp}, \tag{1.239}$$

where $b_\perp$ is the 90° scattering impact parameter defined as

$$\boxed{b_\perp = \frac{Ze^2}{4\pi \varepsilon_0 m v_0^2}}.$$

(1.240)

Alternatively, this also leads to

$$\theta = -2\arctan\left(\frac{b_\perp}{b}\right).$$

(1.241)

The "loss of momentum" of the electron is defined as the variation in longitudinal momentum (with respect to its initial direction) due to collisions, i.e., $\delta v_x = v_0[\cos(\theta) - 1]$. Inserting θ from the expression above and using some trigonometric identities leads to:

$$\delta v_x = -\frac{2v_0}{1 + b^2/b_\perp^2}.$$

(1.242)

In the small angle limit of $\theta \ll 1$, i.e., $b \gg b_\perp$, we have

$$\delta v_x \approx -2v_0 \frac{b_\perp^2}{b^2}.$$

(1.243)

1.4.2   Momentum Loss for a Single Electron Interacting with a Background of Ions

Next we can calculate the loss of momentum for a single electron interacting with a background of ions, still considered immobile. Consider an electron with initial velocity $\mathbf{v} \propto \mathbf{e}_x$: as it propagates along an infinitesimal distance $dx = vdt$, we can estimate the number of ions for which the impact parameter will be between b and $b + db$ as the number of ions in the volume element dV comprised between the two cylinders of radii b and $b + db$ and length dx, as illustrated in Fig. 1.16. Taking db as an infinitesimal element, we can express the volume as $dV = dx\,\pi[(b + db)^2 - b^2] \approx 2\pi b\,db\,dx$, and thus the number of ions in dV as

$$dN_i = 2\pi n_i dx\, b\, db = 2\pi n_i v\, dt\, b\, db,$$

(1.244)

where n_i is the ion density.

The loss of momentum (i.e., reduction in the longitudinal velocity component v_x as derived in the previous section) is then expressed by summing up the contributions from all the ions surrounding the electrons at (almost) all impact parameters—we will quantify what we mean by "almost" in a moment. The reduction in v_x per unit time can then be expressed as

Fig. 1.16 Geometry for a
single electron at velocity **v**
interacting with background
ions during an infinitesimal
time $dt = dx/v$ at impact
parameters comprised in
$[b, b + db]$

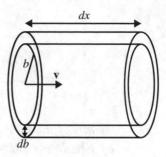

$$\frac{dv_x}{dt} = \int_{b_{min}}^{b_{max}} \frac{dN_i}{dt}\delta v_x \, , \qquad (1.245)$$

with δv_x the loss of momentum from the interaction with a single electron given by
Eq. (1.242). Substituting for the expressions of δv_x and dN_i gives:

$$\frac{dv_x}{dt} = -4\pi n_i v^2 \int_{b_{min}}^{b_{max}} \frac{b}{1 + b^2/b_\perp^2}db = -4\pi n_i v^2 b_\perp^2 \ln(\Lambda) \, , \qquad (1.246)$$

where

$$\ln(\Lambda) = \ln\sqrt{\frac{1 + b_{max}^2/b_\perp^2}{1 + b_{min}^2/b_\perp^2}} \qquad (1.247)$$

is the Coulomb logarithm, obtained from the integration over the impact parameters.
We can now discuss the integration limits: it is common to use hard cutoffs due to
the weak variation of the logarithm function, and to choose $b_{max} = \lambda_{De}$ (as the
electron is not going to interact with ions located beyond the Debye length due to
the screening effect) and $b_{min} = \hbar/(2mv)$, the De Broglie wavelength below which
a quantum treatment is required (cf. Sect. 4.4 footnote 3). Since $b_{max} = \lambda_{De}$ will
typically be $\gg b_\perp$, the expression for Λ simplifies to

$$\Lambda \approx \frac{b_{max}}{\sqrt{b_\perp^2 + b_{min}^2}} \, , \qquad (1.248)$$

with the denominator approximately equal to the maximum of $(b_{min}, b_\perp)$.

This allows us to define a (temporal) rate of loss of momentum $\nu_{/\!/}$ for a single
electron interacting with a background of ions as

$$\frac{dv_x}{dt} = -\nu_{/\!/}v_x \, , \qquad (1.249)$$

with

$$\nu_{/\!/}(v) = \frac{4\pi n_i}{m^2 v^3} \left(\frac{Ze^2}{4\pi \varepsilon_0} \right)^2 \ln(\Lambda) .$$

(1.250)

1.4.3   Momentum Loss for a Drifting Maxwellian Distribution of Electrons Interacting with Ions

In this next step, we are going to estimate the loss of momentum for a Maxwellian distribution whose average velocity is not zero but $\mathbf{v}_d$, a drift velocity assumed small compared to the thermal velocity v_{Te}. The electron-ion collisions will randomize the velocity directions of the electrons and act to negate the drift and re-establish a zero-mean velocity distribution. The rate we are going to derive, which is the rate of loss of momentum of the electron population averaged over the entire distribution, is often simply referred to as the "electron-ion collision frequency."

The distribution function has the form:

$$f_e(\mathbf{v}) = \frac{n_e}{(2\pi)^{3/2} v_{Te}^3} \exp \left[-\frac{(\mathbf{v} - \mathbf{v}_d)^2}{2v_{Te}^2} \right] ,$$

(1.251)

where $\mathbf{v}_d$ is the drift velocity with $v_d \ll v_{Te}$. The average momentum loss for the electrons along x, taken to be the direction of $\mathbf{v}_d$, is estimated by averaging an individual electron's momentum loss over the entire distribution:

$$\frac{d \langle v_x \rangle}{dt} = \frac{dv_d}{dt} = \int \frac{f_e(\mathbf{v})}{n_e} \frac{dv_x}{dt} d^3 v$$

$$= -\frac{1}{n_e} \int f_e(\mathbf{v}) \nu_{/\!/}(v) v_x d^3 v ,$$

(1.252)

where $\nu_{/\!/}$ is the rate of momentum loss for a single electron given by Eq. (1.250) and $d^3 v = dv_x dv_y dv_z$. Technically, the Coulomb logarithm depends on v and makes the integration challenging; this is typically circumvented by taking the value of Λ at the electron thermal velocity and moving the logarithm out of the integral, i.e.,

$$\ln \left[\Lambda(v) \right] \approx \ln \left[\Lambda(v_{Te}) \right] .$$

(1.253)

The other simplification results from our initial hypothesis that $v_d \ll v_{Te}$, which implies that

$$\exp \left[-\frac{(\mathbf{v} - \mathbf{v}_d)^2}{2v_{Te}^2} \right] \approx \left(1 + \frac{\mathbf{v} \cdot \mathbf{v}_d}{v_{Te}^2} \right) \exp \left[-\frac{v^2}{2v_{Te}^2} \right] = \left(1 + \frac{v_x v_d}{v_{Te}^2} \right) e^{-v^2/2v_{Te}^2} .$$

(1.254)

Substituting these and the expression for $v_{//}$ into Eq. (1.252) leads to

$$\frac{dv_d}{dt} = -\frac{2n_i}{\sqrt{2\pi}m^2 v_{Te}^3} \left(\frac{Ze^2}{4\pi\varepsilon_0}\right)^2 \ln(\Lambda) \int \left(1 + \frac{v_x v_d}{v_{Te}^2}\right) e^{-v^2/2v_{Te}^2} \frac{v_x}{v^3} d^3v. \quad (1.255)$$

The 1 in the parenthesis leads to an odd integral over v_x from $-\infty$ to ∞ which is equal to zero. The remaining integral to be calculated is

$$I = v_d \int \frac{u_x^2}{u^3} e^{-u^2/2} d^3u, \quad (1.256)$$

where $u = v/v_{Te}$ and $d^3u = du_x du_y du_z$. Noticing that by symmetry, we would have the same integral by substituting v_x for v_y or v_z, we have

$$I = \frac{v_d}{3} \int \frac{u_x^2 + u_y^2 + u_z^2}{u^3} e^{-u^2/2} d^3u = \frac{v_d}{3} \int \frac{e^{-u^2/2}}{u} d^3u. \quad (1.257)$$

Switching to spherical coordinates leads to

$$I = \frac{v_d}{3} \int_0^{2\pi} d\varphi \int_0^{\pi} d\theta \int_0^{\infty} du \frac{e^{-u^2/2}}{u} u^2 \sin(\theta)$$

$$= \frac{4\pi}{3} v_d, \quad (1.258)$$

where we used $\int_0^{\infty} xe^{-x^2/2} dx = 1$.

Inserting back into the rate of momentum loss leads to

$$\boxed{\frac{dv_d}{dt} = -\nu_{ei} v_d}, \quad (1.259)$$

where the rate of average momentum loss is

$$\boxed{\nu_{ei} = \frac{2n_i}{\sqrt{2\pi}m^2 v_{Te}^3} \left(\frac{Ze^2}{4\pi\varepsilon_0}\right)^2 \frac{4\pi}{3} \ln(\Lambda)}. \quad (1.260)$$

The average rate of momentum loss for electrons interacting with ions is often simply referred to as the electron-ion collision rate. We shall keep its precise definition in mind: in the case where the electron population has a velocity drift along a certain direction, ν_{ei} is the rate at which the collisions with ions will suppress the drift and tend to restore a zero-mean velocity distribution. There are other rates associated with electron-ion collisions, such as the electron energy loss, which we neglected here by assuming elastic collisions. The rate of electron energy loss is much smaller than the rate of momentum loss, by a factor $\propto m/M_i$ (cf. [16] or [15]), but is non-zero.

Note that ν_{ei} is simply equal (up to a constant numerical factor) to the rate of momentum loss for a single electron propagating at v_{Te},

$$\nu_{ei} = \frac{2}{3\sqrt{2\pi}} \nu_{/\!/}(v_{Te}) . \tag{1.261}$$

It can also be expressed in the convenient form:

$$\frac{\nu_{ei}}{\omega_{pe}} = \frac{Z \ln(\Lambda)}{3(2\pi)^{3/2} N_{De}} , \tag{1.262}$$

where $N_{De} = n_e \lambda_{De}^3$ is the number of electrons in the Debye cube. In practical units, we have

$$\nu_{ei}[\text{ns}^{-1}] \approx 9.2 \frac{Z n_e [10^{20}\text{cm}^{-3}]}{T_e^{3/2}[\text{keV}]} \ln(\Lambda). \tag{1.263}$$

The electron-ion collision rate can be used to estimate the collisional absorption of light waves in plasma (cf. Sect. 4.2); however, we will have to remember that the expression for ν_{ei} above is only valid in the limit of velocity drifts that are very small compared to v_{Te}, and as we will discuss later, the Coulomb logarithm will also need to be adjusted when calculating the collisional absorption rate (Sect. 4.4).

1.5 Isothermal Expansion of Plasma in Vacuum (and Ion Acceleration) in 1D

In this section we describe a well-know, simple model of the expansion of plasma in vacuum [17]. This simple model is a fairly good approximation of the type of plasmas produced in high-energy, long-pulse ($\sim$ns) laser experiments. The electron temperature is assumed constant—indeed the electron thermal velocity, which is a typical scale for heat transport, is much larger than the sound speed which constitutes the typical scale for expansion in vacuum as we will see.

We use a fluid approach for both electrons and ions, and make a 1D approximation (which is typically valid for regions within the focal spot of a high-energy laser beam). The plasma is assumed to be initially at constant density n_{e0}, $n_{i0} = n_{e0}/Z$ in the $z < 0$ region, and expands into the vacuum initially located at $z \geq 0$.

Since we are interested in plasma dynamics on the time scale of ion motion, we neglect electron inertia (like we did in our description of IAWs), so the electron fluid equation simplifies to

$$0 = -\frac{e}{m}E - v_{Te}^2\frac{\partial_z n_e}{n_e}, \tag{1.264}$$

where E is the electrostatic field associated with the charge separation (as the electrons move faster and pull the ions behind).

For the ions, the first two fluid equations are

$$\partial_t n_i + \partial_z(n_i v_i) = 0, \tag{1.265}$$

$$(\partial_t + v_i\partial_z)v_i = -3v_{Ti}^2\frac{\partial_z n_i}{n_i} + \frac{Ze}{M_i}E. \tag{1.266}$$

We can slightly recast the first equation and eliminate E from the second via the electron fluid equation to obtain

$$(\partial_t + v_i\partial_z)n_i = -n_i\partial_z v_i, \tag{1.267}$$

$$(\partial_t + v_i\partial_z)v_i = -\frac{c_s^2}{n_i}\partial_z n_i, \tag{1.268}$$

with $c_s^2 = (ZT_e + 3T_i)/M_i$ and where we assumed quasi-neutrality. The system of equations takes an even more symmetric form if we use $\ln(n_i)$ rather than n_i (since $(\partial_z n_i)/n_i = \partial_z(\ln n_i)$, id. with ∂_t) and introduce the Mach number $M = v_i/c_s$:

$$\left(\frac{1}{c_s}\partial_t + M\partial_z\right)\ln(n_i) = -\partial_z M, \tag{1.269}$$

$$\left(\frac{1}{c_s}\partial_t + M\partial_z\right)M = -\partial_z(\ln n_i). \tag{1.270}$$

At this point we introduce the dimensionless variable

$$\xi = \frac{z}{c_s t}, \tag{1.271}$$

which will provide a self-similar solution to the system of equations. Noting that

$$\frac{\partial}{\partial z} = \frac{\partial\xi}{\partial z}\frac{\partial}{\partial\xi} = \frac{1}{c_s t}\partial_\xi, \tag{1.272}$$

$$\frac{\partial}{\partial t} = \frac{\partial\xi}{\partial t}\frac{\partial}{\partial\xi} = -\frac{\xi}{t}\partial_\xi, \tag{1.273}$$

the two coupled equations can then be recast in a matrix form with the self-similar variable ξ replacing z and t:

$$\begin{pmatrix} (M - \xi) & 1 \\ 1 & (M - \xi) \end{pmatrix} \begin{pmatrix} \partial_\xi (\ln n_i) \\ \partial_\xi M \end{pmatrix} = 0 \,. \tag{1.274}$$

The system admits non-trivial solutions if the matrix determinant is zero, meaning

$$M = \xi \pm 1 \,, \tag{1.275}$$

or, expressed in terms of z and t,

$$v_i = \frac{z}{t} \pm c_s \,. \tag{1.276}$$

Inserting back into Eq. (1.270) leads to $\partial_z (\ln n_i) = \mp 1/c_s t$, which can be integrated to yield

$$n_i = K \exp\left[\mp \frac{z}{c_s t} \right], \tag{1.277}$$

where K is a constant of integration.

We need to retain only the solution with "$-$" in the expression above (hence with "$+$" in the expressions for M and v_i), which corresponds to the physically acceptable situation of a decaying density in the vacuum region. The solution indicates that the velocity grows linearly vs. z (at fixed t) in the vacuum region, starting at $z_0(t) = -c_s t$ where $v_i(z_0) = 0$. The density at z_0 is then equal to $n_i(z_0) = K \exp(1)$, which leads us to set the integration constant to $K = n_{i0} \exp(-1)$ so that $n_i(z_0) = n_{i0}$ and $n_i = n_{i0} \exp[-(1 + z/c_s t)]$. We see that $z_0(t)$ moves toward $z < 0$ at c_s (it is a "rarefaction wave" [18]), and corresponds

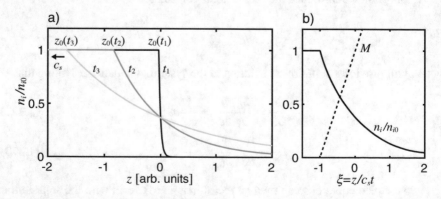

Fig. 1.17 Plasma expansion in vacuum: a) density vs. z at three different times $t_1 < t_2 < t_3$; the rarefaction front, located at $z_0(t)$, moves toward $z < 0$ at c_s. b) Self-similar profiles of density and flow velocity ($M = v_i/c_s$) as a function of the dimensionless parameter $\xi = z/c_s t$. The Mach 1 location stays fixed at the initial location of the vacuum–plasma boundary

to the position from which the plasma starts to flow and expand into vacuum, as represented in Fig. 1.17.

The electric field can then be obtained from the electron fluid equation, leading to $E = T_e/(ec_s t)$. It is uniform in space, and decays in time as $\sim 1/t$.

Since $\int_{z_0(t)}^{\infty} n_i(z)dz = n_{i0}c_s t$ we can verify that the ablated mass during the interval $[0, t]$ is conserved (indeed the mass that was initially in the region $[z_0(t), 0]$ is simply $n_{i0}c_s t$, cf. Fig. 1.17). We also note that the location of Mach 1 flow ($v_i = c_s$) stays fixed at all times, at the initial plasma-vacuum boundary $z = 0$ (cf. Fig. 1.17).

In summary, the key results are:

$$v_i(z, t) = \frac{z}{t} + c_s = \frac{z - z_0(t)}{t}, \tag{1.278}$$

$$n_i(z, t) = n_{i0} \exp[-1 - z/c_s t], \tag{1.279}$$

$$z_0(t) = -c_s t, \tag{1.280}$$

$$E(t) = \frac{T_e}{ec_s t}, \tag{1.281}$$

and of course $n_i(z \leq z_0(t)) = n_{i0}$, $v_i(z < z_0) = 0$ (cf. Fig. 1.17).

These results take a self-similar form via the variable ξ:

$$M = \xi + 1, \tag{1.282}$$

$$n_i = n_{i0} \exp[-(1 + \xi)] = n_{i0} \exp[-(\xi - \xi_0)], \tag{1.283}$$

$$\xi_0 = -1, \tag{1.284}$$

where $\xi_0 = z_0/c_s t$. Note that this simple model ignores the initial behavior—as we can see from the fact that our quantities diverge at $t = 0$. The plasma motion is initially setup by the electrons, which can expand into vacuum by a distance on the order of λ_{De} while the ions still remain immobile with a constant density n_{i0} at $z \leq 0$. The resulting charge separation sets up the ion motion; the self-similar model is only valid at times such that the density spatial scale-length $\sim c_s t \gg \lambda_{De}$.

Note that this self-similar model also provides the energy distribution of accelerated ions in vacuum [19]. Since $n_i = n_{i0} \exp[-(1 + \xi)] = n_{i0} \exp[-M]$, taking the ion kinetic energy $\mathcal{E} = M_i v_i^2/2 = M_i c_s^2 M^2/2$ immediately leads to

$$\frac{\partial n_i}{\partial \mathcal{E}} = \frac{\partial n_i}{\partial M} \frac{\partial M}{\partial \mathcal{E}} = -\frac{n_{i0}}{\sqrt{2\mathcal{E}ZT_e}} \exp[-\sqrt{2\mathcal{E}/ZT_e}]. \tag{1.285}$$

Since the density depends on z but not on the two transverse dimensions x and y (as we assumed at the beginning that the problem was essentially infinite along the transverse directions and only 1D), it makes more sense to integrate along z and define a density of ions per unit surface $N_i = \int n_i dz = n_i z = n_i c_s t$, i.e.,

$$\boxed{\frac{\partial N_i}{\partial \mathcal{E}} = -\frac{n_{i0}c_s t}{\sqrt{2\mathcal{E}ZT_e}} \exp[-\sqrt{2\mathcal{E}/ZT_e}]}. \tag{1.286}$$

This shows that the plasma expansion in vacuum generates a distribution of ions that can reach high energies, on the order of the electron thermal energy T_e (indeed the ions are accelerated by the electric field set by the electrons' displacement, with $E \propto T_e$). This is one of the basic ideas behind laser-driven ion acceleration schemes (cf. [20] and references therein). Clearly, reaching a high electron temperature in the target is paramount to achieving ion acceleration to high energies.

The number of ions in the distribution depends on time ($\propto c_s t$), which simply means that the density of high-energy ions keeps increasing as the rarefaction wave propagates into the target. We can expect the process to stop once the laser, which generates the plasma and feeds the rarefaction wave, is turned off—i.e., the final distribution should be reasonably well-approximated by taking $t \approx \tau_{laser}$ in the expression for $\partial_{\mathcal{E}} N_i$. Laser-solid experiments have measured ion spectra and confirmed that this is indeed the case [21].

A more accurate description of the ion front (launched from the front surface at $t = 0$) including a detailed accounting of charge separation effects was derived by Mora [19]. The analysis also provides an estimate for the maximum energy of the ions or "cutoff" in their energy spectra (indeed the self-similar model predicts acceleration to infinite energies, which is obviously unphysical).

Problems

1.1 Dephasing from a Thin Lens
Express the optical path length of a light ray propagating parallel to z at a given radius $r_{0\perp}$ between the $z = z_0$ and $z = z_0'$ planes, by decomposing into its propagation in air and propagation through the lens' first and second half (path lengths δl_1 and δl_2 in Fig. 1.2). Use a development assuming $r_{0\perp} \ll R_1, |R_2|$ (the paraxial approximation). Verify the dephasing expression from Eq. (1.19).

1.2 The Maxwellian Distribution
Go through the derivations for all the expressions from Sect. 1.2.3. You will make use of the following definite integral formulas:

$$\int_0^\infty x^{2n} e^{-\alpha x^2} dx = \frac{(2n)!}{n!2^{2n+1}} \sqrt{\frac{\pi}{\alpha^{2n+1}}}, \tag{1.287}$$

$$\int_0^\infty x^{2n+1} e^{-\alpha x^2} dx = \frac{n!}{2\alpha^{n+1}}, \tag{1.288}$$

where $\alpha > 0$ and $n \in \mathbb{N}$.

1.3 Some Features and Approximations of IAWs

- Show that for IAWs, the approximation $\chi_e \approx 1/(k\lambda_{De})^2$ is equivalent to assuming $c_s \ll v_{Te}$, and is again equivalent to neglecting $\partial_t \delta \mathbf{v}_e$ in the momentum equation for the electrons. Give an interpretation in terms of electron inertia in a slow IAW.
- Show that for longitudinal waves,

$$\frac{\delta n_e}{n_{e0}} = -\frac{\delta n_i}{n_{i0}} \frac{\chi_e}{\chi_i}, \tag{1.289}$$

$$\delta \mathbf{v}_e = -\frac{\chi_e}{\chi_i} \delta \mathbf{v}_i, \tag{1.290}$$

and that for IAWs we have $-\chi_i/\chi_e = 1 + k^2 \lambda_{De}^2 \approx 1$ for $k\lambda_{De} \ll 1$.
- Combine the two-ion fluid equations by eliminating $\delta \mathbf{v}_i$; eliminate the electric field $\mathbf{E}$ by using the electron momentum equation with $\gamma_e = 1$, $\partial_t \delta \mathbf{v}_e = 0$ and assuming quasi-neutrality, to arrive more rapidly at Eq. (1.129). Try to grasp where the different approximations used for the derivation and simplification of the IAW dispersion relation ($c_s \ll v_{Te}$, $ZT_e \ll T_i$, $k\lambda_{De} \ll 1$) originate in the fluid equations.

1.4 Fluid vs. Kinetic Resonances for EPWs and IAWs

Calculate the estimated resonant frequencies of the EPW and IAW for the conditions of Fig. 1.9; compare to the locations of the two peaks on the figure and comment on the accuracy of the fluid description for the EPW vs. the IAW.

References

1. D. Nicholson, *Introduction to Plasma Theory*, Plasma Physics Series (Wiley, 1983)
2. R.J. Goldston, P. Rutherford, *Introduction to Plasma Physics* (Institute of Physics Pub., 1995)
3. F.F. Chen, *Introduction to Plasma Physics and Controlled Fusion*, 3rd ed. (Springer International Publishing, 2016)
4. P. Mora, *Plasmas créés par laser* (EDP Sciences, 2021)
5. W.L. Kruer, *The Physics of Laser Plasma Interaction* (Addison-Wesley, New York, 1988)
6. B.D. Fried, S.D. Conte, *The Plasma Dispersion Function: The Hilbert Transform of the Gaussian* (Academic Press, 2015)
7. D. Swanson, *Plasma Waves, 2nd Edition*, Series in Plasma Physics (Taylor & Francis, 2003)
8. A. Vedenov, E. Velikhov, R. Sagdeev, Nuclear Fusion **1**, 82 (1961)
9. L. Landau, J. Phys. USSR **10**, 25 (1946)
10. E.A. Williams, R.L. Berger, R.P. Drake, A.M. Rubenchik, B.S. Bauer, D.D. Meyerhofer, A.C. Gaeris, T.W. Johnston, Phys. Plasmas **2**, 129 (1995)
11. I. Alexeff, W.D. Jones, D. Montgomery, Phys. Rev. Lett. **19**, 422 (1967)
12. Y. Nakamura, M. Nakamura, T. Itoh, Phys. Rev. Lett. **37**, 209 (1976)
13. S.H. Glenzer, C.A. Back, K.G. Estabrook, R. Wallace, K. Baker, B.J. MacGowan, B.A. Hammel, R.E. Cid, J.S. De Groot, Phys. Rev. Lett. **77**, 1496 (1996)
14. B.D. Fried, R.B. White, T.K. Samec, Phys. Fluids **14**, 2388 (1971)

15. J.D. Callen, *Fundamentals of Plasma Physics* (University of Wisconsin, 2006). https://cptc. wisc.edu/course-materials/
16. I. Hutchinson, J. Freidberg, 22.611J – *Introduction to Plasma Physics I* (Massachusetts Institute of Technology: MIT OpenCourseWare, 2003). https://ocw.mit.edu
17. A. Gurevich, L. Pariiskaya, L. Pitaevskii, Sov. Phys. JETP **22**, 449 (1966)
18. R.P. Drake, *High-Energy-Density Physics: Fundamentals, Inertial Fusion, and Experimental Astrophysics*, ed. by L. Davison, Y. Horie (Springer, Berlin, Heidelberg, 2006)
19. P. Mora, Phys. Rev. Lett. **90**, 185002 (2003)
20. A. Macchi, M. Borghesi, M. Passoni, Rev. Mod. Phys. **85**, 751 (2013)
21. J. Fuchs, P. Antici, E. d'Humières, E. Lefebvre, M. Borghesi, E. Brambrink, C. A. Cecchetti, M. Kaluza, V. Malka, M. Manclossi, S. Meyroneinc, P. Mora, J. Schreiber, T. Toncian, H. Pépin, P. Audebert, Nature Physics **2**, 48 (2006)

Chapter 2
Single Particle Dynamics in Light Waves and Plasma Waves

In the previous chapter, we saw that the Coulomb gauge conveniently distinguishes longitudinal waves (EPWs or IAWs), described by their electrostatic potential Φ, from the transverse waves (EMWs), described by their vector potential $\mathbf{A}$, such that $\mathbf{E} = -\partial_t \mathbf{A} - \nabla \Phi$. In this chapter, we will describe the motion of a charged particle in either type of wave and see that the nature of the wave (transverse vs. longitudinal) leads to fundamentally different types of dynamics.

In particular, in light waves, we will immediately recover the quiver motion already discussed for the electron fluid in the previous chapter but will see how this motion is perturbed when the laser intensity gets high enough for nonlinear effects to kick in—resulting in the well-known "figure of eight" motion of the electron in an intense laser.

On the other hand, a particle in a longitudinal wave (e.g., an electron in an EPW, or an ion in an IAW) can get trapped by the wave, which can lead to particle acceleration and wave–particle energy exchange (which is fundamentally related to the physics of Landau damping). These processes are of crucial importance for many areas of laser–plasma physics, as they can lead to the formation of suprathermal electrons that play an important role in ICF—or be put to use for plasma-based electron acceleration schemes (cf. Ref. [1] and references therein). The linear and nonlinear evolution of Landau damping due to wave–particle dynamics will also be of great importance for the dynamics of laser–plasma instabilities discussed in later chapters, which strongly depend on the damping of plasma waves.

Finally, we will discuss the single particle dynamics in the presence of a spatial gradient in the intensity of the wave (either light wave or plasma wave), leading to the ponderomotive force. This force is fundamental to laser–plasma physics, and at the origin of all the instabilities and nonlinear processes discussed later in this book. We will distinguish its features for a plasma wave vs. a light wave, and for the latter, see how it is directly connected to the figure of eight motion of an electron in a light wave—i.e., to the onset of nonlinear effects.

© The Author(s), under exclusive license to Springer Nature Switzerland AG 2023
P. Michel, *Introduction to Laser-Plasma Interactions*, Graduate Texts in Physics,
https://doi.org/10.1007/978-3-031-23424-8_2

2.1 Particle Dynamics in a Uniform Light Wave

Let us begin our discussion with the interaction of a single charged particle with a plane and monochromatic light wave. The wave propagates along z and is described by its normalized vector potential $\mathbf{a} = (e/mc)\mathbf{A}$:

$$\mathbf{a}(\mathbf{r}, t) = \frac{1}{2}a_0\mathbf{e}_\pi e^{i\psi_0(z,t)} + c.c.\,, \tag{2.1}$$

where $\psi_0(z, t) = k_0 z - \omega_0 t$ with ω_0 and $\mathbf{k}_0$ the frequency- and wave-vector of the wave, and $c.c.$ means the complex conjugate. Here $\mathbf{e}_\pi$ is a complex unit vector (with $|\mathbf{e}_\pi| = 1$) representing the polarization state of the light wave, e.g., $\mathbf{e}_\pi = \mathbf{e}_x$ for a wave linearly polarized along x and $\mathbf{e}_\pi = (\mathbf{e}_x \pm i\mathbf{e}_y)/\sqrt{2}$ for circularly polarized light, with $\mathbf{a} = (a_0/\sqrt{2})[\mathbf{e}_x \cos(\psi_0) \mp \mathbf{e}_y \sin(\psi_0)]$.

The dynamics of a single electron in the presence of the wave is governed by the Lorentz equation:

$$m\frac{d\mathbf{v}}{dt} = -e(\mathbf{E} + \mathbf{v} \times \mathbf{B})\,, \tag{2.2}$$

or, in terms of the normalized vector potential, since $\mathbf{E} = -\partial\mathbf{A}/\partial t$ and $\mathbf{B} = \nabla \times \mathbf{A}$:

$$\frac{1}{c}d_t\mathbf{v} = \partial_t\mathbf{a} - \mathbf{v} \times (\nabla \times \mathbf{a})\,. \tag{2.3}$$

By expanding the second term on the right-hand side and separating the transverse ($\perp$, along x and y) and longitudinal ($\parallel z$) components (and remembering that $\partial_x a = \partial_y a = 0$), we can see that

$$\frac{d\mathbf{v}_\perp}{cdt} = \partial_t\mathbf{a} + v_z\partial_z\mathbf{a}\,, \tag{2.4}$$

$$\frac{dv_z}{cdt} = -v_x\partial_z a_x - v_y\partial_z a_y\,. \tag{2.5}$$

Since $\mathbf{a} = \mathbf{a}[z(t), t]$, where $z(t)$ is the position of the electron, we recognize that $d_t\mathbf{a} = \partial_t\mathbf{a} + v_z\partial_z\mathbf{a}$, and therefore, the first equation leads to

$$d_t\left(\frac{\mathbf{v}_\perp}{c} - \mathbf{a}\right) = 0\,. \tag{2.6}$$

This is a well-known constant of motion, corresponding to the conservation of canonical momentum [2]. For an electron initially at rest, we simply have $\mathbf{v}_\perp = c\mathbf{a}$. Next, inserting $\mathbf{v}_\perp = c\mathbf{a}$ (i.e., $v_x = ca_x$, $v_y = ca_y$) into Eq. (2.5) leads to

$$d_t v_z = -\frac{c^2}{2}\partial_z\mathbf{a}^2\,. \tag{2.7}$$

Since $\mathbf{a}$ and $\mathbf{a}^2$ are functions of $z - ct$ (assuming propagation in vacuum), we know from the wave equation (or can easily derive from the expression of $\mathbf{a}$ above) that $(\partial_t + c\partial_z)\mathbf{a}^2 = 0$. Furthermore, expressing $d_t\mathbf{a}^2 = (\partial_t + v_z\partial_z)\mathbf{a}^2$, eliminating $\partial_t\mathbf{a}^2$ between these two equations, and inserting into Eq. (2.7) lead to

$$d_t v_z = \frac{1}{2}\frac{c}{1 - v_z/c}d_t\mathbf{a}^2 \,. \tag{2.8}$$

Assuming non-relativistic electron velocities, i.e., $v_z \ll c$, leads to the second constant of the motion:

$$d_t\left(\frac{v_z}{c} - \frac{\mathbf{a}^2}{2}\right) = 0 \,. \tag{2.9}$$

For an electron initially at rest, we have $v_z = c\mathbf{a}^2/2$. Note that this relation indicates the presence of a nonlinear force ($\propto d_t a^2$) pushing the electron toward z; this is the so-called ponderomotive force, which we will study in more detail in Sect. 2.3 (the term "ponderomotive force" is usually associated with the low-frequency part of the force, $\propto \langle\mathbf{a}^2\rangle$).

Equations (2.6) and (2.9) determine the electron trajectory. We first notice that the peak velocities in the transverse and longitudinal directions for an electron initially at rest are $v_\perp/c \sim a_0$ vs. $v_z/c \sim a_0^2/4$, with the normalized vector potential given by

$$\boxed{a_0 = \frac{eA_0}{mc} = \frac{eE_0}{mc\omega_0} \approx 0.855\sqrt{I_{18}\lambda_\mu^2}}\,, \tag{2.10}$$

with I_{18} the intensity in units of 10^{18} W/cm^2 and λ_μ the wavelength in μm. We see that the transverse quiver velocity only approaches c for laser intensities on the order of 10^{18} W/cm^2 for wavelengths in the visible; this is the relativistic regime of laser–matter interaction.[1] For the laser intensities relevant to this book, however, we will typically have $a_0 \ll 1$, which means that the longitudinal velocity is smaller than the transverse one by a factor $\sim a_0$.

Therefore, to first order in a_0, i.e., neglecting the longitudinal motion $\propto a_0^2$, the electron motion is purely transverse, and $z = z_0$ (a constant). For a linear polarization along x, i.e., $\mathbf{e}_\pi = \mathbf{e}_x$ in the definition of $\mathbf{a}$ above with $\mathbf{a} = a_0\cos(k_0 z_0 - \omega_0 t)\mathbf{e}_x$, the electron velocity along x is simply

$$v_x(t) = v_{os}\cos(k_0 z_0 - \omega_0 t) \,, \tag{2.11}$$

---

[1] We can easily perform a similar analysis for the ions, which would show that they quiver as well in the laser electric field, although at a much slower velocity, by a ratio Zm/M_i. Relativistic effects for the ions begin at intensities beyond $\sim 10^{24}$ W/cm^2 for ~ 1 μm wavelengths, well outside the scope of this book.

with the peak velocity amplitude given by

$$v_{os} = ca_0 .$$ (2.12)

The trajectory along x is simply given by, still to first order in a_0:

$$x(t) = x_0 - r_{os} \sin(k_0 z_0 - \omega_0 t)$$ (2.13)

(x_0 is the initial position of the electron before the laser is present), with the excursion amplitude

$$r_{os} = \frac{a_0}{k_0} .$$ (2.14)

Likewise, for a circular polarization, the electron orbit is a circle of radius $r_{os}/\sqrt{2}$:

$$\mathbf{r}_{circ}(t) = \mathbf{r}_0 + \frac{r_{os}}{\sqrt{2}} \left[\sin(k_0 z_0 - \omega_0 t)\mathbf{e}_x + \cos(k_0 z_0 - \omega_0 t)\mathbf{e}_y \right] ,$$ (2.15)

with $\mathbf{r}_0$ the initial position at $t = 0$.

The first-order orbits (i.e., up to terms $\sim a_0$ with $a_0 \ll 1$) of an electron in a laser field for a linear and circular polarization are shown in Fig. 2.1.

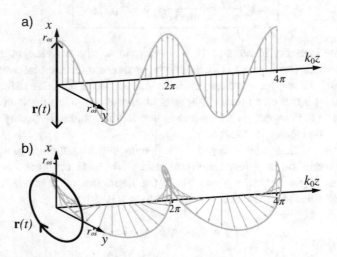

Fig. 2.1 Linear motion (i.e., transverse only and $\sim a_0$, ignoring the longitudinal motion $\sim a_0^2$) of a single electron in a plane light wave propagating along z, shown at a fixed time t: (**a**) linear (along x) and (**b**) circular polarization. The wave's electric field is in gray, and the particle motion $\mathbf{r}(t)$ is the thick black line at $z = 0$: quiver along x in $[-r_{os}, r_{os}]$ in (**a**), and circular motion with radius $r_{os}/\sqrt{2}$ in (**b**)

Let us now look at the longitudinal electron motion. As we just explained, this motion constitutes a small correction to the transverse momentum for non-relativistic intensities; yet it will play a major role in laser–plasma interactions even for intensities well below 10^{18} W/cm^2.

Assuming the electron is initially at rest and $z \approx z_0$, the velocity along z is obtained from Eq. (2.9):

$$v_z(t) = v_d \left[1 + \cos(2k_0z_0 - 2\omega_0t) \right], \qquad (2.16)$$

where v_d is a constant drift velocity of the electron along z given by

$$v_d = c\frac{a_0^2}{4}. \qquad (2.17)$$

The drift is associated with the ponderomotive force on the electron, as will be discussed in Sect. 2.3.4.

In addition to the drift along z, the electron oscillates at $2\omega_0$ along z. The trajectory along z is

$$z(t) = z_0 + v_dt - z_{os} \sin(2k_0z_0 - 2\omega_0t), \qquad (2.18)$$

with

$$z_{os} = r_{os}\frac{a_0}{8} = \frac{a_0^2}{8k_0}. \qquad (2.19)$$

Denoting $\bar{z} = [z(t) - v_dt]/z_{os}$ the trajectory in the frame of reference moving at v_d (normalized to the peak excursion amplitude along z, z_{os}), and $\bar{x} = x(t)/r_{os}$, we see that the trajectory of the electron in that moving frame, as given by $\bar{x}(t)$ and $\bar{z}(t)$, satisfies

$$\bar{z}^2 = 4\bar{x}^2(1 - \bar{x}^2), \qquad (2.20)$$

which describes the well-known "figure of eight" motion of the electron in the laser field, as illustrated in Fig. 2.2.

A similar analysis can be carried out in the relativistic regime, also leading to a figure of eight motion [3, 4]. The longitudinal motion can be interpreted as a relativistic correction to the electron dynamics (even though our analysis is purely non-relativistic). Note that for the case of a circularly polarized laser, the second-order trajectory (including longitudinal motion to order $\sim a_0^2$) is a helix, without any oscillation at $2\omega_0$ like for linearly polarized light—but with the same drift velocity $\mathbf{v}_d$ (Problem 2.1).

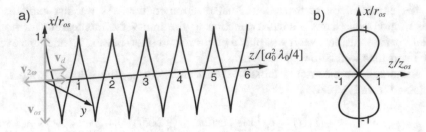

Fig. 2.2 Electron motion in a light wave including nonlinear ($\sim a_0^2$) corrections, appearing in the form of a longitudinal motion. The longitudinal velocity can be separated into a constant drift along z at velocity $\mathbf{v}_d$ and a quiver oscillation along z at $2\omega_0$ (denoted as $\mathbf{v}_{2\omega}$ in the figure). The combination of the transverse quiver at ω_0 and longitudinal at $2\omega_0$ leads to the well-known figure of eight motion in the frame of reference moving at $\mathbf{v}_d$ (**b**)

We brushed over an important question in our derivation, which is: how was the laser turned on? We first note that the monochromatic wave assumption implies that the temporal variations of the envelope (hence, the laser rise time) must remain very slow compared to the laser oscillation period $1/\omega_0$. Furthermore, the presence of a constant drift of the electron indicates that a force must have provided the acceleration from zero initial velocity to $\mathbf{v}_d$. As we already mentioned, this force is the ponderomotive force, which will be presented in Sect. 2.3. The connection between the ponderomotive force (which involves a spatial gradient in the intensity profile) and the dynamics of a single electron in a uniform light wave as was just presented (where no gradient was present, at least not explicitly) will be discussed in Sect. 2.3.4.

2.2 Particle Dynamics in a Uniform Plasma Wave

2.2.1 Phase-Space Analysis: Analogy with the Simple Pendulum

We now consider the dynamics of a charged particle in a uniform longitudinal wave (EPW or IAW) propagating along z. Because the wave is longitudinal, the oscillations of its electric field are also along z, $\mathbf{E} = E\mathbf{e}_z$; therefore, the equation of motion of the particle, $md_t\mathbf{v} = q\mathbf{E}$, is one-dimensional, $md_t v_z = qE$ (velocity components along x and y are unaffected by the wave and remain constant). It will also be convenient to work with the electrostatic potential Φ (such that $\mathbf{E} = -\nabla\Phi$).

We assume that the wave is simply a plane monochromatic wave of peak potential amplitude Φ_0:

$$\Phi(z, t) = \Phi_0 \cos(\psi_p), \tag{2.21}$$

where $\psi_p = k_p(z - v_\phi t) = k_p z - \omega_p t$, with k_p, v_ϕ, and $\omega_p = k_p v_\phi$ the wave's wave-vector, phase velocity, and frequency. The equation of motion is simply

$$d_t^2 z(t) = \frac{q k_p \Phi_0}{m} \sin\left[k_p z(t) - \omega_p t\right] . \tag{2.22}$$

For this analysis, it will be convenient to use Hamiltonian mechanics and derive the particle motion in the frame moving with the wave at v_ϕ. We introduce the new variables in the moving frame:

$$\bar{z} = z - v_\phi t + \delta_{qi}\frac{\pi}{k_p} , \tag{2.23}$$

$$\bar{v} = v - v_\phi . \tag{2.24}$$

Here $\delta_{qi} = 1$ if the particle is an ion ($q = Ze > 0$), and 0 for an electron, which allows us to rewrite the equation of motion as

$$d_t^2 \bar{z}(t) = -\frac{|q| k_p \Phi_0}{m} \sin\left[k_p \bar{z}(t)\right] . \tag{2.25}$$

We now look for the Hamiltonian, a function $H(\bar{z}, \bar{v})$ representing the total energy of the particle (kinetic plus potential) and which must satisfy Hamilton's equations connecting the position and momentum:

$$d_t \bar{z} = \frac{\partial H}{\partial \bar{v}} , \tag{2.26}$$

$$d_t \bar{v} = -\frac{\partial H}{\partial \bar{z}} . \tag{2.27}$$

Finding the expression for H simply consists in integrating these two equations with respect to $\bar{v}$ and $\bar{z}$; since $d_t \bar{z} = \bar{v}$ and $d_t \bar{v} = -(|q| k_p \Phi_0/m) \sin(k_p \bar{z})$, we have

$$H(\bar{z}, \bar{v}) = \int \bar{v} d\bar{v} = \frac{\bar{v}^2}{2} + f(\bar{z}) , \tag{2.28}$$

$$H(\bar{z}, \bar{v}) = \frac{|q| k_p \Phi_0}{m} \int \sin(k_p \bar{z}) d\bar{z} = -\frac{|q| \Phi_0}{m} \cos(k_p \bar{z}) + g(\bar{v}) , \tag{2.29}$$

where f and g are constants of integration—equal to the first term on the right-hand side of the second and first equations, respectively. We obtain the following expression for H:

$$H(\bar{z}, \bar{v}) = \frac{1}{2}\bar{v}^2 - \frac{|q|}{m}\Phi_0 \cos(k_p \bar{z}) . \tag{2.30}$$

We immediately recognize this expression as the sum of the kinetic and potential energies of the particle in the wave (to a factor m), i.e.,

$$mH = W_{tot} = W_k + W_\Phi, \tag{2.31}$$

with $W_k = m\bar{v}^2/2$ and $W_\Phi = -|q|\Phi$. The Hamiltonian is a constant of the motion, independent of time. Note that we could have arrived at this point without resorting to the notion of Hamiltonian, by simply noticing that the equation of motion in the moving frame, Eq. (2.25), is the equation of a simple pendulum, $d_t^2\theta = -(g/l)\sin(\theta)$ with g the gravitational field, l the length of the rod, and θ the angle of the pendulum; saying that H is independent of time is equivalent to saying that the total energy of the pendulum (potential plus kinetic) is conserved.[2]

While the exact particle orbit $\bar{z}(t)$ cannot be expressed analytically in general, its behavior can be investigated in phase-space, i.e., by expressing the velocity as a function of position. From the expression of H, we can extract the phase-space orbit of the particle in the moving frame, i.e., $\bar{v}$ vs. $\bar{z}$:

$$\boxed{\bar{v}(\bar{z}) = \pm\sqrt{2H + \frac{2|q|}{m}\Phi_0\cos(k_p\bar{z})} = \pm\sqrt{\frac{2}{m}(W_{tot} - W_\Phi)}}. \tag{2.32}$$

The energy of a particle in the wave frame is represented in Fig. 2.3a. If $W_{tot} - W_\Phi$ is always positive for any location $\bar{z}$, i.e., if the particle's initial energy is such that $H > |q|\Phi_0/m$, then the velocity is of the form $\bar{v} \propto \pm\sqrt{1 + \varepsilon\cos(k_p\bar{z})}$ with $|\varepsilon| = |q\Phi_0/mH| < 1$: the particle can then travel from $z = -\infty$ to $+\infty$ (or the other way around). The corresponding trajectories are usually called "untrapped" or "passing" orbits.

On the other hand, if $W_{tot} < |q|\Phi_0$, then a solution for $\bar{v}(\bar{z})$ exists only in a restricted region of space in $\bar{z}$ that allows the term in the square root to remain positive ($W_{tot} > W_\Phi$ or $\cos(k_p\bar{z}) > -W_{tot}/|q|\Phi_0$), as seen in Fig. 2.3a. This means that the particle is "trapped" in the potential well of the wave, and its velocity oscillates together with its potential energy in a way that maintains the total energy constant.

The phase-space orbits $\bar{v}(\bar{z})$ are represented in Fig. 2.3b. Trapped particles have closed orbits in phase-space as they oscillate near the wave's phase velocity while being spatially "locked" with the wave, whereas untrapped particles have open orbits and move from right to left on this plot for $v < v_\phi$ as the wave outruns the particle, or left to right if $v > v_\phi$ when the particle outruns the wave.

The physical situation is directly equivalent to the simple pendulum, and it is instructive to exploit the equivalence to gain some intuition of the problem.

---

[2] Like for a pendulum, the choice of the location where the potential energy is zero is arbitrary, and both the potential and total energies can be negative—all that matters is that the total energy is conserved and that the kinetic energy remains positive, i.e., $W_{tot} > W_\Phi$.

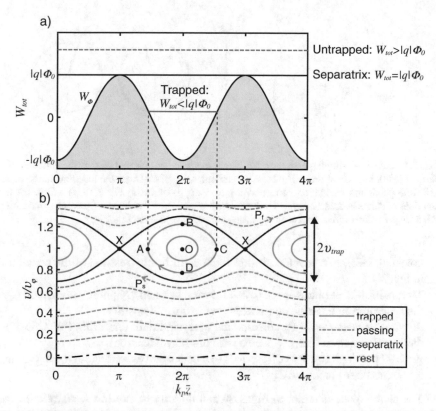

Fig. 2.3 (**a**) Potential well W_Φ and energy W_{tot} of trapped vs. untrapped particles in the moving frame of a plasma wave. The total energy has to remain above the potential energy ($W_{tot} > W_\Phi$) to ensure $W_k = m\bar{v}^2/2 > 0$, which excludes the gray area under the curve of W_Φ. (**b**) Phase-space orbits, showing trapped and untrapped orbits, the separatrix, and the orbit of an electron initially at rest. The locations in phase-space marked by the letters A, B, C, D (along a trapped orbit), and X and O (top and bottom of the potential well) have direct equivalents for a simple pendulum, as illustrated in Fig. 2.4

Figure 2.4 shows a pendulum whose angle θ (chosen positive for counter-clockwise rotation) is equivalent to the particle position in the moving frame $\bar{z}$.

In particular:

- For a trapped orbit, the points A, C correspond to the extrema of the spatial particle excursion where the velocity (in the moving frame) reaches zero before changing sign; this is equivalent to the same points A, C of maximum excursion in a pendulum, Fig. 2.4.

- B and C correspond to the bottom of the potential well where the particle velocity is maximum in either direction—just like when the pendulum reaches the bottom position.

- The most deeply trapped orbit, marked O in Fig. 2.3 (usually identified as an "O-point"), for a particle immobile in the moving frame (trapped at the bottom of the

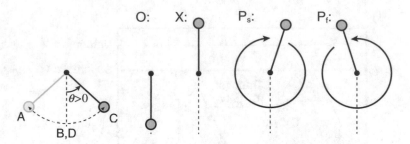

Fig. 2.4 Motion of simple pendulum; the letters mark the correspondence to the situation of a particle in a longitudinal wave, Fig. 2.3b. In particular, A, B, C, D are the locations in the back-and-forth pendulum oscillation corresponding to a trapped orbit in Fig. 2.3b; O and X are the positions of stable and unstable (upright) equilibrium; and P_s, P_f represent a pendulum executing full swings in either direction and are equivalent to passing orbits in Fig. 2.3b

potential well), is equivalent to the stable equilibrium position of the pendulum in Fig. 2.4.

- The position X in phase-space (often referred to as an "X-point") corresponds to the pendulum in unstable equilibrium in the upright position.
- A "slow" passing orbit P_s corresponds to a pendulum executing full swings in the $\theta < 0$ direction (i.e., clockwise with our normalization).
- A "fast" passing orbit P_f corresponds to a pendulum executing full swings along $\theta > 0$ (counter-clockwise).

The phase-space diagram in Fig. 2.3b can directly be used to visually infer the Hamiltonian of any given orbit. Some particular orbits are worth highlighting:

- Particle initially at rest:

For a particle initially at rest in the laboratory frame, we must have $H_{rest} = v_\phi^2/2$ per Eq. (2.30) with $\Phi_0 = 0$ and $v = 0$. In the limit of a small wave amplitude, $|q|\Phi_0 \ll mv_\phi^2$ (i.e., $|\delta n_e/n_{e0}| \ll 1$ in the $k\lambda_{De} \ll 1$ limit for an EPW, per Eq. (1.121)), the orbit of a particle initially at rest can be obtained from Eq. (2.32) with $\bar{v}_{rest} < 0$ for a particle moving slower than the wave (i.e., minus sign in front of the square root):

$$v_{rest} = -\frac{|q|\Phi_0}{mv_\phi} \cos(k_p\bar{z}) . \tag{2.33}$$

Note that we expressed the velocity in the laboratory frame. This is consistent with the approximate solution of the equation of motion in the laboratory frame $d_t^2 z(t) = -(|q|k_p\Phi_0/m)\sin[k_p z(t) - \omega_p t]$ if we assume that the particle excursion during an oscillation in the field is very small compared to the wavelength so that $k_p z(t)$ can be approximated as constant in the phase term, i.e., $k_p z(t) - \omega_p t \approx k_p z_0 - \omega_p t$. The equation of motion can then be integrated vs. time to yield Eq. (2.33).

- Deeply trapped particle:

For a particle near the bottom of the potential well (i.e., near the point O in Fig. 2.3 at $k_p\bar{z} = 0 \mod(2\pi)$, with $H_O = -|q|\Phi_0/m$), we have $k_p\bar{z} \approx 0$, and the situation becomes equivalent to the pendulum in the small angle limit, $d_t^2\bar{z} = -(k_p|q|\Phi_0/m)\sin(k_p\bar{z}) \approx -(k_p^2|q|\Phi_0/m)\bar{z}$, leading to a simple harmonic oscillation:

$$\bar{z} = \frac{\bar{v}(t=0)}{\omega_B} \sin(\omega_B t), \tag{2.34}$$

where we have introduced the "bounce frequency" given by

$$\boxed{\omega_B = \sqrt{\frac{k_p^2|q|\Phi_0}{m}}}. \tag{2.35}$$

The bounce frequency is an important physical quantity as it represents the characteristic time it takes for Landau damping to turn nonlinear, leading to a flattening of the distribution function, as will be discussed in the next sections. Note that for an EPW, using the relationship between potential and density perturbation, Eq. (1.121), we have

$$\omega_{B,\text{EPW}} = \omega_{pe}\sqrt{\frac{\delta n_e}{n_{e0}}}. \tag{2.36}$$

Likewise, for IAWs, using Eq. (1.138), we obtain

$$\omega_{B,\text{IAW}} = \omega_s\sqrt{\frac{\delta n_e}{n_{e0}}}, \tag{2.37}$$

where $\omega_s = k_p\sqrt{ZT_e/M_i}$.

- The "separatrix":

The separatrix separates trapped vs. untrapped orbits; it corresponds to $H_{sep} = |q|\Phi_0/m$ (as can also be inferred from Fig. 2.3a or b). The phase-space orbit of the separatrix is given by

$$\bar{v}_{sep}(\bar{z}) = \pm v_{trap}\cos\left(\frac{k_p\bar{z}}{2}\right), \tag{2.38}$$

where the "trapping width" v_{trap}, defined here as half the maximum velocity separation between the + and − branches of the separatrix at $\bar{z} = 0$ (cf. Fig. 2.3b), is defined as

$$\boxed{v_{trap} = 2\sqrt{\frac{|q|\Phi_0}{m}} = 2\frac{\omega_B}{k_p}}. \tag{2.39}$$

The trapping width is directly related to the wave amplitude: from the expression above, we obtain for the two types of plasma waves:

$$v_{trap} = \begin{cases} 2\dfrac{\omega_{pe}}{k_p}\sqrt{\dfrac{\delta n_e}{n_{e0}}} & \text{(EPWs)}, \\[4mm] 2\sqrt{\dfrac{ZT_e}{M_i}}\sqrt{\dfrac{\delta n_e}{n_{e0}}} & \text{(IAWs)}. \end{cases} \tag{2.40}$$

In the limit of $k_p\lambda_{De} \ll 1$ for EPWs and $ZT_e \ll 3T_i$ for IAWs, we obtain a simple single expression:

$$v_{trap} \approx 2v_\phi\sqrt{\frac{\delta n_e}{n_{e0}}}. \tag{2.41}$$

2.2.2 Transit Time and Bounce Period of Passing and Trapped Particles

The transit time of passing particles (i.e., the time to travel through $k_p\Delta\bar{z} = 2\pi$) and the bounce period of trapped orbits (i.e., the time to perform a full loop along a closed orbit) contain important physics that relates to nonlinear Landau damping.

First, for passing particles, the transit time τ_{pass} is obtained by integrating along a passing orbit over a distance $\Delta\bar{z} = 2\pi/k_p$:

$$\tau_{pass} = \int_0^{2\pi/k_p} \frac{d\bar{z}}{\bar{v}(\bar{z})}. \tag{2.42}$$

Using the trigonometric identity $\cos(k_p\bar{z}) = 1 - 2\sin^2(k_p\bar{z}/2)$, the expression for the velocity, Eq. (2.32), can be recast into

$$\bar{v}(\bar{z}) = \sqrt{2H + \frac{2|q|\Phi_0}{m}}\sqrt{1 - \xi^2\sin^2(k_p\bar{z}/2)}, \tag{2.43}$$

with

$$\xi^2 = \frac{2|q|\Phi_0}{mH + |q|\Phi_0}. \tag{2.44}$$

Inserting into Eq. (2.42) leads to

$$\tau_{pass} = \frac{2}{k_p \sqrt{2H + 2|q|\Phi_0/m}} \int_0^\pi \frac{d\psi}{\sqrt{1 - \xi^2 \sin^2(\psi)}}, \qquad (2.45)$$

with $\psi = k_p \bar{z}/2$. This can be further simplified into

$$\boxed{\tau_{pass}(\xi) = \frac{2\xi}{\omega_B} K(\xi)}, \qquad (2.46)$$

where ω_B is the bounce frequency for a deeply trapped orbit defined above in Eq. (2.35) and $K(\xi)$ is the complete elliptic integral of the first kind [5], defined as

$$K(\xi) = \int_0^{\pi/2} \frac{d\theta}{\sqrt{1 - \xi^2 \sin^2(\theta)}}. \qquad (2.47)$$

Now for trapped orbits, we proceed similarly and express the bounce period (the time it takes to execute a full loop in a closed orbit) as twice the time it takes to go between the two position extrema A and C in Fig. 2.3b:

$$\tau_{trap} = 2 \int_{\bar{z}_A}^{\bar{z}_C} \frac{d\bar{z}}{\bar{v}(\bar{z})} \qquad (2.48)$$

$$= \frac{2\xi}{\omega_B} \int_{\psi_A}^{\psi_C} \frac{d\psi}{\sqrt{1 - \xi^2 \sin^2(\psi)}}, \qquad (2.49)$$

where $\psi = k_p \bar{z}/2$. We cannot immediately identify a special function due to the integration limits, so we are going to look for a change of variable that brings back the expression of the elliptic integrals. We notice the velocity is zero at the two integration limits, $\bar{v}(\bar{z}_A) = \bar{v}(\bar{z}_C) = 0$, which implies $\xi \sin(\psi_{A,C}) = \pm 1$ per Eq. (2.43). Thus we introduce a new variable θ defined as $\sin(\theta) = \xi \sin(\psi)$, which will convert the integration limits to $\theta_A = -\pi/2$ and $\theta_C = \pi/2$. Using the relation $\cos(\theta)d\theta = \xi \cos(\psi)d\psi$ yields

$$\frac{d\psi}{\sqrt{1 - \sin^2(\theta)}} = \frac{d\psi}{|\cos(\theta)|} = \frac{d\theta}{\xi\sqrt{1 - \sin^2(\psi)}} = \frac{d\theta}{\xi\sqrt{1 - \xi^{-2} \sin^2(\theta)}}, \qquad (2.50)$$

finally leading to

$$\boxed{\tau_{trap}(\xi) = \frac{4}{\omega_B} K(1/\xi)}. \qquad (2.51)$$

Fig. 2.5 Transit and bounce
times for passing and trapped
orbits, normalized to the
bounce time of deeply
trapped orbits $\tau_B = 2\pi/\omega_B$.
Passing and trapped orbits
correspond to $\xi < 1$ and
$\xi > 1$, respectively, with
$\xi = 1$ indicating the
separatrix

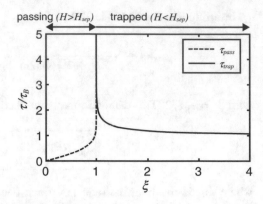

The transit and bounce times for passing and trapped orbits, normalized to the
bounce period $\tau_B = 2\pi/\omega_B$, are represented in Fig. 2.5. From the definition of
ξ above and using the expression for the Hamiltonian of the separatrix $H_{sep} =
|q|\Phi_0/m$, we see that

$$\xi^2 = 2\frac{H_{sep}}{H + H_{sep}} : \tag{2.52}$$

$\xi < 1$ is thus equivalent to $H > H_{sep}$ and vice versa. In other words: the variable ξ
is directly related to the total energy of the particle, and $\xi < 1$ (respectively, $\xi > 1$)
corresponds to passing (respectively, trapped) orbits.

As we can see on the figure, orbits getting close to the separatrix on either side
of it ($|\xi - 1| \rightarrow 0$) take a rapidly increasing amount of time to execute a full
transit or bounce, and a particle on the separatrix will take an infinite amount of time
to execute a cycle. The situation is again directly comparable to the pendulum: a
pendulum whose energy is just above (or just below) the energy required to complete
a full swing will spend an increasingly long amount of time near the upright position
(X-point in Figs. 2.4 or 2.3b), resulting in an increasingly long time to execute a
cycle.

2.2.3 Wave–Particle Energy Conservation: Connection with Landau Damping

Consider an EPW with a phase velocity v_ϕ along z interacting with a Maxwellian
distribution of particles $f(\mathbf{v})$. For velocities near $v_\phi > 0$, we have $\partial_{v_z} f(\mathbf{v})|_{v_\phi} < 0$:
the distribution contains more particles at velocities v_z below v_ϕ than above within
a small range of velocities near v_ϕ. In the following, we assume that $v_{trap} \ll v_\phi$,
i.e., $\delta n_e/n_{e0} \ll 1$ per Eq. (2.41).

If one waits for a time much larger than the bounce period τ_B, then most of the trapped particles (except for those located very close to the separatrix per Fig. 2.5) will have executed at least one full bounce. As a result, the velocity and position of the particles within the trapped orbits inside the separatrix will get randomized,[3] and on average, we will end up with as many particles below v_ϕ as above it. Since we started with more particles below v_ϕ, this results in a net energy increase of the particles, which must be compensated by an energy loss from the wave. This is the basic physical picture of Landau damping.[4]

Heuristic arguments can be developed to recover the scaling of Landau damping. We can reasonably assume that most of the trapped particles take a time $\approx \tau_B$ to execute a full bounce (cf. Fig. 2.5), which means that their initial velocity in the moving frame $\bar{v}$ will switch sign during $\tau_B/2$ (half-cycle along a trapped orbit), i.e., $\bar{v} \rightarrow -\bar{v}$ during $\tau_B/2$. During that time, the particles will gain or lose the energy $\delta W = \frac{1}{2}m[(v_\phi - \bar{v})^2 - (v_\phi + \bar{v})^2] = -mv_\phi\bar{v}$, corresponding to a power gain or loss equal to $\delta P = \delta W/(\tau_B/2) = -2mv_\phi\bar{v}/\tau_B = -2mv_\phi(v - v_\phi)/\tau_B$.

The total power change from the ensemble of the particles within the range $[-v_{trap}, v_{trap}]$ is then

$$\Delta P = \int_{v_\phi - v_{trap}}^{v_\phi + v_{trap}} f(v)\delta P(v)dv. \tag{2.53}$$

Since we assumed that $v_{trap} \ll v_\phi$, the distribution function can be Taylor-expanded in the vicinity of v_ϕ as $f(v) \approx f(v_\phi) + (v - v_\phi)f'(v_\phi)$, where the prime denotes the derivative with respect to velocity (here v denotes the velocity along z). Inserting in the integral and expressing v_{trap} and τ_B in terms of the wave's electric field amplitude $E_0 = -k_p\Phi_0$, we obtain

$$\Delta P \propto -\frac{mv_\phi}{\tau_B}v_{trap}^3 f'(v_\phi) = -\frac{e^2 E_0^2 v_\phi}{mk_p}f'(v_\phi). \tag{2.54}$$

We can then balance this power gain from the particles with the power loss from the field. Taking the field energy density $\epsilon_0 E_0^2/4$, and assuming an exponential decay of the form $E_0(t) \propto \exp[-\nu t]$ (with $\nu > 0$ indicating a damping), the power variation of the wave is given by $-\nu\epsilon_0 E_0^2/2$. Setting $-\nu\epsilon_0 E_0^2/2 + \Delta P = 0$ to impose energy conservation, we obtain the following expression for the damping:

---

[3] Consider an initial and a final time t_i and t_f: since each orbit takes a different time to execute a full bounce, when $t_f - t_i \gg \tau_B$, the final phase-space location of each particle could be anywhere on its orbit.

[4] Note that passing particles also contribute to Landau damping because particles on orbits close to the separatrix spend more time near the X-points—like a pendulum near the upright position, as discussed in the previous section. As a result, when integrating over durations $\gg \tau_B$, the average velocity of the passing particles near the separatrix will move closer to v_ϕ from their initial velocities, which will again lead to a net energy gain if more particles are initially below v_ϕ than above.

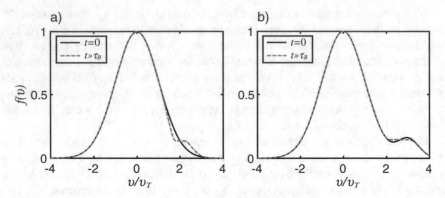

Fig. 2.6 (**a**) Illustration of the flattening of a Maxwellian distribution function near the wave's phase velocity (here at $2v_T$) due to nonlinear Landau damping. (**b**) Illustration of the bump-on-tail instability process, with a beam of particles (here centered at $3v_T$ with a Maxwellian distribution of width $v_T/2$) interacting with a Maxwellian background distribution. As opposed to (**a**), here the particles transfer energy to the wave until the distribution flattens

$$\frac{v}{\omega_p} \propto -\frac{\omega_{pe}^2}{n_e k_p^2} f'(v_\phi).$$
(2.55)

We recover the Landau damping formula (cf. Sect. 1.3.4), up to some numerical constant due to our simplistic approximations.

For time scales comparable to or larger than the bounce period, the phase-space mixing of the particles on their orbits will lead to a flattening of the distribution function around v_ϕ, over a width on the order of the trapping width v_{trap}. At this point, the derivative of the distribution function will drop toward zero, and so will the damping, as illustrated in Fig. 2.6. This corresponds to the onset of nonlinear Landau damping, which has been described in Ref. [6]. This reduction in damping plays an important role in the saturation of parametric instabilities, as will be discussed in Sect. 10.3.

Conversely, for a positive slope of the distribution function at some velocity v_ϕ, v becomes negative, and a wave seeded from noise in the plasma at that phase velocity will be allowed to grow exponentially. This is for example the situation of the "bump-on-tail" instability, where a beam of particles traveling at a velocity in the tail of the bulk plasma distribution function will give rise to a plasma wave that can grow exponentially—eventually leading to a flattening of the distribution function as well. This situation is illustrated in Fig. 2.6b. Like for damping, saturation occurs for time scales on the order of τ_B, beyond which the growth ends and the particle beam will have lost some of its energy to plasma waves.

2.3 Particle Dynamics in a Non-uniform Light or Plasma Wave: The Ponderomotive Force

So far our analyses of single particle dynamics in waves have always assumed that the wave is a plane wave with a uniform amplitude. In this section, we will show that when a particle is interacting with a wave (either longitudinal or transverse) whose envelope is non-uniform, the particle will experience a slow drift on top of the fast oscillations that will push it toward the regions of weaker field amplitude—similar to the "guiding centre" motion of an electron in non-uniform magnetic fields. This effect is known as the ponderomotive force and is at the origin of most laser–plasma interaction processes.

The ponderomotive force exists for both longitudinal plasma waves and transverse light waves, as well as for spatial non-uniformities resulting from the beat between overlapped waves (transverse or longitudinal, in any combination). We will see that there are subtle differences in the ponderomotive force from a plasma wave vs. a light wave: for the former, the force can only be in the propagation direction of the wave, and only the gradient along that direction matters, whereas for a light wave, the force can be in any direction and will just follow the direction of the intensity gradient. The ponderomotive force will also allow us to solve a missing piece of the previous analyses of particle motion in uniform light waves, by establishing the time history of how the wave (and thus the particle motion) was brought to its final state.

2.3.1 Ponderomotive Force from Plasma Waves

We consider a charged particle in a longitudinal wave (EPW or IAW) with a non-uniform envelope. Let z be the direction of the electric field oscillation: the field can be expressed as

$$\mathbf{E}(\mathbf{r}, t) = E_0(\mathbf{r}, t) \cos(\psi)\mathbf{e}_z, \tag{2.56}$$

where $\psi = kz - \omega t$ with k and ω the wave-vector and frequency of the wave. If the particle is initially at rest at $\mathbf{r}(t = 0) = 0$, the equation of motion is 1D:

$$d_t^2 z(t) = \frac{q E_0[x, y, z(t), t]}{m} \cos[kz(t) - \omega t]. \tag{2.57}$$

The particle motion in a uniform plasma wave, with E_0 assumed constant, was investigated in Sect. 2.2. For a particle initially at rest, the motion is described to first order (for small amplitude waves, $|\delta n_e/n_{e0}| \ll 1$) by a simple harmonic oscillation of amplitude $r_{os} = |q E_0/m\omega^2|$, cf. Eq. (2.33). Now, if the wave amplitude is non-uniform, the electric field amplitude may not be the same along the excursion

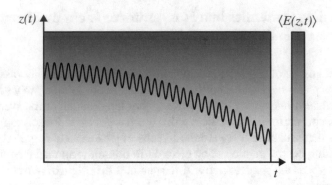

Fig. 2.7 Motion of an electron in an electric field with a gradient in its amplitude. The motion is the sum of fast oscillations and a slow "drift" toward region of smaller field amplitude, coming from the ponderomotive force

distance of the particle r_{os}. As a result, when the particle reaches a location where the field amplitude is lower, the restoring force will be slightly diminished, and the particle will not quite return to its oscillation-averaged position. This will lead to a slow drift of the particle toward the regions of lower field amplitude, as illustrated in Fig. 2.7 (the slow drift is often referred to as the motion in the "oscillation centre"). This is the basic physical idea behind the concept of ponderomotive force—at least for a plasma wave.

Let us now translate this simple physical picture into a physics model. We first assume that the physical scale of the field non-uniformities is much larger than the particle excursion r_{os}. We can then separate the particle motion into a fast oscillation describing the quiver motion at ω and a slow cycle-averaged motion:

$$z(t) = z_0(t) + z_s(t), \tag{2.58}$$

where $z_s(t) = \langle z(t) \rangle$ and the bracket denotes the average over an oscillation period $2\pi/\omega$. Our goal is to find an expression for the particle motion in the oscillation center, i.e., solve

$$m d_t^2 z_s(t) = q \langle E[z(t), t] \rangle . \tag{2.59}$$

Since we are assuming that the field does not vary much over the particle's excursion r_{os} during one cycle, we can Taylor-expand around the cycle-averaged position z_s:

$$E[z(t), t] = E[z_s(t) + z_0(t), t] \approx E_s + z_0(t) \partial_z E_s , \tag{2.60}$$

where $E_s = E[z_s(t), t] = E_{0s} \cos(\psi_s)$ is the field at the cycle-averaged position z_s with $E_{0s} = E_0[z_s(t), t]$ and $\psi_s = k z_s - \omega t$. Taking a time-average over an oscillation cycle gives

$$\langle E[z(t), t]\rangle = \langle z_0(t)\partial_z E_s\rangle \,, \tag{2.61}$$

where we used the fact that $\langle E_s\rangle = 0$.

We now only need to find the expression for $z_0(t)$. Since $md_t^2 z_0(t) = qE_s$, we obtain $z_0(t) = -q/(m\omega^2)E_s$; we also have $\partial_z E_s = \cos(\psi_s)\partial_z E_{0s}$. Inserting into Eq. (2.61) gives

$$\langle E[z(t), t]\rangle = -\frac{q}{4m\omega^2}\frac{\partial E_{0s}^2}{\partial z} \,, \tag{2.62}$$

where we used $\langle \cos^2(\psi_s)\rangle = \frac{1}{2}$.

Inserting in Eq. (2.59) finally gives the desired expression for the slow drift motion of the particle due to the field non-uniformity:

$$md_t^2 z_s(t) = F_p \,, \tag{2.63}$$

where

$$\boxed{F_p = -\frac{q^2}{4m\omega^2}\frac{\partial E_{0s}^2}{\partial z} = -\frac{q^2}{2m\omega^2}\frac{\partial \langle E^2\rangle}{\partial z}} \tag{2.64}$$

is known as the ponderomotive force. It will push all charged particles (since $F_p \propto q^2$) away from the region of high-field intensity. Note that for this particular case of a longitudinal plasma wave, only the gradient along z matters: if the field envelope E_0 had a spatial non-uniformity along x or y, there would be no ponderomotive force and no slow drift. The situation is different for a non-uniform light wave, which we analyze below.

2.3.2 Ponderomotive Force from Light Waves

A charged particle in a non-uniform light wave will also experience a slow drift or oscillation center motion due to the ponderomotive force. The derivation is similar to the case of a longitudinal wave, except that the equation of motion now includes the $\mathbf{v} \times \mathbf{B}$ term that will slightly change the physical picture.

Assuming that the light is linearly polarized along x and propagating toward z, the electric field of the light wave is

$$\mathbf{E}(\mathbf{r}, t) = E_0(\mathbf{r}, t)\mathbf{e}_x \cos(\psi) \,, \tag{2.65}$$

with $\psi = kz - \omega t$ and E_0 is a slowly varying envelope.

Like in the previous discussion, we are looking for particle motion of the form:

$$\mathbf{r}(t) = \mathbf{r}_0(t) + \mathbf{r}_s(t) , \tag{2.66}$$

where $\mathbf{r}_s(t) = \langle \mathbf{r}(t) \rangle$ is the slow drift component.

The electric field is assumed to have small variations over the particle excursion $\mathbf{r}_0$ during one oscillation cycle and can thus be Taylor-expanded near the cycle-averaged location of the particle:

$$\mathbf{E}\left[\mathbf{r}(t), t\right] = \mathbf{E}[\mathbf{r}_s(t) + \mathbf{r}_0(t), t] \approx \mathbf{E}_s + [\mathbf{r}_0(t) \cdot \nabla]\mathbf{E}_s , \tag{2.67}$$

where $\mathbf{E}_s = \mathbf{E}[\mathbf{r}_s(t), t] = E_{0s}\mathbf{e}_x \cos(\psi_s)$ with $\psi_s = kz_s - \omega t$. Likewise, the magnetic field is taken as approximately equal to its cycle-averaged value at $\mathbf{r}_s$, $\mathbf{B} \approx B_{s0}\mathbf{e}_y \sin(\psi)$ and can be connected to the electric field via the Maxwell–Faraday equation $\nabla \times \mathbf{E} = -\partial \mathbf{B}/\partial t$, leading to

$$\mathbf{B}_s = \frac{1}{\omega}\nabla \times (E_{0s}\mathbf{e}_x \sin(\psi)) . \tag{2.68}$$

We are again looking for the cycle-averaged equation of motion of the particle:

$$md_t^2\mathbf{r}_s = md_t^2 \langle \mathbf{r}(t) \rangle = q \langle \mathbf{E}(\mathbf{r}, t) \rangle + q \langle \mathbf{v} \times \mathbf{B}(\mathbf{r}, t) \rangle . \tag{2.69}$$

The cycle-averaged electric field is obtained from the Taylor expansion:

$$\langle \mathbf{E}(\mathbf{r}, t) \rangle \approx \langle (\mathbf{r}_0 \cdot \nabla)\mathbf{E}_s \rangle , \tag{2.70}$$

where we used $\langle \mathbf{E}_s \rangle = 0$. Next we need the expressions for the quiver motion of the particle at $\mathbf{r}_s$, under the influence of $\mathbf{E}_s$. For non-relativistic intensities, to first order, we can neglect the contribution of the magnetic field to $\mathbf{r}_0$ and thus get

$$\mathbf{r}_0(t) = \frac{-q}{m\omega^2}E_{0s}\mathbf{e}_x \cos(\psi) , \tag{2.71}$$

$$\mathbf{v}_0(t) = \frac{-q}{m\omega}E_{0s}\mathbf{e}_x \sin(\psi) . \tag{2.72}$$

It is now obvious from these expressions and the Taylor expansion of the electric field that since $\mathbf{r}_0$ and $\mathbf{v}_0$ are both first order in E_{0s}, the two terms on the right-hand side in Eq. (2.69) will be second order in E_{0s} and both need to be retained. Inserting the expressions for $\mathbf{r}_0$ and $\mathbf{v}_0$ into the averaged fields, we obtain

$$\langle \mathbf{E} \rangle \approx -\frac{q}{m\omega^2} \langle E_s \partial_x \mathbf{E}_s \rangle = -\frac{q}{4m\omega^2} \begin{pmatrix} \partial_x E_{0s}^2 \\ 0 \\ 0 \end{pmatrix} , \tag{2.73}$$

$$\langle \mathbf{v} \times \mathbf{B} \rangle \approx \langle \mathbf{v}_0 \times \mathbf{B}_s \rangle = -\frac{q}{4m\omega^2} \begin{pmatrix} 0 \\ \partial_y E_{0s}^2 \\ \partial_z E_{0s}^2 \end{pmatrix}. \tag{2.74}$$

Note that the expression for the cycle-averaged electric field is exactly the same that was obtained for the case of a longitudinal plasma wave (Eq. (2.62)); now, we see that the $\mathbf{v} \times \mathbf{B}$ term contributes to adding the derivatives of the square of the envelope in the other two directions as well. Inserting into the cycle-averaged equation of motion finally gives

$$md_t^2 \mathbf{r}_s(t) = \mathbf{F}_p, \tag{2.75}$$

with the expression for the ponderomotive force from a light wave:

$$\boxed{\mathbf{F}_p = \frac{-q^2}{4m\omega^2} \nabla E_{0s}^2 = \frac{-q^2}{2m\omega^2} \nabla \langle \mathbf{E}_s^2 \rangle.} \tag{2.76}$$

Like in the case of a plasma wave, since the ponderomotive force is proportional to $-q^2$, it will push all charged particles (electrons and ions) away from the regions of high electric field amplitude. However, for a light wave, the ponderomotive force will push the particles in any spatial direction as long as it follows the gradient of $\langle \mathbf{E}^2 \rangle$—whereas the ponderomotive force in a plasma wave will only push the particles along the direction of the wave. More specifically, we have seen that the push along the electric field's polarization direction is coming from the $q\mathbf{E}$ term in the Lorentz force (like for a plasma wave), while the $\mathbf{v} \times \mathbf{B}$ term provides the push in the other two directions.

Because the ponderomotive force is inversely proportional to the mass of the particle, it will mostly affect the electrons and not the ions; in fact, as we will see in the later chapters, the electrostatic potential generated by the displacement of electrons will be a lot more efficient at moving ions than the ponderomotive force from a laser, so the interaction between a laser and the plasma ions is often mediated by the ponderomotive displacement of electrons by the laser.

Furthermore, because the force is proportional to the *gradient* of $\langle \mathbf{E}^2 \rangle$, it will be most effective for steep spatial variations of laser intensity; two examples that will be discussed in great detail later are the cases of a tightly focused laser beam and the interference pattern between two light waves, generating a modulation over a distance of the order of an oscillation wavelength only. This will be discussed in more detail below.

Note that the Taylor expansion used in the ponderomotive force derivation (Eq. (2.67)) becomes incorrect when the variations of the field amplitude cannot be considered small compared to the maximum excursion of an electron executing its quiver motion. For an electron in a light wave, the maximum quiver excursion is related to the field amplitude via $r_{os} = a_0/k$, so the ponderomotive force expression remains valid as long as the gradient of the laser amplitude remains small compared to a_0/k. In the following chapters, we will encounter situations (three-wave instabilities) where the field amplitude will vary over a distance comparable to λ due to the interference pattern between two light waves; the ponderomotive force expression above will thus remain valid as long as $a_0 \ll 1$, i.e., for non-relativistic laser intensities.

Because the ponderomotive force can lead to charge displacements in a plasma and setup longitudinal waves in response, it is natural to express it as a potential, i.e., $\mathbf{F}_p = -q\nabla\Phi_p$ with

$$\Phi_p = \frac{q}{4m\omega^2}E_{0s}^2 . \tag{2.77}$$

Using Eq. (2.72), we can easily see that the ponderomotive potential energy $q\Phi_p$ is equal to the average kinetic energy of the oscillations:

$$q\Phi_p = \frac{1}{2}m\left\langle \mathbf{v}_0^2 \right\rangle . \tag{2.78}$$

For the case of an electron in a light wave, using the normalized vector potential $\mathbf{a} = (e/mc)\mathbf{A}$ and normalized ponderomotive potential $\varphi_p = (e/mc^2)\Phi_p$ yields the simple expression:

$$\varphi_p = -\frac{\left\langle \mathbf{a}^2 \right\rangle}{2} . \tag{2.79}$$

Alternatively, we may also write

$$\mathbf{F}_p = -\frac{1}{2}mc^2\nabla\left\langle \mathbf{a}^2 \right\rangle . \tag{2.80}$$

2.3.3 Beat Pattern Between Two Overlapped Waves

We now consider the case of two overlapped oscillating electric fields. The beat pattern created by their interference can create strong modulations over distances on the order of the wavelengths of the waves, typically much smaller than the typical spatial variations of the envelope of either beam alone and thus potentially prone to exerting a significant ponderomotive force on the electrons. Therefore, in the following, we assume that the two overlapped waves are spatially uniform and concentrate on their beat pattern only.

We write the total electric field as

$$\mathbf{E}(\mathbf{r}, t) = \frac{1}{2}\mathbf{E}_0 e^{i\psi_0} + \frac{1}{2}\mathbf{E}_1 e^{i\psi_1} + c.c., \tag{2.81}$$

where $\psi_0 = \mathbf{k}_0 \cdot \mathbf{r} - \omega_0 t$ and $\psi_1 = \mathbf{k}_1 \cdot \mathbf{r} - \omega_1 t$. The two envelopes $\mathbf{E}_0$ and $\mathbf{E}_1$ are assumed to be complex, to "absorb" any arbitrary phase shift between the two waves.[5]

The main assumption for this situation is that the sum of the frequencies is much larger than their difference, i.e.,

$$\omega_0 + \omega_1 \gg \omega_0 - \omega_1 , \tag{2.82}$$

where we have assumed that $\omega_0 > \omega_1$. Next we define $\psi_\Sigma = \psi_0 + \psi_1$ and $\psi_\Delta = \psi_0 - \psi_1$, with similar definitions for ω_Σ, ω_Δ, $\mathbf{k}_\Sigma$, and $\mathbf{k}_\Delta$.

The electric field can then be expressed as follows:

$$\mathbf{E}(\mathbf{r}, t) = \frac{1}{2}\mathbf{E}_0 e^{i(\psi_\Sigma + \psi_\Delta)/2} + \frac{1}{2}\mathbf{E}_1 e^{i(\psi_\Sigma - \psi_\Delta)/2} + c.c. \tag{2.83}$$

$$= \frac{1}{2}\left(\mathbf{E}_0 e^{i\psi_\Delta/2} + \mathbf{E}_1 e^{-i\psi_\Delta/2} \right) e^{i\psi_\Sigma/2} + c.c.. \tag{2.84}$$

The entire term in parenthesis in this last expression can be considered a slowly varying envelope compared to the rapid oscillation at $\psi_\Sigma/2$; to simplify our notations and make a connection with the previous subsections, we denote it $\mathbf{E}_s$, i.e.,

$$\mathbf{E}_s = \mathbf{E}_0 e^{i\psi_\Delta/2} + \mathbf{E}_1 e^{-i\psi_\Delta/2} , \tag{2.85}$$

such that $\mathbf{E} = \frac{1}{2}\mathbf{E}_s \exp[i\psi_\Sigma/2] + c.c..$

From here, the analysis of a particle motion in the total field from the two waves takes the exact same form as in the previous subsection. We can again separate the particle motion following $\mathbf{r}(t) = \mathbf{r}_0(t) + \mathbf{r}_s(t)$, where $\mathbf{r}_0(t)$ is the fast oscillation at $\omega_\Sigma/2$ and $\mathbf{r}_s(t) = \langle \mathbf{r}(t) \rangle$, where the average is taken over $4\pi/\omega_\Sigma$. Skipping over the details of the derivation (which is exactly the same as before), we obtain the following result:

$$m d_t^2 \mathbf{r}_s = -\frac{q^2}{4m(\omega_\Sigma/2)^2} \nabla |\mathbf{E}_s|^2 = \mathbf{F}_p . \tag{2.86}$$

Replacing $\mathbf{E}_s$ by its expression as a function of $\mathbf{E}_0$ and $\mathbf{E}_1$, and assuming that the spatial variations of each individual wave's envelope are much larger than $1/k_\Delta$, we get

$$\mathbf{F}_p = -\frac{q^2}{4m\bar{\omega}^2} \nabla \left(\mathbf{E}_0 \cdot \mathbf{E}_1^* e^{i\psi_\Delta} + c.c. \right) , \tag{2.87}$$

---

[5] This is equivalent to writing $\mathbf{E} = |\mathbf{E}_0| \cos(\psi_0 + \varphi_0) + |\mathbf{E}_1| \cos(\psi_1 + \varphi_1)$, with $\mathbf{E}_0 = |\mathbf{E}_0| \exp[i\varphi_0]$ and $\mathbf{E}_1 = |\mathbf{E}_1| \exp[i\varphi_1]$; the two arbitrary constant phase terms φ_0 and φ_1 are unknown and do not impact the results as we will see.

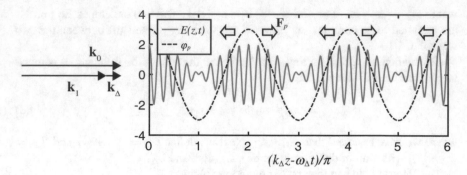

Fig. 2.8 Ponderomotive force from a beat wave between two waves' electric fields. In this example, the two waves have the same amplitude ($E_0 = E_1$), both propagate along z and have a spatial frequency shift $\mathbf{k}_\Delta = \mathbf{k}_0/10$. The ponderomotive potential leads to a bunching of the particles in the regions of lower field intensity

where $\bar{\omega} = \omega_\Sigma/2$ is the averaged frequency between the two waves. We can take out the phase factor of the two waves to simplify this expression: writing the field envelopes as $\mathbf{E}_{0,1} = |E_{0,1}| \exp[i\varphi_{0,1}]\mathbf{e}_{0,1}$ with φ_0 and φ_1 arbitrary phase factors and $\mathbf{e}_{0,1}$ the unit vectors along the polarization direction of the two waves, we get $\mathbf{E}_0 \cdot \mathbf{E}_1^* e^{i\psi_\Delta} + c.c. = |E_0||E_1|\mathbf{e}_0 \cdot \mathbf{e}_1 \cos(\psi_\Delta + \varphi_0 - \varphi_1)$. Since $\varphi_0 - \varphi_1$ is a constant phase shift that is arbitrary and does not change any of the results here, we can drop it and get the following expression for the ponderomotive force and potential Φ_p, defined as $\mathbf{F}_p = -q\nabla\Phi_p$:

$$\Phi_p = \frac{q}{4m\bar{\omega}^2}|E_0||E_1|\mathbf{e}_0 \cdot \mathbf{e}_1 \cos(\psi_\Delta)\,, \tag{2.88}$$

$$\mathbf{F}_p = \frac{q^2}{4m\bar{\omega}^2}|E_0||E_1|\mathbf{e}_0 \cdot \mathbf{e}_1 \sin(\psi_\Delta)\mathbf{k}_\Delta\,. \tag{2.89}$$

This shows that the ponderomotive potential is modulated at the beat wave frequency and wave-number ω_Δ, $\mathbf{k}_\Delta$. The force acts to bunch the charged particles into the regions where the beat wave amplitude is lowest—for example, in the case of two light waves producing interference fringes as they overlap, the particles will bunch into the "dark" fringes, as illustrated in Fig. 2.8.

We should reiterate here that in general the ponderomotive force exists regardless of the nature of either of the two oscillating electric fields beating together—longitudinal or transverse. The particle bunching will create a traveling density modulation in the plasma that can resonantly excite a natural plasma wave, EPW, or IAW. This is at the origin of the three-wave instabilities that will be studied in later chapters.

2.3.4   Connection Between the Ponderomotive Force and the Single Particle Motion in Uniform Light or Plasma Waves

In this section, we wish to clarify the connection between the ponderomotive force and some of the results from our previous analysis of single particle dynamics in a light wave or plasma wave.

First, recall that an electron in a plane, monochromatic light wave (with linear polarization), has a motion characterized by three main velocity components: a quiver velocity $v_{os} \propto a_0$ at ω_0 along the electric field direction, a quiver velocity at $2\omega_0$ and $\propto a_0^2$ along z (laser propagation direction) and a constant drift velocity $\propto a_0^2$ along z (cf. Fig. 2.2 and Eqs. (2.11), (2.16)), all leading to the figure of eight particle motion:

$$\frac{\mathbf{v}(t)}{c} = a_0 \cos(\omega_0 t)\mathbf{e}_x + \frac{a_0^2}{4} \cos(2\omega_0 t)\mathbf{e}_z + \frac{a_0^2}{4}\mathbf{e}_z . \tag{2.90}$$

As we briefly mentioned in Sect. 2.1, the drift velocity is in fact associated with the ponderomotive force. Indeed, let us look for the velocity along z averaged over the laser oscillation period, starting from the force equation with the ponderomotive force (Eq. (2.80)):

$$d_t \langle v_z \rangle = \frac{F_p}{m} = -\frac{1}{2}c^2 \partial_z \langle \mathbf{a}^2 \rangle . \tag{2.91}$$

Using the relations $d_t \langle \mathbf{a}^2 \rangle = (\partial_t + v_z \partial_z) \langle \mathbf{a}^2 \rangle$ and $(\partial_t + c \partial_z) \langle \mathbf{a}^2 \rangle = 0$, and eliminating $\partial_t \langle \mathbf{a}^2 \rangle$ between the two yields, $d_t \langle \mathbf{a}^2 \rangle = (v_z - c)\partial_z \langle \mathbf{a}^2 \rangle \approx -c \partial_z \langle \mathbf{a}^2 \rangle$, i.e.,

$$d_t \langle v_z \rangle = \frac{c}{2} d_t \langle \mathbf{a}^2 \rangle . \tag{2.92}$$

Time integrating from $-\infty$ (when the laser is turned off and the initial electron velocity $\langle v_z(-\infty) \rangle = 0$) to the time considered in the plane wave analysis (when the laser has reached its final amplitude) leads to $\langle v_z \rangle = c \langle \mathbf{a}^2 \rangle /2 = v_d$. In other words: the ponderomotive force associated with turning the laser on (which acts via a negative gradient along z along the rising edge of the pulse) is responsible for the drift velocity of the electron.

The fact that the drift is directed toward z in the analysis of the electron in a plane laser wave comes from the implicit hypothesis that the wave remains plane and without any gradients in amplitude along the transverse directions but does have a gradient along z along its rising edge, associated with the temporal rise in intensity. In reality, the drift will be directed toward the intensity gradient, which may not

Fig. 2.9 Motion of an
electron in a finite laser pulse
moving toward $z > 0$. The
ponderomotive force on the
rising edge pushes the
electron toward z, whereas
the ponderomotive force from
the dropping edge accelerates
it toward $-z$. The two net
accelerations cancel, leaving
the electron with the same
initial velocity (but with a
finite displacement along z)

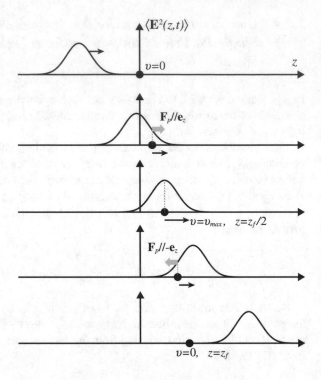

be along z if the laser has a finite spatial envelope in the transverse directions; for
example, if the electron is initially located slightly off-axis near the best focus of
a tightly focused and slowly rising laser pulse, then it will be pushed sideways (as
long as the scale-length of the rise, $L_{rise} = \tau_{rise}/c$, where τ_{rise}, is the rise time,
remains much larger than the laser focal spot w_0, $L_{rise} \gg w_0$). In any case, the rise
time must remain large compared to the laser oscillation period $2\pi/\omega_0$ in order to
not violate the monochromatic wave assumption.

If the laser pulse is finite, then the dropping edge of the pulse will exert a
ponderomotive force in the opposite direction from the one on the rising edge. This
will lead to a deceleration of the electron (i.e., acceleration toward $-z$), which will
eventually bring the electron back to zero velocity—but will leave it displaced from
its initial location along z, as illustrated in Fig. 2.9. This means that an electron
interacting with a plane wave cannot gain energy, which is often referred to as the
Lawson–Woodward theorem [7]. This criterion is easily violated in reality as laser
pulses always have finite sizes, which can lead to "vacuum acceleration" as the
electrons can be expelled forward *and sideways* by the rising laser pulse and thus not
be subject to an equivalent deceleration, as the electron can be away from the best
focus region by the time the dropping edge passes by (cf. Ref. [4] and references
therein). An example of electron dynamics in a finite light pulse is described in
Problem 2.2.

The situation is in fact similar for an electron in an EPW. In Sect. 2.2, we described the dynamics of an electron initially at rest in a plasma wave. The Hamiltonian is given by $H_{rest} = v_\phi^2/2$, which once inserted in Eq. (2.32) gives the following expression for the velocity in the frame of the wave ($\bar{v} = v - v_\phi$):

$$\bar{v}(t) = -v_\phi \sqrt{1 + 2\frac{\delta n_e}{n_{e0}} \cos(\omega t)}, \tag{2.93}$$

where we used $e\Phi_0/m = v_\phi^2 \delta n_e/n_{e0}$ and assumed that the electron velocity remains much smaller than the phase velocity of the wave, so that the phase term can be assumed to be dependent on time only, $\psi = kz(t) - \omega t \approx kz(0) - \omega t$ (we can then set $z(0) = 0$ without loss of generality).[6] Expanding to first order in $\delta n_e/n_{e0}$ yields Eq. (2.33), i.e., a simple harmonic oscillation in the wave; however, expanding to second order leads to

$$\frac{v}{v_\phi} \approx -\frac{\delta n_e}{n_{e0}} \cos(\omega t) + \frac{1}{4}\left(\frac{\delta n_e}{n_{e0}}\right)^2 \cos(2\omega t) + \frac{1}{4}\left(\frac{\delta n_e}{n_{e0}}\right)^2. \tag{2.94}$$

Note the similarity with the motion in a light wave above in Eq. (2.90): we recognize again a harmonic quiver velocity in the EPW at ω and $\propto \delta n_e/n_{e0}$ (but this time along z, vs. $/\!/ x$ for an EMW), a quiver velocity at 2ω and $\propto (\delta n_e/n_{e0})^2$, and a drift velocity $\propto (\delta n_e/n_{e0})^2$ (all three along z).

Proceeding like for the EMW, we can also write the force equation for the oscillation-averaged velocity using the ponderomotive force, Eq. (2.64), as

$$d_t \langle v_z \rangle = \frac{F_p}{m} = -\frac{1}{2}v_\phi^2 \partial_z \left\langle (\delta n_e/n_{e0})^2 \right\rangle, \tag{2.95}$$

where we used $|E| \approx m v_\phi \omega (\delta n_e/n_{e0})/e$ in the limit of $k\lambda_{De} \ll 1$, per Eq. (1.124). Using again $d_t \left\langle (\delta n_e/n_{e0})^2 \right\rangle \approx -v_\phi \partial_z \left\langle (\delta n_e/n_{e0})^2 \right\rangle$ and integrating from $-\infty$ (when the plasma wave amplitude and the electron velocity are both 0) to the time considered in the plane wave analysis lead to

$$\langle v_z \rangle = \frac{1}{2}v_\phi \left\langle (\delta n_e/n_{e0})^2 \right\rangle. \tag{2.96}$$

Having used the ponderomotive force to describe the oscillation-averaged motion, we recovered the drift velocity from the single particle analysis in a plane wave, Eq. (2.94)—just like we did above for an EMW. This shows again that the drift velocity that appears in the analysis of particle dynamics in an EPW is in fact coming

---

[6] For example: consider two times t_1 and t_2 such that $\delta t = t_2 - t_1$ and $\delta z = z(t_2) - z(t_1)$: we then have $\psi(t_2) - \psi(t_1) = k\delta z - \omega\delta t = -\omega\delta t(1 - v_z/v_\phi) \approx -\omega t$ for $v_z \approx \delta z/\delta t$.

Fig. 2.10 Field $a(z,t)$ of the pulse used for Problem 2.2, with a triangular intensity profile a^2, shown here at $t = 0$. The electron is initially located at $z = L$

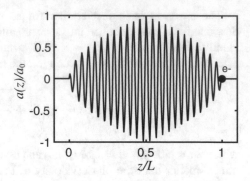

from the ponderomotive force in the rising edge of the wave as it was "turned on" adiabatically to reach its final amplitude.

Problems

2.1 Helicoidal Electron Motion in a Circularly Polarized Light Wave

Derive the longitudinal motion of an electron in a circularly polarized light wave, starting from Eq. (2.9) and taking $\mathbf{e}_\pi = (\mathbf{e}_x + i\mathbf{e}_y)/\sqrt{2}$ in the definition of $\mathbf{a}$. Show that the second harmonic oscillation from the linear polarization disappears, leaving only the drift velocity $\mathbf{v}_d$.

2.2 Electron Motion in a Finite Laser Pulse

- Consider a laser pulse linearly polarized along the x-direction and propagating toward z, with a triangular temporal profile for its intensity envelope:

$$\mathbf{a}(z,t) = a_0\sqrt{T_L(z,t)}\mathbf{e}_x \cos(kz - \omega t), \qquad (2.97)$$

  where T_L is a triangular function of length L, defined as $T_L(z,t) = 1 - |2(z - ct)/L - 1|$ for $|z - ct| \in [0, L]$ and 0 elsewhere (cf. Fig. 2.10). We will assume that $a_0 \ll 1$. Take the electron initial velocity as zero, its initial position before the pulse arrives as $z = L$, and define $t = 0$ as the time when the front edge of the pulse reaches the electron (i.e., $z(0) = L$ and $T(z, t = 0) = 1 - |2z/L - 1|$). Write down the force equation for the oscillation-averaged velocity v_s along z in the presence of the ponderomotive force.
- Solve for the electron trajectory along the rising edge of the pulse, up to its peak intensity. In particular, show that the electron reaches the peak laser intensity at a time $t_m = (L/2c)(1 + a_0^2/8)$, corresponding to a peak velocity $v_m = ca_0^2/4$ and location $z(t_m) = L + La_0^2/16$.

- Calculate the trajectory on the falling edge, and show that the electron velocity returns to zero once the laser has passed. Show that the electron has been displaced from its initial position by a distance $\Delta z = La_0^2/8$.
- Solve the full trajectory numerically, i.e., the exact Lorentz force equation (including the **B**-field), not just the oscillation-averaged velocity like in the previous questions. Use a Runge–Kutta integration scheme (use an existing routine from a scientific software, or even better, code one yourself), and verify the results above after taking the oscillation-average quantities (you might want to normalize the time by ω, distance by k, velocity by c).

References

1. E. Esarey, C.B. Schroeder, W.P. Leemans, Rev. Mod. Phys. **81**, 1229 (2009)
2. L.D. Landau, E.M. Lifshitz, *The classical theory of fields, by L. D. Landau and E. M. Lifshitz*, rev. 2d ed. ed. (Pergamon Press; Addison-Wesley Pub. Co Oxford, Reading, 1962), p. 404 p.
3. A. Macchi, *A Superintense Laser-Plasma Interaction Theory Primer* (Springer, Berlin, 2013)
4. P. Gibbon, *Short Pulse Laser Interactions with Matter: An Introduction* (World Scientific, Singapore, 2005)
5. M. Abramowitz, I.A. Stegun, *Handbook of Mathematical Functions with Formulas, Graphs, and Mathematical Tables*, vol. 55 (US Government Printing Office, 1948)
6. T. O'Neil, Phys. Fluids **8**, 2255 (1965)
7. J.D. Lawson, IEEE Trans. Nucl. Sci. **26** (1979)

Chapter 3
Linear Propagation of Light Waves in Plasmas

In this chapter, we present a description of linear propagation of light waves in plasmas, by which we mean that the plasma is simply treated as a linear medium with an index of refraction that is independent of the light wave intensity. We begin with the WKB method, which allows the description of light propagation when the plasma index (i.e., the electron density) varies on scales that are larger than the oscillation wavelength. We then introduce the well-known Airy function continuation, which describes the transition from the oscillatory behavior of the light wave for densities below critical to the evanescent behavior of the wave past n_c. The analysis introduces the so-called Airy skin depth, which is a scale-length of important meaning as it represents both the size of the region in the vicinity of n_c where an Airy function is required (as opposed to an oscillatory WKB solution) and the scale over which the field exponentially decays past the critical density.

Finally, we will present the ray-tracing method, which is a simple yet powerful tool to describe light propagation in plasmas with density variations in space and in time and emphasize the main results of this model to describe both the refraction of light in plasmas and its frequency shift due to time-varying density.

3.1 The WKB Method

We now wish to describe the propagation of a light wave in a plasma with a varying electron density, i.e., refractive index. For now, we assume that the refractive index is constant in time and that the light is monochromatic. The plasma density variation is taken to be along the z-direction only, with the light wave incident at normal incidence (i.e., propagating along z). We write the following expression for the light wave's normalized vector potential $\mathbf{a}$:

$$a(z, t) = \frac{1}{2} a_0 e^{-i\omega_0 t} f(z) + c.c., \tag{3.1}$$

where $c.c.$ stands for complex conjugate, which means that the light is monochromatic at the frequency ω_0, but with an unknown dependence on z since the plasma density also depends on z. The polarization does not matter here since the light propagates along the density gradient (it is thus necessarily s-polarized). Our goal is to find an expression for $f(z)$.

Taking the wave equation $(\partial_t^2 - c^2 \nabla^2 + \omega_{pe}^2)a = 0$ and inserting the expression for a above yield

$$\frac{c^2}{\omega_0^2} f''(z) = -\varepsilon(z) f(z) , \tag{3.2}$$

where $\varepsilon(z) = 1 - \omega_{pe}^2(z)/\omega_0^2$ is the plasma dielectric constant (cf. Sect. 1.3.1.1, Eq. (1.101)), and the prime denotes the derivative with respect to z.

We are now going to assume that the spatial variations in $\varepsilon(z)$ occur over distances that are much larger than the laser wavelength. The equation above then naturally lends itself to the WKB method (for Wentzel, Kramers and Brillouin), due to the clear separation of scales between the ("small") coefficient $c^2/\omega_0^2 \sim \lambda_0^2$ in front of the second derivative on the left-hand side (with λ_0 the vacuum wavelength), and ε in front of f on the right-hand side. Following Bender and Orszag [1], we make the following ansatz:

$$f(z) = e^{i\psi(z)} , \tag{3.3}$$

where the function ψ is expressed as a perturbative series:

$$\psi(z) = \frac{1}{\delta} \sum_{n=0}^{\infty} \delta^n S_n(z) , \quad \delta \to 0 . \tag{3.4}$$

The small parameter δ can be chosen arbitrarily as we shall see below; the validity of the perturbation method will be verified and quantified below as well.

Inserting the expression for f into Eq. (3.2) leads to

$$\frac{c^2}{\omega_0^2} \left[i\psi'' - (\psi')^2 \right] = -\varepsilon , \tag{3.5}$$

and inserting the perturbative expansion of ψ now leads to

$$\frac{c^2}{\omega_0^2} \left[i\frac{S_0''}{\delta} + iS_1'' - \left(\frac{S_0'}{\delta} + S_1' \right)^2 + O(\delta^0) \right] = -\varepsilon . \tag{3.6}$$

The term $\propto \delta^{-2}$ on the left-hand side is dominant, leading to

$$\frac{c^2}{\omega_0^2} \frac{(S_0')^2}{\delta^2} = \varepsilon . \tag{3.7}$$

We thus take $\delta = c/\omega_0$ (which we had already identified as a "small" parameter), leading to $S'_0 = \pm\sqrt{\varepsilon} = \pm n(z)$, with $n = \sqrt{1 - n_e(z)/n_c}$ the plasma refractive index, and therefore,

$$S_0(z) = \pm \int^z n(\bar{z})d\bar{z}, \qquad (3.8)$$

where $\bar{z}$ is a dummy integration variable (which we drop in the following, simply using z for simplicity).

We see that the dominant term in the perturbative expansion is simply $S_0/\delta = \pm(\omega_0/c) \int n(\bar{z})d\bar{z}$, leading to

$$f(z) \approx \exp\left[\pm i\frac{\omega_0}{c} \int^z n(\bar{z})d\bar{z}\right] = \exp\left[\pm \int^z k_0(z)dz\right], \qquad (3.9)$$

where $k_0(z) = \omega n(z)/c$ is the local wave-number. Obviously, if the density is uniform, i.e., $k_0(z) = cst.$, then we simply recover the plane wave solution $f(z) = \exp[\pm ik_0 z]$ (with the $+$ and $-$ sign representing a wave propagating toward $z > 0$ or $z < 0$, respectively, both of which being solutions to the wave equation under WKB).

At the next order, collecting the terms $\propto \delta^{-1}$ in Eq. (3.6) leads to $iS''_0 = 2S'_0 S'_1$, or

$$\frac{S''_0}{S'_0} = -2i S'_1. \qquad (3.10)$$

Integrating on both sides with $S'_0 = n(z)$ leads to

$$S_1(z) = i \ln \sqrt{n(z)} + \varphi, \qquad (3.11)$$

where φ is an integration constant. Inserting the expressions for S_0 and S_1 in $f(z)$ now leads to the following expression for the light wave's vector potential at the next order:

$$\boxed{a(z,t) = \frac{1}{2}\frac{a_0}{\sqrt{n(z)}} \exp\left[\pm i \int k_0(z)dz - i\omega_0 t + i\varphi\right] + c.c.,} \qquad (3.12)$$

where the integration constant φ can be set to zero without loss of generality. The WKB method can be pushed to higher-order (i.e., S_2, S_3, etc.), but the expression above is sufficient in most cases.

Note that the effect of the higher-order term S_1 compared to keeping only S_0 like in Eq. (3.9) is the appearance of the $1/\sqrt{n(z)}$ term. Physically, this last term enables $|a|\sqrt{n(z)} = cst.$: as the light propagates toward higher intensities, the decrease in refractive index leads to an increase in the field amplitude, usually referred to as "field swelling". Since $n(z)|a(z)|^2 \propto I = c\varepsilon_0 n|E_0|^2/2$, this simply expresses the

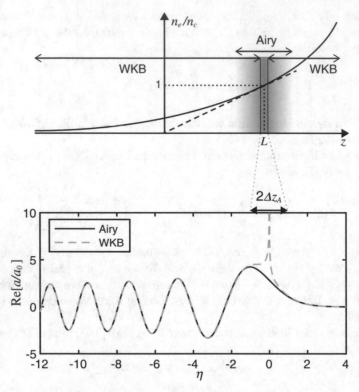

Fig. 3.1 Light propagation models including the vicinity of the turning point, here at $n_e = n_c$ and $z = L$ (or $\eta = 0$) for normal incidence. The Airy solution from Eq. (3.24) is valid in the region around the turning point where the density is approximately linear, $n_e/n_c \approx z/L$. WKB is valid all the way up to the Airy skin depth, defined as the region where $|\eta| \leq 1$ or $|z - L| \leq \Delta z_A$, cf. Eq. (3.28); the WKB field is oscillating before the turning point (Eq. (3.12), or (3.21) with $R = -i$), and evanescent (exponentially decaying) beyond n_c (Eq. 3.26)

conservation of the intensity, i.e., the energy flux. This simultaneous increase in oscillation wavelength and field amplitude is visible in Fig. 3.1, which compares the WKB approximation to the exact analytical solution for a linear density ramp up to (and beyond) the critical density where $n(z) = 0$ and the WKB approximation fails, as described in the following subsection. We can see that as the density increases (and the index decreases), the oscillation wavelength increases.

We shall now quantify the condition for validity of the WKB expansion. Comparing the terms of order $\propto \delta^{-2}$ and $\propto \delta^{-1}$ in Eq. (3.6) leads to the condition:

$$\frac{|S_0''|}{\delta} \ll \frac{(S_0')^2}{\delta^2}$$

$$\Leftrightarrow \frac{|n'|}{n^2} \ll \frac{\omega_0}{c}. \tag{3.13}$$

3.2 Airy Description at the Turning Point

3.2.1 Connection Between WKB and the Airy Description Up to the Turning Point

The WKB method relies on the approximation of a slowly varying envelope compared to a rapid phase oscillation; however, as the density approaches n_c (referred to as the "turning point" in the case of normal incidence), the oscillation wavelength $\lambda(z) = 2\pi/k(z)$ goes to infinity and that approximation becomes invalid (i.e., $\varepsilon \to 0$ in Eq. (3.2), which invalidates the conditions to use WKB). In other words, the refractive index goes to zero and the field expression in Eq. (3.12) diverges.

However, the wave equation can actually be solved analytically as long as the density profile can be Taylor-expanded to first order in the vicinity of the turning point, i.e., if the density profile is approximately linear in a small region near the turning point located at $z = z_t$ such that $n_e(z_t) = n_c$:

$$n_e(z) \approx n_c + (z - z_t)n'_e(z_t),\tag{3.14}$$

where the prime denotes again the derivative along z.

We define L as the gradient scale-length at the turning point:

$$L \equiv \frac{n_c}{n'_e(z_t)}.\tag{3.15}$$

For simplicity, we will choose the origin of the z-axis such that $z_t = L$, i.e., $n_e(z)/n_c \approx z/L$ in the vicinity of the turning point ($z \approx L$), as illustrated in Fig. 3.1. The refractive index in that region is $n^2(z) \approx 1 - z/L$.

According to the previous section and Eq. (3.13), the WKB approximation should remain valid as long as $|n'|/n^2 \ll \omega_0/c$. Replacing with the linearized expression for $n(z)$ near z_t indicates that the WKB approximation is valid for z such that

$$(\omega_0 L/c)^{2/3} \left(1 - \frac{z}{L}\right) \gg 1,\tag{3.16}$$

where we have assumed that the density is increasing toward $z > 0$ and the light wave also propagates toward $z > 0$.

We go back to the general expression of the light wave-vector potential from Eq. (3.1), $a = \frac{1}{2}a_0 \exp[-i\omega_0 t] f(z) + c.c.$, and the wave equation for $f(z)$ from Eq. (3.2), $f'' = -(\omega_0/c)^2 \varepsilon f$. Using a change of variable:

$$\eta = \left(\frac{\omega_0 L}{c}\right)^{2/3} \left(\frac{z}{L} - 1\right),\tag{3.17}$$

with $\eta < 0$ up to the turning point ($z < L$), transforms the wave equation into an Airy equation:

$$(\partial_\eta^2 - \eta)f = 0.\tag{3.18}$$

The Airy equation has a general solution of the form $f(\eta) = \alpha A_i(\eta) + \beta B_i(\eta)$, where A_i and B_i are the Airy functions of the first and second kinds, respectively, and α and β are constants. Since $A_i(\eta) \to 0$ and $B_i(\eta) \to \infty$ when $\eta \to \infty$, we need to have $\beta = 0$ in order to ensure a physical solution where the field is evanescent past the critical density.

At this point, we already see that the validity condition for WKB from Eq. (3.16) is exactly equivalent to $-\eta \gg 1$ (with $\eta < 0$), for which the Airy function has the following asymptotic expansion [2]:

$$A_i(\eta) \approx \pi^{-1/2}(-\eta)^{-1/4} \cos\left(\frac{2}{3}(-\eta)^{3/2} - \frac{\pi}{4}\right).\tag{3.19}$$

Therefore, we should expect that this asymptotic expression matches the WKB solution for locations z that are sufficiently far from the turning point—and we will proceed to show that it is indeed the case.

First, we write the WKB solution for the linear expansion of the density; here we have to recognize that the field must be composed of the incoming wave as well as a reflected wave (which both appear in WKB, per the $\pm$ sign in front of the local wave-vector in Eq. (3.12)). For now, we attribute an arbitrary (complex) reflectivity coefficient R to the reflected wave (such that $|R|^2$ represents the reflected energy coefficient):

$$a(z,t) = \frac{a_0 e^{-i\omega_0 t}}{2\sqrt{n(z)}}\left\{\exp\left[i\int k_0(z)dz\right] + R\exp\left[-i\int k_0(z)dz\right]\right\} + c.c..\tag{3.20}$$

Since $k_0(z) = (\omega_0/c)\sqrt{1 - z/L}$ in the region of interest ($z \approx L$), we have $\int k_0(z)dz = -\frac{2}{3}(-\eta)^{3/2}$. Also noting that $\sqrt{n} = (-\eta)^{1/4}(\omega_0 L/c)^{-1/6}$, the WKB field can be expressed as a function of η as follows:

$$a(\eta,t) = \frac{a_0 e^{-i\omega_0 t}}{2(-\eta)^{1/4}}\left(\frac{\omega_0 L}{c}\right)^{1/6}\left[\exp\left(-i\frac{2}{3}(-\eta)^{3/2}\right) + R\exp\left(i\frac{2}{3}(-\eta)^{3/2}\right)\right] + c.c..$$
$$\tag{3.21}$$

Second, recasting the Airy function asymptotic expansion from Eq. (3.19) in $f(\eta) = \alpha A_i(\eta)$ (with $a(\eta,t) = \frac{1}{2}a_0\exp[-i\omega_0 t]f(\eta) + c.c.$) gives

$$a(\eta,t) = \frac{a_0 e^{-i\omega_0 t}}{2}\frac{\alpha e^{i\pi/4}}{2\sqrt{\pi}(-\eta)^{1/4}}\left[\exp\left(-i\frac{2}{3}(-\eta)^{3/2}\right) - i\exp\left(i\frac{2}{3}(-\eta)^{3/2}\right)\right] + c.c..$$
$$\tag{3.22}$$

A comparison of Eqs. (3.21) and (3.22) shows that as expected, the Airy function expansion can be matched to the WKB solution, with

$$\alpha = 2\sqrt{\pi}e^{-i\pi/4}\left(\frac{\omega_0 L}{c}\right)^{1/6} \tag{3.23}$$

and $R = -i$, i.e., the wave is fully reflected at the critical density with $|R|^2 = 1$ and a $-\pi/2$ phase shift. Note that the matching requires that the linearization (first-order Taylor expansion) of the density near n_c stays valid over a distance sufficiently large so that $|\eta| \gg 1$. We will see below that this is easily satisfied for typical laser–plasma environments, at least for typical HED or ICF conditions.

The full expression for the field in the vicinity of the turning point is then

$$a(\eta, t) = \frac{a_0 e^{-i\omega_0 t}}{2} 2\sqrt{\pi}e^{-i\pi/4}\left(\frac{\omega_0 L}{c}\right)^{1/6} A_i(\eta) + c.c. \tag{3.24}$$

3.2.2   Field Evanescence Past the Turning Point

Beyond the turning point, far enough from n_c for $n_e/n_c > 1$ (or $z > L$, or $\eta > 0$ for the same linear profile as before), WKB becomes valid again, and the exact same analysis as in Sect. 3.1 applies up to the final result, Eq. (3.12) —except that we now have $\varepsilon < 0$, so the refractive index must be imaginary, $n(z) = \sqrt{\varepsilon} = i\sqrt{n_e/n_c - 1} = i|n(z)|$. The expression for the field becomes, assuming propagation toward $z > 0$ for clarity and using $\sqrt{n} = \sqrt{|n|}\exp[i\pi/4]$:

$$a(z, t) = \frac{a_0}{\sqrt{|n(z)|}}\exp\left[-\frac{2L\omega_0}{3c}\left(\frac{z}{L} - 1\right)^{3/2}\right]\cos(\omega_0 t + \varphi), \tag{3.25}$$

where φ is an arbitrary phase, and we used $\int |n(z)|dz = (2L/3)(z/L - 1)^{3/2} = (2L/3)|n|^3$. We obtain an evanescent behavior of the light wave (i.e., exponential decay) past the turning point, as expected.

Note that a change of variable from z to η leads to the following expression:

$$a(\eta) = a_0\left(\frac{\omega_0 L}{c}\right)^{1/6}\eta^{-1/4}\exp\left[-\frac{2}{3}\eta^{3/2}\right]\cos(\omega_0 t + \varphi). \tag{3.26}$$

If we use the following asymptotic expansion for the Airy function for $\eta \gg 1$ [2]:

$$A_i(\eta) \approx \frac{1}{2\sqrt{\pi}}\eta^{-1/4}\exp\left[-\frac{2}{3}\eta^{3/2}\right], \tag{3.27}$$

then we see that using this expansion in the expression for the field near the turning point, Eq. (3.24), gives the exact same result as Eq. (3.26). In other words, the Airy and WKB solutions match again on the other side of the turning point, where $\eta \gg 1$, and show an exponential decay (evanescence) of the field.

3.2.3 Regions of Validity

Figure 3.1 shows the Airy solution from Eq. (3.24) compared to the WKB model (Eqs. (3.21), (3.26) for $n_e < n_c$ and $n_e > n_c$, respectively) near a turning point, located at $\eta = 0$. We can see that the WKB expression is an excellent approximation on either side of the turning point up to $|\eta| \approx 1$.

Figure 3.1 also summarizes the different regions of validity of our model:

- The Airy solution, Eq. (3.24), is valid in the region around the turning point where the density is approximately linear, i.e., the first-order Taylor expansion $n_e/n_c \approx z/L$ is valid.
- The WKB model is valid everywhere except in a narrow region near the turning point (at the critical density for normal incidence, or a lower density for oblique incidence—see next section) whose width we shall call the "Airy skin depth" Δz_A; the Airy skin depth is defined as the distance on either side of the turning point such that $|\eta| \leq 1$ or $|z - L| \leq \Delta z_A$, with

$$\boxed{\Delta z_A = \left(\frac{Lc^2}{\omega_0^2} \right)^{1/3} \approx 0.3\lambda_0^{2/3} L^{1/3}.} \tag{3.28}$$

The Airy skin depth also corresponds to the location before the turning point ($\eta = -1$) where the field is maximum, as well as the characteristic decay length past the critical density ($\eta = 1$). Since the scaling with L is weak, the Airy skin depth will be on the order of a micron for typical ICF or HED conditions where the gradients are typically on the order of tens to hundreds of microns: for example, for a laser wavelength of 0.351 μm, we have $\Delta z_A = 0.3$ μm and 1.5 μm for $L = 10$ μm and 1 mm, respectively. Since WKB is valid up to the Airy skin depth, the analysis above is valid as long as the density is approximately linear over the Airy skin depth.

Finally, the amplification or "swelling" of the field due to the drop in group velocity near the turning point (cf. Sect. 3.1) depends on the density scale-length L at n_c. Since $A_i(-1) \approx 0.536$, the peak intensity amplification at the turning point $a_{max} \equiv a(\eta = -1)$ compared to the vacuum intensity can be expressed as

$$\frac{I_{max}}{I_0} = \left| \frac{a_{max}}{a_0} \right|^2 = 4\pi(\omega_0 L/c)^{1/3} A_i^2(-1) \approx 6.6(L/\lambda_0)^{1/3}. \tag{3.29}$$

3.3   Oblique Incidence (s-Polarization)

The case of oblique incidence for s-polarized light waves can be treated in a similar manner. By s-polarized, we mean that the light is linearly polarized with its polarization vector perpendicular to the plane of incidence, i.e., the plane formed by the wave-vector $\mathbf{k}_0$ and the density gradient ∇n_e. Using the notations from Fig. 3.2, we assume that the light wave is linearly polarized along y and propagates in the (x, z) plane with a density variation along z. The vacuum–plasma interface is at $z = 0$, and the light wave has an incidence angle θ_0 at $z = 0$. The case of p-polarized light (i.e., polarization vector in the (x, z) plane) is more complicated since the electric field will have a component in the direction of the density gradient and therefore have a longitudinal (electrostatic) component as it contributes to charge separation oscillations along z. This situation can lead to resonance absorption of the light wave, as will be discussed in Sect. 4.8.

We can repeat the WKB analysis from the previous section but extend it to two dimensions (x, z) to find the local wave-vector. We now have the general expression for the light wave's vector potential:

$$a(x, z, t) = \frac{1}{2} a_0 e^{-i\omega_0 t} f(x, z) + c.c., \tag{3.30}$$

which after insertion in the wave equation $(\partial_t^2 - c^2 \nabla^2 + \omega_{pe}^2) a = 0$ yields

$$\frac{c^2}{\omega_0^2} (\partial_x^2 + \partial_z^2) f(x, z) = -\varepsilon(z) f(x, z) \tag{3.31}$$

with $\varepsilon(z) = 1 - n_e(z)/n_c$. Taking the WKB ansatz

$$f(x, z) = e^{i\psi(x, z)} \ , \quad \psi(x, z) = \frac{1}{\delta} \sum_{n=0}^{\infty} \delta^n S_n(x, z) \tag{3.32}$$

with $\delta = c/\omega_0$, inserting into the wave equation for f above, and keeping only the leading term $\propto \delta^{-2}$ lead to the so-called eikonal equation:

$$(\partial_x S_0(x, z))^2 + (\partial_z S_0(x, z))^2 = \varepsilon(z) . \tag{3.33}$$

Fig. 3.2 Geometry of light propagation into an increasing density profile at oblique incidence. The turning point is at z_t such that $n_e(z_t) = \cos^2(\theta_0) n_c$

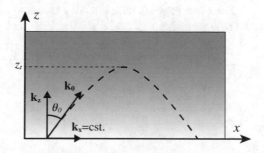

We take the plane wave-like solution $\partial_x S_0 = 0$, meaning that the x-component of the wave-vector is constant and therefore equal to its value at the vacuum boundary:

$$k_x = \frac{\omega_0}{c} \sin(\theta_0) . \tag{3.34}$$

The rest of the WKB procedure for the z-direction is exactly the same as before, leading to the WKB expression for the vector potential:

$$a(z) = \frac{a_0}{2\sqrt{n(z)}} \exp\left[i\frac{\omega_0}{c}x \sin\theta_0 - i\omega_0 t + i \int k_z(z)dz\right] + c.c.. \tag{3.35}$$

The local wave-number is $\mathbf{k}_0(z) = k_x \mathbf{e}_x + k_z(z)\mathbf{e}_z$ and must still satisfy the dispersion relation of EMWs $\omega_0^2 = \omega_{pe}^2(z) + k_0^2(z)c^2$, from which we obtain

$$k_z^2(z) = \frac{\omega_0^2}{c^2}\left(\cos^2(\theta_0) - \frac{n_e(z)}{n_c}\right). \tag{3.36}$$

The turning point is reached when the z-component of the wave-vector vanishes, which occurs at $z = z_t$ such that

$$\boxed{n_e(z_t) = \cos^2(\theta_0)n_c}. \tag{3.37}$$

Note that this is independent of the exact shape of the density profile, as long as the profile remains 1D: since this is where the z-component of the wave-vector vanishes, $\cos^2(\theta_0)n_c$ is the highest density the light will reach before turning back. Some examples of light trajectory in 1D profiles are discussed in Problem 3.1.

The behavior at the turning point z_t can be derived following the same analysis as in Sect. 3.2. Separating the function f as

$$f(x, z) = e^{ik_x x}g(z), \tag{3.38}$$

with $k_x = (\omega_0/c)\sin(\theta_0)$ (such that $a = \frac{1}{2}a_0 \exp[ik_x x - i\omega_0 t]g(z) + c.c.$), and inserting into Eq. (3.31), leads to

$$g''(z) = -\frac{\omega_0^2}{c^2}\left(\cos^2(\theta_0) - \frac{z}{L}\right)g. \tag{3.39}$$

The same analysis with the Airy function can thus be repeated by simply using the new variable:

$$\eta = (\omega_0 L/c)^{2/3}\left(\frac{z}{L} - \cos^2(\theta_0)\right), \tag{3.40}$$

which transforms the wave equation above for g into the Airy equation,

$$d_\eta^2 g = \eta g \, . \tag{3.41}$$

The expression for the field near the turning point is the same as for normal incidence (Eq. 3.24) with the new variable η above valid for any incidence angle.

Note that the Airy skin depth defined earlier in Eq. (3.28) remains the same regardless of the incidence angle, i.e., the penetration depth, and distance to the peak of the electric field is independent of the incidence angle. Even though we used a linear density ramp, $n_e/n_c = z/L$, it is only necessary that the density be approximately linear near the turning point (for $|z - L \cos^2(\theta_0)| < \Delta z_A$) for the Airy function solution to be valid; WKB can be used elsewhere.

3.4   Ray-Tracing

We have seen that the WKB method provides an excellent description of the light wave's electric field except in a narrow region at the turning point. We will now introduce the concept of ray-tracing, which essentially extends the WKB approach presented earlier to three spatial dimensions as well as to time variations, i.e., allowing the index to vary in both time and space, "slowly" enough to recover a local wave-vector and frequency.

We express the field as

$$\mathbf{a}(\mathbf{r}, t) = \frac{1}{2} a(\mathbf{r}, t) \mathbf{e}_\pi \exp[i\psi(\mathbf{r}, t)] + c.c. \, , \tag{3.42}$$

where $\mathbf{e}_\pi$ is the wave's polarization vector. The phase ψ is oscillating rapidly in both time and space compared to a, such that the wave can be locally described by a frequency and wave-vector defined as

$$\mathbf{k} = \nabla \psi \, , \quad \omega = -\frac{\partial \psi}{\partial t} \, . \tag{3.43}$$

In other words, we assume that at any point in time and space, the wave can locally (and in at least a small interval of time) be approximated by a plane wave of frequency and wave-vector ω and $\mathbf{k}$. Both ω and $\mathbf{k}$ are allowed to vary slowly in space and time, due to spatial and temporal variations of the plasma refractive index—i.e., its electron density.

The frequency and wave-vector can be directly connected by taking the time derivative and gradient of the two equations in Eq. (3.43), respectively: since $\partial_t \nabla \psi = \nabla \partial_t \psi$, we obtain

$$\nabla \omega = -\frac{\partial \mathbf{k}}{\partial t} \, . \tag{3.44}$$

Furthermore, ω and $\mathbf{k}$ must still satisfy the dispersion relation of EMWs in plasmas at any point in time and space, $\omega = \Omega(\mathbf{k}, \mathbf{r}, t)$; here Ω is a function, $\Omega(\mathbf{k}(\mathbf{r}, t), \mathbf{r}, t) = [\omega_{pe}(\mathbf{r}, t)^2 + k(\mathbf{r}, t)^2 c^2]^{1/2}$. Since $\mathbf{k}$ is a function of space and time, taking partial derivatives leads to

$$\frac{\partial \omega}{\partial t} = \Omega_t + \Omega_{\mathbf{k}} \cdot \partial_t \mathbf{k}, \tag{3.45}$$

$$\nabla \omega = \Omega_{\mathbf{r}} + \Omega_{\mathbf{k}} \cdot \nabla \mathbf{k}, \tag{3.46}$$

where the subscripts indicate partial derivatives, e.g., $\Omega_t = \partial_t \Omega$, $\Omega_{\mathbf{k}} = (\partial \Omega / \partial k_x) \mathbf{e}_x + (\partial \Omega / \partial k_y) \mathbf{e}_y + (\partial \Omega / \partial k_z) \mathbf{e}_z$, etc.

We now introduce the concept of rays or trajectories of wave packets, defined by their position $\mathbf{r}(t)$ and the group velocity $\mathbf{v}_g = \Omega_{\mathbf{k}}$ such that:

$$\dot{\mathbf{r}}(t) = \frac{d\mathbf{r}}{dt} = \mathbf{v}_g = \Omega_{\mathbf{k}}. \tag{3.47}$$

This is in fact the first ray-tracing equation, connecting the velocity of the ray to the function $\Omega_{\mathbf{k}}$ that can be obtained from the dispersion relation.

Next we will be looking for expressions of $\dot{\mathbf{k}}$ and $\dot{\omega}$, in order to get a complete set of coupled ordinary differential equations describing the position, velocity (or wave-vector), and frequency of the wave packet as a function of time. Since $\mathbf{k} = \mathbf{k}(\mathbf{r}, t)$, we have $\dot{\mathbf{k}} = \partial_t \mathbf{k} + \dot{\mathbf{r}} \cdot \nabla \mathbf{k} = \partial_t \mathbf{k} + \Omega_{\mathbf{k}} \cdot \nabla \mathbf{k}$; substituting using Eqs. (3.44) and (3.46), we obtain $\dot{\mathbf{k}} = -\Omega_{\mathbf{r}}$. This is the second ray-tracing equation.

Finally, we have $\dot{\omega} = \Omega_t + \Omega_{\mathbf{k}} \dot{\mathbf{k}} + \Omega_{\mathbf{r}} \dot{\mathbf{r}}$, where we immediately see that the last two terms cancel due to the first two ray-tracing equations ($\dot{\mathbf{k}} = -\Omega_{\mathbf{r}}$ and $\dot{\mathbf{r}} = \Omega_{\mathbf{k}}$); we thus obtain $\dot{\omega} = \Omega_t$ and now have the final set of ray-tracing equations parametrized with time:

$$\dot{\mathbf{r}} = \Omega_{\mathbf{k}} \tag{3.48}$$

$$\dot{\mathbf{k}} = -\Omega_{\mathbf{r}} \tag{3.49}$$

$$\dot{\omega} = \Omega_t. \tag{3.50}$$

This constitutes a coupled set of ordinary differential equations that can be readily integrated numerically (e.g., using a Runge–Kutta method) and sometimes solved analytically. It describes the evolution of the position, group velocity (or wave-vector), and frequency of wave packets as a function of time, and like WKB (but extended to three dimensions and time) is valid as long as spatial and temporal variations of the envelope are slow compared to $\mathbf{k}$ and ω.

Since the function $\Omega(\mathbf{r}, \mathbf{k}, t)$ comes from the dispersion relation, and taking partial derivatives of the dispersion relation itself will often be easier than having to take the extra step of expressing Ω explicitly, the ray-tracing equations are often recast using the implicit dependence of $\Omega(\mathbf{r}, \mathbf{k}, t)$ on the dispersion relation, i.e.,

$$D(\Omega(\mathbf{r}, \mathbf{k}, t), \mathbf{r}, \mathbf{k}, t) = 0. \tag{3.51}$$

Using the rules of implicit partial differentiation, we get $\Omega_t = -D_t/D_\omega$, $\Omega_\mathbf{r} = -D_\mathbf{r}/D_\omega$, and $\Omega_\mathbf{k} = -D_\mathbf{k}/D_\omega$, so the ray-tracing equations can be expressed as

$$\dot{\mathbf{r}} = -\frac{D_\mathbf{k}}{D_\omega}, \tag{3.52}$$

$$\dot{\mathbf{k}} = \frac{D_\mathbf{r}}{D_\omega}, \tag{3.53}$$

$$\dot{\omega} = -\frac{D_t}{D_\omega}. \tag{3.54}$$

These equations are valid for any type of wave in plasmas, i.e., EMW, EPW, or IAW. For a light wave, $D = -\omega^2 + \omega_{pe}^2 + k^2 c^2$, and therefore,

$$\dot{\mathbf{r}} = \mathbf{v}_g = \frac{c^2}{\omega}\mathbf{k}, \tag{3.55}$$

$$\dot{\mathbf{k}} = -\frac{1}{2\omega}\nabla(\omega_{pe}^2) \approx -\frac{1}{2}\omega\nabla(n_e/n_c), \tag{3.56}$$

$$\dot{\omega} = \frac{1}{2\omega}\partial_t(\omega_{pe}^2) \approx \frac{\omega}{2}\partial_t(n_e/n_c). \tag{3.57}$$

Note that the time scales for the plasma density evolution in most conditions are typically much longer than $1/\omega$ (which is on the order of a femtosecond for a light wave), resulting in a very small frequency shift compared to the light wave frequency. The frequency shifts are thus typically too small to directly affect the ray trajectories, and the first two equations can be treated independently of the third. These frequency shifts have been used as a diagnostic of the plasma evolution for ICF implosions in spherical "direct-drive" geometry [3, 4].

If the third ray-tracing equation is neglected (i.e., ignoring the small frequency variations), the ray equations resemble the equations of motion of classical mechanics. In fact, we can see that Eqs. (3.48), (3.49) are similar to Hamilton's equations used in Chap. 2, $\dot{\mathbf{q}} = \partial_\mathbf{p}H$, $\dot{\mathbf{p}} = -\partial_\mathbf{q}H$, where $\mathbf{q}$ and $\mathbf{p}$ are the generalized coordinate and momentum and H is the Hamiltonian, with ω taking the role of the Hamiltonian [5]. Since $\dot{\mathbf{v}}_g = -\frac{1}{2}c^2\nabla(n_e/n_c)$, the system is equivalent (with similar solutions) to the situation of an object moving in a conservative force field with $\mathbf{F}/m = \dot{\mathbf{v}} = -(1/m)\nabla\Phi$, with n_e being equivalent to the potential. Therefore, analogies with well-known problems in classical mechanics will sometimes allow rapid assessments of ray trajectories, as seen in the problems below.

An example of ray-tracing for a spherical density profile of the form $n_e(r)/n_c = \exp[-(r - r_c)/L]$ is shown in Fig. 3.3, based on a numerical integration of Eqs. (3.55)–(3.56). It is known from classical mechanics that only a handful of central potentials lead to integrable solutions for the particle trajectory; an exponential profile, as is typical from density profiles in laser–plasma experiments

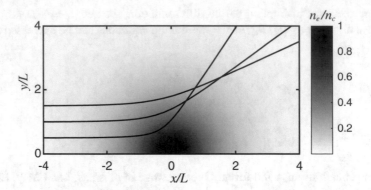

Fig. 3.3 Ray propagation in a spherical, exponential density profile of the form $n_e(r)/n_c = \exp[-(r - r_c)/L]$. Shown here are three rays with impact parameters $b/L = 0.5$, 1, and 1.5, for $r_c = 0.4L$

as we saw in Chap. 1, is not one of them [6]. Therefore, this type of density profile does not lead to analytical solutions for the ray propagation; however, their numerical integration is straightforward.

3.5   Frequency Shift of a Light Wave Reflecting Off an Expanding Density Profile

The frequency shift of a light wave propagating in a plasma, while typically small, can be a valuable diagnostic of the plasma dynamics. Here we give the expression for the frequency shift of a wave reflected off a 1D rarefaction profile (cf. Sect. 1.5) at normal incidence. This analysis can easily be generalized to arbitrary incidence or to other types of idealized plasma profiles (cf. Problem 3.4).

We consider a light wave with vacuum frequency ω_0 incident onto the rarefaction density profile derived in Sect. 1.5, Eq. 1.279, $n_e = n_0 \exp[-(z/c_s t + 1)]$. The light propagates up to the turning point, here corresponding to $n_e = n_c$ due to the normal incidence and located at z_T, from where it is reflected back (Fig. 3.4).

The frequency shift can be calculated from the third ray-tracing equation, Eq. (3.57). Noting that $d\omega/dt = v_g d\omega/dz$, with $|v_g| = nc$ and $n = \sqrt{1 - n_e/n_c}$, the frequency shift between incident and reflected light can be expressed by integrating Eq. (3.57) along the ray propagation,

$$\Delta\omega = \omega_r - \omega_0 = 2 \int_{\infty}^{z_T} \frac{d\omega}{dz} dz = 2\frac{\omega}{c} \int_{z_T}^{\infty} \frac{\partial_t n_e/n_c}{2n} dz , \qquad (3.58)$$

where the factor of 2 comes from the symmetry of the frequency shift going from $z \to \infty$ to z_T and back (paying attention to the fact that $v_g < 0$ on the way in).

Fig. 3.4 Frequency shift for
a light wave reflected off a
rarefaction profile at normal
incidence

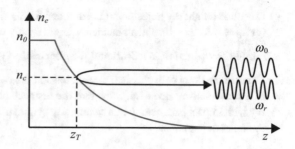

Using the expression of the density profile, we can do a change of variable
from z to n, the refractive index, to facilitate the integration. Noticing that $dz = 2nc_s t\,dn/(1-n^2)$ and that $z = -c_s t[1 + \ln(n_e/n_0)]$, we finally arrive at

$$\boxed{\Delta\omega = 2\omega_0 \frac{c_s}{c}\left[1 - 2\ln(2) + \ln(n_0/n_c)\right]}.\qquad(3.59)$$

Note that this expression was obtained under the WKB approximation, even
though the light propagates up to the turning point where WKB isn't valid. A more
rigorous analysis was derived in Ref. [7]: the field can be solved in the vicinity of
the turning point using the Airy function like we did in Sec. 3.2, and phase matched
to the WKB solution near the "Airy skin depth". The phase contribution from the
Airy skin depth vanishes, and one gets the same result as Eq. (3.59) above.

Finally, we can notice that the turning point location can be expressed (from
the definition of the density profile) as $z_T = c_s t[\ln(n_0/n_c) - 1]$; its velocity is thus
$v_T = c_s[\ln(n_0/nc) - 1]$. Expressing $c_s = dL/dt$ with $L = c_s t$ the scale-length of the
density profile, we can then recast Eq. (3.59) into

$$\frac{\Delta\omega}{\omega_0} = 2\frac{v_T}{c} + \frac{4}{c}(1 - \ln 2)\frac{dL}{dt};\qquad(3.60)$$

the first term is analogous to a Doppler shift from the "moving mirror" at the turning
point, while the second is associated with the variation of the optical path length due
to the change in plasma density [7].

Problems

3.1 Light Propagation in a Stationary 1D Density Profile

- Calculate a light ray trajectory in a linear density profile defined as $n_e(z)/n_c = z/L$ for
  $z > 0$ (and $n_e = 0$ for $z < 0$), with θ_0 the incidence angle of the light at $z = 0$: first,
  derive the expressions for $x(t)$ and $z(t)$ (the light propagates in the (x, z) plane), then
  express $z(x)$, and show that the trajectory is a parabola.

- Calculate a light ray trajectory $z(x)$ for a density profile defined by $n_e(z)/n_c = (z/L)^2$ (or $n_e = 0$ for $z < 0$), and an incidence angle θ_0 at $z = 0$.

3.2 Light Propagation in a Stationary, Spherically Symmetric Density Profile

- Using an analogy with classical mechanics and the conservation of angular momentum in a conservative force field, show that the distance of closest approach r_{min} for a light ray with impact parameter b in a spherically symmetric density profile $n_e(r)$ is given by the following relation:

$$b = r_{min}\sqrt{1 - n_e(r_{min})/n_c}. \tag{3.61}$$

- Take the density profile used for Fig. 3.3, $n_e(r)/n_c = \exp[-(r - r_c)/L]$ with $r_c = 0.4L$. Defining a vector of possible r_{min} (e.g., linearly distributed values from $0.4L$ to $2L$, to ensure $n_e < n_c$), calculate the corresponding vector of impact parameters b's from the expression above in a computing software, and plot r_{min} vs. b. Check that the distances of closest approaches for the three values of b used in Fig. 3.3 are consistent with the plot of r_{min} vs. b.

3.3 Ray-Tracing for EPWs

Derive ray-tracing equations for EPWs in an isothermal plasma with non-uniform density. Show that EPWs follow exactly the same trajectory as light waves.

3.4 Frequency shifts from expanding plasma profiles

- Fill in the steps leading to Eq. (3.59). List the assumptions used to derive this expression.
- Generalize the derivation for an arbitrary incidence angle θ.
- Estimate the relative frequency shift $\Delta\omega/\omega$ for a 351 nm laser at normal incidence on an isothermal aluminum plasma with an electron temperature of 1 keV and a negligible ion temperature. Assume that the initial density n_0 of the rarefaction profile corresponds to the solid mass density of aluminum at room temperature, with $\rho = 2.7$ g/cm^3.
- The plasma density profile from an exploding thin solid foil is crudely estimated as $n_e(z)/n_c = L/c_s t$ for $|z| \leq c_s t/2$, and $n_e = 0$ for $|z| > c_s t/2$. Consider a light wave propagating along z through the entire expanding plasma (for times t satisfying $n_e < n_c$): derive the expression for the relative frequency shift $\Delta\omega/\omega$ experienced by the light after propagating through the plasma.

References

1. C. Bender, S. Orszag, S. Orszag, *Advanced Mathematical Methods for Scientists and Engineers I: Asymptotic Methods and Perturbation Theory, Advanced*. Mathematical Methods for Scientists and Engineers (Springer, 1999)
2. M. Abramowitz, I.A. Stegun, *Handbook of Mathematical Functions with Formulas, Graphs, and Mathematical Tables*, vol. 55 (US Government Printing Office, 1948)
3. W. Seka, D.H. Edgell, J.P. Knauer, J.F. Myatt, A.V. Maximov, R.W. Short, T.C. Sangster, C. Stoeckl, R.E. Bahr, R.S. Craxton, J.A. Delettrez, V.N. Goncharov, I.V. Igumenshchev, D.

Shvarts, Phys. Plasmas **15**, 056312 (2008)
4. I.V. Igumenshchev, D.H. Edgell, V.N. Goncharov, J.A. Delettrez, A.V. Maximov, J.F. Myatt, W. Seka, A. Shvydky, S. Skupsky, C. Stoeckl, Phys. Plasmas **17**, 122708 (2010)
5. L.D. Landau, E.M. Lifshitz, *The Classical Theory of Fields*, vol. 2 (Elsevier, 2013)
6. H. Goldstein, C. Poole, J. Safko, *Classical Mechanics* (Addison Wesley, 2002)
7. T. Dewandre, J.R. Albritton, E.A. Williams, Phys. Fluids **24**, 528 (1981)

Chapter 4
Absorption of Light Waves (and EPWs) in Plasmas

In this chapter, we will derive the well-known collisional absorption rate and give simple estimates for the absorption of light waves (or EPWs) propagating in idealized plasma profiles. This process corresponds to the absorption of the energy of an external oscillating electric field due to electron–ion collisions: during each collision, some of the oscillation velocity of the electrons is randomly "scattered" in velocity space. The back-and-forth energy exchange between the wave and the electrons executing their quiver motion is then disrupted, and the particles' energy gain is not fully restored to the wave due to the collisions. This leads to an overall energy increase of the electrons—primarily from the "slow" electrons, as we will see. The net energy gained by the electrons is lost by the wave.

This process is often referred to as inverse bremsstrahlung, since it can be pictured as a photon absorption during the collision of an electron with an ion.

We will also introduce the resonance absorption mechanism, whereby a p-polarized light wave incident at a finite angle onto a density gradient partly converts into a longitudinal mode that will subsequently be damped in the plasma, leading to irreversible energy transfer from the light wave to the plasma.

4.1 Ballistic Model of Collisional (or Inverse Bremsstrahlung) Absorption

In this section, we will use a ballistic model to derive the collisional electron heating and corresponding wave absorption. Our derivation, which provides an intuitive physical picture of the collisional absorption mechanism and its subtleties, is similar to the ones of Bunkin and Pert in Refs. [1] and [2]. It is reminiscent of the derivation of the electron loss of momentum from Sect. 1.4, except we are now including the effect of an external oscillating electric field on the electron motion. A similar way of deriving the absorption coefficient, used by Mulser [3, 4],

© The Author(s), under exclusive license to Springer Nature Switzerland AG 2023
P. Michel, *Introduction to Laser-Plasma Interactions*, Graduate Texts in Physics,
https://doi.org/10.1007/978-3-031-23424-8_4

consists in using a Drude model ansatz (which we will introduce in Sect. 4.2) before calculating the electron–ion collision frequency using a ballistic approach. The absence of collisional absorption from electron–electron collisions is the subject of Problem 4.2.

The ballistic model approach only considers electron–ion interactions with a simple Coulomb potential; shielding and collective effects are thus ignored. As we will see in our discussion of the Coulomb logarithm in Sect. 4.4, most of the electron–ion collisions responsible for the absorption mechanism occur over distances that are much shorter than the Debye length—especially for densities well below critical. As a result, this approach provides the same result as the well-known kinetic theory from Dawson–Oberman–Johnston[1] in the limit of $\omega \gg \omega_{pe}$ [5, 6] and recovers the kinetic results from Silin [7] in the high-field limit of $v_{os} \gg v_{Te}$ as well—but with a substantially simpler derivation.

4.1.1   Physical Picture

Consider a plasma with a finite electron temperature and an external oscillating electric field that is locally approximated by a plane wave,

$$\mathbf{E}(\mathbf{r}, t) = \mathbf{E}_0 \cos(\psi) \tag{4.1}$$

with $\psi = \mathbf{k} \cdot \mathbf{r} - \omega t$.

The external field can be from an EPW or a light wave: the derivation that follows is valid for both, giving the same absorption coefficient.

We consider a single electron whose velocity $\mathbf{w}$ just before the collision consists in a time-average velocity $\mathbf{v} = \langle \mathbf{w} \rangle$ (typically on the order of the thermal velocity v_{Te}) and a quiver velocity $\mathbf{u}$ in the external electric field, given by (cf. Sect. 2.1)

$$\mathbf{u}(\mathbf{r}, t) = v_{os} \sin(\psi) \ , \quad v_{os} = \frac{e E_0}{m \omega} \ . \tag{4.2}$$

The ions are assumed stationary, since the time scales we are considering are much faster than the ion motion, and we neglect ponderomotive effects and assume non-relativistic intensities. Another key assumption is that the electron–ion interaction time is much faster than the external field oscillation period $\sim 1/\omega$ and occurs over a distance much smaller than the wavelength of the external field, which allows us to ignore the spatial dependence in the oscillation phase ψ (i.e., we can assume $\mathbf{k} \cdot \mathbf{r} \approx cst.$). More details on these assumptions will be given in the discussion on the Coulomb logarithm (Sect. 4.4). Denoting t_0 the "collision time"

---

[1] Reference [5] includes numerical errors, which are corrected in Ref. [6].

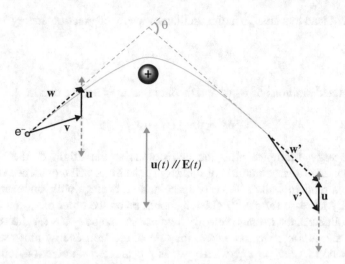

Fig. 4.1 Electron–ion collision in the presence of the external oscillating electric field. The electron velocity immediately before and after the collision is the same since the collision is elastic, i.e., $w = w'$; however, the *time-averaged* velocities before and after the collision v and v' are different, since the electron resumes its quiver motion after the collision (assuming the collision time is shorter than the oscillation period). This results in a (small) random variation in the average velocity after each collision, i.e., a random walk in velocity space, which eventually leads to an increase in velocity and thus in electron energy. The electron energy increase is balanced by an energy loss from the wave: this constitutes the basic physics of the collisional absorption mechanism

(i.e., the time of closest approach between the electron and the ion), and t_0^-, t_0^+ the times just before and just after the collision, we have then

$$\mathbf{w}(t_0^-) = \mathbf{v}(t_0^-) + \mathbf{u}(t_0^-) \approx \mathbf{v}(t_0^-) + \mathbf{u}(t_0) \,, \tag{4.3}$$

$$\mathbf{w}(t_0^+) = \mathbf{v}(t_0^+) + \mathbf{u}(t_0^+) \approx \mathbf{v}(t_0^+) + \mathbf{u}(t_0) \tag{4.4}$$

(with $u(t_0^-) \approx u(t_0^+) \approx u(t_0)$ per our assumption that the collision time is short compared to an oscillation period of the electric field) or, simply denoting the velocities just after the collision with a prime and those just before the collision without a prime,

$$\mathbf{w} = \mathbf{v} + \mathbf{u} \,, \tag{4.5}$$

$$\mathbf{w}' = \mathbf{v}' + \mathbf{u} \,. \tag{4.6}$$

Since the collision is elastic, we also have $w = w'$. The situation is illustrated in Fig. 4.1.

As the electron resumes its oscillation in the external field after the collision, its oscillation-averaged velocity $\mathbf{v}' = \mathbf{w}' - \mathbf{u}$ will have a different amplitude from before, $v \neq v'$.

This will lead to a change in the oscillation-averaged electron energy

$$\delta\epsilon = \frac{m}{2}(v'^2 - v^2),$$ (4.7)

and from the definitions of $\mathbf{w}$, $\mathbf{w}'$ and the fact that $w = w'$, we obtain

$$\delta\epsilon = m\mathbf{u} \cdot (\mathbf{w} - \mathbf{w}').$$ (4.8)

That energy difference after the collision is at the origin of the collisional absorption: after a large number of collisions, electrons will experience something similar to a random walk in velocity space and on average will gain energy. Since the velocity increase for every collision should be on the order of v_{os}, and because after N collisions, the average velocity increase should be $\sim\sqrt{N}v_{os}$, i.e., an energy increase of $\sim Nmv_{os}^2$, we can expect the rate of electron energy increase per unit time to scale like $\sim\nu_{col}mv_{os}^2$, where ν_{col} is a characteristic electron–ion collision frequency to be determined (since electrons interact with multiple ions in a plasma, there is no single definition for an electron–ion collision frequency). We are going to see that it is indeed the case and that ν_{col} happens to be very close (but not exactly equal) to the momentum loss rate introduced in Sect. 1.4.

In the following sections, we will obtain the rate of electron energy increase by averaging the single electron–ion energy difference $\delta\epsilon$ over: (i) the collisions with all neighboring ions; (ii) an equilibrium (Maxwellian) distribution of electron velocities; and (iii) the oscillation phase ψ of the electric field. The absorption rate will be obtained by balancing the electron energy increase with the external wave energy loss.

4.1.2   Low-Field Limit ($v_{os} \ll v_{Te}$)

The low-field limit of collisional absorption corresponds to $v_{os} \ll v_{Te}$, or, in practical units, when the external wave is a light wave:

$$\frac{v_{os}}{v_{Te}} \approx 1.93\lambda_0[\mu m]\sqrt{\frac{I_{16}}{T_e[\text{keV}]}},$$ (4.9)

where I_{16} is the intensity in units of 10^{16} W/cm^2. This is the limit where $u \ll v$ in Fig. 4.1, i.e., the initial electron velocity is dominated by its "thermal" contribution, while the quiver motion represents only a small correction. Note that the present analysis of collisional absorption is valid for EPWs as well as light waves; for EPWs, the collisional absorption process will practically always be in the low-field limit, as discussed in Problem 4.4.

Fig. 4.2 Geometry used in the derivation of the collisional absorption via a ballistic model

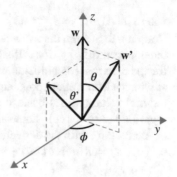

In the following, we will use the notations and geometry from Fig. 4.2. Using $\mathbf{u} \cdot \mathbf{w} = uw \cos(\theta')$ (here and in the following, u represents $|\mathbf{u}|$, i.e., $u = v_{os} |\sin(\psi)|$, where v_{os} is assumed positive) and $w = w'$, and expressing $\mathbf{w}'$ in the spherical coordinates from Fig. 4.2, yields the expression for the single-collision energy variation from Eq. (4.8):

$$\delta \epsilon = m \left[\mathbf{u} \cdot \mathbf{w} (1 - \cos \theta) - uw \sin \theta' \sin \theta \cos \phi \right] . \tag{4.10}$$

The following steps are similar to the derivation of the electron–ion collision in Sect. 1.4. We first want to estimate the rate of variation in a single electron energy from colliding with all neighboring ions,

$$\frac{d\epsilon}{dt} = \int_{ions} \frac{\delta N_i \, \delta \epsilon}{dt} , \tag{4.11}$$

where $\delta N_i = 2\pi n_i w \, dt \, b \, db$ is the number of ions in the volume element comprised between the two cylinders of radii b and $b + db$ and length $w \, dt$ (cf. Eq. (1.244) and Fig. 1.16).

The scattering angle θ is directly related to the impact parameter b via $\theta = -2\arctan(b_\perp/b)$ (Eq. (1.241)), where $b_\perp$ is the 90° scattering impact parameter defined as $b_\perp = Ze^2/(4\pi\varepsilon_0 m w^2)$ for the pre-collision velocity w (Eq. (1.240); keep in mind that the collision time is assumed very short compared to $1/\omega$, the oscillation period of the light wave). These expressions lead to, after using simple trigonometric identities, $1 - \cos(\theta) = 2b_\perp^2/(b^2 + b_\perp^2)$.

Recognizing that the second term in the brackets in Eq. (4.10) averages to zero when averaging over all azimuthal angles ϕ, the expression for the rate of energy change $d\epsilon/dt$ above becomes

$$\frac{d\epsilon}{dt} = \int_{b_{min}}^{b_{max}} 4\pi n_i wbm\mathbf{u} \cdot \mathbf{w} \frac{b_\perp^2}{b^2 + b_\perp^2} db$$

$$= \frac{4\pi n_i}{m} \left(\frac{Ze^2}{4\pi\varepsilon_0} \right)^2 \frac{\mathbf{u} \cdot \mathbf{w}}{w^3} \ln(\Lambda) , \tag{4.12}$$

where $\ln[\Lambda(w)] = \int_{b_{min}}^{b_{max}} db\, b/(b^2 + b_\perp^2) = \ln\sqrt{(b_\perp^2 + b_{max}^2)/(b_\perp^2 + b_{min}^2)}$ is the Coulomb logarithm that we already encountered in Sect. 1.4 (Eq. (1.247)). However, the upper cutoff b_{max} for collisional absorption is usually different from the one used for the electron–ion collision rate in Sect. 1.4, which specifically describes the rate at which a Maxwellian population with a velocity drift that is small compared to the thermal velocity goes back to a zero-average velocity Maxwellian. The Coulomb logarithm used for collisional absorption will be discussed in Sect. 4.4.

Proceeding with the derivation, we express $w^2 = u^2 + v^2 + 2uv\cos\chi$ and $\mathbf{u}\cdot\mathbf{w} = \mathbf{u}\cdot(\mathbf{u}+\mathbf{v}) = u(u + v\cos\chi)$, where χ is the angle between $\mathbf{u}$ and $\mathbf{v}$, and obtain

$$\frac{d\epsilon}{dt} = K\ln(\Lambda)u(u + v\cos\chi)(u^2 + v^2 + 2uv\cos\chi)^{-3/2}, \tag{4.13}$$

where the constant k is defined as

$$K = \frac{4\pi n_i}{m}\left(\frac{Ze^2}{4\pi\varepsilon_0}\right)^2. \tag{4.14}$$

Now that we have an expression for the energy gain per unit time for a single electron at $\mathbf{w}$, the average rate of electron energy increase is obtained by summing up the energy increase of all plasma electrons, which are assumed to follow a distribution function $f(\mathbf{v})$, and average over the phase of the oscillation ψ:

$$\frac{dU}{dt} = \int d^3v \int_0^{2\pi} \frac{d\psi}{2\pi}\frac{d\epsilon}{dt} f(\mathbf{v}) \tag{4.15}$$

$$= \int_0^\infty v^2 dv \int_0^{2\pi} d\phi' \int_{-1}^1 dC \int_0^{2\pi} \frac{d\psi}{2\pi}\frac{d\epsilon}{dt}(\psi, v, C) f(v), \tag{4.16}$$

where $d^3v = v^2\sin(\chi)d\chi\, d\phi'\, dv$. The polar coordinate (in spherical) was chosen with respect to the axis parallel to $\mathbf{u}$, and ϕ' is the azimuthal coordinate in this basis (the reference does not matter since the integrand is independent of ϕ'). We also used $C = \cos(\chi)$, so that $\int_0^\pi \sin(\chi)d\chi = \int_{-1}^1 dC$.

Using the standard integral:

$$\int \frac{\alpha + x}{(\alpha^2 + 1 + 2\alpha x)^{3/2}}dx = \frac{\alpha x + 1}{\alpha^2\sqrt{\alpha^2 + 2\alpha x + 1}} + cst., \tag{4.17}$$

we get

$$\int_0^{2\pi} d\phi' \int_{-1}^1 dC\frac{d\epsilon}{dt} = \frac{2\pi}{u}K\ln(\Lambda_{lf})\left[\frac{1 + u/v}{|1 + u/v|} - \frac{1 - u/v}{|1 - u/v|}\right]$$

$$= \frac{4\pi}{u}K\ln(\Lambda_{lf})H\left(\frac{u}{v} - 1\right), \tag{4.18}$$

where H is the Heaviside step function.

Note that the Coulomb logarithm $\ln(\Lambda_{lf})$ (the subscript lf for *low-field* absorption indicates that it is different from the value defined in Sect. 1.4) will be completely taken out of the entire integration, even though it depends on w (via $b_\perp \propto w^{-2}$), i.e., on all the integration variables (since w depends on u, hence, ψ, on χ and on v). This will be discussed in more detail in Sect. 4.4: for now, we assume that the velocity dependence is weak due to the logarithm and that a good approximation can be obtained by taking a "characteristic" value for the integration cutoffs that will be specified later.

Next we integrate over the velocity amplitudes v assuming a Maxwellian distribution function

$$\int d^3v \frac{d\epsilon}{dt} f(\mathbf{v}) = \frac{4\pi K}{u} \ln(\Lambda_{lf}) \int_0^u v^2 f(v)dv\,,\tag{4.19}$$

with $f(v) = n_e(2\pi)^{-3/2}v_{Te}^{-3}\exp[-v^2/2v_{Te}^2]$. In the low-field limit $v_{os} \ll v_{Te}$, we have *a fortiori* $u \ll v_{Te}$ at any oscillation phase ψ, and therefore, the values of v in the integral all satisfy $v \ll v_{Te}$. We can then expand the integrand by approximating $f(v) \approx n_e(2\pi)^{-3/2}v_{Te}^{-3}$, which eventually leads to

$$\int d^3v \frac{d\epsilon}{dt} f(v) = \frac{4}{3}\pi K \ln(\Lambda_{lf}) \frac{n_e}{(2\pi)^{3/2}v_{Te}^3}u^2$$

$$= n_e m u^2 v_{ei}\,,\tag{4.20}$$

where v_{ei} takes the same form as the electron–ion collision frequency (i.e., the rate of momentum loss) that we already derived in Sect. 1.4, except for the Coulomb logarithm:

$$v_{ei} = \frac{4\sqrt{2\pi}}{3} \frac{n_i}{m^2 v_{Te}^3} \left(\frac{Ze^2}{4\pi\varepsilon_0}\right)^2 \ln(\Lambda_{lf}) = \omega_{pe}\frac{Z}{3(2\pi)^{3/2}}\frac{\ln(\Lambda_{lf})}{N_{De}}\,,\tag{4.21}$$

where $N_{De} = n_e\lambda_{De}^3$ is the number of electrons in the Debye cube. The expression for $\ln(\Lambda_{lf})$ will be discussed in Sect. 4.4.

It can easily be verified that in the case of multi-species plasmas, the expression for v_{ei} remains the same except that one needs to replace Z by $Z^* = \langle Z^2 \rangle / \langle Z \rangle$ as introduced in Sect. 1.3.6 (cf. Problem 4.3).

All that is left is to perform the averaging over the phase of the oscillations ψ, which leads to the average rate of electron energy increase from electron–ion collisions in an external oscillating field:

$$\frac{dU}{dt} = \frac{1}{2}mv_{os}^2 n_e v_{ei}\,.\tag{4.22}$$

The temporal absorption rate of the external field (either light wave or EPW) is obtained by balancing the plasma heating rate per unit volume above against an energy loss per unit time and volume for the field, equal to $\nu_c \varepsilon_0 E_0^2/2$, where ν_c is the collisional absorption rate. We obtain

$$\nu_c = \frac{\omega_{pe}^2}{\omega^2} \nu_{ei}, \tag{4.23}$$

where we used $\nu_{os} = eE_0/(m\omega)$ and $\omega_{pe}^2 = e^2 n_e/(\varepsilon_0 m)$.

This absorption rate can also be expressed as a spatial rate κ_c for a wave propagating at the group velocity ν_g, such that the absorption after propagation from $z = 0$ to L is given by $\exp[-\int_0^L \kappa_c(z)dz]$; we obtain

$$\kappa_c = \frac{\omega_{pe}^2}{\omega^2} \frac{\nu_{ei}}{\nu_g}, \tag{4.24}$$

or, in practical units,

$$\kappa_c[\text{m}^{-1}] \approx 2.75 \frac{Z^* \ln(\Lambda_{lf})}{T_{ek}^{3/2} \sqrt{1 - n_e/n_c}} \lambda_\mu^2 n_{e20}^2, \tag{4.25}$$

where T_{ek} is the electron temperature in keV, λ_μ the wavelength in μm, n_{e20} the electron density in 10^{20} cm^{-3}, and the Coulomb logarithm will be given in practical units in the next section and is typically on the order of $\sim[1 - 10]$. As mentioned above, $Z^* = \langle Z^2 \rangle / \langle Z \rangle$, where the average is over the atomic fraction of ion species for mixed-ion-species plasma (for single species, $Z^* \rightarrow Z$).

This is a crucial scaling for ICF and HED experiments: it shows that the laser absorption, and therefore the energy deposition and heating rate, increases:

- With the ionization state Z (because the deflection experienced by an electron as it "collides" with an ion will be stronger if the ion charge state is higher)
- With colder electron temperatures T_e (because the thermal velocity of the electrons is lower, so they will tend to be more deflected as they pass near an ion)
- When the density is higher (because it increases the probability of electron–ion collision)
- For longer laser wavelengths (because it increases the oscillation energy $\frac{1}{2}m\nu_{os}^2 \propto \lambda_0^2$, which is what gets transferred to the electrons via their collisions with ions)

This expression applies when the laser intensity is moderate and the plasma is hot, such that $\nu_{os} \ll \nu_{Te}$. In fact, it has been shown to be valid up to $\nu_{os}/\nu_{Te} \approx 1$ [8]. However for higher intensity lasers and/or colder plasma temperatures, the electron quiver velocity will typically dominate over its thermal velocity, and the collisional heating and absorption rates must be calculated in the high-field limit, as described in the next section.

4.1.3 High-Field Limit ($v_{os} \gg v_{Te}$)

We now turn our attention to the high-field limit, $v_{os} \gg v_{Te}$, or, in practical units:

$$I_{16}\lambda_\mu^2 \gg 0.27 T_{ek} , \tag{4.26}$$

where I_{16} is the intensity in units of 10^{16} W/cm^2, λ_μ the wavelength in μm and T_{ek} the electron temperature in keV.

The derivation in the high-field limit follows the same steps as for the low-field limit up to the integration of the energy gain rate for a single electron $d\epsilon/dt$ over the solid angle, i.e., Eq. (4.18). However, integrating over the velocity distribution next would lead to a problem, since $v_{os} \gg v_{Te}$ does not necessarily lead to $u \gg v_{Te}$, since there are times when the oscillation phase $\sin(\omega_0 t) \approx 0$ and thus $u < v_{Te}$.

The problem is avoided by performing the averaging over the phase of the oscillation *before* integrating over the velocity amplitudes. Going back to Eqs. (4.15) and (4.18), the rate of electron energy increase is

$$\frac{dU}{dt} = \int d^3v \int_0^{2\pi} \frac{d\psi}{2\pi} \frac{d\epsilon}{dt} f(v) \tag{4.27}$$

$$= \int_0^\infty v^2 dv f(v) \int_0^{2\pi} \frac{d\psi}{2\pi} \frac{4\pi}{u} K \ln(\Lambda_{hf}) H\left(\frac{u}{v} - 1\right) , \tag{4.28}$$

where we introduced the Coulomb logarithm for the high field limit $\ln(\Lambda_{hf})$, which will again be taken out of the integration entirely and will be specified in Sect. 4.4. We take care of the integration over the phase ψ first, i.e., $\int_0^{2\pi} d\psi \, H(u/v - 1)/u$ with $u = |\mathbf{u}| = v_{os}|\sin \psi|$ and $v_{os} > 0$. The Heaviside function can be rewritten as $H(u/v-1) = H(|\sin(\psi)| - v/v_{os})$. We also notice that the integrand is π-periodic, since $|\sin(\psi)| = |\sin(\psi + \pi)|$, and symmetric around $\pi/2$, since $|\sin(\pi/2 - \psi)| = |\sin(\pi/2 + \psi)|$; therefore, the integral simplifies to

$$\int_0^{2\pi} \frac{H(u/v - 1)}{u} d\psi = \frac{1}{v_{os}} \int_0^{2\pi} \frac{H(|\sin(\psi)| - v/v_{os})}{|\sin(\psi)|} d\psi$$

$$= \frac{4}{v_{os}} \int_0^{\pi/2} \frac{H(|\sin(\psi)| - v/v_{os})}{|\sin(\psi)|} d\psi$$

$$= \frac{4}{v_{os}} \int_{\arcsin(v/v_{os})}^{\pi/2} \frac{1}{\sin(\psi)} d\psi . \tag{4.29}$$

Using the standard integral

$$\int \frac{d\psi}{\sin(\psi)} = \ln\left[\frac{\sin \psi}{1 + \cos \psi}\right], \tag{4.30}$$

we obtain

$$\int_0^{2\pi} \frac{H(u/v - 1)}{u} d\psi = \frac{4}{v_{os}} \ln\left[\frac{1}{v}\left(v_{os} + \sqrt{v_{os}^2 - v^2}\right)\right]$$

$$\approx \frac{4}{v_{os}} \ln\left(\frac{2v_{os}}{v}\right) \quad , \quad v_{os} \gg v . \quad (4.31)$$

We now have to perform the final integration over the velocity to get the rate of electron energy increase in the high-field limit:

$$\frac{dU}{dt} = \ln(\Lambda_{hf}) \int_0^\infty dv v^2 f(v) \frac{8K}{v_{os}} \ln\left(\frac{2v_{os}}{v}\right). \quad (4.32)$$

We take the logarithm out of the integration, replacing v by $\sqrt{2}v_{Te}$ (as justified in Problem 4.1). Noticing that:

$$\int_0^\infty v^2 f(v) dv = \frac{1}{4\pi} \int d^3 v f(v) = \frac{n_e}{4\pi}, \quad (4.33)$$

we finally obtain the expression for the rate of electron energy increase in the high-field limit:

$$\boxed{\frac{dU}{dt} = \frac{8n_i n_e}{m v_{os}} \left(\frac{Ze^2}{4\pi\varepsilon_0}\right)^2 \ln(\Lambda_{hf}) \ln\left(\frac{\sqrt{2}v_{os}}{v_{Te}}\right).} \quad (4.34)$$

Balancing again the electron energy increase per unit time and volume above with the associated energy decrease of the external field, we can express the absorption rate as

$$\nu_c = \frac{\omega_{pe}^2}{\omega^2} \nu_s, \quad (4.35)$$

where the "effective collision frequency" ν_s in the strong-field limit is defined as

$$\boxed{\nu_s = \frac{16 n_i}{m^2 v_{os}^3} \left(\frac{Ze^2}{4\pi\varepsilon_0}\right)^2 \ln(\Lambda_{hf}) \ln\left(\frac{\sqrt{2}v_{os}}{v_{Te}}\right).} \quad (4.36)$$

The "effective" collision rate now takes the form:

$$\nu_s = \nu_{ei} \frac{12}{\sqrt{2\pi}} \frac{v_{Te}^3}{v_{os}^3} \ln\left(\frac{\sqrt{2}v_{os}}{v_{Te}}\right) \ln(\Lambda_{hf})/\ln(\Lambda_{lf}), \quad (4.37)$$

i.e., it is similar to the collision rate in the low-field limit but with the thermal velocity replaced by the quiver velocity, i.e., $\nu \propto v_{os}^{-3}$ instead of v_{Te}^{-3}—as well as a constant numerical factor $12/\sqrt{2\pi}$ and a slowly varying logarithm in front, and a different expression for the Coulomb logarithm. This means that the high quiver velocity leads to a decrease of the effective electron–ion collision frequency, and the collisional heating and absorption rates will decrease accordingly.

This expression is consistent with the results from Silin [9] obtained via a Fokker–Planck model, up to a very small numerical coefficient inside the logarithm.[2] Mulser [4] used a similar ballistic model to derive the electron–ion collision frequency in the high-field limit. An expression valid for arbitrary values of v_{os}/v_{Te} was obtained by Bunkin [1] in terms of elliptic integrals. Other authors often treat the intermediate regime of $v_{os} \sim v_{Te}$ by replacing v_{Te} in the expression for ν_{ei} by $(v_{Te}^2 + v_{os}^2)^{1/2}$, which is only correct up to the numerical pre-factor and logarithm terms of Eq. (4.36). In fact, it has been shown via numerical integrations that the high-field expression above is valid starting around $v_{os}/v_{Te} \geq 2$, while the low-field expression was found to be satisfactory up to $v_{os}/v_{Te} \leq 1$, leaving only the intermediate range $\sim[1 - 2]$ uncovered [8]. This intermediate range can be well-approximated by extending the two expressions. Asymptotic expansions for the intermediate regime are also derived in [3].

Like for the low-field limit, we can also define the spatial absorption rate $\kappa_c = \nu_c/v_g$. We obtain, in practical units,

$$\kappa_c[\text{m}^{-1}] \approx \frac{Z \ln(\Lambda_{hf})}{I_{16}^{3/2}\lambda_\mu \sqrt{1 - n_e/n_c}} n_{e20}^2 \ln\left(7.4\lambda_\mu^2 I_{16}/T_{ek}\right), \qquad (4.38)$$

where like earlier, I_{16} denotes the intensity in units of 10^{16} W/cm^2, λ_μ the wavelength in μm, and n_{e20} the density in 10^{20} cm^{-3}. Comparing to the low-field absorption rate, Eq. (4.25) shows that the general scalings are similar, except for the temperature that has been replaced by the laser intensity—or to be more precise, the thermal velocity $v_{Te} \propto \sqrt{T_e}$ has been replaced by the quiver velocity in the laser field $v_{os} \propto \lambda_0\sqrt{I}$, which also explains why the absorption rate in the high-field limit is now $\propto 1/\lambda_0$ as opposed to λ_0^2 in the low-field limit.

4.2 Collisional Absorption from a Fluid Model: Dielectric Constant with Absorption

The collisional absorption can also be derived from the plasma fluid description. While shorter, this derivation also ignores some of the underlying physics and makes it easier to miss the relevant hypotheses and their validity range.

---

[2] Ref. [9] Eq. (3.13) is similar except for the term in brackets, $1 + \ln(v_{os}/2v_{Te}) = \ln\left(\frac{e^1}{2}v_{os}/v_{Te}\right)$, vs. $\ln(\sqrt{2}v_{os}/v_{Te})$ with our ballistic model; since $e^1/2 \approx 1.36$, this only represents a 4% difference in front of v_{os}/v_{Te} inside the logarithm.

The physical picture is now that of a plasma whose electron fluid collectively quivers in the external electric field of a wave (again, either a light wave or an EPW) while experiencing a friction force from electron–ion collisions. Recall that by definition, the electron–ion collision frequency ν_{ei} is defined as the relaxation rate (or rate of momentum loss) for a drifting Maxwellian distribution of electrons,

$$\frac{d\mathbf{v}_d}{dt} = -\mathbf{v}_d \nu_{ei} \,, \tag{4.39}$$

where $\mathbf{v}_d$ is the drift velocity of the electrons—i.e., ν_{ei} is the characteristic rate at which a population of Maxwellian electrons with a drift velocity compared to the background ions will return to zero-average velocity due to collisions.

In our case, the drift is due to the quiver motion in the external electric field, i.e., $v_d = eE/(m\omega)$ (we will drop the subscript d in the following and simply use $v_d = v$).

Here one also needs to recall that the usual derivation of ν_{ei} (cf. Sect. 1.4) is based on an expansion in $v_d/v_{Te} \ll 1$, i.e., the electrons' drift velocity has to remain much smaller than their thermal velocity. Since the "drift" from the electron fluid is now due to their quiver motion in the external field, this implies that the following derivation is only valid when the quiver amplitude is much smaller than the thermal velocity, i.e., $v_{os} \ll v_{Te}$: in other words, this is the "low field" limit of the ballistic derivation from the previous section.

We write the equation of motion for the electron fluid in the presence of collisions with immobile ions and an external harmonic electric field $\mathbf{E}(\mathbf{r}, t) = \frac{1}{2}\mathbf{E}(\mathbf{r})\exp[-i\omega t] + c.c.$ (with $c.c.$ the complex conjugate),

$$\frac{\partial \mathbf{v}}{\partial t} = -\frac{e}{m}\mathbf{E} - \nu_{ei}\mathbf{v} \,, \tag{4.40}$$

where we neglected the higher-order terms $\propto \nabla\mathbf{v}$, $\mathbf{B}$ that lead to the nonlinear ponderomotive force (cf. Sect. 5.1). With our expression of the electric field, we can look for a harmonic response for the plasma electron current (i.e., $\mathbf{v}$) of the form $\mathbf{v}(\mathbf{r}, t) = \frac{1}{2}\mathbf{v}(\mathbf{r})\exp[-i\omega t] + c.c.$. We obtain

$$\mathbf{v}(\mathbf{r}) = \frac{-ie}{m(\omega + i\nu_{ei})}\mathbf{E}(\mathbf{r}) \,. \tag{4.41}$$

From the definition of the dielectric susceptibility $\chi_e(\omega)$, $\mathbf{j}(\mathbf{r}) = -i\omega\varepsilon_0\chi_e(\omega)\mathbf{E}(\mathbf{r})$, we obtain

$$\chi_e(\omega) = -\frac{\omega_{pe}^2}{\omega(\omega + i\nu_{ei})} \,. \tag{4.42}$$

The expression for the complex dielectric constant $\varepsilon = 1 - \chi_e$ in the presence of dissipation from collisions immediately follows:

$$\boxed{\varepsilon(\omega) = 1 - \frac{\omega_{pe}^2}{\omega(\omega + i\nu_{ei})}}.$$ (4.43)

Since the refractive index is $n(\omega) = \sqrt{\varepsilon(\omega)}$ and the wave-number $k = (\omega/c)n$, we obtain, in the limit of small absorption ($\nu_{ei} \ll \omega$),

$$\text{Re}(n) = \sqrt{1 - \frac{n_e}{n_c}},$$ (4.44)

$$\text{Im}(n) = \frac{1}{2}\frac{\nu_{ei}}{\omega}\frac{n_e}{n_c}\left(1 - \frac{n_e}{n_c}\right)^{-1/2},$$ (4.45)

or, equivalently,

$$\text{Re}(k) = \frac{\omega}{c}\sqrt{1 - \frac{n_e}{n_c}},$$ (4.46)

$$\text{Im}(k) = \frac{1}{2}\frac{\nu_{ei}}{v_g}\frac{n_e}{n_c}$$ (4.47)

with $v_g = c\sqrt{1 - n_e/n_c}$.

If we denote z the propagation direction of the wave, then the wave's electric field will get damped as it propagates following $\exp[-\text{Im}(k)z]$, i.e., the wave intensity will have the spatial damping rate $\kappa_c = 2\text{Im}(k)$ or

$$\boxed{\kappa_c = \frac{\nu_{ei}}{v_g}\frac{n_e}{n_c}}.$$ (4.48)

This is the same expression obtained using our ballistic model in Sect. 4.1, Eq. (4.24). Likewise we can also write the same temporal absorption rate $\nu_c = \nu_{ei}n_e/n_c$.

Note again that this derivation is only valid in the low-field limit, i.e., $v_{os} \ll v_{Te}$, and that the Coulomb log in ν_{ei} should be modified from the usual expression used in the derivation of ν_{ei} following the discussion of Sect. 4.4. In order to apply this derivation to the high-field limit, one has to separately derive a new electron–ion collision frequency for the high-field limit—see Mulser et al., Refs. [3, 4].

4.3 Collisional Heating Rate

The ballistic model of collisional absorption directly provides the rate at which the plasma electron energy increases due to electron–ion collisions when an oscillating electric field is present (Eq. (4.22) for the low-field limit). Equating U with the total

Fig. 4.3 Collisional heating rate T_e/T_{e0} vs. time normalized to $\tau = \frac{6}{5}v_{Te0}^2/(v_{os}^2 v_{ei0})$ in the low-field limit, $v_{os} \ll v_{Te}$

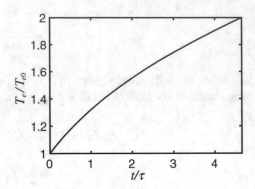

internal energy of the electrons $\frac{3}{2}n_e T_e$ and assuming a constant density give the rate of temperature increase:

$$\frac{dT_e}{dt} = \frac{1}{3}mv_{os}^2 v_{ei}.$$

(4.49)

As the temperature rises, the electron–ion collision frequency will decrease, since $v_{ei} \propto T_e^{-3/2}$. Allowing v_{ei} to vary in time due to heating turns this last equation into a nonlinear first-order differential equation that is separable and can thus be readily solved, giving the following equation for the temperature evolution:

$$T_e(t) = T_{e0}\left[1 + \frac{5}{6}v_{ei0}\frac{v_{os}^2}{v_{Te0}^2}t\right]^{2/5},$$

(4.50)

where T_{e0}, v_{Te0}, and v_{ei0} are the electron temperature, thermal velocity, and electron–ion collision at time 0 when the laser is turned on (cf. Fig. 4.3). The heating rate decreases with time because the collisionality (v_{ei}) decreases as the electrons get hotter, slowing down the heating process.

The relation can be expressed in practical units as

$$T_e(t) = T_{e0}\left(1 + 28.6Z\ln(\Lambda)n_{e20}T_{ek}^{-5/2}I_{16}t[\text{ns}]\right)^{2/5},$$

(4.51)

with n_{e20} the density in 10^{20} cm^{-3}, T_{ek} the temperature at $t = 0$ in keV, and I_{16} the intensity in 10^{16} W/cm^2. Note that in general, this heating will need to be balanced against heat conduction.

In the high-field limit, ignoring the slow variations of the logarithm term in Eq. (4.34) gives a heating rate $dT_e/dt = (2/3n_e)dU/dt$ that is essentially independent of T_e and therefore approximately constant with time, leading to

$$T_e(t) = T_{e0} + \frac{16}{3}\frac{n_e}{mZv_{os}}\left(\frac{Ze^2}{4\pi\varepsilon_0}\right)^2 \ln(\Lambda)\ln\left(\frac{\sqrt{2}v_{os}}{v_{Te}}\right)t.$$

(4.52)

In practical units,

$$T_e[\text{keV}] \approx T_{e0} + 7.6 n_{e20} \frac{Z}{\sqrt{I_{16}}\lambda_\mu} \ln(\Lambda) \ln\left(2.7\lambda_\mu \sqrt{\frac{I_{16}}{T_{e0}[\text{keV}]}}\right) t[\text{ns}], \quad (4.53)$$

with n_{e20} the electron density in units of 10^{20} cm^{-3}, I_{16} the laser intensity in 10^{16} W/cm^2, and λ_μ the laser wavelength in μm.

This heating rate is an overestimate as it does not include heat conduction. This expression stops being valid when the temperature rises enough to be comparable to mv_{os}^2, at which point the condition for the high-field limit (4.26) does not apply anymore and one should go back to the low-field limit.

4.4 Coulomb Logarithm

Broadly speaking, the Coulomb logarithm is a function that originates from the integration of a single electron–ion collision over all the ions within some range of interaction distances. This range of interaction distances, usually specified via minimum and maximum impact parameters b_{min} and b_{max}, varies depending on the physics problem under consideration: as we are about to see, it is for example different between the problem of the momentum relaxation of an electron population with a drift velocity (like we discussed in Sect. 1.4) and the problem of collisional absorption of a light wave that we are considering here. In both cases, the Coulomb logarithm takes the same form (cf. Eqs (1.247) and (4.12)—for a single electron with velocity v only, i.e., before the integration over a velocity distribution):

$$\ln(\Lambda) = \ln\left[\sqrt{\frac{b_\perp^2 + b_{max}^2}{b_\perp^2 + b_{min}^2}}\right], \quad (4.54)$$

where $b_\perp = Ze^2/(4\pi\varepsilon_0 mv^2)$ is the 90° scattering impact parameter for the single electron with velocity v (as we saw in Sect. 1.4, this expression does not require a small angle approximation, as b_{min} can in principle be smaller than $b_\perp$).

The two important points that were left off from our discussion in the earlier sections of this chapter and will now be addressed are: (i) the choice of the cutoffs b_{min} and b_{max} and (ii) the choice of a characteristic value for the velocity v, to justify taking $\ln[\Lambda(v)]$ out of the integration over the electron velocity distribution (cf. Eq. (4.18)).

For the latter point, it is customary to use v_{Te} and v_{os} as the characteristic velocity for the low-field and high-field limits, respectively. Now for the cutoffs, one usually uses half the de Broglie wavelength for the minimum impact parameter, which has been shown to recover the exact quantum-mechanical results [10] in the limit of low-Z plasmas and Maxwellian distributions:[3]

$$b_{min} = \frac{\hbar}{2m\bar{v}},$$ (4.55)

where $\bar{v}$ is the characteristic velocity, i.e., v_{Te} or v_{os}; it has been verified in molecular dynamics simulations [12] that the following expression for $\bar{v}$ provides the best match to simulations when transitioning from the low-field to the high-field limit:

$$\bar{v} = \sqrt{v_{Te}^2 + v_{os}^2/6},$$ (4.56)

indicating that the characteristic velocity is related to the internal energy of the electrons, i.e., $m\langle v^2\rangle/2 = 3mv_{Te}^2/2$ for the low-field (where the averaging is over the velocity distribution) vs. $m\langle v^2\rangle/2 = mv_{os}^2/4$ for the high-field limit, where the averaging is over the oscillation period (with $\langle(v_{os}\sin\omega t)^2\rangle = v_{os}^2/2$), leading to the factor 1/6 in front of v_{os}^2 in the expression of $\bar{v}$.

Regarding the maximum impact parameter, we have seen in our ballistic model that energy gain in an electron–ion collision event can only occur if the collision occurs over a duration much shorter than the external field period $1/\omega$; no energy exchange can occur if the electron experiences many oscillations in the external field during its interaction with an ion. Since the interaction length of the electron with the ion is on the order of the impact parameter (the electron trajectory as it passes by the ion is a hyperbola with the ion at its focus), the interaction time is $\sim b/\bar{v}$. The requirement that this time be much shorter than the oscillation period for energy exchange to be possible, i.e., $b/\bar{v} \ll 1/\omega$, leads to the choice of the upper cutoff:

$$b_{max} = \frac{\bar{v}}{\omega}.$$ (4.57)

In other words: for an electron with velocity $\bar{v}$, collisions with ions located at impact parameters larger than b_{max} will not contribute to collisional heating and can be discarded. Note that light propagation in underdense plasmas requires $\omega > \omega_{pe}$, which in turns implies (in the low-field limit of $v_{os} \ll v_{Te}$) $b_{max} = v_{Te}/\omega < \lambda_{De} = v_{Te}/\omega_{pe}$. This means that the electron–ion interactions contributing to collisional absorption are not subject to screening (or collective) effects, at least as long as the absorption occurs at densities well below critical. This upper cutoff, which we

---

[3] A heuristic argument relies on the uncertainty principle, which states that the uncertainty in transverse momentum of the electron $\Delta p_\perp$ ("transverse" with respect to the initial velocity before collision) and the uncertainty in impact parameter Δb must verify $\Delta p_\perp \Delta b \geq \hbar/2$, i.e., $\Delta b \geq \hbar/2\Delta p$; our classical treatment requires that the initial velocity be well known, i.e., $\Delta p \ll p$, which in turn implies that $\Delta b \gg \hbar/2p$. This condition sets the minimum value for the impact parameter b_{min}, cf. [11] p. 629.

introduced somewhat ad hoc, can be reproduced exactly by kinetic calculations without invoking a cutoff at all [5, 6].

In summary, a reasonable expression for the Coulomb logarithm for collisional absorption valid in underdense plasmas ($\omega \gg \omega_{pe}$) in the low-field or high-field regime is Eq. (4.54), with the following averaged cutoffs:

$$\bar{b}_{min} = \frac{\hbar}{2m\bar{v}}, \tag{4.58}$$

$$\bar{b}_{max} = \frac{\bar{v}}{\omega}, \tag{4.59}$$

$$\bar{b}_\perp = \frac{Ze^2}{4\pi\varepsilon_0 m\bar{v}^2} \tag{4.60}$$

with

$$\bar{v} = \sqrt{v_{Te}^2 + v_{os}^2/6}. \tag{4.61}$$

Note that when $b_{max} \gg b_{min}, b_\perp$, Eq. (4.54) is consistent with

$$\Lambda = \frac{b_{max}}{\text{Max}[b_{min}, b_\perp]}, \tag{4.62}$$

which is the small angle scattering limit usually found in textbooks for the derivation of ν_{ei} with $b_{max} = \lambda_{De}$.

In practical units, we can use

$$\frac{b_{max}}{b_\perp} \approx 4.9 \times 10^3 \frac{\lambda_\mu}{Z} \left[0.6 I_{16}\lambda_\mu^2 + T_{ek}\right]^{3/2}, \tag{4.63}$$

$$\frac{b_{min}}{b_\perp} \approx \frac{3}{Z}\sqrt{0.6 I_{16}\lambda_\mu^2 + T_{ek}}, \tag{4.64}$$

$$\ln(\Lambda) = \ln\left(\sqrt{\frac{1 + (b_{max}/b_\perp)^2}{1 + (b_{min}/b_\perp)^2}}\right), \tag{4.65}$$

where T_{ek} is the temperature in keV, λ_μ is the wavelength in μm, and I_{16} is the intensity in 10^{16} W/cm^2.

4.5 Collisional Damping of EPWs

Since we never had to specify whether the external electric field in the derivation of the collisional absorption rate was longitudinal or transverse, the entire derivation is valid for both light waves and electron plasma waves. The collisional damping rate is then usually simply added to the Landau damping to give the total wave absorption rate, i.e.,

$$\nu_{EPW} = \nu_c + 2\nu_L, \tag{4.66}$$

where ν_L is the Landau damping rate (defined in Sect. 1.3.4 for the field amplitude, hence the factor of 2 for the rate of energy loss) and $\nu_c = \nu_{ei}\omega^2/\omega_{pe}^2$ is the collisional damping with ω the EPW frequency. Inserting ω from the EPW dispersion relation $\omega^2 = \omega_{pe}^2 + k^2 v_{Te}^2$ gives the following expression for the EPW collisional damping rate:

$$\nu_c = \frac{\nu_{ei}}{1 + 3(k\lambda_{De})^2}. \tag{4.67}$$

Like for light waves, we can also express the absorption as a spatial rate κ, given by

$$\kappa_{EPW} = \frac{\nu_c}{v_g} \approx \nu_{ei}\frac{\omega_{pe}}{3kv_{Te}^2}, \tag{4.68}$$

where we assumed $\omega \approx \omega_{pe}$ for the EPW frequency.

As we will see in later chapters, EPWs encountered in laser–plasma experiments are typically created by three-wave instabilities such as stimulated Raman scattering (SRS) or two plasmon decay (TPD). In most cases, these EPWs have much slower group velocities than the laser, $v_{g,EPW} = 3kv_{Te}^2/\omega \ll v_{g,EMW} = k_0c^2/\omega_0$. Since the temporal absorption rates are roughly on the same order for the densities of relevance, this shows that the spatial absorption rate of EPWs for most laser–plasma experiments is usually much larger than that for light waves, due to the slower group velocity of EPWs (by a factor $3v_{Te}^2/c^2 \approx T_e[keV]/170$).

Unlike Landau damping, collisional damping for EPWs is only weakly dependent on $k\lambda_{De}$. Since EPWs typically exist in the regime of $k\lambda_{De} \ll 1$, their collisional damping rate is then $\approx\nu_{ei}$.

Figure 4.4 represents the total EPW damping as a function of plasma density and temperature for a $Z = 2$ plasma and an EPW wave-number $k = 1.8 \times 10^7$ cm^{-3}, corresponding to $k = k_0 = 2\pi/\lambda_0$ for $\lambda_0 = 351$ nm (most EPWs relevant to laser–plasma interactions typically have wave-numbers on the order of the laser wave-number). The drop in damping in the upper-left corner comes from the fact that Landau damping goes back down for $k\lambda_{De} > 0.6$.

Our main take-away is that even at high densities and low temperatures, when Landau damping becomes negligible, EPW damping will not vanish as collisional damping will take over and dominate. As we will see in later chapters, laser–plasma instabilities involving EPWs typically have spatial amplification rates that are inversely proportional to the EPW damping rate. Collisional damping will put a finite limit on the damping and, hence, the instability growth in the limit of $k\lambda_{De} \ll 1$ where Landau damping vanishes.

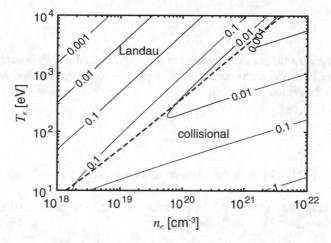

Fig. 4.4 EPW damping $v/\omega_{pe} = v_c/\omega_{pe} + v_L/\omega_{pe}$ (collisional + Landau) for a $Z = 2$ plasma and $k = 1.8 \times 10^7$ cm^{-3}, corresponding to $k = 2\pi/\lambda_0$ for $\lambda_0 = 351$ nm. The dashed line corresponds to $v_c = v_L$; Landau and collisional damping dominate in the upper-left and lower-right regions, respectively

As a final note, we should keep in mind that the ballistic derivation used here neglected shielding and thus collective effects (since we only considered binary electron–ion collisions with a Coulomb potential). While this is arguably a reasonable assumption for light waves, at least in the underdense region of $n_e/n_c \ll 1$, this might seem more questionable for EPWs for which $\omega \approx \omega_{pe}$. Kinetic theory of collisional absorption of EPWs has shown that the unshielded results presented here are nonetheless in reasonable agreement with the kinetic treatment as long as v_{ei}/ω_{pe} remains small enough [13].

4.6 Non-Maxwellian Electron Distributions from Collisional Heating: The Langdon Effect

We saw that in the low-field limit, collisions preferentially heat low-energy electrons, since their collision frequency ($\propto v^{-3}$) is larger. This can lead to the formation of non-Maxwellian distribution functions, more specifically super-Gaussian functions, a phenomenon known as the Langdon effect after Ref. [14]. The condition for this process to occur is that the rate of energy increase of the plasma electrons relative to their initial total energy must be faster than the time it takes for electron–electron collisions to thermalize the distribution and bring it back to a Maxwellian,[4] which we can write as

---

[4] Note that electron–electron collisions do not lead to collisional heating of the plasma electrons, cf. Problem 4.2.

$$\frac{dU/dt}{3n_e T_e/2} \geq \nu_{ee}. \tag{4.69}$$

Replacing dU/dt by its expression from Eq. (4.22), and using the fact that $\nu_{ei} = \sqrt{2}Z\nu_{ee}$, leads to the following condition for the formation of non-Maxwellian distributions in the low-field limit:

$$\alpha = Z^* \frac{v_{os}^2}{v_{Te}^2} \gg 1, \tag{4.70}$$

where $Z^* = \langle Z^2 \rangle / \langle Z \rangle$ with the average taken over ion species for mixed-species plasmas.

Fokker–Plank analyses [15] have confirmed that the resulting distribution functions are super-Gaussian

$$f(v) = c_m \frac{n_e}{4\pi} \exp\left[-(v/v_m)^m\right] , \quad c_m = \frac{m}{\Gamma(3/m)v_m^3} , \quad v_m^2 = 3v_{Te}^2 \frac{\Gamma(3/m)}{\Gamma(5/m)} ,$$

$$\tag{4.71}$$

with the super-Gaussian order m given by

$$m(\alpha) \approx 2 + \frac{3}{1 + 1.66/\alpha^{0.724}} . \tag{4.72}$$

The existence of super-Gaussian distribution functions from collisional heating has recently been confirmed in the laboratory in conditions relevant to ICF and HED experiments [16]. The deviation of the electron distribution function from a Maxwellian under the influence of collisional heating can change the plasma response to a laser propagation beyond the absorption rate itself. As we will see in later chapters, the coupling coefficients for most laser–plasma instabilities are sensitive to the details of the distribution function; collisional absorption is one mechanism (among others) that can lead to non-Maxwellian distribution functions and impact laser–plasma instabilities.

Finally, note that super-Gaussian distributions have also been observed in simulations in high-field conditions with $v_{os} \geq v_{Te}$, but the very high-field limit when $v_{os}/v_{Te} \gg 1$ typically sees a return of the distribution function to a Maxwellian. Indeed, while the high quiver velocity reduces the effective electron–ion collision frequency (cf. Eq. (4.36)), the electron–electron frequency, which is responsible for the thermalization of the distribution, is not impacted (for the same reason that electron–electron collisions do not lead to absorption or heating, cf. Problem 4.2). As a result, for a high enough v_{os}/v_{Te} ratio, the thermalization from electron–electron collisions will eventually happen at a faster rate than the collisional heating and prevent the deviation from Maxwellian [17].

4.7 Estimating Collisional Laser Absorption in Idealized Plasma Profiles

4.7.1 Laser Absorption at a Turning Point

In this section, we will show that the laser absorption in the turning point region can be derived analytically and that it yields the same result as the WKB approximation—even though WKB is invalid at the turning point. In other words, the absorption can be calculated via a line integral along the light propagation path (i.e., along the rays) even as it goes through a turning point, just like we saw for the frequency shift discussed in Sect. 3.5.

Let us consider a light wave at normal incidence along z on a density gradient (the case of oblique incidence can be derived in a similar fashion), which is assumed to be approximately linear near the turning point, i.e., $n_e(z)/n_z \approx z/L$ for a region that is at least equal to the Airy skin depth Δz_A defined in Sect. 3.2. We wish to calculate the absorption of a light wave from a location z as it gets to the turning point at $z = L$ and then back to z. We will assume that z is sufficiently close to the turning point that the electron–ion collision frequency can be assumed constant throughout the propagation path.[5]

First, the absorption can be estimated using the line integral along the light propagation path (i.e., along a ray, under the WKB approximation):

$$A = 1 - \exp\left[-2\int_z^L \kappa_c(z')dz'\right],\tag{4.73}$$

where $\kappa_c = (n_e/n_c)\nu_{ei}/\nu_g$ is the collisional absorption rate (in energy) per unit length. The factor 2 in the exponential represents the round trip of the light from z to the turning point and back.

Inserting the expression for the group velocity $\nu_g = c\sqrt{1 - n_e/n_c}$ and using the linear approximation for $n_e(z) \approx n_c z/L$, the integral can be solved and yields

$$A(z) = 1 - \exp\left[-\frac{4}{3}\frac{\nu_{ei}L}{c}\sqrt{1 - \frac{z}{L}}\left(2 + \frac{z}{L}\right)\right].\tag{4.74}$$

Second, let us go back to the exact expression for the light wave's electric field in the turning point region based on the Airy function. In Sect. 4.2, we derived the expression for the complex dielectric constant in the presence of absorption, Eq. (4.43):

---

[5] A rather dubious assumption, since $\nu_{ei} \propto n_e$; however, it is the only way to get back to an Airy equation that can be solved without WKB and then compared to WKB—while making the same approximation in both cases.

$$\varepsilon = 1 - \frac{\omega_{pe}^2}{\omega_0^2(1 + i\nu_{ei}/\omega_0)} . \tag{4.75}$$

If we use a linear density profile with $\omega_{pe}^2/\omega_0^2 = z/L$, then we can introduce a complex variable

$$\tilde{L} = L\left(1 + i\frac{\nu_{ei}}{\omega_0}\right), \tag{4.76}$$

so that the entire analysis of light propagation near the turning point can be conducted identically to Sect. 3.2 by introducing the complex variable

$$\eta = -\left(\frac{\tilde{L}\omega_0}{c}\right)^{2/3}\left(1 - \frac{z}{\tilde{L}}\right). \tag{4.77}$$

The wave equation takes again the form of an Airy equation, $(\partial_\eta^2 - \eta)f = 0$, the asymptotic expansion of the Airy function used in Sect. 3.2, Eq. (3.19) is still valid for complex variables, and so Eq. (3.22) is still valid as well.

Assuming that the absorption is small during a laser oscillation period, i.e., $\nu_{ei}/\omega_0 \ll 1$, gives

$$(-\eta)^{3/2} \approx (-\eta_R)^{3/2} + i\frac{L\nu_{ei}}{2c}\sqrt{1 - \frac{z}{L}}\left(2 + \frac{z}{L}\right), \tag{4.78}$$

where $\eta_R = \mathrm{Re}[\eta]$ is the real part as was used in Sect. 3.2. We can then rewrite Eq. (3.22) as

$$f(\eta) = \frac{\alpha e^{i\pi/4}}{2\sqrt{\pi}(-\eta)^{1/4}}e^{g(z)}\left[\exp\left(-i\frac{2}{3}(-\eta_R)^{3/2}\right) - ie^{-2f(z)}\exp\left(i\frac{2}{3}(-\eta_R)^{3/2}\right)\right], \tag{4.79}$$

where $g(z) = \frac{2}{3}\mathrm{Im}[(-\eta)^{3/2}]$. Proceeding like in Sect. 3.2, we can identify the two exponential terms as the incoming and the reflected wave components, with the complex pre-factor in front of the reflected wave corresponding to a reflectivity coefficient R, i.e., $R = -i\exp[-2f(z)]$ (i.e., an energy reflection coefficient equal to $|R|^2$); inserting the expression for $f(z)$ from Eq. (4.78), we easily verify that the energy absorption coefficient $A = 1 - |R|^2$ gives the same expression as in Eq. (4.74) above.

The conclusion is that the energy absorption A of a light wave in a non-uniform plasma profile can simply be calculated by performing a line integral along the light propagation path, regardless of whether the light goes through a turning point, i.e.,

$$A = 1 - \exp\left[-\int_P \kappa_c ds\right], \tag{4.80}$$

where P is the path followed by the light wave, governed by the ray-tracing equations derived earlier. More explicitly, replacing κ_c by its expression gives

$$A = 1 - \exp\left[-\int_P \frac{n_e(s)}{n_c} \frac{\nu_{ei}(s)}{c n(s)} ds\right],$$ (4.81)

where ν_{ei} varies along the path via its linear dependence on $n_e(s)$ and $n = \sqrt{1 - n_e/n_c}$.

4.7.2 Absorption in Isothermal 1D Density Profiles

Here we consider the case of light propagation in a 1D density profile varying along z; we will provide a general expression for the collisional absorption and derive simple analytic formulae for ideal profiles.

In the general case of oblique incidence, where the laser propagates in the 2D plane (x, z), we have already seen in Sect. 3.3 that the x-component of the wave-vector is conserved, i.e., $k_x = (\omega_0/c) \sin(\theta_0)$, where θ_0 is the incidence angle at the vacuum–plasma boundary, and

$$k_z(z) = \frac{\omega_0}{c}\sqrt{\cos^2(\theta_0) - \frac{n_e(z)}{n_c}}.$$ (4.82)

The expression for the dielectric constant from Eq. (4.43) shows that in the limit of $\nu_{ei} \ll \omega_0$, collisional absorption can be modeled by simply substituting n_e/n_c by $(n_e/n_c)(1 - i\nu_{ei}/\omega_0)$. Since k_x is conserved, the absorption in 1D profiles can be calculated via the z-component of the wave-vector only, i.e.,

$$A = 1 - \exp\left[-4\int_{-\infty}^{z_t} \mathrm{Im}\,(k_z(z))\,dz\right],$$ (4.83)

where the factor 4 in the exponential accounts for the absorption in energy (i.e., factor 2) and for the round trip of the light wave to the turning point at z_t and back (another factor 2). Replacing by the expression for k_z above, we obtain

$$A = 1 - \exp\left[-\frac{2}{c}\int_{-\infty}^{z_t} \nu_{ei}\frac{n_e}{n_c}\left(\cos^2(\theta_0) - \frac{n_e}{n_c}\right)^{-1/2} dz\right].$$ (4.84)

We can now proceed to derive the absorption formula for a light wave in a few typical density profiles. We assume that the plasma is isothermal, meaning that the only spatial dependence in the absorption coefficient is via the electron density. We define $\nu_{ei,c}$ as the electron–ion collision frequency at the critical density, such that:

$$v_{ei} = v_{ei,c} \frac{n_e}{n_c},$$ (4.85)

where $v_{ei,c} \propto T_e^{-3/2}$. We take the simple examples of a linear density ramp and an exponential profile.

- Linear density ramp:

Here we assume that $n_e/n_c = z/L$. The turning point is located at $z_t = L \cos^2(\theta_0)$. The absorption can be readily calculated from Eq. 4.84, using the standard integral:

$$\int \frac{x^2}{\sqrt{a-x}} dx = -\frac{2}{15}\sqrt{a-x}(8a^2 + 4ax + 3x^2).$$ (4.86)

We get

$$A = 1 - \exp\left[-\frac{32}{15}\frac{v_{ei,c}L}{c}\cos^5(\theta_0)\right].$$ (4.87)

- Exponential density profile:

This type of profile is characteristic of an isothermal expansion of plasma in vacuum. We define $n_e/n_c = \exp[z/L]$, with the laser originating from $z = -\infty$ and reaching the turning point at $z_t = 2L \ln[\cos(\theta_0)]$. We use the integral:

$$\int \frac{e^{2x}}{\sqrt{a-e^x}} dx = -\frac{2}{3}\sqrt{a-e^x}(2a + e^x)$$ (4.88)

and immediately get:

$$A = 1 - \exp\left[-\frac{8}{3}\frac{v_{ei,c}L}{c}\cos^3(\theta_0)\right].$$ (4.89)

4.8   Resonance Absorption of p-Polarized Light Waves

4.8.1   Physical Picture

Resonance absorption can occur during the propagation of a p-polarized light wave in a density gradient. By p-polarized, we mean that the wave is linearly polarized with its polarization vector in the plane of incidence, as shown in Fig. 4.5. As we have already seen in Chap. 3, the light reaches its turning point at $n_e = n_c \cos^2(\theta)$. When the light is s-polarized, i.e., polarization normal to the plane of incidence

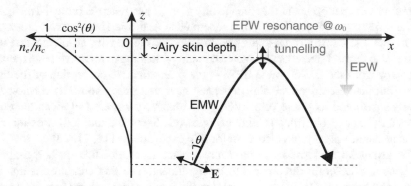

Fig. 4.5 Resonance absorption of a p-polarized light wave at oblique incidence in a density gradient: the EMW's electric field component along the density gradient "tunnels" (i.e., exponentially decays) from the turning point at $n_e = n_c \cos^2(\theta)$ to the critical density, where it can resonantly drive an EPW at the laser frequency ω_0. Since it is created at its turning point, the EPW will propagate down the density ramp (i.e., toward $-z$). Landau damping will prevent the EPW from propagating beyond $n_e/n_c \lesssim 0.8$ (see text). Energy transfer from the light wave to the EPW and ultimately to the plasma electrons via the EPW damping (irreversible) leads to a net absorption of the EMW. Resonance absorption only occurs when the tunneling distance from the turning point to the critical density is on the order of the Airy skin depth defined in Eq. (3.28), i.e., when $\tau = (\omega_0 L/c)^{1/3} \sin(\theta) \sim 1$

(along y in Fig. 4.5), it can be described by the WKB approximation except in the immediate vicinity of the turning point in the region we described as the "Airy skin depth" in Sect. 3.2, where the light electric field is described by an Airy function. The wave becomes evanescent past the turning point, i.e., it exponentially decays over a distance equal to the Airy skin depth—cf. Eq. (3.28). The light wave remains purely transverse as long as it is s-polarized, because the electron quiver motion in the light's electric field is always orthogonal to the density variations.

On the other hand, when the light is p-polarized, its electric field's z-component is aligned with the density gradient and cannot be purely transverse anymore, since the resulting quiver motion along z will lead to charge separation due to the density variation along that same direction. Note that a finite incidence angle is required for a z-component of the electric field to exist; for normal incidence ($\theta = 0$), **E** is perpendicular to ∇n_e and remains purely transverse (in other words, p-polarization is only defined for oblique incidence: normal incidence is always s-polarized).

Since these longitudinal oscillations are driven at the laser frequency ω_0, they can resonantly excite an electrostatic mode if the EPW dispersion relation is satisfied for the laser frequency, i.e., if $\omega_0^2 = \omega_{pe}^2 + 3k^2 v_{Te}^2 = \omega_{pe}^2[1 + 3(k\lambda_{De})^2]$. Since Landau damping restricts EPWs to the $k\lambda_{De} \ll 1$ regime, this means that resonant excitation of an EPW at ω_0 can only occur in the region where $\omega_0 \approx \omega_{pe}$, i.e., near the critical density $n_e = n_c$.

Resonance absorption is the transfer of energy from the transverse to the longitudinal component of the field, i.e., a "mode conversion" process from the EMW to an EPW [18, 19]. It is called "resonant" since the excitation of an EPW can be resonant (or near-resonant) in the vicinity of ω_{pe}, i.e., near the critical density for the laser frequency. Since that EPW is generated at its turning point, it will propagate back down the density ramp along the direction of the density gradient,[6] and will then vanish due to Landau or collisional damping,[7] leading to an irreversible transfer of energy from the EMW to the EPW—i.e., to an effective absorption of the light wave. Because the driven EPW is generated near its turning point, its wave-vector is very small, and its phase velocity ω_0/k is thus very large,[8] and it can therefore accelerate trapped electrons to high energies (cf. Sects. 2.2 and 8.4); this initially made resonance absorption a concern for ICF experiments [19, 21, 22].

We just mentioned two conditions for resonance absorption that can help us guess the general criteria for the process to be significant. The first condition is that the wave must have a finite incidence angle, as the p-component disappears for normal incidence. The second condition is that the light wave must reach densities close to critical so that its electric field can drive resonant oscillations near n_c; this means that the turning point must approximately lie within one Airy skin depth from the critical density (since the field is evanescent past the turning point, cf. Sect. 3.2), and thus that the incidence angle must generally be very small. We can already infer that the only way to satisfy these two seemingly contradictory conditions is if the critical density is located at approximately one Airy skin depth from the turning point, which maximizes the incidence angle before the field decays too much past the turning point for larger angles (remember that the Airy skin depth is independent of the incidence angle). For the simple case of a linear density ramp with $n_e(z)/n_c = z/L$ in the vicinity of n_c, this condition is equivalent to $L - L\cos^2(\theta) \sim \Delta z_A$, or, after inserting the expression for Δz_A from Eq. (3.28),

$$\tau^2 \sim 1, \tag{4.90}$$

where $\tau = (\omega_0 L/c)^{1/3} \sin(\theta)$. The qualitative analysis below will confirm this rough guess and show that resonance absorption is maximum for $\tau \approx 0.7$.

---

[6] We saw in Sect. 3.4 Problem 3.3 that EMWs and EPWs follow the same trajectories in plasmas: therefore, n_c is the turning point for either an EMW or an EPW with normal incidence.

[7] The EPW picks up a k-vector as it propagates back down the density ramp, following the dispersion relation $\omega_0^2 = \omega_{pe}^2[1 + 3(k\lambda_{De})^2]$ (assuming an isothermal plasma). Landau damping will effectively terminate the EPW when $k\lambda_{De}$ reaches $\sim$0.3, i.e., $n_e/n_c = 1/[1+3(k\lambda_{De})^2] \approx 0.8$, and collisional damping might even prevent it from going that far (cf. Sect. 4.5).

[8] The z-component of the electric field in the tunneling region with $n_e/n_c > 0.8$ will be the superposition of the EMW (with both incoming and reflected components) and the EPW. Since their wave-numbers are, respectively, given by $k_{EMW} = n\omega_0/c$ and $k_{EPW} = n\omega_0/(\sqrt{3}v_{Te})$, with $n = \sqrt{1 - n_e/n_c}$ the plasma refractive index, the EPW will have a much shorter wavelength than the EMW due to $v_{Te} \ll c$ (for example, cf. Ref. [20] for illustrations of the fields from numerical simulations).

4.8.2 Qualitative Analysis

The analytical analysis of resonance absorption is difficult; various theoretical formulas and approximations have been obtained through the years but always require considerable mathematical treatment and only exist for highly simplified situations, like the linear density ramp considered in the following. For a complete treatment, one needs to solve coupled equations for the light wave's magnetic field (representing the purely electromagnetic component), the electric field along the density gradient (consisting of electromagnetic and electrostatic parts), and the density perturbation associated with the EPW. Here we will merely provide a few qualitative observations to capture the key physics of the process and quote the expression for the absorption coefficient from one of the most thorough analytical analyses by Hinkel et al. [20, 23].

We use the geometry of Fig. 4.5, where z is the direction of the density gradient and x is such that (x, z) is the plane of incidence of the laser. As we have already seen in Sect. 3.3, the x-component of the light wave's wave-vector is conserved, whereas its z-component will vanish at the turning point, so we can envelope the field at k_x but not k_z. Since the system is infinite in x, variations in the amplitude are only possible along z, so we get the following enveloped expression for the electric and magnetic fields with $k_v = \omega_0/c$ the vacuum wave-vector:[9]

$$\mathbf{E}(\mathbf{r}, t) = \frac{1}{2}\left[E_x(z)\mathbf{e}_x + E_z(z)\mathbf{e}_z\right]\exp[i\,(k_v x \sin\theta - \omega_0 t)] + c.c. \,, \quad (4.91)$$

$$\mathbf{B}(\mathbf{r}, t) = \frac{1}{2}B_y(z)\mathbf{e}_y \exp[i\,(k_v x \sin\theta - \omega_0 t)] + c.c.. \qquad (4.92)$$

Using the relation $\nabla \cdot (\varepsilon \mathbf{E}) = 0$ with $\varepsilon(z) = 1 + \chi_e(z) = 1 - \omega_{pe}^2(z)/\omega^2$ and the vector identity $\nabla \cdot (\psi \mathbf{A}) = \psi \nabla \cdot \mathbf{A} + \mathbf{A} \cdot \nabla \psi$ immediately confirms that the field is not purely transverse but has a longitudinal component, since

$$\nabla \cdot \mathbf{E} = -\frac{\partial_z \varepsilon}{\varepsilon} E_z \,. \qquad (4.93)$$

Next we take the time derivative of Ampère's law $\nabla \times \mathbf{B} = \mu_o \mathbf{j} + \partial_t \mathbf{E}/c^2$, substitute the current $\mathbf{j} = -e n_e \mathbf{v}$ and $m d\mathbf{v}/dt = -e\mathbf{E}$, and get

$$\partial_t (\nabla \times \mathbf{B}) = \frac{\omega_{pe}^2}{c^2}\mathbf{E} + \frac{1}{c^2}\frac{\partial^2 \mathbf{E}}{\partial t^2} \,. \qquad (4.94)$$

---

[9] Even though the magnetic field is always along y, we will keep the subscript in B_y to remind ourselves that it is an enveloped quantity, with the fast oscillations in time and in space along x taken out.

Inserting the enveloped expressions for the electric and magnetic fields above gives

$$E_x(z) = -i\frac{c}{k_v}\frac{B_y'}{\varepsilon}, \tag{4.95}$$

$$E_z(z) = -c\sin(\theta)\frac{B_y}{\varepsilon}, \tag{4.96}$$

where the prime denotes the derivative with respect to z. These equations can provide some qualitative insight into resonance absorption; it can be shown (cf. Problem 4.5) that at the critical density the magnetic field $B_y(z_c)$ is finite, but its derivative is zero, $B_y'(z_c) = 0$, where $n_e(z_c) = n_c$. The divergence of E_z at the critical density, due to $\varepsilon(z_c) = 0$, illustrates the resonant excitation of the EPW by the light wave tunneling from the turning point to the critical density. To avoid the divergence, we introduce a finite collisional damping, i.e., we use the complex expression of ε from Eq. (4.43); as long as the electron temperature is small, then the electromagnetic and electrostatic dielectric functions $\varepsilon = 1 + \chi_e$ are approximately the same (cf. Eqs. (1.102), (1.116)), including their imaginary part (i.e., collisional absorption) that is the same for EPWs and EMWs as we saw in Sect. 4.5. The existence of an irreversible damping process of the EPW is required to get absorption of the EMW; Landau damping would play a similar role but would require a different treatment since the corresponding dielectric function ε is not a function anymore but an operator in Fourier space (cf. Sect. 1.3.3).

Since the EPW power will be entirely converted into thermal energy in plasma electrons, the power of E_z in the vicinity of the critical density represents the fraction of laser power lost to resonance absorption. It can be estimated as

$$P_a = \int_{-\infty}^{\infty} \nu_{ei}(z)\varepsilon_0 \left\langle \mathbf{E}_z^2 \right\rangle dz = \frac{1}{2}\int_{-\infty}^{\infty} \nu_{ei}(z)\varepsilon_0 |E_z|^2 dz. \tag{4.97}$$

Defining a "driver field" at the critical density:

$$E_d = \varepsilon E_z = -c\sin(\theta)B_y(z_c), \tag{4.98}$$

we obtain

$$P_a = \frac{\varepsilon_0}{2}\int_{-\infty}^{\infty} \nu_{ei}(z)\frac{|E_d|^2}{|\varepsilon|^2} dz. \tag{4.99}$$

Assuming a linear density profile $n_e(z)/n_c = z/L$ and defining $\nu_{ei,c} = \nu_{ei}(z = L)$ as the electron–ion collision frequency at the critical density, for a sufficiently small $\nu_{ei,c}/\omega_0 \ll 1$, the integrand is a Lorentzian function

$$P_a \approx \frac{\varepsilon_0}{2}|E_d|^2\omega_0 L \int_{-\infty}^{\infty} \frac{\nu_{ei,c}/\omega_0}{\zeta^2 + (\nu_{ei,c}/\omega_0)^2} d\zeta = \frac{\pi}{2}\varepsilon_0|E_d|^2 L\omega_0, \tag{4.100}$$

with $\zeta = z/L - 1$. Since E_z is strongly peaked at the critical density (in the limit of $\nu_{ei,c}/\omega_0 \ll 1$), its power integrated throughout that narrow resonance region is proportional to $|E_d|^2$.

The fraction of absorbed power via resonance absorption is obtained by taking the ratio of P_a and the incident laser intensity in vacuum $I_0 = \varepsilon_0|E_0|^2 c/2$, i.e.,

$$f_a = \pi k_v L \frac{|E_d|^2}{|E_0|^2} , \qquad (4.101)$$

where $k_v = \omega_0/c$ is the light wave-vector in vacuum. The important point here is that the absorption is dictated by the driver field E_d, i.e., the value of the magnetic field at the critical density.

Evaluating E_d is a significant task; here we simply show that the driver field follows the asymptotic expansion (cf. Problem 4.5):

$$|E_d| \approx |E_0| 2\sqrt{\pi} A_i'(0)\tau(k_0 L)^{-1/2} , \quad \tau \ll 1, \qquad (4.102)$$

where $\tau = \sin(\theta)(k_v L)^{1/3}$ and the numerical factor can be evaluated as $2\sqrt{\pi} A_i'(0) \approx 0.92$. Therefore, when $\tau \ll 1$, the absorption coefficient only depends on the variable τ, with

$$f_a \approx 4\pi^2 A_i'^2(0)\tau^2 \approx 2.7\tau^2 , \quad \tau \ll 1. \qquad (4.103)$$

On the other hand, for large angles of incidence—or to be more precise for $\tau \gg 1$ (as we shall confirm just below)—the behavior of E_z can be roughly estimated via a WKB approximation past the turning point, i.e., with a purely imaginary wave-number and an evanescent field. The behavior of E_z from the turning point at $z = L\cos^2(\theta)$ to the critical density at $z = L$ can be estimated as (cf. Sect. 3.1)

$$E_z \propto \exp\left[i\int_{L\cos^2(\theta)}^{L} k_z dz\right], \qquad (4.104)$$

with k_z purely imaginary. This expansion is valid beyond the Airy skin depth, i.e., for $z > L\cos^2(\theta) + \Delta z_A$ with $\Delta z_A = (k_v L)^{1/3}/k_v$ (cf. Eq. (3.28)). It is thus only relevant to our situation if the separation between the turning point and the critical density is larger than the Airy skin depth, i.e., if $L - L\cos^2(\theta) \gg \Delta z_A$; substituting for Δz_A, we immediately see that this condition is exactly equivalent to $\tau^2 \gg 1$.

From the EMW dispersion relation $\omega_0^2 = \omega_{pe}^2(z) + k_x^2 c^2 + k_z^2(z)c^2$, where $k_x^2 = k_v^2 \sin^2(\theta)$ is conserved, we obtain $k_z = k_v\sqrt{\cos^2(\theta) - n_e/n_c}$ (cf. Sect. 3.3). Past the turning point, i.e., for $n_e/n_c = z/L > \cos^2(\theta)$, k_z is imaginary and given by

$$k_z = -ik_v\sqrt{\frac{n_e}{n_c} - \cos^2(\theta)}, \qquad (4.105)$$

Fig. 4.6 Fraction of incident
laser power absorbed via
resonance absorption, from
Eq. (4.108) (Ref. [20]). The
dashed and dotted lines
correspond to the asymptotic
expansions
Eqs. (4.103), (4.107)

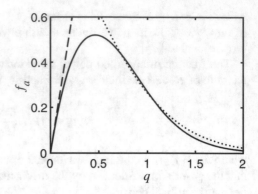

which when inserted in Eq. (4.104) gives after integration

$$E_z \propto \exp\left[-\frac{2}{3}k_v L \sin^3(\theta)\right] = \exp\left[-\frac{2}{3}\tau^3\right]. \tag{4.106}$$

Therefore, for $\tau \gg 1$, i.e., when the critical density is sufficiently far from the turning point that the field is well-approximated by the evanescent WKB solution, we expect $|E_d| \propto \exp[-2\tau^3/3]$, and hence,

$$f_a \propto \exp\left[-\frac{4}{3}\tau^3\right], \quad \tau \gg 1. \tag{4.107}$$

Note that the proportionality term in the last two equations depends on the value of E_z at the turning point, which cannot be easily estimated.[10]

The two asymptotic limits for $\tau \ll 1$ and $\tau \gg 1$ are plotted in Fig. 4.6 together with the analytic solution obtained by Hinkel in Ref. [20] Eq. (56), valid for non-relativistic electron temperatures and showing excellent agreement with numerical solutions:

$$f_a(q) = \frac{(2\pi)^2 q A_i'^2(q)}{[1 + q\pi^2 A_i'^2(q)]^2 + q^2\pi^4 A_i'^2(q)B_i'^2(q)}, \tag{4.108}$$

where $q = \tau^2 \approx 3.4(L/\lambda_0)^{2/3}\sin^2(\theta)$ and A_i', B_i' are the derivatives of the Airy functions.

---

[10] It can be tempting to use the expression from Eq. (4.102), valid for $\tau \to 0$, and then multiply it by $\exp[-2\tau^3/3]$ to impose the correct asymptotic behavior for large values of τ [24]. However, doing so relies on two incompatible assumptions of $\tau \gg 1$ *and* $\tau \ll 1$, and the resulting absorption function, while showing the correct qualitative trends of $f_a \to 0$ for $\tau \ll 1$ and $\tau \gg 1$, is inaccurate by ~100% over most of the relevant range of $\tau \sim 1$ compared to numerical simulations or the expression from Eq. (4.108).

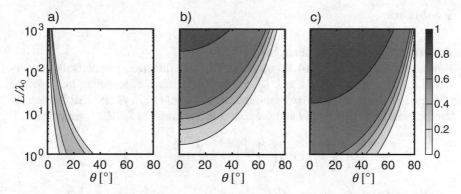

Fig. 4.7 Fraction of absorbed power for a 351 nm laser incident at angle θ onto a linear density ramp with profile $n_e(z) = n_c z/L$, for: (**a**) resonance absorption (independent of plasma temperature); (**b**) collisional absorption, $T_e = 1$ keV; (**c**) collisional absorption, $T_e = 0.1$ keV

The fraction of power absorbed by resonance absorption is at most 50%. The result is independent of the plasma temperature, and the laser wavelength only enters through the $k_v L$ term. For density gradient scale-lengths much greater than the laser wavelength, $k_v L \gg 1$, resonance absorption will only occur for a narrowly peaked incidence angle close to zero. Resonance absorption will typically be significant (compared to collisional absorption) only in very steep density gradients such as those produced by short pulse lasers, before the plasma expansion smoothes out the density profile over scale-lengths longer than the laser wavelength.

To illustrate this last point, a comparison between the fraction of absorbed laser power for resonance and collisional absorption is shown in Fig. 4.7. Here the density profile is assumed purely linear along z, and we are comparing Eqs. (4.108) and (4.87) for two different electron temperatures, $T_e = 1$ and 0.1 keV, and for a laser wavelength of 351 nm. Resonance absorption only dominates over collisional absorption for sufficiently high temperatures (such that collisional absorption becomes ineffective) and steep density gradients on the order of a few (≤ 10) laser wavelengths. These two conditions are often mutually exclusive in the long-pulse ($\sim$ns) regime but can simultaneously occur in the interaction of intense, short pulse lasers with solid targets. However in this case, things can rapidly become a lot more complicated: collisional absorption will be in the high-field regime if $v_{os} \gg v_{Te}$ (in Fig. 4.7, we took the low-field limit), the EPW driven by resonance absorption can become nonlinear (invalidating the theory presented here), and many other nonlinear processes will kick in.

Problems

4.1 Integration of the ln Term in Eq. (4.32)

Rewrite the integral in Eq. (4.32) as the sum of two integrals, respectively, proportional to $\int_0^\infty x^2 \exp(-x^2)dx$ and $\int_0^\infty x^2 \exp(-x^2)\ln(x)dx$. Knowing that the first of these two terms is a standard definite integral equal to $\sqrt{\pi}/4$ and that the second can be evaluated numerically and is approximately equal to 8×10^{-3}, show that

$$\int d^3 v f(v) \ln\left(\frac{2v_{os}}{v}\right) = \ln\left(\frac{2v_{os}}{\sqrt{2}v_{Te}}\right) + \varepsilon, \qquad (4.109)$$

where $\varepsilon < 2\%$.

4.2 The Absence of Collisional Heating from Electron–Electron Collisions

Consider the collision of two electrons in an oscillating electric field, with cycle-averaged velocities $\mathbf{v}_1$ and $\mathbf{v}_2$ before impact, i.e., total velocities $\mathbf{w}_1 = \mathbf{v}_1 + \mathbf{u}$ and $\mathbf{w}_2 = \mathbf{v}_2 + \mathbf{u}$, where $\mathbf{u} = \mathbf{v}_{os} \sin\psi$ is the quiver velocity (and ψ its phase) at the time of impact. The velocities right after the collision (assuming that the collision time is much shorter than the external field's oscillation period) are $\mathbf{w}_1' = \mathbf{v}_1' + \mathbf{u}$ and $\mathbf{w}_2' = \mathbf{v}_2' + \mathbf{u}$.

Write down the conservation of energy and momentum for the system consisting of the two electrons, and show that the cycle-averaged energy gain $\Delta\epsilon$ from an electron–electron collision is zero—and that therefore, electron–electron collisions do not lead to particle heating and collisional absorption of the external electric field.

4.3 Collisional Absorption in Mixed-Ion-Species Plasmas

Re-derive the collisional absorption coefficient in multi-ion-species plasma, and show that the final expression is the same except that Z is replaced by $Z^* = \langle Z^2 \rangle / \langle Z \rangle$. You will define the fraction of each ion species as f_j such that the ion density for that species is $n_{i,j} = f_j n_i$, where n_i is the global ion density with $n_e = \langle Z \rangle n_i$.

4.4 Low-Field vs. Strong-Field Limit for Collisional Absorption of EPWs

Express v_{os}/v_{Te}, the quiver velocity of an electron in an EPW (not to be confused with the quiver velocity in a light wave) divided by the thermal velocity, as a function of the EPW amplitude $\delta n_e/n_e$, its phase velocity $v_\phi = \omega/k$, and the thermal velocity v_{Te} (cf. Sect. 1.3). Discuss the conditions under which an EPW can drive electrons in the strong-field limit, with $v_{os} \gg v_{Te}$.

4.5 Asymptotic Expansion of Resonance Absorption for Small τ

*   Following similar steps as for the derivation of the general wave equation for an electric field in Sect. 1.3.1.1, Eq. (1.79), show that the general wave equation for the magnetic field is

$$\nabla^2 \mathbf{B} + \frac{\nabla \varepsilon}{\varepsilon} \times (\nabla \times \mathbf{B}) + \frac{\omega_0^2}{c^2} \varepsilon \mathbf{B} = 0. \qquad (4.110)$$

- Rewrite this wave equation for the magnetic field for a 1D profile and the enveloped expression for the magnetic field from Eq. (4.92) and get to the following expression:

$$\left[\partial_z^2 + \frac{\omega_0^2}{c^2} \left(\varepsilon - \sin^2(\theta) \right) - \frac{1}{\varepsilon} \frac{\partial \varepsilon}{\partial z} \partial_z \right] B_y = 0 \,. \tag{4.111}$$

- Assuming a linear density profile $n_e(z) = n_c z/L$ and using the new variables $\tau = (k_v L)^{1/3} \sin(\theta)$ and $\eta = k_v (k_v L)^{-1/3} [z - L \cos^2(\theta)]$ (like in Sect. 3.3, Eq. (3.40)), rewrite the previous equation as

$$\left[d_\eta^2 - \frac{1}{\eta - \tau^2} d_\eta - \eta \right] B_y(\eta) = 0 \,. \tag{4.112}$$

This is one of the key equations that need to be solved for the complete analysis of resonance absorption, together with an equation for E_z and another one for δn, the density perturbation from the EPW.

- Taking the limit of small incidence angle, $\tau \ll 1$, show that the previous equation simplifies to

$$\beta'' - \eta\beta = 0 \,, \tag{4.113}$$

where $B_y(\eta) = d\beta/d\eta$. This shows that in the small angle limit, B_y is proportional to the derivative of an Airy function.

- Using Ampère's law and Eq. (4.96), show that

$$B_y = -\frac{i}{\omega_0} \frac{\varepsilon}{\varepsilon - \sin^2(\theta)} \frac{dE_x}{dz} \,. \tag{4.114}$$

- Using the new variables η, τ in that last equation, show that $E_x(\eta) \propto A_i(\eta)$; following the same analysis as in Sect. 3.2 to find the constant of integration in front of A_i, show that

$$|B_y(\eta = 0)| = |B_0| 2\sqrt{\pi} A_i'(0) (k_v L)^{-1/6} \,, \tag{4.115}$$

and finally arrive at Eq. (4.102).

4.6 Alternate Derivation for the Absorption in a Linear Density Profile

Consider a linear density ramp $n_e/n_c = z/L$. Establish that the trajectory of a light wave incident at θ_0 at $z = 0$ can be expressed as $x(t) = c \sin(\theta_0) t$, $z(t) = -c^2 t^2/4L + c \cos(\theta_0) t$ (Problem 3.1). Using the temporal expression for the collisional absorption rate $\nu_c = \nu_{ei} n_e/n_c$, use a parametrization vs. time to derive the total absorption in the linear ramp. Verify that the result is the same as derived in Eq. (4.87).

References

1. F.B. Bunkin, A.E. Kazakov, M.V. Fedorov, Sov. Phys. Uspekhi **15**, 416 (1973)
2. G.J. Pert, J. Phys. A: Gener. Phys. **5**, 506 (1972)
3. P. Mulser, F. Cornolti, E. Bésuelle, R. Schneider, Phys. Rev. E **63**, 016406 (2000)
4. P. Mulser, R. Schneider, J. Phys. A: Math. Theor. **42**, 214058 (2009)
5. J. Dawson, C. Oberman, Phys. Fluids **5**, 517 (1962)
6. T.W. Johnston, J.M. Dawson, Phys. Fluids **16**, 722 (1973)
7. V. Silin, S. Uryupin, Sov. Phys. JETP **54**, 485 (1981)
8. C.D. Decker, W.B. Mori, J.M. Dawson, T. Katsouleas, Phys. Plasmas **1**, 4043 (1994)
9. V. Silin, Sov. Phys. JETP **20**, 1510 (1965)
10. S. Skupsky, Phys. Rev. A **36**, 5701 (1987)
11. J.D. Jackson, *Classical Electrodynamics*, 3rd edn. (Wiley, London, 1999)
12. R. Devriendt, O. Poujade, Phys. Plasmas **29**, 073301 (2022)
13. J.W. Banks, S. Brunner, R.L. Berger, T.M. Tran, Phys. Plasmas **23**, 032108 (2016)
14. A.B. Langdon, Phys. Rev. Lett. **44**, 575 (1980)
15. J.P. Matte, M. Lamoureux, C. Moller, R.Y. Yin, J. Delettrez, J. Virmont, T.W. Johnston, Plasma Phys. Controlled Fusion **30**, 1665 (1988)
16. D. Turnbull, A. Colaïtis, A.M. Hansen, A.L. Milder, J.P. Palastro, J. Katz, C. Dorrer, B.E. Kruschwitz, D.J. Strozzi, D.H. Froula, Nat. Phys. **16**, 181 (2020)
17. S.-M. Weng, Z.-M. Sheng, J. Zhang, Phys. Rev. E **80**, 056406 (2009)
18. N. Denisov, Sov. Phys. JETP **7**, 364 (1958)
19. J.P. Freidberg, R.W. Mitchell, R.L. Morse, L.I. Rudsinski, Phys. Rev. Lett. **28**, 795 (1972)
20. D.E. Hinkel-Lipsker, B.D. Fried, G.J. Morales, Phys. Fluids B: Plasma Phys. **4**, 559 (1992)
21. D.W. Forslund, J.M. Kindel, K. Lee, Phys. Rev. Lett. **39**, 284 (1977)
22. J.P. Palastro, J.G. Shaw, R.K. Follett, A. Colaïtis, D. Turnbull, A.V. Maximov, V.N. Goncharov, D.H. Froula, Phys. Plasmas **25**, 123104 (2018)
23. D.E. Hinkel-Lipsker, B.D. Fried, G.J. Morales, Phys. Rev. Lett. **62**, 2680 (1989)
24. W.L. Kruer, *The Physics of Laser Plasma Interactions* (Westview Press, 2003)

Chapter 5
Nonlinear Self-action Effects in Light Propagation in Plasmas

In this chapter we will investigate how plasmas respond to the presence of a ponderomotive force and will in particular describe how the electron density—and hence the refractive index—is modified due to the presence of an intense electric field with spatial non-uniformities. The non-uniformity could be associated with the natural intensity profile of a laser (e.g., a Gaussian profile near best focus) or with the beat wave pattern between overlapping waves—light waves or plasma waves or a combination of both.

This nonlinear index modification can in turn lead to "self-action" effects on a single beam propagating in a plasma, i.e., a modification of the beam propagation due to the modification of the index induced by the intensity profile. The most famous example is the self-focusing effect, which has been extensively studied in nonlinear optics; another related process we are going to present in this chapter is the phenomenon of nonlinear beam deflection in a flowing plasma. This analysis will establish the physical basis for the description of wave coupling instabilities in Chap. 6. Note that self-focusing is often presented in association with the filamentation instability, as the two processes are closely related; however, since filamentation is a wave-mixing process, we will describe it (and explain its connection to self-focusing) in a later chapter (Chap. 7), after having introduced some general concepts about wave-mixing in plasmas in Chap. 6.

5.1 The Ponderomotive Force: Fluid Approach

The ponderomotive force, introduced for a single particle in Sect. 2.3, can also be derived for a plasma under the fluid description; as we will see, the expression for the force ends up being the same, but it is useful to be aware of the similarities and differences in the derivation between the two descriptions. Like for the single particle derivation, the force will apply primarily to the plasma electrons, as the ions

© The Author(s), under exclusive license to Springer Nature Switzerland AG 2023
P. Michel, *Introduction to Laser-Plasma Interactions*, Graduate Texts in Physics,
https://doi.org/10.1007/978-3-031-23424-8_5

are too massive to respond; however, the resulting displacement of the electrons will generate an electrostatic field that can pull the ions and set them in motion as well on longer time scales.

We shall try not to confuse the Eulerian description from the fluid equations and the Lagrangian approach used in the single-particle derivation of the ponderomotive force: for the latter, $\mathbf{r}(t)$ described the position of the particle at time t, subjected to an external electric field $\mathbf{E}[\mathbf{r}(t), t]$. On the other hand, in the fluid description, $\mathbf{r}$ and t are independent variables, and the fluid velocity $\mathbf{v}(\mathbf{r}, t)$ is a *field* quantity (like density, temperature, and the other fluid quantities) describing the fluid velocity at the position and time $\mathbf{r}, t$. Therefore, the Taylor expansion method used in Sect. 2.3 is not applicable here.

We consider the case of an "external"[1] electric field $\mathbf{E}(\mathbf{r}, t)$ oscillating at the frequency ω_0. The field can be from a light wave (in which case it has an associated magnetic field $\mathbf{B}$) or a plasma wave (in which case $\mathbf{B} = 0$) or a superposition of waves overlapping in the plasma. We are only concerned with the response of the plasma electron fluid to this field, so we write the force equation, ignoring the pressure term for simplicity (cf. Eq. (1.91)), as

$$\frac{\partial \mathbf{v}_e}{\partial t} + (\mathbf{v}_e \cdot \nabla)\mathbf{v}_e = -\frac{e}{m}(\mathbf{E} + \mathbf{v}_e \times \mathbf{B}). \tag{5.1}$$

Recall that in our analysis of plasma waves in Sect. 1.3.1, we assumed that the fluid velocity consisted in a linear or "first-order" response $\mathbf{v}_e^{(1)}$ of the plasma to the electric and magnetic fields:

$$v_e^{(1)} \sim E \sim B. \tag{5.2}$$

Keeping only first-order terms in Eq. (5.1) yields the simple relation between the first-order fluid velocity and the electric field:

$$\mathbf{E} = -\frac{m}{e}\partial_t \mathbf{v}_e^{(1)}, \tag{5.3}$$

i.e., $\mathbf{v}_e^{(1)}$ is simply the quiver motion of the plasma electrons in the oscillating electric field, just like for the single particle case from Sect. 2.1. In case the wave is an EMW, its magnetic field can be related to $\mathbf{E}$ (and thus $\mathbf{v}_e^{(1)}$) via Faraday's law: applying a curl to the equation above and using Faraday's law lead to

$$\mathbf{B} = \frac{m}{e}\nabla \times \mathbf{v}_e^{(1)}. \tag{5.4}$$

---

[1] By "external," we really mean that the field is assumed to be unaffected by the plasma dynamics. A self-consistent treatment leads to nonlinear field evolution processes like self-focusing or beam bending, described in later sections of this chapter.

By limiting the expansion of the electron fluid velocity to first order, we had to eliminate the quadratic terms $(\mathbf{v}_e^{(1)} \cdot \nabla)\mathbf{v}_e^{(1)}$ and $\mathbf{v}_e^{(1)} \times \mathbf{B}$, which are both second order, $\sim E^2$. Let us now expand the fluid velocity to second order to investigate the role of these quadratic terms: we write

$$\mathbf{v}_e = \mathbf{v}_e^{(1)} + \mathbf{v}_e^{(2)}, \tag{5.5}$$

where $\mathbf{v}_e^{(2)} \sim E^2$ is a second-order perturbation. Collecting the second-order terms in Eq. (5.1) leads to

$$m\partial_t \mathbf{v}_e^{(2)} = -m(\mathbf{v}_e^{(1)} \cdot \nabla)\mathbf{v}_e^{(1)} - e\mathbf{v}_e^{(1)} \times \mathbf{B}. \tag{5.6}$$

Inserting the expression for $\mathbf{B}$ above and using the vector identity $\frac{1}{2}\nabla(\mathbf{A} \cdot \mathbf{A}) = (\mathbf{A} \cdot \nabla)\mathbf{A} + \mathbf{A} \times (\nabla \times \mathbf{A})$ finally lead to

$$m\partial_t \mathbf{v}_e^{(2)} = \mathbf{F}_p, \tag{5.7}$$

where $\mathbf{F}_p$ is the ponderomotive force defined as

$$\boxed{\mathbf{F}_p = -\frac{m}{2}\nabla(\mathbf{v}_e^{(1)})^2}. \tag{5.8}$$

Usually only the low-frequency part of the force is retained, i.e.,

$$\mathbf{F}_p = -\frac{m}{2}\nabla\left\langle(\mathbf{v}_e^{(1)})^2\right\rangle, \tag{5.9}$$

where the angular brackets denote the time average over the oscillation period $T_0 = 2\pi/\omega_0$. The force can also be expressed as a function of the electric field, which yields the same expression we already derived for the single particle case in Eq. (2.76):

$$\mathbf{F}_p = -\frac{e^2}{2m\omega_0^2}\nabla\left\langle\mathbf{E}^2\right\rangle. \tag{5.10}$$

Let us now go back and quantify the conditions of validity of our perturbation analysis. We write the electric field as

$$\mathbf{E} = \frac{1}{2}\mathbf{e}_E \tilde{E} e^{i(kz-\omega t)} + c.c., \tag{5.11}$$

where $\mathbf{e}_E$ is the polarization unit vector and $\tilde{E}$ the field's slowly varying envelope (likewise $\tilde{v}_e^{(1)}$ is the envelope of the fluid velocity perturbation); c.c. stands for complex conjugate. The assumption that first-order quantities are greater than the second-order ones requires that $|F_p| \ll |e\tilde{E}|$, where F_p is the slowly varying ponderomotive force envelope. Expressing F_p and $\tilde{E}$ in terms of $\tilde{v}_e^{(1)}$, we obtain

$$|\nabla(\tilde{v}_e^{(1)})^2| \ll |\partial_t \tilde{v}_e^{(1)}|$$

$$\Leftrightarrow |\tilde{v}_e^{(1)}||\nabla|\tilde{v}_e^{(1)}| \ll \omega_0|\tilde{v}_e^{(1)}|$$

$$\Leftrightarrow \frac{|\tilde{v}_e^{(1)}|}{\omega_0} \ll \frac{|\tilde{v}_e^{(1)}|}{\nabla|\tilde{v}_e^{(1)}|} = \frac{|\tilde{E}|}{\nabla|\tilde{E}|} . \tag{5.12}$$

Defining a characteristic scale-length of the spatial variations of the electric field amplitude $L_E = |\tilde{E}|/\nabla|\tilde{E}|$ and defining $r_{os} = \int |\tilde{v}_e^{(1)}(t)|dt$ as the spatial excursion of the electron fluid under the quiver motion in the electric field, the last relation is then equivalent to

$$r_{os} \ll L_E . \tag{5.13}$$

In other words, our separation between "first-order" and "second-order" quantities is valid as long as the spatial excursion of the electron fluid quivering in the electric field remains much smaller than the scale of the spatial variations of the electric field amplitude. This is exactly the same condition we used in our treatment of the ponderomotive force on a single particle (cf. Sect. 2.3). This also means that the ponderomotive force is intrinsically low frequency since the assumption of a small excursion compared to the electric field envelope scale-length implies that the time-average position over an oscillation period must be a small quantity. In particular, if $\mathbf{E}$ is purely monochromatic, $\propto e^{-i\omega_0 t}$, then all the first-order quantities average to zero over an oscillation period, $\langle \mathbf{E} \rangle = \langle \mathbf{B} \rangle = \left\langle \mathbf{v}_e^{(1)} \right\rangle = 0$, and keeping only the second-order components of the force equation is equivalent to taking a time-average over an oscillation period.

Let us go back to the full expression for the force equation on the electron fluid in the presence of the ponderomotive force from an external field. We obtain the following relation for the second-order (i.e., low-frequency) fluid velocity $\mathbf{v}_e^{(2)}$, after inserting back the pressure force:

$$\frac{\partial \mathbf{v}_e^{(2)}}{\partial t} = \frac{1}{m}\mathbf{F}_p - \frac{e}{m}\mathbf{E}^{(2)} - \gamma_e v_{Te}^2 \left(\frac{\nabla n_e^{(2)}}{n_{e0}} - \frac{n_e^{(1)}\nabla n_e^{(1)}}{n_{e0}^2} \right) , \tag{5.14}$$

where $\mathbf{E}^{(2)}$ is the remaining "internal" (second-order, low-frequency, electrostatic) electric field resulting from the charge displacement due to the ponderomotive force. We used the second-order development of the pressure term, with $n_e = n_{e0} + n_e^{(1)} + n_e^{(2)}$ where the background density n_{e0} is assumed uniform, which yields

$$\frac{\nabla n_e}{n_e} = \frac{\nabla(n_{e0} + n_e^{(1)} + n_e^{(2)})}{n_{e0} + n_e^{(1)} + n_e^{(2)}} \approx \frac{\nabla n_e^{(1)}}{n_{e0}} + \frac{\nabla n_e^{(2)}}{n_{e0}} - \frac{n_e^{(1)}\nabla n_e^{(1)}}{n_{e0}^2} . \tag{5.15}$$

It is convenient to introduce the "ponderomotive potential" Φ_p such that $\mathbf{F}_p = e\nabla\Phi_p$, i.e.,

$$\Phi_p = -\frac{e}{2m\omega_0^2}\langle\mathbf{E}^2\rangle.$$
(5.16)

The low-frequency force equation can then be recast as

$$\frac{\partial\mathbf{v}_e^{(2)}}{\partial t} = \frac{e}{m}\nabla(\Phi_p + \Phi^{(2)}) - \gamma_e v_{Te}^2\left(\frac{\nabla n_e^{(2)}}{n_{e0}} - \frac{n_e^{(1)}\nabla n_e^{(1)}}{n_{e0}^2}\right),$$
(5.17)

where $\Phi^{(2)}$ is the "internal" self-consistent potential, $\mathbf{E}^{(2)} = -\nabla\Phi^{(2)}$.

Note that a general derivation of the first two fluid equations (continuity and force) to second order for arbitrary species will be presented in detail in Sect. 6.3.

As a final note, the ponderomotive force on the ion fluid can be derived in a similar manner and remains much smaller than the force on the electrons due to the ions' higher inertia:

$$\mathbf{F}_{pi} = Ze\mathbf{v}_i^{(1)}\times\mathbf{B} - M_i(\mathbf{v}_i^{(1)}\cdot\nabla)\mathbf{v}_i^{(1)}$$
(5.18)

$$= -\frac{M_i}{2}\nabla(\mathbf{v}_i^{(1)})^2$$
(5.19)

$$= -\frac{Z^2e^2}{M_i\omega_0^2}\nabla\mathbf{E}^2$$
(5.20)

$$= Z^2\frac{m}{M_i}\mathbf{F}_{pe} \ll \mathbf{F}_{pe}.$$
(5.21)

In the following sections, we are going to investigate the plasma response to the ponderomotive force from an external source, such as a tightly focused laser beam or the beat wave intensity pattern from two overlapped waves. We will then turn our attention to the case of nonlinear laser propagation and the self-consistent coupling between the propagation in the plasma and the plasma response, i.e., the self-action effects, among which are self-focusing and beam deflection.

5.2   Linear Plasma Response to a Laser's Ponderomotive Force: Transient Stage

Now that we have established the expression of the ponderomotive force in a plasma, we are going to investigate how the plasma responds to this force. In this section we assume that the ponderomotive force is not affected by the plasma response.

Two different time scales will be considered: the fast "electron" time scale where ions are assumed immobile (i.e., the time scale of EPWs) and the slower "ion" time scale (e.g., IAWs) corresponding to the ion motion. On the EPW time scale, the electron fluid response is described by a driven EPW wave equation derived similarly to the linear EPW from Sect. 1.3.2.

We expand the electron density and velocity in a perturbation series, $n_e = n_{e0} + n_e^{(1)} + n_e^{(2)}$ and $\mathbf{v}_e = 0 + \mathbf{v}_e^{(1)} + \mathbf{v}_e^{(2)}$ (i.e., no background plasma flow), where, like in the previous section, the first-order terms correspond to the linear response of the plasma to the electric field, and the second order will describe the plasma response to the ponderomotive force.

We assume that the electric field is from an EMW (the laser): therefore, we have $\mathbf{E} = -(m/e)\partial_t \mathbf{v}_e^{(1)}$ and $n_e^{(1)} = 0$, i.e., the linear response of the plasma to the laser is a simple quiver motion at $\mathbf{v}_e^{(1)}$ and has no associated density modulation, as can be seen from the continuity equation (cf. Sect. 1.3.1.1). The continuity equation, $\partial_t n_e + \nabla \cdot (n_e \mathbf{v}_e) = 0$, yields at the second order:

$$\frac{\partial n_e^{(2)}}{\partial t} + n_{e0}\nabla \cdot \mathbf{v}_e^{(2)} = 0. \tag{5.22}$$

Proceeding like in Sect. 1.3.2, i.e., taking the time derivative of Eq. (5.22), the divergence of Eq. (5.17) and using Poisson's equation $\nabla^2 \Phi^{(2)} = en_e^{(2)}/\varepsilon_0$ — or in normalized units for the potentials, $\nabla^2 \varphi^{(2)} = (\omega_{pe0}^2/c^2)n_e^{(2)}/n_{e0}$, where $\omega_{pe0}^2 = n_{e0}e^2/(m\varepsilon_0)$ and $\varphi = (e/mc^2)\Phi$—to connect the density perturbation to the electrostatic potential finally gives the following wave equation for EPWs with a ponderomotive source term:

$$\boxed{(\partial_t^2 + \omega_{pe0}^2 - 3v_{Te}^2\nabla^2)\frac{n_e^{(2)}}{n_{e0}} = -c^2\nabla^2\varphi_p}. \tag{5.23}$$

This equation describes the response of the electron fluid to the ponderomotive force on short time scales, when ions have not started to move yet.

On longer time scales, which are more relevant to the ICF and HED conditions that we are primarily concerned about, the motion of electrons sets up a charge separation which can drag the ions. In this case, we consider again the electron fluid in the presence of a ponderomotive force as described in Eq. 5.17; we assume an isothermal equation of state for the electrons, $\gamma_e = 1$, and neglect the electrons inertia (i.e., we set $\partial_t(m\mathbf{v}_e^{(2)}) \approx 0$) since we are investigating ion motion (cf. Chap. 1.3.1.1, and Problem 1.3). We get

$$\frac{e}{m}\nabla(\Phi_p + \Phi^{(2)}) = v_{Te}^2\frac{\nabla n_e^{(2)}}{n_{e0}}. \tag{5.24}$$

Next, we write the continuity and force equations for the ions; here the ponderomotive force does not appear explicitly but will be experienced by the ions indirectly, via the electrostatic potential setup by the electrons' displacement due to the ponderomotive force. We have, using an adiabatic equation of state for the ions,

$$\frac{\partial n_i}{\partial t} + \nabla \cdot (n_i \mathbf{v}_i) = 0, \tag{5.25}$$

$$\frac{\partial \mathbf{v}_i}{\partial t} + (\mathbf{v}_i \cdot \nabla)\mathbf{v}_i = \frac{Ze}{M_i}(\mathbf{E} + \mathbf{v}_i \times \mathbf{B}) - 3v_{Ti}^2 \frac{\nabla n_i}{n_i}. \tag{5.26}$$

Like for the electrons, we expand the ion density and velocity in a perturbation series, $n_i = n_{i0} + n_i^{(1)} + n_i^{(2)}$ and $\mathbf{v}_i = 0 + \mathbf{v}_i^{(1)} + \mathbf{v}_i^{(2)}$ ($n_{i0} = n_{e0}/Z$ is the uniform background density, and we assume no background flow velocity). Again, like for the electrons, since the electric field is from an EMW, there is no first-order ion density modulation, $n_i^{(1)} = 0$. The second-order fluid equations are in this case

$$\frac{\partial n_i^{(2)}}{\partial t} + n_{i0}\nabla \cdot \mathbf{v}_i^{(2)} = 0, \tag{5.27}$$

$$\frac{\partial \mathbf{v}_i^{(2)}}{\partial t} + (\mathbf{v}_i^{(1)} \cdot \nabla)\mathbf{v}_i^{(1)} = \frac{-Ze}{M_i}\nabla \Phi^{(2)} + \frac{Ze}{M_i}\mathbf{v}_i^{(1)} \times \mathbf{B} - 3v_{Ti}^2 \frac{\nabla n_i^{(2)}}{n_{i0}}. \tag{5.28}$$

Combining the time derivative of the continuity equation and the divergence of the force equation and eliminating $\Phi^{(2)} = (mc^2/e)\psi^{(2)}$ using Eq. (5.24) finally give the IAW wave equation with a ponderomotive source term:

$$(\partial_t^2 - c_s^2\nabla^2)\frac{n_e^{(2)}}{n_{e0}} = -\frac{Zm}{M_i}c^2\nabla^2\varphi_p, \tag{5.29}$$

with $c_s^2 = (ZT_e + 3T_i)/M_i$ and $\varphi_p = -\langle a^2 \rangle/2$, where $\mathbf{a}$ is the laser vector potential. We neglected the ponderomotive force on the ions as justified in the previous section (cf. Eq. (5.18)).

Since the first-order density perturbation in this case is zero, we are going to simplify our notations throughout the rest of this section and use

$$\delta n_e = n_e^{(2)}. \tag{5.30}$$

It should always be obvious from the context that the density perturbation is second order with respect to the electric field.

The driven IAW equation from a laser ponderomotive force takes the form:

$$\boxed{(\partial_t^2 - c_s^2\nabla^2)\frac{\delta n_e}{n_{e0}} = \frac{Zm}{2M_i}c^2\nabla^2 \langle \mathbf{a}^2 \rangle}. \tag{5.31}$$

Ignoring the time derivative in Eq. (5.31) gives the steady-state expression for the density perturbation:[2]

$$\boxed{\frac{\delta n_e(\mathbf{r}, t \to \infty)}{n_{e0}} = \frac{Z}{ZT_e + 3T_i} mc^2 \varphi_p(\mathbf{r}) = -\frac{Zm}{2M_i}\frac{c^2}{c_s^2}\left\langle \mathbf{a}(\mathbf{r})^2 \right\rangle.} \tag{5.32}$$

In the limit where $ZT_e \gg T_i$ and for a laser with peak amplitude a_0, $\mathbf{a} = \frac{1}{2}\mathbf{a}_0 \exp[i(k_0 z - \omega_0 t)] + c.c.$, we have

$$\frac{\delta n_e(\mathbf{r}, t \to \infty)}{n_{e0}} \approx -\frac{a_0^2(\mathbf{r})}{4v_{Te}^2}. \tag{5.33}$$

To illustrate the plasma response to a laser ponderomotive force, let us consider the simple case of a laser pulse with a finite profile in space in the transverse direction, in 1D along x (the laser is assumed uniform and infinite along its propagation direction z and the other transverse dimension y). This would correspond to the ponderomotive force from a tightly focused laser beam, perpendicular to its propagation axis. We first assume that the laser turns on very rapidly compared to the plasma response time and approximate the temporal envelope by a step function. The potential vector takes the simple form:

$$\mathbf{a}(\mathbf{r}, t) = \mathbf{a}(x)H(t)\sin(k_0 z - \omega_0 t), \tag{5.34}$$

where $H(t)$ is the Heaviside step function and $|\mathbf{a}(x)| = e|\mathbf{E}(x)|/(mc\omega_0)$ is the laser vector potential with a finite spatial profile along x (e.g., Gaussian, as illustrated in Fig. 5.1).

Using the expression for $\mathbf{a}$ into Eq. (5.31), we obtain

$$(\partial_t^2 - c_s^2 \nabla^2)\frac{\delta n_e}{n_{e0}} = \frac{c_s^2 c^2}{4v_{Te}^2}H(t)\frac{\partial^2 a^2(x)}{\partial x^2}, \tag{5.35}$$

where we have assumed that $T_i \ll T_e$ so that $c_s^2 \approx ZT_e/M_i$. Performing a Laplace transform in time (with variable s) and a Fourier transform in space (with variable k), we obtain

---

[2] In the steady-state limit, the adiabatic equation of state for the ions may not be appropriate anymore, depending on the time scale under consideration. Indeed, the adiabatic approximation used in the ions equation of motion essentially means that in IAWs, where $v_{Ti} \ll c_s$, the ion temperature cannot equilibrate as the wave propagates since the thermal ions are much slower than the wave. Whereas for stationary perturbations, once a steady-state has been reached the ion temperature will eventually equilibrate (typically on time scales on the order of w/v_{Ti}, where w is the laser spot size), and an isothermal equation of state is required. The resulting correction to the density perturbation is typically small, i.e., $3T_i$ should be replaced by T_i..

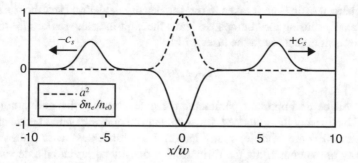

Fig. 5.1 Plasma response to the ponderomotive force from a laser in 1D, with instantaneous rise time (arbitrary units), at $t = 6w/c_s$, where w is the width of the (Gaussian) intensity profile along x

$$(s^2 + k^2 c_s^2)\frac{\delta \hat{n}_e(s, k)}{n_{e0}} = -\frac{c_s^2 c^2}{4v_{Te}^2}k^2\frac{\widehat{a^2}(k)}{s},$$ (5.36)

where the hat denotes a Laplace and/or Fourier transform. Performing a partial fraction decomposition leads to

$$\frac{\delta \hat{n}_e(s, k)}{n_{e0}} = -\frac{c^2\widehat{a^2}(k)}{4v_{Te}^2}\left[\frac{1}{s} - \frac{1}{2(s + ikc_s)} - \frac{1}{2(s - ikc_s)}\right].$$ (5.37)

Taking the inverse Laplace transform and inverse Fourier transform immediately gives the solution:

$$\frac{\delta n_e(x, t)}{n_{e0}} = \frac{c^2}{4v_{Te}^2}H(t)\left[-a^2(x) + \frac{1}{2}a^2(x - c_s t) + \frac{1}{2}a^2(x + c_s t)\right].$$ (5.38)

This shows that when a laser pulse turns on rapidly on a time scale compared to the ion acoustic transit time through the spot size, the plasma response has the form of a density depression with the same exact profile as the laser intensity profile (but with opposite sign) and two "bumps" propagating away from the laser at the sound speed; each bump also has the same profile as the laser pulse and is half the height of the density depression, ensuring mass conservation. The depression (the first term in the bracket) corresponds to the steady-state solution after the bumps have propagated to infinity, as can be immediately seen by taking $\partial_t = 0$ in Eq. (5.35). The solution is illustrated in Fig. 5.1.

From Eq. (5.38), we can also extract the time it takes to "dig" the density depression at $x = 0$. Assuming a laser with Gaussian spatial profile $a^2(x) = a_0^2 \exp[-x^2/w^2]$, we can express the density at $x = 0$ and for $t > 0$ as

$$\frac{\delta n_e(0, t)}{n_{e0}} = -\frac{a_0^2 c^2}{4v_{Te}^2}\left[1 - \exp\left(-(c_s t/w)^2\right)\right].$$ (5.39)

The characteristic time it takes to dig the density depletion (i.e., the time for the excess density causing the "bumps" to leave the high-intensity region) is simply the acoustic transit time through the laser width,

$$\tau_{ac} = w/c_s . \tag{5.40}$$

This will be an important parameter for understanding the mitigating role of temporal beam smoothing; indeed, focal spots from optically smoothed beams used in ICF and HED are made of many uncorrelated "speckles" with finite lifetimes which may or may not allow the formation of density perturbations, depending on the ratio of lifetime to acoustic transit time. This will be discussed in Sect. 9.5.1.

It is also interesting to consider the effect of a finite rise time for the laser intensity, which is the subject of Problem 5.1. In this problem, we investigate a rise time of the form $(1 - \exp[-t/\tau])H(t)$ instead of simply $H(t)$ as above. It can then be shown that the expression for the density perturbation is

$$
\frac{\delta n_e(x, t)}{n_{e0}} = \frac{c^2}{4v_{Te}^2} H(t) \Big[(1 - e^{-t/\tau}) a^2(x) - \frac{1}{2\tau c_s} e^{-t/\tau} e^{-|x|/\tau c_s} * a^2(x)
$$

$$
+ \frac{1}{2\tau c_s} \left(e^{x/\tau c_s} H(-x) \right) * a^2(x - c_s t) + \frac{1}{2\tau c_s} \left(e^{-x/\tau c_s} H(x) \right) * a^2(x + c_s t) \Big],
$$

$$\tag{5.41}$$

where $*$ denotes a spatial convolution. The critical parameter is the ratio between the intensity rise time τ and the acoustic transit time across the laser spot size w/c_s. Figure 5.2 shows the density profile from Eq. (5.41) calculated again with a Gaussian laser profile and varying the ratio $w/\tau c_s$.

For $w/\tau c_s \gg 1$, i.e., a fast (small) laser rise time compared to the acoustic transit time, the solution looks like the one obtained before for a step function in Fig. 5.1. In the opposite limit, when the rise time is much larger than the acoustic transit time, the mass is expelled adiabatically from the high-intensity region as the intensity slowly rises, and the density bumps become very small.

We should also note that plasma wave damping will damp the bumps on a time scale $\sim 1/\nu_2$, where ν_2 is the damping rate of the IAW—but it will not affect the amplitude of the density depression in steady-state.

5.3   Nonlinear Plasma Response to a Laser Ponderomotive Force in Steady-State

In this section we will investigate the nonlinear plasma response to a ponderomotive driver in steady-state, i.e., at long time scales, which is often the most relevant situation in ICF or HED conditions where the laser pulse durations (typically several

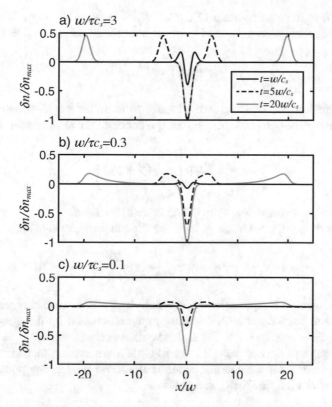

Fig. 5.2 Plasma response to the ponderomotive force from a laser in 1D, with a laser intensity profile time dependence $\propto (1 - \exp[-t/\tau])H(t)$

nanoseconds) are much longer than the typical time scales of the transient regime of the plasma response (such as the acoustic transit time discussed in the previous section). The nonlinear expression, valid for arbitrary density perturbations (and not just $|\delta n_e/n_{e0}| \ll 1$ like in the previous section), will be needed to describe the self-focusing process later in the chapter.

For the electrons, we start from the fluid equation for the momentum:

$$\frac{\partial \mathbf{v}_e}{\partial t} + (\mathbf{v}_e \cdot \nabla)\mathbf{v}_e = -\gamma_e v_{Te}^2 \frac{\nabla n_e}{n_e} - \frac{e}{m}(\mathbf{E} + \mathbf{v}_e \times \mathbf{B}). \qquad (5.42)$$

We decompose the electric field into a transverse component associated with the laser (with vector potential $\mathbf{A}$) and a longitudinal electrostatic potential Φ corresponding to the restoring force in response to the electrons' displacement under the ponderomotive force: $\mathbf{E} = -\partial_t \mathbf{A} - \nabla\Phi$. We are interested in the steady-state behavior, including ion motion, so we take $\partial_t = 0$ (which effectively removes the fast-oscillating and transverse component of this equation, $\partial_t \mathbf{v}_e = c\partial_t \mathbf{a}$ with $\mathbf{a} = e\mathbf{A}/mc$) and make an isothermal electron assumption ($\gamma_e = 1$). Identifying the ponderomotive force (or potential) $\mathbf{F}_p = e\nabla\Phi_p = -e\mathbf{v}_e \times \mathbf{B} - m(\mathbf{v}_e \cdot \nabla)\mathbf{v}_e$ leads to the following relation, expressed in terms of normalized potentials $\varphi = (e/mc^2)\Phi$:

$$c^2\nabla(\varphi + \varphi_p) = v_{Te}^2 \frac{\nabla n_e}{n_e} . \tag{5.43}$$

This equation can be integrated, leading to a Boltzmann distribution for the electrons under the potential energy from the ponderomotive and restoring forces:

$$n_e = n_{e0} \exp\left[\frac{c^2}{v_{Te}^2}(\varphi + \varphi_p)\right] , \tag{5.44}$$

with n_{e0} the background density in the absence of the laser.

Now for the ions, we start again from the fluid momentum equation:

$$\frac{\partial \mathbf{v}_i}{\partial t} + (\mathbf{v}_i \cdot \nabla)\mathbf{v}_i = -\gamma_i v_{Ti}^2 \frac{\nabla n_i}{n_i} + \frac{Ze}{M_i}(\mathbf{E} + \mathbf{v}_i \times \mathbf{B}) . \tag{5.45}$$

We can again decompose the electric field into a transverse component (from the laser) and the longitudinal, restoring potential created by the electrons' displacement, $\mathbf{E} = -\partial_t \mathbf{A} - \nabla\Phi$. For a steady-state analysis, we set $\partial_t = 0$ and use an isothermal equation of state for the ions (i.e., we assume that the time scale considered allows for ion thermal equilibration). We neglect the ponderomotive force on the ions (cf. Eq. (5.18)) and obtain

$$Zmc^2\nabla\varphi = -T_i \frac{\nabla n_i}{n_i} , \tag{5.46}$$

which can be integrated to yield

$$n_i = n_{i0} \exp\left[-\frac{Zmc^2}{T_i}\varphi\right] , \tag{5.47}$$

where $n_{i0} = n_{e0}/Z$ is the background ion density in the absence of the ponderomotive and restoring potentials.

If one assumes quasi-neutrality ($n_e = Zn_i$), which is valid for spatial scales greater than the Debye length (i.e., $k\lambda_{De} \ll 1$, cf. Sect. 1.3.1.5), it is possible to derive the nonlinear solution for the density perturbation [1]; this will be useful to study the self-focusing process in a later section, where the full nonlinear expression for the density perturbation allows the electron density to smoothly drop to zero when the ponderomotive force becomes too strong. Assuming quasi-neutrality together with Eqs. (5.44), (5.47) leads to the expression of the electrostatic potential as a function of the ponderomotive driver:

$$\varphi = -\frac{\varphi_p}{1 + ZT_e/T_i} . \tag{5.48}$$

In the limit of $ZT_e/T_i \gg 1$, we have $|\varphi^{(2)}| \ll |\varphi_p|$: this illustrates that the ions have moved to join the electrons in the regions of low intensity, thus (almost) eliminating the electrostatic potential that was initially set up by the electrons' displacement under the ponderomotive potential.

Substituting this expression in Eq. (5.44) leads to the expression for the stationary density perturbation resulting from the ponderomotive potential:

$$n_e = n_{e0} \exp\left[\frac{Z}{ZT_e + T_i} mc^2 \varphi_p\right]. \tag{5.49}$$

In the linear limit where the term in brackets is much smaller than one (i.e., small density perturbations), we have

$$n_e \approx n_{e0}\left[1 + \frac{Z}{ZT_e + T_i} mc^2 \varphi_p\right] = n_{e0} + \delta n_e, \tag{5.50}$$

which is identical to Eq. (5.32) derived in the previous section except for the numerical factor in front of T_i, which was 3 in the previous section (adiabatic ion fluid) vs. 1 here (isothermal ions), cf. footnote 2 in the previous section.

To illustrate the nonlinear density depression from the ponderomotive force, let us consider the simple case of a tightly focused laser beam at best focus, with a vector potential expressed in cylindrical coordinates as $\mathbf{a}(z, r, t) = a_0 \mathbf{e}_x \exp[-r^2/w^2] \cos(k_0 z - \omega_0 t)$. The ponderomotive potential is then $\varphi_p = -\langle \mathbf{a}^2\rangle/2 = -\frac{1}{4}a_0^2 \exp[-2r^2/w^2]$, and in the limit of $ZT_e \gg T_i$, we have

$$n_e \approx n_{e0} \exp\left[-\frac{a_0^2 c^2}{4 v_{Te}^2} \exp(-2r^2/w^2)\right]. \tag{5.51}$$

The resulting density depression at equilibrium is shown in Fig. 5.3: when $a_0 c/v_{Te} \gg 1$, the depression completely depletes the electron density. Note that the mass is not conserved here, since the density is everywhere smaller than the initial background density: this is because the excess particles that have been removed from the depression have propagated outward toward infinity in the form of a wave traveling at the sound speed, as we saw in Sect. 5.2 and Fig. 5.1.

5.4 Plasma Response to a Ponderomotive Driver: Linear Kinetic Model

Let us now derive the plasma response to an external ponderomotive force using a linear kinetic model. This model describes the plasma response to a stationary driver (i.e., $\omega = 0$, like in Sect. 5.3 in the linear limit) or a driver near an EPW

Fig. 5.3 Density depression at equilibrium for a Gaussian ponderomotive potential $\varphi_p = -\frac{1}{4}a_0^2 \exp[-2r^2/w^2]$, for three values of $a_0^2 c^2/4v_{Te}^2 = 0.1, 1$, and 10

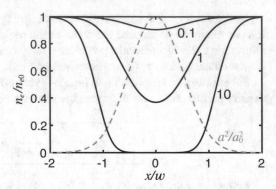

or IAW resonance or anything in between, i.e., off-resonance. It also lends itself to the description of non-Maxwellian plasmas and to multi-ion species and will be particularly useful for the description of crossed-beam energy transfer (CBET, cf. Sect. 7.2).

Like in Sect. 1.3.3, we start from the Vlasov equation for the particle species s ($\in \{i, e\}$ for ions or electrons):

$$\left(\partial_t + \mathbf{v} \cdot \nabla + \frac{\mathbf{F}_s}{m_s} \cdot \partial_{\mathbf{v}}\right) f_s(\mathbf{r}, \mathbf{v}, t) = 0, \tag{5.52}$$

where $\mathbf{F}_s$ is the force exerted on the particle species s. Like in the previous sections, we assume that two forces are at play: an externally driven ponderomotive force with potential Φ_p acting only on the electrons and the self-consistent electrostatic force with potential Φ resulting from the charge separation acting on both the electrons and the ions:

$$\mathbf{F}_e = e\nabla(\Phi_p + \Phi), \tag{5.53}$$

$$\mathbf{F}_i = -Ze\nabla\Phi. \tag{5.54}$$

Like in the previous sections, we have removed the high-frequency electron quiver in the external electric field (or, equivalently, we keep only the longitudinal component of the equations and ignore the transverse component, which describes the quiver motion).

Proceeding like in Sect. 1.3.3, we linearize the distribution function into a background distribution which depends only on velocity and a small perturbation, $f_s(\mathbf{v}, \mathbf{r}, t) = f_{s0}(\mathbf{v}) + \delta f_s(\mathbf{v}, \mathbf{r}, t)$. Assuming that the ponderomotive driver and therefore the other quantities of interest (Φ, δf_s, etc.) are plane waves oscillating at the phase $\exp[i\psi]$ with $\psi = \mathbf{k} \cdot \mathbf{r} - \omega t$, e.g.,

$$\Phi_p = \frac{1}{2}\tilde{\Phi}_p e^{i\psi} + c.c., \tag{5.55}$$

etc. (the tilde denotes envelopes), leads to the expression for the density perturbation:

$$\frac{\delta\tilde{n}_s}{n_{s0}} = \frac{1}{n_{s0}} \int \delta\tilde{f}_s(\mathbf{v}, \mathbf{r}, t) d^3v = -\frac{i\tilde{\mathbf{F}}_s}{n_{s0}m_s} \cdot \int \frac{\partial f_{s0}/\partial \mathbf{v}}{\omega - \mathbf{k} \cdot \mathbf{v}} d^3v . \qquad (5.56)$$

Inserting the susceptibility for the particle species s, defined from Eq. (1.160) as

$$\chi_s = \frac{\omega_{ps}^2}{n_{s0}k^2} \int \frac{\mathbf{k} \cdot \partial f_{s0}/\partial \mathbf{v}}{\omega - \mathbf{k} \cdot \mathbf{v}} d^3v , \qquad (5.57)$$

leads to

$$\frac{\delta\tilde{n}_s}{n_{s0}} = -i \frac{\mathbf{k} \cdot \tilde{\mathbf{F}}_s}{m_s \omega_{ps}^2} \chi_s . \qquad (5.58)$$

Remember that when the background distribution function is Maxwellian, the susceptibility can be expressed using the plasma dispersion function Z [2], cf. Eq. (1.162).

Using the expressions for the forces exerted on both species from Eqs. (5.53), (5.54) leads to

$$\frac{\delta\tilde{n}_s}{n_{s0}} = -\frac{q_s k^2}{m_s \omega_{ps}^2} \chi_s (\tilde{\Phi} + \delta_{es}\tilde{\Phi}_p) , \qquad (5.59)$$

where δ_{es} is the Kronecker delta.

Finally, Poisson's equation connects the potential and the charge separation via $k^2\tilde{\Phi} = \Sigma_s q_s \delta\tilde{n}_s/\varepsilon_0$, leading to the kinetic dispersion relation of the driven plasma under the ponderomotive force:

$$\boxed{(1 + \chi_e + \chi_i)\tilde{\Phi} = -\chi_e\tilde{\Phi}_p} \qquad (5.60)$$

and the expression for the density perturbation:

$$\boxed{\frac{\delta\tilde{n}_e}{n_{e0}} = \frac{k^2 c^2}{\omega_{pe0}^2} \frac{\chi_e(1 + \chi_i)}{1 + \chi_e + \chi_i} \tilde{\varphi}_p} , \qquad (5.61)$$

where $\varphi_p = (e/mc^2)\Phi_p$ is the normalized ponderomotive potential and $\omega_{pe0}^2 = n_{e0}e^2/(m\varepsilon_0)$.

Note that when the driver's frequency and wave-number are near an EPW or IAW resonance, the relations above lead back to the wave equations for EPWs and IAWs with a ponderomotive source term obtained in Sect. 5.2 (cf. Problem 5.2). Likewise, taking the limit of a stationary ponderomotive driver (i.e., $\omega = 0$) and using the appropriate asymptotic limits for the susceptibilities lead to the fluid expressions for the density perturbation and electrostatic potential derived in Sect. 5.3 (cf. Problem 5.3).

5.5   The Nonlinear Refractive Index of Plasmas

The electron density perturbation resulting from a laser ponderomotive force leads to an associated refractive index perturbation. Decomposing the electron density and index into a uniform "background" value and a small perturbation, $n_e = n_{e0} + \delta n_e$ and $n = n_0 + \delta n$, using the expression for the index $n = \sqrt{1 - n_e/n_c}$, we get

$$\delta n = -\frac{1}{2}\frac{\delta n_e}{n_0 n_c} = -\frac{1}{2}\frac{\delta n_e}{n_{e0}}\frac{n_{e0}}{n_c}\frac{1}{n_0}. \tag{5.62}$$

On long time scales and in the limit where $ZT_e \gg T_i$, the density perturbation $\delta n_e \ll n_{e0}$ due to the ponderomotive force is given by Eq. (5.33):

$$\frac{\delta n_e}{n_{e0}} = -\frac{1}{2}\frac{c^2}{v_{Te}^2}\left\langle \mathbf{a}^2 \right\rangle, \tag{5.63}$$

with $\mathbf{a}$ the normalized vector potential such that $\mathbf{E} = -(mc/e)\partial_t \mathbf{a}$.

Inserting into Eq. (5.62) leads to the expression for the refractive index perturbation δn due to the laser ponderomotive force (with $k_0 = n_0 \omega_0/c$):

$$\boxed{\frac{\delta n}{n_0} = \frac{\left\langle \mathbf{a}^2 \right\rangle}{4(k_0 \lambda_{De})^2}}. \tag{5.64}$$

To facilitate the connection with the well-established results from nonlinear optics, it will be convenient to introduce the n_2 parameter, defined as

$$n = n_0 + n_2 I_0, \tag{5.65}$$

where $I_0 = \varepsilon_0 c n_0 \left\langle \mathbf{E}^2 \right\rangle$ is the laser intensity. Using $a_0 = e E_0/(mc\omega_0)$, where a_0 and E_0 denote the peak oscillation amplitudes of $|\mathbf{a}|$ and $|\mathbf{E}|$, we get

$$\boxed{n_2 = \frac{1}{4}\frac{1}{(k_0 \lambda_{De})^2}\frac{1}{n_c mc^3}}, \tag{5.66}$$

where $n_c = m\varepsilon_0 \omega_0^2/e^2$ is the critical density for the frequency ω_0.

Note that the ponderomotive force is not the only physical effect leading to a Kerr effect (i.e., a variation in refractive index proportional to the intensity) in a plasma: the process may also occur due to relativistic effects when a_0 becomes comparable to or larger than one [3–5] (via the relativistic electron mass increase, which reduces $\omega_{pe} \propto 1/\sqrt{m_e}$) or thermal effects when heat transport is responsible for the density depression (cf. Sect. 7.3.7). The analysis that follows can be carried out using the alternative expressions for n_2 from relativistic or thermal effects.

5.6 Self-Focusing

In the previous sections we established that the ponderomotive force from a laser pushes the plasma electrons (and the ions as well, on a longer time scale) out of the regions of high-intensity; as a result, the electron density locally drops, i.e., the refractive index increases. For the simple case of a single laser beam propagating in a plasma, the density perturbation takes the same shape as the transverse intensity profile of the beam once a steady-state equilibrium has been reached (cf. Sect. 5.2). For a simple transverse laser intensity profile (e.g., Gaussian), the index perturbation acts like a focusing lens and leads to self-focusing of the beam. Light self-focusing in nonlinear media has been extensively investigated for more than half a century (for an early review on the subject, see [6]). Here we first describe the process from the point of view of geometric optics and will then provide a more detailed description using wave optics.

5.6.1 Geometric Optics Description

The simplest description of the process relies on a geometric optics argument [6, 7]. While very non-rigorous, it provides a good metric for estimating a threshold for self-focusing in a laser in a plasma or any other nonlinear medium with a "Kerr" nonlinearity of the type $\delta n = n_2 I_0$, where n_2 depends on the medium conditions and I_0 is the laser intensity.

For this simple description, we will use Snell's law with the rather crude representation of a Gaussian beam as a simple flat-top cylinder of diameter $d = 2w_0$, propagating toward z (Fig. 5.4). The plasma refractive index is simply described as being equal to n_0 outside the beam envelope and $n_0 + \delta n$ inside the beam envelope.

The beam contains a finite range of transverse spatial frequencies, corresponding to a finite divergence due to the diffraction limit for realistic beams. For a Gaussian

Fig. 5.4 Derivation of the self-focusing power threshold using Snell's law

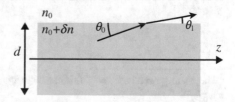

beam, the divergence is $\theta_G = \lambda_0/(\pi n_0 w_0)$, where w_0 is twice the Gaussian waist such that the intensity $I_0(r) \propto \exp[-2r^2/w_0^2]$ at best focus.[3] This means that at any position z along the beam propagation, the light contains a finite range of divergence angles up to $\sim \theta_G$ (i.e., the width of the Fourier transform of the Gaussian beam).

Snell's law implies that a "ray" inside the beam envelope will be refracted as it leaves the envelope following

$$(n_0 + \delta n) \cos(\theta_0) = n_0 \cos(\theta_1), \tag{5.67}$$

with the conventions from Fig. 5.4. Since $\delta n > 0$ (see previous section), the rays at grazing incidence will not be able to leave the envelope due to the index difference (equivalent to the total internal reflection). The self-focusing threshold is defined as the conditions under which the entire beam up to the natural divergence angle θ_G undergoes the total internal reflection, i.e.,

$$(n_0 + \delta n_c) \cos(\theta_G) = n_0, \tag{5.68}$$

where δn_c is the critical index perturbation causing self-focusing of the beam. When this condition is met, the nonlinear index variation balances the natural diffraction of the beam.

For a Gaussian beam, using $\delta n_c = n_2 I_c$ with I_c the critical intensity and using a small angle expansion in Eq. (5.68) to second order in θ_G (and neglecting the term $\propto \delta n_c \theta_G^2$), we immediately get

$$I_c = \frac{n_0 \theta_G^2}{2n_2}. \tag{5.69}$$

Since $\theta_G = \lambda_0/(\pi n_0 w_0)$ and $I = 2P/\pi w_0^2$, we immediately see that the beam area disappears and that the threshold is really a threshold in power, not intensity; defining $P_c = \pi w_0^2 I_c/2$, we get

$$\boxed{P_c = K \frac{\lambda_0^2}{4\pi n_0 n_2}}. \tag{5.70}$$

This is the standard definition of the critical power for self-focusing found in many nonlinear optics textbooks. It is only accurate up to a numerical factor K which depends on the exact shape of the beam, since the choice of w_0 as the beam radius to apply Snell's law was arbitrary. For Gaussian beams, simulations have shown that $K \approx 2$.

---

[3] The beam power is $P = \int I_0(r)ds$, with the surface element $ds = rdrd\phi$ such that $P = \pi w_0^2 I_0/2$.

Fig. 5.5 Derivation of the self-focusing distance using Fermat's principle

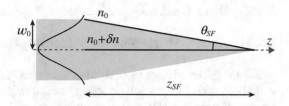

Inserting the expression for n_2 for the steady-state ponderomotive force (Eq. 5.66) and using $K = 2$ in Eq. (5.70) for Gaussian beams, we get the following expressions for the critical power for self-focusing in plasmas as a function of various physical constants or in more physical units:

$$P_c = 8\pi n_0 \lambda_{De}^2 n_c mc^3 , \tag{5.71}$$

$$= \frac{2c}{r_e} n_0 \frac{T_e}{n_{e0}/n_c} , \tag{5.72}$$

$$= 3.4 \times 10^7 n_0 \frac{T_e[keV]}{n_{e0}/n_c} [W], \tag{5.73}$$

where $r_e = e^2/(4\pi\varepsilon_0 mc^2)$ is the classical electron radius and $n_0 = \sqrt{1 - n_{e0}/n_c}$ is the background plasma refractive index. This expression shows in particular that the critical power for self-focusing is in fact very easily exceeded in many laser–plasma environments.

A typical self-focusing distance z_{SF} can be derived using similar arguments. We consider the situation of a beam undergoing self-focusing as depicted in Fig. 5.5. Considering a ray at the outer edge of the beam, where the index is approximately n_0, and a central ray for which the index is $n_0 + \delta n$, Fermat's principle states that the optical path for the two rays to the focus must be the same, i.e., $n_0 z_{SF}/\cos(\theta_{SF}) = (n_0 + \delta n)z_{SF}$.

Taking a small angle expansion for θ_{SF} yields the following expressions for θ_{SF} and z_{SF}:

$$\theta_{SF} = \sqrt{2\frac{\delta n}{n_0}} , \tag{5.74}$$

$$z_{SF} = \frac{w_0}{\sqrt{2\delta n/n_0}} , \tag{5.75}$$

where w_0 is the beam waist. These expressions are very crude, as they assume a constant index variation δn, i.e., a constant intensity. In reality the self-focusing process raises the intensity (like a regular lens would do), which increases δn and thus focuses the beam even more, which exacerbates the intensity increase and density depletion in a feedback loop, which can lead to a catastrophic collapse of the beam. We discuss this more realistic situation in more detail in the next section.

5.6.2   Wave Optics Description

5.6.2.1   Envelope Equations and Soliton Solution

While the geometric optics derivation above provides a reasonable account of the conditions under which self-focusing occurs, in particular the power threshold, it fails to capture the self-consistent evolution of the process. Indeed, as the wave starts to self-focus, its intensity increases, which reinforces the nonlinear refractive index increase in a feedback loop.

The self-focusing process is difficult to address analytically without making some rather strong simplifying assumptions. In the following we use a Gaussian ansatz, i.e., we will assume that the beam follows a Gaussian profile whose waist and radius of curvature are allowed to vary along z depending on the nonlinear response; on top of that we will also have to limit ourselves to the central portion of the beam (along the axis) and ignore the "wings" of the beam. We will also see that a simple n_2 nonlinearity (i.e., $n \approx n_0 + n_2 I$, as we saw in Sect. 5.5) leads to catastrophic collapse of the beam, as the shrinking of the beam leads to an ever-increasing intensity increase, the process reinforcing itself until unphysical breakdown. To recover a physically valid description, we will need to introduce a more complete description that includes saturation via depletion of the electron density from the ponderomotive force; indeed it has been known for a long time in the nonlinear optics community that an n_2 nonlinearity alone leads to non-physical collapse but is avoided by introducing saturation processes such as an intensity-dependent absorption or a higher-order "n_4" nonlinearity (such that $n \approx n_0 + n_2 I + n_4 I^2$) [6, 8]. For plasmas, when considering long time scales, it is natural to introduce the density equilibration between the ponderomotive and thermal pressures, as was derived in Sect. 5.3. This was first used by Kaw et al. in 2D [1] and then by Max in 3D geometry (i.e., 2D cylindrical coordinates r, z) [9].

Despite all these simplifying assumptions, the analytical results below provide a good intuitive understanding of the various aspects of self-focusing, which can readily be observed in numerical simulations. The beginning of our analysis closely follows Max [9].

We start as usual from the wave equation for light waves in plasmas,

$$(\partial_t^2 + \omega_{pe}^2 - c^2 \nabla^2)\mathbf{a} = 0, \qquad (5.76)$$

where we describe the light wave via its normalized potential vector $\mathbf{a} = e\mathbf{A}/mc$, assuming a steady-state (time-independent) envelope:

$$\mathbf{a}(\mathbf{r}, t) = \frac{1}{2}\mathbf{e}_y a(\mathbf{r}) e^{i\psi} + c.c., \qquad (5.77)$$

where $\psi = k_0 z - \omega_0 t$. Next we assume thermal equilibrium for the plasma density depression, as seen in Sect. 5.3 Eq. (5.49) where we assume that $T_e \gg T_i$ for simplicity:

$$\omega_{pe}^2 = \omega_{pe0}^2 \exp\left[-\frac{c^2 \langle \mathbf{a}^2 \rangle}{2v_{Te}^2}\right] \tag{5.78}$$

$$= \omega_{pe0}^2 \exp[-\beta a^2], \quad \beta = c^2/4v_{Te}^2. \tag{5.79}$$

Inserting into the wave equation and taking a paraxial approximation such that $|\partial_z^2 a| \ll k_0 |\partial_z a|, |\nabla_\perp^2 a|$ (cf. Sect. 1.1.1) lead to

$$\left(2ik_0\partial_z + \nabla_\perp^2 - \frac{\Gamma^2}{c^2}\right) a = -\frac{\omega_{pe0}^2}{c^2}\left(1 - \exp[-\beta a^2]\right) a, \tag{5.80}$$

where $\Gamma^2 = -\omega_0^2 + \omega_{pe0}^2 + k_0^2 c^2$ is a phase mismatch, which can give a dispersion relation for the self-guided wave into the plasma channel (in the case of an infinite plane wave, $\Gamma = 0$).

Before we introduce the Gaussian ansatz, it will be convenient to first write down equations for the amplitude and phase of a under the assumption of azimuthal symmetry. We define the (real) amplitude and phase as α and s (with cylindrical coordinates) such that

$$a(r, z) = \alpha(r, z)e^{ik_0 s(r,z)}. \tag{5.81}$$

Inserting this expression into the paraxial equation with the cylindrical expression for $\nabla_\perp^2 f = (1/r)\partial_r[r\partial_r f]$ (assuming $\partial_\varphi a = 0$) and separating the real and imaginary parts lead to the following pair of coupled equations for the amplitude and phase:

$$k_0^2\left[2s' + (\partial_r s)^2\right] - \frac{\partial_r \alpha}{\alpha r} - \frac{\partial_r^2 \alpha}{\alpha} + \frac{\Gamma^2}{c^2} - \frac{\omega_{pe0}^2}{c^2}\left(1 - \exp[-\beta\alpha^2]\right) = 0, \tag{5.82}$$

$$2\alpha' + \alpha\frac{\partial_r s}{r} + 2\partial_r s \partial_r \alpha + \alpha\partial_r^2 s = 0, \tag{5.83}$$

where the prime denotes a derivative along z.

We can now introduce our Gaussian ansatz and express the field as

$$a(r, z) = a_0 \frac{w_0}{w(z)} \exp\left[-\frac{r^2}{w^2(z)}\right] \exp\left[ik_0\left(\frac{r^2}{2}\frac{w'(z)}{w(z)} + \varphi\right)\right], \tag{5.84}$$

where $w(0) = w_0$, the phase front curvature satisfies $R(z) = w(z)/w'(z)$ and φ is the Gouy phase. In the following we assume that $z = 0$ represents the plasma boundary, which is not necessarily at the best focus; therefore we have the initial conditions $w(0) = w_0$ and $R(0) = R_0$ (with $1/R_0 = 0$ only if $z = 0$ is the best focus location).

Inserting the resulting expressions for α and s into Eq. (5.83) yields $0 = 0$. Next, we make our second strong assumption, which is to consider only the fraction of the beam concentrated near axis, and take

$$\exp\left[-\frac{r^2}{w^2(z)}\right] \approx 1 - \frac{r^2}{w^2(z)} . \tag{5.85}$$

Inserting the Gaussian ansatz into Eq. (5.82), using the expansion above and collecting the terms in $O(r^0)$ and $O(r^2)$, respectively, lead to the following two equations:

$$2\varphi' = -\frac{4}{k_0^2 w^2} - \frac{\Gamma^2}{k_0^2 c^2} + \frac{\omega_{pe0}^2}{k_0^2 c^2}\left(1 - \exp[-\beta a_0^2 w_0^2/w^2]\right) , \tag{5.86}$$

$$w'' = \frac{4}{k_0^2 w^3}\left(1 - \frac{w_0^2 a_0^2}{8\lambda_{De}^2}\exp[-\beta a_0^2 w_0^2/w^2]\right) . \tag{5.87}$$

Equation (5.86) dictates the evolution of the phase and eventually leads to the dispersion relation for the light wave propagating in the density channel dug by the ponderomotive force (cf. [9]). In the following we focus on Eq. (5.87), which describes the self-consistent evolution of the beam waist as it self-focuses.

Setting $w'' = 0$ and $w = cst.$ in Eq. (5.87) gives the condition for the formation of a beam whose size is constant due to the exact balancing of the diffraction and self-focusing effects. Such a mode is often referred to as a soliton; it corresponds to a beam whose power is exactly at the critical power for self-focusing that was heuristically derived in the previous section. Denoting w_s the soliton waist, which must satisfy $w(z) = w_0 = w_s$, we obtain

$$\boxed{w_s = \frac{c}{\omega_{pe0}}\sqrt{\frac{8v_{Te}^2}{a_0^2 c^2}}\exp\left[\frac{a_0^2 c^2}{8v_{Te}^2}\right]} . \tag{5.88}$$

The soliton waist size's dependence on the intensity normalized to the thermal pressure, i.e., $\propto v_{os}^2/v_{Te}^2$, is represented in Fig. 5.6.

The minimum soliton waist size, satisfying $\partial w_s/\partial a_0 = 0$, occurs at the vector potential value $a_{0,min}$ and takes the value $w_{s,min}$ such that

$$a_{0,min} = 2\frac{v_{Te}}{c} , \tag{5.89}$$

$$w_{s,min} = \sqrt{2}\frac{c}{\omega_{pe0}}e^{1/2} . \tag{5.90}$$

We note that the minimum soliton waist size is equal, up to a numerical factor, to the plasma skin depth $\lambda_s = c/\omega_{pe}$. It can be shown (cf. [9]) that the paraxial

Fig. 5.6 Waist size of a "soliton" mode (propagating at a constant beam radius, exactly at the critical power) as a function of normalized intensity (Eq. 5.88)

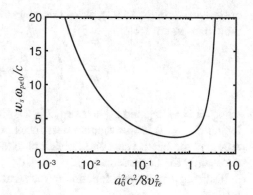

approximation breaks down when $ca_0/v_{Te} \geq 1$. In the following we will therefore stay in the regime of $ca_0/v_{Te} \ll 1$.

The soliton also corresponds to the critical power threshold that was introduced via geometric optics in the previous section. The critical power $P_c = \pi w_s^2 I_0 / 2$ can be inferred from Eq. (5.88), assuming $\exp[a_0^2 c^2 / 8 v_{Te}^2] \approx 1$. Using $I_0 = \varepsilon_0 c n_0 E_0^2 / 2 = c n_0 n_c a_0^2 m c^2 / 2$ yields

$$P_c = 2\pi \lambda_{De}^2 n_0 n_c m c^3 , \tag{5.91}$$

where $\lambda_{De} = v_{Te}/\omega_{pe0}$. This can also be expressed more generally in terms of the nonlinear index coefficient n_2 defined in Eq. (5.66); we obtain

$$P_c = \frac{\lambda_0^2}{8\pi n_0 n_2} . \tag{5.92}$$

This expression is almost identical to what we obtained using geometric optics (Eq. 5.70), except for a factor 4; keep in mind that both derivations use rather strong simplifications (arbitrary choice of beam radius in the geometric optics model to apply Snell's law and restriction to near-axis regions for the wave model with Gaussian ansatz), so finding a different numerical factor is not unexpected.

From Eq. (5.88) (still in the limit of $\exp[a_0^2 c^2 / 8 v_{Te}^2] \approx 1$), we can also easily express the ratio of incident power at the plasma boundary to the critical power, $P/P_c = w_0^2/w_s^2$ as

$$\frac{P}{P_c} = \frac{w_0^2 a_0^2}{8 \lambda_{De}^2} . \tag{5.93}$$

5.6.2.2 Behavior at $P > P_c$

The behavior of the beam when the power exceeds the critical power for self-focusing is again dictated by the envelope equation, Eq. (5.87). Having defined the critical power in Eq. (5.93), we can start by re-expressing the equation in a somewhat more transparent form:

$$\tilde{w}'' = \frac{1}{\tilde{w}^3 Z_d^2} \left(1 - \frac{P}{P_c} \exp[-\beta a_0^2/\tilde{w}^2] \right), \tag{5.94}$$

where $\tilde{w} = w/w_0$ and $Z_d = k_0 w_0^2/2$ is the diffraction length, equal to the Rayleigh length if $z = 0$ is the vacuum best focus position (keep in mind that w_0 and a_0 represent the beam waist and field amplitude at the plasma boundary, which is not necessarily the beam focus).

Defining a temporary function U such that $\tilde{w}'' = -dU/d\tilde{w}$, an integration easily leads to

$$U = \frac{1}{2Z_d^2 \tilde{w}^2} + \frac{w_0^2 \omega_{pe0}^2}{4Z_d^2 c^2} \exp[-\beta a_0^2/\tilde{w}^2]. \tag{5.95}$$

Multiplying the relation $d^2\tilde{w}/dz^2 + dU/d\tilde{w} = 0$ by $d\tilde{w}/dz$ leads to the following conservation law:

$$\tilde{w}'^2 + 2U = K, \tag{5.96}$$

where K is a constant given by the boundary conditions at $z = 0$, $\tilde{w}(0) = 1$, and $\tilde{w}'(0)/\tilde{w}(0) = 1/R(0) = 1/R_0$. We obtain

$$\tilde{w}'^2 = \frac{1}{R_0^2} + \frac{1}{Z_d^2}\left(1 - \frac{1}{\tilde{w}^2}\right) + \frac{w_0^2 \omega_{pe0}^2}{2c^2 Z_d^2}\left(\exp[-\beta a_0^2] - \exp[-\beta a_0^2/\tilde{w}^2] \right). \tag{5.97}$$

This equation cannot be solved analytically, but some interesting insight can be gained from deriving solutions in the limit $\beta a_0^2 \ll 1$ (i.e., $v_{os} \ll v_{Te}$) and expanding the exponential terms.

First, we can verify that the zero-order expansion, $\exp[-\beta a_0^2] \approx \exp[-\beta a_0^2/\tilde{w}^2] \approx 1$, describes the natural diffraction of the Gaussian beam in vacuum.[4]

Next, it can be shown (cf. Problem 5.4) that a first-order expansion, $\exp[-\beta a_0^2] \approx 1 - \beta a_0^2$ or equivalently $n \approx n_0 + n_2 I$, is also insufficient and leads to catastrophic collapse of the beam. This is a well-known result of nonlinear optics, which comes from the fact that nothing saturates the instability, with the index perturbation $n_2 I$ growing indefinitely as the self-focusing develops. This situation is unphysical in a plasma, as the growth in $n_2 I$ must be limited by the electron density depletion.

---

[4] The waist of a Gaussian beam propagating in vacuum with its best focus at $z = 0$ (i.e., $1/R_0 = 0$) is $\tilde{w} = \sqrt{1 + (z/Z_d)^2}$, which satisfies $\tilde{w}'^2 = (1 - 1/\tilde{w}^2)/Z_d^2$ in accordance with Eq. (5.97) with $a_0 = R_0 = 0$.

We must therefore expand the density expression to fourth order in $a_0 c/v_{Te}$:
$\exp[-\beta a_0^2] \approx 1 - \beta a_0^2 + \beta^2 a_0^4/2$ (and likewise, $\exp[-\beta a_0^2/\tilde{w}^2] \approx 1 - \beta a_0^2/\tilde{w}^2 + \beta^2 a_0^4/2\tilde{w}^4$). This is equivalent to introducing an "n_4" optical nonlinearity, i.e., $n \approx n_0 + n_2 I + n_4 I^2$ with $|n_4| I^2 \ll |n_2| I \ll n_0$. For simplicity, we introduce the normalized coefficients v_2 and v_4 defined as follows:

$$n \approx n_0 + v_2 \left\langle \mathbf{a}^2 \right\rangle + v_4 \left\langle \mathbf{a}^2 \right\rangle^2 = v_0 + v_2 \frac{a_0^2}{2} + v_4 \frac{a_0^4}{4} . \tag{5.98}$$

Expanding the expression for the density $n_e = n_{e0} \exp[-\beta a_0^2]$ used above, with $\beta = c^2/4v_{Te}^2$, in powers of βa_0^2, gives the following normalized nonlinear coefficients:

$$v_2 = \frac{n_0}{4k_0^2 \lambda_{De}^2} , \tag{5.99}$$

$$v_4 = -\frac{n_0 c^2}{16 k_0^2 \lambda_{De}^2 v_{Te}^2} . \tag{5.100}$$

Self-focusing in the presence of an n_4 nonlinearity has been studied in nonlinear optics; it has been shown that a negative n_4 coefficient, like we have here, prevents catastrophic collapse and can lead to self-trapping behavior [8]. Physically, this corresponds to the situation where the laser intensity increases up to the point where the ponderomotive force removes all the electrons from the high-intensity region (as well as the ions, since we assume thermal equilibrium on the ion motion time scale), saturating the process and letting diffraction take over again.

The self-focusing behavior can be investigated by looking for the turning points of the beam waist, i.e., by looking for solutions of $\tilde{w}' = 0$ in Eq. (5.97). Expanding the exponential terms in that equation to second order ($\sim \beta^2 a_0^4/2\tilde{w}^4$) and setting $\tilde{w}'^2 = 0$ lead to a second-order polynomial in $\tilde{w}^2$, whose solutions are

$$w_m^2 = \frac{w_0^2}{2C} \left[-\frac{1}{R_{nl}^2} \pm \sqrt{\frac{1}{R_{nl}^4} + 4C \frac{|v_4| a_0^4}{n_0 w_0^2}} \right] , \tag{5.101}$$

with the definitions:

$$\frac{1}{R_{nl}^2} = \frac{1}{Z_d^2} \left(\frac{P}{P_c} - 1 \right) , \tag{5.102}$$

$$C = \frac{1}{R_0^2} - \frac{1}{R_{nl}^2} + \frac{|v_4| a_0^4}{n_0 w_0^2} . \tag{5.103}$$

Here R_{nl} can be interpreted as a nonlinear curvature due to self-focusing.

From this, we can distinguish two types of behavior:

- Beams that are "nearly collimated" at the plasma boundary:

By "nearly collimated," we mean that $C < 0$, i.e., $R_0^2 \gg R_{nl}^2$ (ignoring the higher-order term $\propto a_0^4$). This corresponds to a beam entering the plasma close to its best focus. In this case, Eq. (5.101) has two solutions satisfying $w_m^2 > 0$, which means that the beam waist oscillates between a minimum and a maximum value; using an expansion up to order βa_0^2 leads to the following expressions for these two radii:

$$w_{max} \approx w_0, \tag{5.104}$$

$$w_{min} \approx \lambda_s \frac{P/P_c}{\sqrt{P/P_c - 1}}, \tag{5.105}$$

where $\lambda_s = c/\omega_{pe0}$ is the plasma skin depth.

In other words, the beam waist oscillates between its initial value at the plasma boundary w_0 and w_{min}, which can be down to the plasma skin depth. This is the so-called trapped solution, where the beam is self-guided with its waist undergoing oscillations between w_{max} and w_{min} as it propagates, without ever diffracting out.[5]

- Beams "strongly focused" on the plasma boundary:

By "strongly focused," we now mean that $C > 0$, i.e., $R_0^2 \ll R_{nl}^2$. In other words, the initial curvature from the beam is stronger than the nonlinear one. In this case Eq. (5.101) can only admit one solution that satisfies $w_m^2 > 0$ (corresponding to the "plus" sign in front of the square root). Going through a similar expansion as for the nearly collimated case leads to the same result for the minimum beam waist, Eq. (5.105). Physically, the beam will focus only once to a smaller diameter than its vacuum focal spot and will then diffract without self-trapping.

The two self-focusing behaviors, self-trapped and untrapped, are represented in Fig. 5.7a and b, respectively. These results were obtained by integrating Eq. (5.94) numerically.

At this point, we should check the validity of our perturbative development and the resulting equation (5.105). Since power is conserved in our problem, i.e., $w^2 a^2 = cst.$, the normalized intensity at the minimum waist w_{min} is

$$a_{0,max}^2 = a_0^2 \frac{w_0^2}{w_{min}^2}. \tag{5.106}$$

However we still need to satisfy $\beta a_{0,max}^2 \ll 1$, i.e.,

$$\frac{c^2}{4v_{Te}^2} \frac{a_0^2 w_0^2}{w_{min}^2} \ll 1. \tag{5.107}$$

---

[5] In reality absorption and nonlinear mechanisms will keep the trapping distance finite.

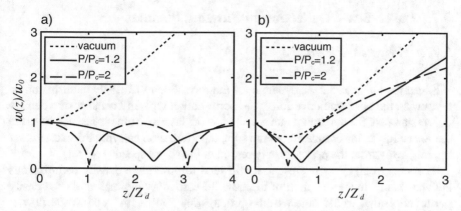

Fig. 5.7 Self-focusing behavior for $P > P_c$: (**a**) collimated beam at the plasma boundary (i.e., the laser enters the plasma at its best focus), exhibiting self-trapping, and (**b**) convergent beam at the plasma boundary: the beam is not self-trapped. Here the initial curvature at $z = 0$ was $R_0 = -Z_d = -k_0 w_0^2/2$ (corresponding to a Rayleigh length $z_R = Z_d/2$ and a vacuum best focus located at $z_f = Z_d/2$, and a vacuum waist size at best focus $w_{min,vac} = w_0/\sqrt{2}$)

Inserting the expression for w_{min} and using the definition of P/P_c from Eq. (5.93) lead to $P/P_c \ll 2$, so our development of the index nonlinearity to n_4 and the resulting expression for w_{min} are only valid for

$$ 1 < \frac{P}{P_c} \ll 2, \tag{5.108} $$

i.e., for beams that are only slightly above the critical self-focusing power.

Even though the validity range of this analysis and the resulting analytic formulas is rather limited, the general self-focusing behavior (self-trapped beams vs. untrapped) is easily observed in numerical simulations beyond this validity range (e.g., non-paraxial conditions or $\beta a_0 \geq 1$) and has been observed in experiments for various nonlinear media.

Finally, we should note that for beams whose waist sizes are large compared to the laser wavelength or whose phase fronts have poor quality, the filamentation instability might develop first, as will be discussed in Sect. 7.3. As we will see, "large" beams (this condition will be quantified) will tend to break up into multiple filaments, which will then self-focus according to the general behavior presented above. Even though the filamentation instability is closely related to self-focusing, it is technically a four-wave instability, and therefore its analysis is deferred to Chap. 7, after we introduce some general concepts of wave-mixing instabilities in Chap. 6.

5.7 Laser Beam Deflection in Flowing Plasmas

5.7.1 Physical Picture

We described in Sect. 5.2 the dynamics of the formation of a density perturbation by a laser via the ponderomotive force; the perturbation typically consists of a density depression near the maximum intensity and, if the laser rise time is fast compared to the ion acoustic transit time across the spot, a perturbation carrying the excess mass moving away from the peak intensity region at the sound speed (Fig. 5.1).

If a background flow is present, it will tend to push these density perturbations "downstream," in the direction of the flow. In particular, the region of the density perturbation that would otherwise be propagating "upstream" against the flow is going to be pushed back, which will lead to a pile-up of mass on the upstream side of the laser spot—and conversely, to a larger depression on the downstream side, resulting in a tilt of the density profile that will bend the laser beam downstream due to refraction. This is particularly important when the flow velocity is close to the sound speed, in which case the density pile-up on the upstream side cannot evacuate as it is being pushed by the laser's ponderomotive force on the one side and by the flow on the other side and will then reach its peak amplitude (as we will describe in detail below).

To illustrate the process, let us go back to a simple ray optics model as presented in Sect. 3.4. The ray direction, given by $\mathbf{k}$, is given by Eq. (3.56). Taking z as the initial propagation direction of the laser with $\perp$ referring to the directions transverse to z, we can recast Eq. (3.56) as

$$\frac{d\mathbf{k}_\perp}{dz} = -\frac{\omega_0}{2v_g}\nabla_\perp\frac{n_e}{n_c}, \tag{5.109}$$

where we used $d/dt = (dz/dt)d/dz$. Decomposing the density into a uniform background and the perturbation from the laser ponderomotive force δn_e, we obtain, in the low density limit where $n_0 = \sqrt{1 - n_{e0}/n_c} \approx 1$,

$$\frac{1}{k_0}\frac{d\mathbf{k}_\perp}{dz} = -\frac{1}{2}\frac{n_{e0}}{n_c}\nabla_\perp\frac{\delta n_e}{n_{e0}}. \tag{5.110}$$

The term on the left-hand side represents the ray deflection rate vs. z in the small angle limit; e.g., for a flow direction along x, we have $(1/k_0)dk_x/dz = d\varphi/dz$, where $\varphi \approx k_x/k_0$ is the deflection angle (cf. Fig. 5.8). Note that the process is cumulative: as the flow constantly keeps displacing the density perturbation, the deflection angle will keep growing for as long as the laser propagates in the flowing plasma region (assuming that the laser remains intense enough to keep driving the density perturbation—the process is nonlinear, and the deflection disappears for $\delta n \to 0$).

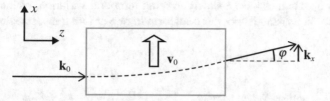

Fig. 5.8 Illustration of the deflection of a laser beam propagating in a plasma (bounded by the gray box in this illustration) with a uniform background flow $\mathbf{v}_0$

In the next sections we will derive the expression for the density perturbation in steady-state, similar to what we did in Sect. 5.2, but with a background flow, and derive expressions for the deflection rate as the beam propagates in a region of flowing plasma. Our derivation follows the theoretical framework originally developed by Rose [10] and extended by Ghosal and Rose in [11]. Nonlinear beam deflection by plasma flow was observed experimentally [12–14] and in simulations [15] and is particularly important in indirect-drive ICF as it can potentially modify the pointing of laser beams inside the target due to flows near the laser entrance holes.

5.7.2 Nonlinear Density Perturbation Displacement by Plasma Flow

The starting point to study this process is the ion acoustic wave equation with a ponderomotive driver, Eq. (5.31), while accounting for a finite background velocity $\mathbf{v}_0$ of the ion fluid as we derived in Sect. 1.3.2, Eq. (1.149), which gives

$$\left((\partial_t + \nu + \mathbf{v}_0 \cdot \nabla)^2 - c_s^2 \nabla^2\right) \frac{\delta n_e}{n_{e0}} = \frac{Zmc^2}{2M_i} \nabla^2 \langle \mathbf{a}^2 \rangle ; \qquad (5.111)$$

the damping term ν was introduced to represent IAW Landau damping.

In the following we assume a Gaussian profile near best focus for the laser,

$$a(\mathbf{r}, t) = \frac{1}{2} a_0 e^{-r^2/w^2} e^{i\psi_0} + c.c. \qquad (5.112)$$

with $\psi_0 = k_0 z - \omega_0 t$ and $r = \sqrt{x^2 + y^2}$ the radius in the transverse direction, and define the normalized intensity J as

$$J_0(\mathbf{r}) \equiv 2\langle \mathbf{a}^2 \rangle = a_0^2 e^{-2r^2/w^2} \qquad (5.113)$$

(in general J_0 can also have a slowly varying temporal variation, which we ignore here for simplicity) and its Fourier transform in one or two transverse directions:[6]

$$\hat{J}_0(k_x) = \frac{w}{2}a_0^2 e^{-k_x^2 w^2/8} \quad \text{(1D } \perp \text{)}, \tag{5.114}$$

$$\hat{J}_0(k_x, k_y) = \frac{w^2}{4}a_0^2 e^{-k^2 w^2/8} \quad \text{(2D } \perp, k^2 = k_x^2 + k_y^2\text{)}. \tag{5.115}$$

To derive the expression for the density perturbation in the presence of flow, it will be easier to assume a 1D geometry along x (the direction of the flow), i.e., we assume that the laser profile is very wide (quasi-infinite) along y, like we did in Sect. 5.2.

Taking the steady-state limit $\partial_t = 0$ in Eq. (5.111) and neglecting terms of order v^2 lead to

$$\left((\mathbf{v}_0 \cdot \nabla)^2 + 2v(\mathbf{v}_0 \cdot \nabla) - c_s^2 \nabla^2\right)\frac{\delta n_e}{n_{e0}} = \frac{Zmc^2}{4M_i}\nabla^2 J_0. \tag{5.116}$$

Introducing the Mach number $\mathbf{M}_0 = \mathbf{v}_0/c_s$ and taking a Fourier transform lead to the steady-state expression for the density perturbation in Fourier space:

$$\frac{\delta\hat{n}_e}{n_{e0}} = -\frac{c^2}{4v_{Te}^2}\frac{\hat{J}_0}{1 - M_0^2 \cos^2(\theta) + 2i\tilde{v}M_0\cos(\theta)}, \tag{5.117}$$

where $\cos(\theta) = \mathbf{k}_\perp \cdot \mathbf{v}_0/(|k_\perp \mathbf{v}_0|)$, $\tilde{v} = v/k_\perp c_s$ [which is usually independent of k in the case of ion Landau damping, cf. Eq. (1.189)[7]] and where we assumed $ZT_e \gg T_i$ for simplicity.

In 1D, we have $\cos(\theta) = k_x/|k_x| = \text{sign}(k_x)$; taking the expression for the 1D Gaussian beam above in the small Mach number limit, $M_0 \ll 1$, thus leads to

$$\frac{\delta\hat{n}_e}{n_{e0}} \approx -\frac{c^2 a_0^2}{8v_{Te}^2}we^{-k_x^2 w^2/8}\left[1 - 2i\tilde{v}M_0\text{sign}(k_x)\right]. \tag{5.118}$$

The expression for the density perturbation in real space is obtained after an inverse Fourier transform, which involves the following relation:

$$\mathcal{F}^{-1}\left[\text{sign}(k)e^{-k^2 w^2/8}\right] = -\frac{2i}{w}e^{-2x^2/w^2}\text{erfi}\left(\frac{\sqrt{2}x}{w}\right), \tag{5.119}$$

---

[6] We use the following definition of the Fourier transform, with unitary normalization and angular frequency: $\mathcal{F}[f(x)] = \hat{f}(k) = (2\pi)^{-1/2}\int_{-\infty}^{\infty} f(x)e^{-ikx}dx$, and likewise for the inverse Fourier transform: $f(x) = \mathcal{F}^{-1}[\hat{f}(k)] = (2\pi)^{-1/2}\int_{-\infty}^{\infty} \hat{f}(k)e^{ikx}dk$.

[7] A more rigorous approach consists in using a kinetic model, which provides the right damping in all cases; cf. [16].

Fig. 5.9 Steady-state density
perturbation in 1D,
normalized to the peak
perturbation amplitude
without flow

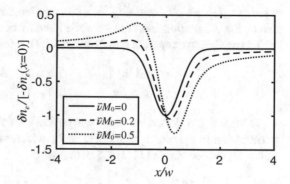

where $\mathrm{erfi}(x) = -i\,\mathrm{erf}(ix)$ is the imaginary error function. This leads us to the final
expression for the steady-state density perturbation in real space:

$$\frac{\delta n_e}{n_{e0}} = -\frac{c^2 a_0^2}{4 v_{Te}^2} e^{-2x^2/w^2} \left[1 - 2\tilde{v} M_0 \mathrm{erfi}\left(\frac{\sqrt{2}x}{w}\right) \right]. \tag{5.120}$$

The density perturbation is represented in Fig. 5.9 (note that the case $\tilde{v} M_0 = 0.5$
does not quite satisfy our assumptions that $\tilde{v} \ll 1$ and $M_0 \ll 1$ but is shown
anyway to illustrate the evolution of the density in the more pronounced situations).
As described earlier, as the flow gets closer to Mach 1, we observe a density pile-
up on the upstream side (left in the figure) and a depression on the downstream
side. The density develops a gradient near the center of the beam intensity profile
($x = 0$ on the figure), which will deflect the beam downstream due to refraction as
described in the next section.

5.7.3   Expression for the Beam Bending Rate

To quantify the beam deflection rate as the beam propagates in the flowing plasma,
we extend the simple ray optics model from Eq. (5.110) to the case of a Gaussian
beam. To do so, we introduce the intensity-weighted averages in real and Fourier
space for a function $f(r)$ (real space) or $f(k)$ (Fourier):

$$\langle f \rangle_\perp = \frac{\iint dr_\perp J_0(\mathbf{r}_\perp) f(\mathbf{r}_\perp)}{\iint dr_\perp J_0(\mathbf{r}_\perp)}, \tag{5.121}$$

$$\langle f \rangle_{k_\perp} = \frac{\iint dk_\perp \hat{J}_0(\mathbf{k}_\perp) f(\mathbf{k}_\perp)}{\iint dk_\perp \hat{J}_0(\mathbf{k}_\perp)}, \tag{5.122}$$

where $dk_\perp = dk_x dk_y$ and $dr_\perp = dx dy$ (unlike for the previous section, here we can do the calculation analytically while keeping both transverse components).

The ray-based deflection rate from Eq. (5.110) becomes for a Gaussian beam

$$\frac{1}{k_0}\frac{d\langle \mathbf{k}_\perp \rangle_{k_\perp}}{dz} = -\frac{1}{2}\frac{n_{e0}}{n_c}\left\langle \nabla_\perp \frac{\delta n_e}{n_{e0}}\right\rangle_\perp . \tag{5.123}$$

To derive the deflection rate, we go back to the expression for the density perturbation in Fourier space, Eq. (5.117), which we insert in the right-hand side of the expression above [11, 16]. Starting from

$$\nabla_\perp \frac{\delta n_e}{n_{e0}} = \mathcal{F}^{-1}\left[i\mathbf{k}_\perp \frac{\delta \hat{n}_e}{n_{e0}}\right] = \frac{i}{2\pi}\iint \mathbf{k}_\perp \frac{\delta \hat{n}_e}{n_{e0}}e^{i\mathbf{k}_\perp \cdot \mathbf{r}_\perp}dk_\perp , \tag{5.124}$$

we see that the average in real space, $\langle . \rangle_\perp$, only applies to the term $\exp[i\mathbf{k}_\perp \cdot \mathbf{r}_\perp]$, i.e.,

$$\left\langle \nabla_\perp \frac{\delta n_e}{n_{e0}}\right\rangle_\perp = \frac{i}{2\pi}\iint \mathbf{k}_\perp \frac{\delta \hat{n}_e}{n_{e0}}\left\langle e^{i\mathbf{k}_\perp \cdot \mathbf{r}_\perp}\right\rangle_\perp dk_\perp . \tag{5.125}$$

Using standard Fourier transform formulas (being careful about the conventions, cf. footnote 6) gives

$$\left\langle e^{i\mathbf{k}_\perp \cdot \mathbf{r}_\perp}\right\rangle_\perp = e^{-k^2 w^2/8} . \tag{5.126}$$

Inserting back into the expression above with the full expression for $\delta \hat{n}_e$ gives

$$\left\langle \nabla_\perp \frac{\delta n_e}{n_{e0}}\right\rangle_\perp = -\frac{a_0^2 c^2 w^2}{32\pi v_{Te}^2}i\iint \frac{\mathbf{k}_\perp e^{-k^2 w^2/4}}{1 - M_0^2 \cos^2(\theta) + 2i\tilde{\nu}M_0 \cos(\theta)}dk_\perp . \tag{5.127}$$

We assume a flow in the x-direction and look for deflection φ along x such that $(1/k_0)dk_x/dz = d\varphi/dz$ like in the previous section (cf. Fig. 5.8), i.e.,

$$\frac{d\varphi}{dz} = -\frac{1}{2}\frac{n_{e0}}{n_c}\left\langle \nabla_\perp \frac{\delta n_e}{n_{e0}}\right\rangle_\perp \cdot \mathbf{e}_x . \tag{5.128}$$

To perform the integral in 2D Fourier space we switch from the Cartesian coordinates, (k_x, k_y) to polar, (k, θ) where $\cos(\theta) = k_x/k$ per our definition of θ with a flow along x. After integrating along k from 0 to ∞, we arrive at

$$\frac{d\varphi}{dz} = i\frac{n_{e0}}{n_c}\frac{a_0^2 c^2}{32\sqrt{\pi}v_{Te}^2 w}G(M_0, \tilde{\nu}) , \tag{5.129}$$

$$G(M_0, \tilde{\nu}) \equiv \int_{-\pi}^{\pi} \frac{\cos(\theta)}{1 - M_0^2 \cos^2(\theta) + 2i\tilde{\nu}M_0 \cos(\theta)} d\theta . \tag{5.130}$$

We can simplify the expression of the function G by first noting that it is even with respect to θ, so the integral from $-\pi$ to π is twice the integral from 0 to π. Next we expand to more clearly show the real and imaginary parts:

$$G(M_0, \tilde{\nu}) = 2 \int_0^{\pi} \cos(\theta) \frac{1 - M_0^2 \cos^2(\theta) - 2i\tilde{\nu}M_0 \cos(\theta)}{(1 - M_0^2 \cos^2(\theta))^2 + 4\tilde{\nu}^2 M_0^2 \cos^2(\theta)} d\theta . \tag{5.131}$$

Using the symmetries or anti-symmetries from $\pi/2$, stemming from $\cos^2(\pi - \theta) = \cos^2(\theta)$ and $\cos(\pi - \theta) = -\cos(\theta)$, we immediately see that Re[G]=0, and therefore

$$G(M_0, \tilde{\nu}) = -8if(M_0, \tilde{\nu}) , \tag{5.132}$$

$$f(M_0, \tilde{\nu}) = \int_0^{\pi/2} \frac{\cos^2(\theta)\tilde{\nu}M_0}{(1 - M_0^2 \cos^2(\theta))^2 + 4\tilde{\nu}^2 M_0^2 \cos^2(\theta)} d\theta . \tag{5.133}$$

Our final expression for the deflection rate is then

$$\boxed{\frac{d\varphi}{dz} = \frac{1}{4\sqrt{\pi}w} \frac{n_{e0}}{n_c} \frac{a_0^2 c^2}{v_{Te}^2} f(M_0, \tilde{\nu})} . \tag{5.134}$$

This expression is valid for any value of M_0, up to 1 and beyond. The function f can be integrated analytically [17], but the derivation is tedious and the analytical expression is cumbersome. However, some insightful asymptotic properties can easily be obtained (as shown below), and it is straightforward to integrate the function numerically, yielding the curves shown in Fig. 5.10.

As we can see from the figure, a resonance appears when the flow is equal to c_s and is increasingly peaked and narrow as the damping gets smaller.

Fig. 5.10 Deflection rate $f(M_0, \tilde{\nu}) \propto d\varphi/dz$ as a function of flow Mach number

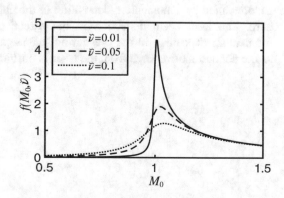

For small flow velocities, $M_0 \ll 1$, it is easy to show that the function $f(M_0, \tilde{\nu}) \approx \pi \tilde{\nu} M_0/4$: in this case the deflection rate is simply

$$\left. \frac{d\varphi}{dz} \right|_{M_0 \ll 1} \approx \frac{\sqrt{\pi}}{16w} \frac{n_{e0}}{n_c} \frac{a_0^2 c^2}{v_{Te}^2} \tilde{\nu} M_0 . \tag{5.135}$$

We can also extract a useful asymptotic limit for the case of a Mach 1 flow; the function f becomes

$$f(1, \tilde{\nu}) = \int_0^{\pi/2} \frac{\tilde{\nu}}{\sin^4(\theta)/\cos^2(\theta) + 4\tilde{\nu}^2} d\theta . \tag{5.136}$$

The denominator in the integrand becomes infinitely small at $\theta = 0$ as $\tilde{\nu} \to 0$: in other words, the whole integrand has a strong peak at $\theta = 0$, which extends up to $\sin^4(\theta)/\cos^2(\theta) \approx 4\tilde{\nu}^2$. For $\tilde{\nu} \ll 1$, which is generally the case, the corresponding width of the peak is $\theta_m^4 \approx 4\tilde{\nu}^2$, i.e., $\theta_m \approx \sqrt{2\tilde{\nu}} \ll 1$.

Since the integrand is only "significant" up to $\theta_m \ll 1$, we can approximate the integral by

$$f(1, \tilde{\nu}) \approx \int_0^\infty \frac{\tilde{\nu}}{\theta^4 + 4\tilde{\nu}^2} d\theta = \int_0^\infty \frac{\tilde{\nu}}{\theta^4 + \theta_m^4} d\theta = \frac{\pi}{8\sqrt{\tilde{\nu}}} , \tag{5.137}$$

where we used the definite integral

$$\int_0^\infty \frac{1}{x^n + a^n} dx = \frac{\pi}{n a^{n-1} \sin(\pi/n)} . \tag{5.138}$$

The deflection rate for a Mach 1 flow is then

$$\left. \frac{d\varphi}{dz} \right|_{M_0=1, \tilde{\nu} \ll 1} \approx \frac{\sqrt{\pi}}{32\sqrt{\tilde{\nu}} w} \frac{n_{e0}}{n_c} \frac{a_0^2 c^2}{v_{Te}^2} . \tag{5.139}$$

Note that by momentum conservation, the laser beam deflection is always accompanied by a momentum deposition in the plasma, which we did not consider here. The laser beam deflection in the direction of the flow implies that the momentum deposition in the plasma must be against the flow and act as a drag as the ponderomotive force from the laser is pushing against the flow [11].

Problems

5.1 Plasma Response to a Laser's Ponderomotive Force with a Finite Rise Time

- Derive Eq. (5.41) using the same procedure leading to Eq. (5.38) (Fourier and Laplace transforms in space and time, respectively). You will use the Laplace transform of the "exponential approach":

$$\mathcal{L}\left[(1 - e^{-\alpha t})H(t)\right] = \frac{\alpha}{s(s + \alpha)}. \tag{5.140}$$

- Check that the density is conserved at any time, i.e.,

$$\int_{-\infty}^{\infty} \frac{\delta n_e(x)}{n_{e0}} dx = 0. \tag{5.141}$$

- Check that in the limit of a very fast rise time compared to the acoustic transit time, $\tau \ll w/c_s$, Eq. (5.41) gives back Eq. (5.38). Hint: you will use

$$\frac{\alpha}{2} \exp[-\alpha|x|] \rightarrow \delta(x) , \quad \alpha \rightarrow \infty. \tag{5.142}$$

- Check that the opposite limit of a slow rise time compared to the acoustic transit time, $\tau \gg w/c_s$, gives the following expression for the density perturbation, where we used a Gaussian laser spatial profile, $a^2(x) = a_0^2 \exp[-x^2/w^2]$:

$$\frac{\delta n_e(x,t)}{n_{e0}} = \frac{c^2}{4v_{Te}^2} H(t) \left[(e^{-t/\tau} - 1)a^2(x) + \frac{\sqrt{\pi} w a_0^2}{2\tau c_s} e^{-t/\tau + |x|/\tau c_s} H(c_s t - |x|)\right]. \tag{5.143}$$

Hint: you will use

$$\frac{1}{w\sqrt{\pi}} \exp[-x^2/w^2] \rightarrow \delta(x) , \quad w \rightarrow 0. \tag{5.144}$$

Give a physical interpretation of this result.

5.2 Plasma Response to a Ponderomotive Driver at (or Near) a Plasma Resonance: Kinetic to Fluid

Assuming a Maxwellian background distribution of electrons and ions and a ponderomotive driver of the form $\varphi_p = \frac{1}{2}\tilde{\varphi}_p \exp[i(kz - \omega t)] + c.c.$, show that Eq. (5.61) leads back to Eqs. (5.23) and (5.31). You will use the asymptotic expansions of the Z function given in Sect. 1.3.3.

5.3 Plasma Response to a Stationary Ponderomotive Driver: Kinetic to Fluid

Assuming a Maxwellian background distribution of electrons and ions and a stationary driver (i.e., $\omega = 0$ in the Fourier analysis of Sect. 5.4) and using $Z'(0) = -2$, expand the kinetic expression for the electrostatic potential and density perturbation, Eqs. (5.60) and (5.61), and recover the fluid expressions from Sect. 5.3 for the self-consistent potential and density perturbation, Eqs. (5.48) and (5.50).

5.4 Beam Collapse from Self-Focusing with an n_2 Nonlinearity

Using the development $\exp[-\beta a_0^2] \approx 1 - \beta a_0^2$ in Eq. (5.97) and assuming $1/R_0 = 0$ (i.e., plasma boundary at best focus), analyze the solutions of that equation for $\tilde{w}' = 0$ and show that the only two solutions are for $P = P_c$ (use Eq. (5.93)), i.e., the soliton solution, or $\tilde{w} = 1$, meaning that the only inflection point is at the plasma boundary, $z = 0$ (i.e., catastrophic collapse).

References

1. P. Kaw, G. Schmidt, T. Wilcox, Phys. Fluids **16**, 1522 (1973)
2. B.D. Fried, S.D. Conte, *The Plasma Dispersion Function: The Hilbert Transform of the Gaussian* (Academic Press, 2015)
3. P. Kaw, J. Dawson, Phys. Fluids **13**, 472 (1970)
4. C.E. Max, J. Arons, A.B. Langdon, Phys. Rev. Lett. **33**, 209 (1974)
5. P. Gibbon, *Short Pulse Laser Interactions with Matter: An Introduction* (World Scientific, 2005)
6. S.A. Akhmanov, A.P. Sukhorukov, R.V. Khokhlov, Sov. Phys. Uspekhi **10**, 609 (1968)
7. R.Y. Chiao, E. Garmire, C.H. Townes, Phys. Rev. Lett. **13**, 479 (1964)
8. S. Akhmanov, A. Sukhorukov, R. Khokhlov, Sov. Phys. JETP **23**, 1025 (1966)
9. C.E. Max, Phys. Fluids **19**, 74 (1976)
10. H.A. Rose, Phys. Plasmas **3**, 1709 (1996)
11. S. Ghosal, H.A. Rose, Phys. Plasmas **4**, 2376 (1997)
12. J.D. Moody, B.J. MacGowan, D.E. Hinkel, W.L. Kruer, E.A. Williams, K. Estabrook, R.L. Berger, R.K. Kirkwood, D.S. Montgomery, T.D. Shepard, Phys. Rev. Lett. **77**, 1294 (1996)
13. P.E. Young, C.H. Still, D.E. Hinkel, W.L. Kruer, E.A. Williams, R.L. Berger, K.G. Estabrook, Phys. Rev. Lett. **81**, 1425 (1998)
14. D.S. Montgomery, R.P. Johnson, H.A. Rose, J.A. Cobble, J.C. Fernández, Phys. Rev. Lett. **84**, 678 (2000)
15. D.E. Hinkel, E.A. Williams, C.H. Still, Phys. Rev. Lett. **77**, 1298 (1996)
16. C. Ruyer, A. Debayle, P. Loiseau, M. Casanova, P.E. Masson-Laborde, Phys. Plasmas **27**, 102105 (2020)
17. D.E. Hinkel, E.A. Williams, R.L. Berger, L.V. Powers, A.B. Langdon, C.H. Still, Phys. Plasmas **5**, 1887 (1998)

Chapter 6
Introduction to Three-Wave Instabilities

This chapter will provide a simple physical picture of wave coupling in plasmas, and introduce all the possible three-wave instabilities that can exist between the three types of waves in non-magnetized plasmas, i.e., EMWs, EPWs, and IAWs.

The perturbative method that we used to derive wave equations for all three types of waves in Sect. 1.3 will be extended to second-order perturbations in the fluid quantities (fluid velocity or density modulations), which will lead to a new set of wave equations for each type of wave with second-order coupling terms on the right-hand side (corresponding to the beating between other waves). We will give the general form of the coupled mode equations for any given three-wave instability (the specific systems of equations for each instability will be derived in the later chapters); this standard form of the wave-coupling equations is sufficient to extract some of the most important quantities and physical concepts of instabilities, such as the growth rate, threshold, spatial amplification—as well as the "convective" vs. "absolute" nature of the instability, which is the subject of Sect. 6.6.

Each of the possible three-wave instabilities that can exist in (non-magnetized) plasmas will be described individually and in more detail in the subsequent chapters: all the "primary" instabilities, where an intense laser decays into a pair of "daughter waves," will be treated in Chap. 7 and 8, whereas the "secondary" instabilities, where a high-amplitude plasma wave (EPW or IAW) decays into a pair of daughter waves, will be discussed in Sect. 10.2.

6.1 Physical Picture

In this section we will provide simple, intuitive arguments describing how several waves can couple and exchange energy in plasmas. Consider the situation illustrated in Fig. 6.1. Two waves (shown as light waves in the figure, but as we will see plasma waves can be considered as well) with wave-vectors and frequencies $(\mathbf{k}_0, \omega_0)$ and

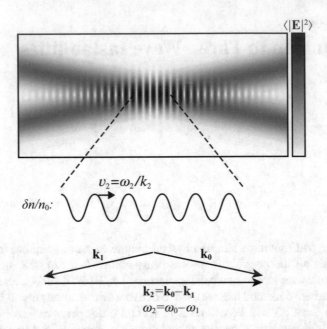

Fig. 6.1 Illustration of three-wave mixing in plasmas: two overlapped waves (here represented as light waves of equal intensity) with electric fields $\mathbf{E}_0$ and $\mathbf{E}_1$, wave-vectors and frequencies $(\mathbf{k}_0, \omega_0)$, $(\mathbf{k}_1, \omega_1)$ generate a beat wave in the overlap region with wave-vector $\mathbf{k}_2 = \mathbf{k}_0 - \mathbf{k}_1$ and frequency $\omega_2 = \omega_0 - \omega_1$. This beat wave imprints a density modulation in the plasma via the ponderomotive force, also oscillating at $\mathbf{k}_2$, ω_2, and can resonantly drive a plasma wave if $\mathbf{k}_2$ and ω_2 satisfy the dispersion relation of either an EPW or an IAW

$(\mathbf{k}_1, \omega_1)$ overlap in a plasma. For the sake of simplicity we will consider a region of space within the overlap volume which is small enough that the plasma conditions are approximately uniform and the waves can be approximated as plane waves, with electric fields $\mathbf{E}_0$ and $\mathbf{E}_1$ given by:

$$\mathbf{E}_0(\mathbf{r}, t) = \frac{1}{2} E_0 \mathbf{e}_0 e^{i\psi_0} + c.c. \ , \quad \psi_0 = \mathbf{k}_0 \cdot \mathbf{r} - \omega_0 t \ , \tag{6.1}$$

$$\mathbf{E}_1(\mathbf{r}, t) = \frac{1}{2} E_1 \mathbf{e}_1 e^{i\psi_1} + c.c. \ , \quad \psi_1 = \mathbf{k}_1 \cdot \mathbf{r} - \omega_1 t \ , \tag{6.2}$$

where $c.c.$ stands for complex conjugate.

We already described the ponderomotive force from two overlapped plane waves in Sect. 2.3.3:

$$\mathbf{F}_p = -\frac{e^2}{4m\bar{\omega}^2} \nabla \left(\mathbf{E}_0 \cdot \mathbf{E}_1^* e^{i\psi_2} + c.c. \right) \ , \tag{6.3}$$

where $\bar{\omega} = (\omega_0 + \omega_1)/2$ and $\psi_2 = \psi_0 - \psi_1$. The ponderomotive force will lead to a bunching of the plasma electrons into the "dark" fringes of the beat wave pattern, as the electrons are pushed out of the high-intensity regions, and to a density (or refractive index) modulation with the same wave-number and frequency as the beat wave, and the same phase velocity $v_2 = \omega_2/k_2$ (cf. Fig. 6.1).

If the beat wave's frequency and wave-vector satisfy the dispersion relation of a plasma wave (either EPW or IAW), then a plasma wave is resonantly driven by the beat wave, and the process turns into a feedback loop: the amplitude of the driven wave grows, which increases the scattering of one of the first two waves into the other, which increases the product $\mathbf{E}_0 \cdot \mathbf{E}_1$ and thus the ponderomotive force, etc.

Although the two beating waves can be of any nature (transverse or longitudinal), it is clear from this physical picture that the third wave, excited by the beat, has to be longitudinal, as it corresponds to particle bunching with an associated density modulation δn (cf. Fig. 1.7)—whereas transverse waves do not lead to density modulations, cf. Sect. 1.3.1.

In the following section we list all the possible three-wave couplings that can exist in plasmas; we will then derive driven wave equations for each type of wave (EMW, EPW, and IAW) as it is coupled to other waves existing in the plasma, and introduce some general concepts of three-wave instabilities. The detailed study of the different instabilities will be presented in the following chapters.

6.2   List of Possible Three-Wave Coupling in Plasma

As we saw above, the frequency- and wave-vector-matching conditions for three-wave coupling are

$$\omega_2 = \omega_0 - \omega_1 , \tag{6.4}$$

$$\mathbf{k}_2 = \mathbf{k}_0 - \mathbf{k}_1 , \tag{6.5}$$

where $(\omega_2, \mathbf{k}_2)$ satisfy the dispersion relation of either an EPW or an IAW. This can be interpreted as the exchange (and conservation) of quanta of energy and momentum between the three waves, i.e.,

$$\hbar\omega_0 = \hbar\omega_1 + \hbar\omega_2 , \tag{6.6}$$

$$\hbar\mathbf{k}_0 = \hbar\mathbf{k}_1 + \hbar\mathbf{k}_2 . \tag{6.7}$$

The wave 0 with the highest frequency, usually called the "pump," loses energy to waves 1 and 2, the "daughter" waves; in Sect. 6.4.6 we will give a more quantitative analysis of the energy conservation process in three-wave couplings via the Manley–Rowe relations. Three-wave couplings are usually referred to as "decay" instabilities, since the pump wave is depleted of its energy to feed the growth of the two daughter waves.

We can now look for all possible three-wave coupling in plasmas, where a "pump" wave with frequency ω_0 couples to two daughter waves with frequency

Table 6.1 Summary of all three-wave instabilities occurring in laser–plasma interactions. The "primary" instabilities have a laser beam as the "pump"; secondary instabilities have a plasma wave (EPW or IAW) as the pump, presumably driven to high enough amplitude by one of the primary processes above

Pump	Daughter 1	Daughter 2 (EPW/IAW)	Instability name
Primary processes			
EMW (ω_0)	EMW ($\omega_1 \lesssim \omega_0$)	IAW ($\omega_2 \ll \omega_0$)	Stimulated Brillouin scattering (SBS), Crossed-beam energy transfer (CBET)
EMW (ω_0)	EMW ($\omega_1 < \omega_0$)	EPW ($\omega_2 < \omega_0, \omega_1$)	Stimulated Raman scattering (SRS)
EMW (ω_0)	EPW ($\omega_1 \approx \omega_0/2$)	EPW ($\omega_2 \approx \omega_0/2$)	Two plasmon decay (TPD)
EMW (ω_0)	EPW ($\omega_1 \lesssim \omega_0$)	IAW ($\omega_2 \ll \omega_0$)	Ion acoustic decay (IAD)
Secondary processes			
EPW (ω_0)	EPW ($\omega_1 \lesssim \omega_0$)	IAW ($\omega_2 \ll \omega_0$)	Langmuir decay instability (LDI)
EPW (ω_0)	EMW ($\omega_1 \lesssim \omega_0$)	IAW ($\omega_2 \ll \omega_0$)	Electromagnetic decay instability (EDI)
IAW (ω_0)	IAW ($\omega_1 < \omega_0$)	IAW ($\omega_2 < \omega_0$)	Two-ion decay (TID)

ω_1 and ω_2, under the conditions that: i) $\omega_0 > \omega_1, \omega_2$ (to ensure energy conservation as mentioned above) and ii) wave #2 is a plasma wave, EPW, or IAW. Given these conditions, and the usual relation between the frequencies of waves in plasmas:

$$\omega_{IAW} \ll \omega_{EPW} \leq \omega_{EMW} , \qquad (6.8)$$

the possible three-wave couplings reduce to the list of processes shown in Table 6.1 (cf. Problem 6.1).

In this table, the "primary" processes refer to those where a light wave (i.e., a laser) acts as the pump, since a high-amplitude pump is generally required for these instabilities to develop (we will quantify the conditions in detail later). The "secondary" processes are those where the pump is a plasma wave: generally, these processes occur after one of the primary instabilities has created a plasma wave of high enough amplitude that it can subsequently behave as a pump and decay into other daughter waves. Secondary processes can act as saturation mechanisms for primary instabilities, by limiting the amplitude that the daughter waves can reach in the primary process. All primary instabilities in this table will be discussed in detail in Chap. 7 and 8; secondary instabilities will be described in Sect. 10.2.

The terminology for some of these processes is sometimes misleading and not always consistent through the literature; for example, what we call here ion acoustic decay (IAD) is sometimes called the parametric decay instability (PDI) or the electromagnetic decay instability (EDI) [1]—which is not the same as what we call EDI in the table. Throughout this book we adopt the most-commonly used terminology as summarized in Table 6.1—but one should be aware of these inconsistencies and remember to verify that the name actually refers to the correct process.

Note that the one exception to the relation in Eq. (6.8) in Table 6.1 is the electromagnetic decay instability (EDI): here, having $\omega_{EPW} > \omega_{EMW}$ is only possible if $3k_{EPW}^2 v_{Te}^2 > k_{EMW}^2 c^2$, i.e., the wave-vector of the pump EPW has to be much larger than the one from the light wave, by a factor of at least $c/\sqrt{3}v_{Te}$. Each three-wave instability in this table, either primary or secondary, will be discussed in greater detail in later chapters.

The regions of existence of the three-wave (decay) instabilities as a function of plasma density are summarized in Fig. 6.2. These can be inferred directly from the frequency matching conditions above, in particular:

- SRS: the daughter waves dispersion relations require that $\omega_1 > \omega_{pe}$ and $\omega_2 > \omega_{pe}$; therefore, $\omega_0 = \omega_1 + \omega_2 > 2\omega_{pe}$, i.e., $n_e/n_c = \omega_{pe}^2/\omega_0^2 < 1/4$: SRS is thus limited to electron densities lower than $n_c/4$;
- TPD: since both daughter waves are EPWs, their frequencies ω_1, ω_2 are both $\approx \omega_{pe}$, therefore $\omega_0 \approx 2\omega_{pe}$ and $n_e/n_c \approx 1/4$: TPD only occurs in the vicinity of $n_c/4$ (we will see in Sect. 8.2 that the region of existence depends on the plasma electron temperature, but is typically $[0.2 - 0.25]n_c$ for keV plasmas);
- IAD: since $\omega_2 \ll \omega_0$, we must have $\omega_0 \approx \omega_1 \approx \omega_{pe}$, i.e., $n_e \approx n_c$: IAD only occurs near the critical density (likewise this will be quantified in Sect. 8.3, where we will see that IAD can occur in the range $[0.8 - 1]n_c$).

6.3   Derivation of the Second-Order Coupled Mode Equations

In Sect. 1.3.1 we discussed the different types of waves that can exist in a plasma: light waves (EMWs), electron plasma waves (EPWs) and ion acoustic waves (IAWs). We saw that these waves can be described by perturbing the density and

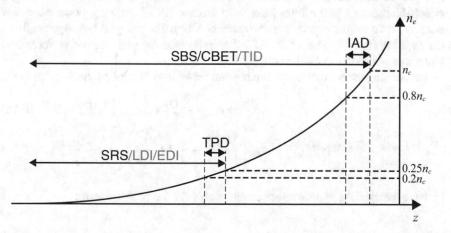

Fig. 6.2 Regions of existence of primary (black) or secondary (gray) three-wave instabilities as a function of electron density

velocity to first order around equilibrium, i.e., $n_s = n_{s0} + \delta n_s$, $\mathbf{v}_s = 0 + \delta\mathbf{v}_s$, where n_s and $\mathbf{v}_s$ are the density and fluid velocity for the particle $s \in \{e, i\}$, n_{s0} is the background (equilibrium) density assumed to be uniform, and δn_s, $\delta\mathbf{v}_s$ correspond to the linear response of the plasma to a field $\mathbf{E}$, i.e., $\delta n_s \sim \delta v_s \sim E$.

When multiple waves are simultaneously present in a plasma, we can have the formation of *second-order* density and velocity perturbations, corresponding to the product of first-order quantities. Let us look again at the fluid equations:

$$\partial_t n_s + \nabla(n_s \mathbf{v}_s) = 0, \tag{6.9}$$

$$[\partial_t + (\mathbf{v}_s \cdot \nabla)]\,\mathbf{v}_s = -\gamma_s v_{Ts}^2 \frac{\nabla n_s}{n_s} + \frac{q_s}{m_s}(\mathbf{E} + \mathbf{v}_s \times \mathbf{B})\,. \tag{6.10}$$

These equations contain nonlinear terms, which appear as products (or ratios) of fluid quantities; these nonlinear terms are $\nabla(n_s \mathbf{v}_s)$, $(\mathbf{v}_s \cdot \nabla)\mathbf{v}_s$, $\nabla n_s / n_s$ and $\mathbf{v}_s \times \mathbf{B}$ (since the fields $\mathbf{E}$ and $\mathbf{B}$ are first-order quantities like we saw in Sect. 1.3.1).

Let us rewrite these equations with the following perturbative development:

$$n_s = n_{s0} + n_s^{(1)} + n_s^{(2)} \tag{6.11}$$

$$\mathbf{v}_s = 0 \quad + \mathbf{v}_s^{(1)} + \mathbf{v}_s^{(2)} \tag{6.12}$$

$$\mathbf{E} = 0 \quad + \mathbf{E}^{(1)} + \mathbf{E}^{(2)} \tag{6.13}$$

$$\mathbf{B} = 0 \quad + \mathbf{B}^{(1)} + \mathbf{B}^{(2)}\,. \tag{6.14}$$

Here the first-order quantities ($n_s^{(1)}$, $\mathbf{v}_s^{(1)}$, etc.) correspond to linear waves present in the plasma (any type of wave—EMW, EPW, or IAW); the second-order quantities correspond to the product(s) of first-order quantities, such as the beat between two waves. Examples will be given below, but include processes like three-wave coupling as described in the previous section (i.e., a pair of waves beating together in a plasma and whose ponderomotive force, $\propto \nabla(\mathbf{E}_0 \cdot \mathbf{E}_1)$, drives a third wave), or self-focusing (where a light wave with electric field $\mathbf{E}$ drives its own nonlinear coupling term via the ponderomotive force $\propto \nabla\mathbf{E}^2$). The analysis below generalizes the approach used in Sects. 5.1 and 5.2 to introduce the ponderomotive force and plasma response to a ponderomotive driver.

Introducing the second-order expansions above into the fluid equations leads to:

$$\partial_t n_s^{(2)} + n_{s0}\nabla \cdot \mathbf{v}_s^{(2)} + \nabla \cdot (n_s^{(1)}\mathbf{v}_s^{(1)}) = 0, \quad (6.15)$$

$$\partial_t \mathbf{v}_s^{(2)} = -\frac{1}{2}\nabla(\mathbf{v}_s^{(1)})^2 - \gamma_s v_{Ts}^2\left[\frac{\nabla n_s^{(2)}}{n_{s0}} - \frac{1}{2}\nabla\left(\frac{n_s^{(1)}}{n_{s0}}\right)^2\right] + \frac{q_s}{m_s}\mathbf{E}^{(2)}\,. \quad (6.16)$$

Here we used the second-order development for the pressure term:

$$\frac{\nabla n_s}{n_s} = \frac{\nabla(n_s^{(1)} + n_s^{(2)})}{n_{s0} + n_s^{(1)} + n_s^{(2)}} \approx \frac{\nabla n_s^{(1)}}{n_{s0}} + \frac{\nabla n_s^{(2)}}{n_{s0}} - \frac{n_s^{(1)} \nabla n_s^{(1)}}{n_{s0}^2} + O\left(n_s^{(3)}\right), \quad (6.17)$$

as well as the development used for the derivation of the ponderomotive force in Sect. 5.1, $-(\mathbf{v}_s^{(1)} \cdot \nabla)\mathbf{v}_s^{(1)} + (q_s/m_s)\mathbf{v}_s^{(1)} \times \mathbf{B}^{(1)} = -\frac{1}{2}\nabla(\mathbf{v}_s^{(1)})^2$.

In the following sections we use these fluid equations to re-derive the second-order wave equations for EMWs, EPWs, and IAWs. The method is the same as in Sect. 1.3.2, but we are now going to see coupling terms on the right-hand side of the wave equations (instead of 0) in the form of products of first-order terms (i.e., linear waves). The resulting equations can describe a wide range of nonlinear effects, including the self-action effects that we already presented in Chap. 5 but also nonlinear couplings between multiple waves like the three-wave processes introduced in the previous section. They will constitute the starting point for the study of three-wave instabilities, including the primary processes from Table 6.1 in Chaps. 7 (when one of the daughter waves is an IAW) and 8 (with an EPW as daughter wave) and the secondary processes as well in Chap. 10 (when the primary wave is an EPW or IAW).

6.3.1 Driven Light Wave Equation

The driven light wave equation can be obtained almost immediately by going back to Eq. (1.140), which does not assume any linearization of the fluid quantities, taking the second-order components of that equation leads to

$$(\partial_t^2 - c^2\nabla^2)\mathbf{A}^{(2)} = \frac{1}{\varepsilon_0}\mathbf{j}_\perp^{(2)}, \quad (6.18)$$

where $\mathbf{A}^{(2)}$ is the second-order vector potential.

Since $\mathbf{j}_{e\perp} = -en_e\mathbf{v}_{e\perp}$, the second-order term for the current is:

$$\mathbf{j}_{e\perp}^{(2)} = -en_{e0}\mathbf{v}_{e\perp}^{(2)} - en_e^{(1)}\mathbf{v}_{e\perp}^{(1)}. \quad (6.19)$$

The first term on the right-hand side is related to $\mathbf{a}^{(2)} = (e/mc)\mathbf{A}^{(2)}$ (normalized vector potential) via $\mathbf{v}_{e\perp}^{(2)} = c\mathbf{a}^{(2)}$, which leads to the driven EMW equation:

$$\boxed{\left[\partial_t^2 + \omega_{pe0}^2 - c^2\nabla^2\right]\mathbf{a}^{(2)} = -\omega_{pe0}^2 \frac{n_e^{(1)}}{n_{e0}}\frac{\mathbf{v}_{e\perp}^{(1)}}{c}}. \quad (6.20)$$

This equation describes any of the three-wave coupling instabilities from Table 6.1 involving a light wave, the terms $n_e^{(1)}$ and $\mathbf{v}_e^{(1)}$ on the right-hand side corresponding to the other two waves involved in the process (note that one of these two waves is described via a density perturbation $n_e^{(1)}$, so has to be longitudinal, as we had already mentioned in our introduction on three-wave coupling in Sect. 6.1).

Since $(\partial_t^2 + \omega_{pe0}^2 - c^2\nabla^2)\mathbf{a}^{(1)} = 0$ (linear EMW wave equation), with $\mathbf{a} = \mathbf{a}^{(1)} + \mathbf{a}^{(2)}$, we will usually drop the $^{(1)}$ and $^{(2)}$ superscripts, and just use $\mathbf{a}$ to describe the EMW and δn_e, $\delta \mathbf{v}_{e\perp}$ (with $n_e = n_{e0} + \delta n_e$ and $\mathbf{v}_e = \delta \mathbf{v}_e$) for the fluid perturbations on the right-hand side:

$$\left[\partial_t^2 + \omega_{pe0}^2 - c^2\nabla^2\right]\mathbf{a} = -\omega_{pe0}^2 \frac{\delta n_e}{n_{e0}} \frac{\delta \mathbf{v}_{e\perp}}{c}. \tag{6.21}$$

Let us illustrate how to use this equation to describe three-wave instabilities. We first consider the example of SBS or SRS: these instabilities correspond to the coupling between two light waves, described by their vector potentials $\mathbf{a}_0 = \delta \mathbf{v}_0/c$ and $\mathbf{a}_1 = \delta \mathbf{v}_1/c$, and a plasma wave (EPW or IAW) described by its density modulation δn_e. Per Eq. (6.20), the wave equations describing the two light waves are then:

$$\left[\partial_t^2 + \omega_{pe0}^2 - c^2\nabla^2\right]\mathbf{a}_0 = -\omega_{pe0}^2 \frac{\delta n_e}{n_{e0}}\mathbf{a}_1, \tag{6.22}$$

$$\left[\partial_t^2 + \omega_{pe0}^2 - c^2\nabla^2\right]\mathbf{a}_1 = -\omega_{pe0}^2 \frac{\delta n_e}{n_{e0}}\mathbf{a}_0. \tag{6.23}$$

The plasma wave δn_e will be described by a driven (i.e., second-order) EPW or IAW wave equation, which we derive below.

Equation (6.20) also describes self-action effects such as self-focusing, described earlier in Sect. 5.6. In this case $\delta \mathbf{v}_e = c\mathbf{a}_0$ on the right-hand side corresponds to the main light wave itself, and δn_e is the density perturbation caused by the ponderomotive force from the spatial intensity gradient and provided by Eq. (5.63).

6.3.2   Driven EPW Equation

The wave equation for EPWs is obtained from the electron fluid equations, ignoring ion motion. Combining the time derivative of Eq. (6.15) with the divergence of Eq. (6.16) (by eliminating the term $\partial_t \nabla \cdot \mathbf{v}_e^{(2)}$) with $s = e$ for both, and eliminating the term $\nabla \cdot \mathbf{E}^{(2)}$ via Poisson's equation, $\nabla \cdot \mathbf{E}^{(2)} = -en_e^{(2)}/\varepsilon_0$, leads to:

$$\boxed{\left[\partial_t^2 + \omega_{pe0}^2 - 3v_{Te}^2\nabla^2\right]\frac{n_e^{(2)}}{n_{e0}} = \frac{1}{2}\nabla^2\left(\mathbf{v}_e^{(1)}\right)^2 - \partial_t\nabla\cdot\left(\frac{n_e^{(1)}}{n_{e0}}\mathbf{v}_e^{(1)}\right) - \frac{\gamma_e v_{Te}^2}{2}\nabla^2\left(\frac{n_e^{(1)}}{n_{e0}}\right)^2}.$$

$$\tag{6.24}$$

The nonlinear coupling terms on the right-hand side, respectively, correspond to the ponderomotive force, the nonlinearity coming from the $\nabla \cdot (n_e \mathbf{v}_e)$ term in the continuity equation, and the nonlinearity from the pressure term $\nabla n_e / n_e$. Like we discussed for the driven EMW equation, since $[\partial_t^2 + \omega_{pe0}^2 - 3v_{Te}^2 \nabla^2] n_e^{(1)} = 0$ we can simply write $n_e = n_{e0} + \delta n_e$, and likewise $\mathbf{v}_e = \delta \mathbf{v}_e$ and drop the first- and second-order superscripts (the ordering should be obvious):

$$\left[\partial_t^2 + \omega_{pe0}^2 - 3v_{Te}^2 \nabla^2\right] \frac{\delta n_e}{n_{e0}} = \frac{1}{2} \nabla^2 \delta \mathbf{v}_e^2 - \partial_t \nabla \cdot \left(\frac{\delta n_e}{n_{e0}} \delta \mathbf{v}_e\right) - \frac{\gamma_e v_{Te}^2}{2} \nabla^2 \left(\frac{\delta n_e}{n_{e0}}\right)^2.$$

(6.25)

This expression can be used for any of the three-wave instabilities in Table 6.1 involving an EPW. We can already notice that the second term will only intervene if one of the other waves is a longitudinal wave (since it involves a density modulation $n_e^{(1)}$). Likewise, the third term will only come into play if both the other waves are longitudinal, which only leaves the Langmuir decay instability (LDI)—and in fact, we will see that this term is then negligible compared to the others (cf. Sect. 10.2.1), so in reality, this last term will never be retained for the treatment of three-wave instabilities.

6.3.3 Driven IAW Equation

For driven IAWs, we need fluid equations for the electrons and the ions. The main steps of the derivation are similar to the derivation for the driven EPW wave equation above: we start again with the time derivative of Eq. (6.15) for the ions, $s = i$ (assuming only one ion species):

$$\partial_t^2 \frac{n_i^{(2)}}{n_{i0}} + \partial_t \nabla \cdot \mathbf{v}_i^{(2)} + \partial_t \nabla \cdot \left(\frac{n_i^{(1)}}{n_{i0}} \mathbf{v}_i^{(1)}\right) = 0.$$

(6.26)

Next we take the divergence of Eq. (6.15) for $s = i$:

$$\partial_t \nabla \cdot \mathbf{v}_i^{(2)} = -\frac{1}{2} \nabla^2 \left(\mathbf{v}_i^{(1)}\right)^2 - 3v_{Ti}^2 \nabla^2 \frac{n_i^{(2)}}{n_{i0}} + \frac{1}{2} \gamma_i v_{Ti}^2 \nabla^2 \left(\frac{n_i^{(1)}}{n_{i0}}\right)^2 + \frac{Ze}{M_i} \nabla \cdot \mathbf{E}^{(2)}.$$

(6.27)

And finally, we take the divergence of Eq. (6.15) for the electron fluid, $s = e$, with the approximation $\partial_t \mathbf{v}_e^{(2)} \approx 0$ based on the same arguments as for the linear IAW wave equation derivation discussed in Sect. 1.3.1 (see also Problem 1.3). We obtain:

$$-\frac{1}{2}\nabla^2\left(\mathbf{v}_e^{(1)}\right)^2 - v_{Te}^2\nabla^2\frac{n_e^{(2)}}{n_{e0}} + \frac{1}{2}\gamma_e v_{Te}^2\nabla^2\left(\frac{n_e^{(1)}}{n_{e0}}\right)^2 - \frac{e}{m}\nabla\cdot\mathbf{E}^{(2)} = 0. \quad (6.28)$$

We now combine these three equations, eliminating $\partial_t\nabla\cdot\mathbf{v}_i^{(2)}$ between Eqs. (6.26) and (6.27), then eliminating $\nabla\cdot\mathbf{E}^{(2)}$ from Eq. (6.28) to finally arrive at:

$$\boxed{\begin{aligned}\left(\partial_t^2 - c_s^2\nabla^2\right)\frac{n_e^{(2)}}{n_{e0}} &= \frac{Zm}{2M_i}\nabla^2\left(\mathbf{v}_e^{(1)}\right)^2 + \frac{1}{2}\nabla^2\left(\mathbf{v}_i^{(1)}\right)^2 - \partial_t\nabla\left(\frac{n_i^{(1)}}{n_{i0}}\mathbf{v}_i^{(1)}\right) \\ &\quad - \frac{1}{2}\left[\gamma_i v_{Ti}^2\nabla^2\left(\frac{n_i^{(1)}}{n_{i0}}\right)^2 + \frac{Zm}{M_i}\gamma_e v_{Te}^2\nabla^2\left(\frac{n_e^{(1)}}{n_{e0}}\right)^2\right]\end{aligned}}, \quad (6.29)$$

with $c_s^2 = (ZT_e + 3T_i)/M_i$ and where we assumed quasi-neutrality ($n_e^{(2)}/n_{e0} = n_i^{(2)}/n_{i0}$, $k\lambda_{De} \ll 1$).

As discussed for the driven EMW and EPW, since $(\partial_t^2 - c_s^2\nabla^2)n_e^{(1)} = 0$ we will usually drop the ordering superscripts and simply write $n_s = n_{s0} + \delta n_s$, and likewise $\mathbf{v}_s = \delta\mathbf{v}_s$ for the two species, yielding:

$$\begin{aligned}\left(\partial_t^2 - c_s^2\nabla^2\right)\frac{\delta n_e}{n_{e0}} &= \frac{Zm}{2M_i}\nabla^2\delta\mathbf{v}_e^2 + \frac{1}{2}\nabla^2\delta\mathbf{v}_i^2 - \partial_t\nabla\left(\frac{\delta n_i}{n_{i0}}\delta\mathbf{v}_i\right) \\ &\quad - \frac{1}{2}\left[\gamma_i v_{Ti}^2\nabla^2\left(\frac{\delta n_i}{n_{i0}}\right)^2 + \frac{Zm}{M_i}\gamma_e v_{Te}^2\nabla^2\left(\frac{\delta n_e}{n_{e0}}\right)^2\right].\end{aligned} \quad (6.30)$$

We will see in later chapters that the first term on the right-hand side is the dominant one and the only one that needs to be accounted for in all three-wave processes except the two-ion decay (TID), i.e., the coupling of an IAW to two other IAWs, in which case all the right-hand side terms count *except* the first one (cf. Sect. 10.2.2).

6.4   General Concepts and Methods for Three Wave Coupling

In the previous section we derived the general expressions for EMW, EPW, and IAW wave equations with second-order driving terms corresponding to other waves existing in the plasma. When three waves are present in the plasma, with $\mathbf{E}_{j,k,l}$ the electric field associated with the wave #j, k or l, each wave equation can take the following form, as can be inferred from Eqs. (6.20), (6.24), and (6.29) (and will be clarified in examples in the following chapters):

$$D_j(\nabla, \partial_t)\mathbf{E}_j = C_j\mathbf{E}_k\cdot\mathbf{E}_l, \quad (6.31)$$

where $D_j(\nabla, \partial_t)$ is the propagation operator for the wave j and C_j is a coupling term (or an operator) whose expression depends on the process being considered and will be elaborated upon in the later chapters describing three-wave instabilities. Note that the electric field will rarely be the quantity of choice to describe a wave: as explained before, it will be convenient to work with the normalized vector potential **a** for EMWs (connected to **E** via $\mathbf{E} = -\partial_t[mc\mathbf{a}/e]$), and with the electron density perturbations δn_e for EPWs or IAWs (connected to **E** via Poisson's equation).

A mathematical framework for the fully unified description of all possible three-wave instabilities was introduced in Ref. [2] and expanded in Ref. [3]; specific examples of parametric instabilities in plasmas were also the subject of review papers such as Refs. [4, 5]. Here we will only give a few simple definitions and methods that will then be applied to each of the three-wave processes in later chapters.

6.4.1 The Slowly Varying Envelope Approximation

The wave propagation operator can be written in the following general form, valid for EMWs, EPWs or IAWs:

$$D_j(\nabla, \partial_t) = \partial_t^2 + (1 - \delta_{ja})\omega_{pe}^2 - c_j^2\nabla^2, \tag{6.32}$$

where $\delta_{ja} = 1$ if j is an ion acoustic wave ("a") and 0 otherwise (Kronecker delta), and $c_j^2 = c^2$, $3v_{Te}^2$ or c_s^2 for an EMW, EPW, or IAW, respectively. a_j is a physical quantity used to represent the wave, for example, the vector potential for an EMW or the plasma density perturbation for an EPW or IAW, and will often be treated as a wave with a slowly varying envelope,

$$a_j = \frac{1}{2}\tilde{a}_j \exp[i\psi_j] + c.c \ , \quad \psi_j = k_jz_j - \omega_jt, \tag{6.33}$$

where z_j is the propagation direction of the wave (along $\mathbf{k}_j$) and

$$|\partial_z^2\tilde{a}_j| \ll k_j^2|\tilde{a}_j| \ , \quad |\partial_t^2\tilde{a}_j| \ll \omega_j^2|\tilde{a}_j|. \tag{6.34}$$

These conditions constitute the slowly varying envelope approximation, and once included in the expression for the propagation operator lead to the following simplification:

$$\boxed{D_j(\nabla, \partial_t)a_j \approx -2i\omega_j\frac{e^{i\psi_j}}{2}[\partial_t + v_{gj}\partial_{zj}]\tilde{a}_j + c.c.} \ , \tag{6.35}$$

where $v_{gj} = c_j^2k_j/\omega_j$ is the group velocity of the wave. Here we used the dispersion relation of the wave in $(\mathbf{k}, \omega)$, i.e.,

$$\hat{D}_j(\mathbf{k}_j, \omega_j) = -\omega_j^2 + (1 - \delta_{ja})\omega_{pe}^2 + k_j^2 c_j^2 = 0, \tag{6.36}$$

with the hat representing a Fourier transform (technically Fourier in space and Laplace in time with $\omega = -is$ used in lieu of the Laplace variable s, cf. Sect. 1.3.1).

Note that the enveloped propagator, Eq. (6.35), can also be obtained via a first-order expansion in Fourier space: taking $\omega = \omega_j + \delta\omega$ and $k = k_j + \delta k$ with $|\delta\omega/\omega_j|, |\delta k/k_j| \ll 1$ and inserting into $\hat{D}(k, \omega)$ with $\hat{D}(k_j, \omega_j) = 0$ gives:

$$\hat{D}_j(k, \omega) \approx -2\omega_j(\delta\omega - v_{gj}\delta k), \tag{6.37}$$

which is the equivalent of Eq. (6.35) in Fourier space.

6.4.2 Coupled Equations with Wave Damping

Wave damping is often introduced phenomenologically in the wave equation for any kind of plasma wave. It will typically account for collisional damping for EMWs and EPWs, and Landau damping for EPWs and IAWs.

The damping coefficient v_j comes into the wave equation for the wave j as:

$$D_j(\nabla, \partial_t) = \partial_t^2 + 2v_j\partial_t + (1 - \delta_{ja})\omega_{pe}^2 - c_j^2\nabla^2. \tag{6.38}$$

Proceeding as before with the slowly varying envelope (SVE) approximation, and making the assumption that the damping is small compared to the resonant frequency of the wave, $v_j \ll \omega_j$, gives the SVE propagator:

$$D_j(\nabla, \partial_t) \approx -i\omega_j e^{i\psi_j}[\partial_t + v_j + v_{gj}\partial_{zj}]\tilde{a}_j + c.c.. \tag{6.39}$$

Note that the Green's function of the operator $(\partial_t + v_j)$ is simply $H(t)\exp[-v_j t]$, with H the Heaviside step function, as required from our definition of the damping.

The slowly varying envelope approximation will allow us to reduce the three coupled equations for the three waves in decay instabilities, based on Eqs. (6.20), (6.24), and (6.29), to a system of three coupled equations with first-order derivatives in space and time. As we will see when discussing the specific instabilities, applying the phase-matching conditions between the three waves, $\psi_0 = \psi_1 + \psi_2$ will lead to a set of three coupled equations for any instability listed in Table 6.1 with the following form:

$$(\partial_t + \nu_0 + v_{g0}\partial_{z0})\tilde{a}_0 = iK_0\tilde{a}_1\tilde{a}_2 \,,$$

$$(\partial_t + \nu_1 + v_{g1}\partial_{z1})\tilde{a}_1 = iK_1\tilde{a}_0\tilde{a}_2^* \,, \qquad (6.40)$$

$$(\partial_t + \nu_2 + v_{g2}\partial_{z2})\tilde{a}_2 = iK_2\tilde{a}_0\tilde{a}_1^* \,,$$

where the coefficients K_j are real quantities.

6.4.3   Temporal Growth Rate

The temporal growth rate is an important quantity that provides a typical time scale associated with each of the three-wave instabilities—even though the actual temporal behavior of the waves in three-wave instabilities is more complicated than a simple growth at this rate, as we will discuss later. The growth rate is also an important quantity because it can be used as the single coupling term on the right-hand side of Eqs. (6.40), after some normalization of the wave quantities a_j, as will be shown in Sect. 6.4.6.

The growth rate is obtained by assuming a spatially uniform (i.e., temporal only) problem, i.e., setting the spatial derivatives ∂_{zj} in Eqs. (6.40) to zero, and neglecting the depletion of the pump wave 0 (i.e., taking a_0 as a constant). Starting with the undamped problem ($\nu_j = 0$), the two daughter waves are coupled to each other and the pump via:

$$\partial_t\tilde{a}_1 = iK_1\tilde{a}_0\tilde{a}_2^* \,, \qquad (6.41)$$

$$\partial_t\tilde{a}_2^* = -iK_2\tilde{a}_0^*\tilde{a}_1 \,. \qquad (6.42)$$

Taking the time-derivative of Eq. (6.41) and inserting into Eq. (6.42), and vice versa, immediately gives the solution:

$$\tilde{a}_{1,2}(t) = \tilde{a}_{1,2}(0)e^{\gamma_0 t} \,, \qquad (6.43)$$

where γ_0 is the growth rate, given by:

$$\boxed{\gamma_0^2 = K_1K_2|\tilde{a}_0|^2} \,. \qquad (6.44)$$

Therefore, once we establish a system of coupled equations for the wave envelopes of the form of Eqs. (6.40) for a given instability, we can immediately obtain the instability growth rate by identification of the right-hand side terms with the expression above.

Finite damping of the daughter waves reduces—and can even eliminate—the temporal growth; looking for solutions of the type $\tilde{a}_{1,2}(t) = \tilde{a}_{1,2}(0)\exp[\gamma t]$ and inserting into the coupled equations with damping present leads to:

$$(\gamma + \nu_1)\tilde{a}_1 = iK_1\tilde{a}_0\tilde{a}_2^* , \tag{6.45}$$

$$(\gamma + \nu_2)\tilde{a}_2^* = -iK_2\tilde{a}_0^*\tilde{a}_1 . \tag{6.46}$$

Eliminating $\tilde{a}_1$ (or $\tilde{a}_2$) from the system of equations leads to the following expression for the growth rate γ in the presence of damping:

$$(\gamma + \nu_1)(\gamma + \nu_2) = \gamma_0^2 . \tag{6.47}$$

This shows that as expected, the growth rate is reduced due to the presence of damping of the daughter waves; in particular, looking for positive solutions $\gamma > 0$ to this equation gives the threshold condition for the instability:

$$\boxed{\gamma_0^2 > \nu_1\nu_2} . \tag{6.48}$$

6.4.4 Spatial Amplification Rate and Definition of the Strongly Damped Regime of Instability

As we mentioned earlier, since the density perturbation driven by the beat wave between two other waves is longitudinal in nature, one of the two daughter waves (say, wave #2) is necessarily a plasma wave, EPW, or IAW.

As we shall see in later chapters on three-wave instabilities, we will often encounter situations where the plasma wave is in the "strongly damped regime" for the instability, meaning that $|\nu_2\tilde{a}_2| \gg |\nu_{g2}\partial_z\tilde{a}_2|$ (we will come back to the meaning of this approximation in a moment). It is important to keep in mind that this criterion is different from the condition for strong Landau damping of plasma waves, which is about the phase velocity of the wave relative to the thermal velocity of the background particle distribution, ω_2/k_2 vs. v_{Ts} where $s = e, i$.

Ignoring again the depletion of the pump wave #0, and assuming that the daughter wave #1 is an EMW (i.e., SRS or SBS/CBET) whose absorption can be neglected, the first-order coupled equations for the daughter waves simplify to:

$$v_{g1}\partial_z\tilde{a}_1 = iK_1\tilde{a}_0\tilde{a}_2^* , \tag{6.49}$$

$$v_2\tilde{a}_2^* = -iK_2\tilde{a}_0^*\tilde{a}_1 , \tag{6.50}$$

where the direction $z = z_1$ is chosen along $\mathbf{k}_1$.

Physically, this corresponds to the case where the plasma wave #2 immediately takes the form of the beat pattern between 0 and 1 (since $\tilde{a}_2 \propto \tilde{a}_0\tilde{a}_1^*$).[1] Wave #2 is basically an imprint of the beat pattern between 0 and 1.

---

[1] Since here we only look for steady-state solutions we are ignoring the transient regime associated with the "turning on" of 0 and 1.

This system of equations is readily solved for $\tilde{a}_1$:

$$\tilde{a}_1(z) = \tilde{a}_1(0)e^{\Gamma z}, \tag{6.51}$$

with the spatial amplification rate in amplitude Γ given by:

$$\boxed{\Gamma = \frac{\gamma_0^2}{v_{g1}v_2}} \tag{6.52}$$

(the amplification rate in intensity is 2Γ).

This expression connects the spatial growth rate Γ to the temporal growth rate γ_0. The spatial growth rate (assuming a steady-state response of the EPW or IAW daughter wave) is a physical quantity that is frequently used to characterize three-wave instabilities in laser–plasma interactions.

The physical quantity for the wave #2, which will typically be taken as the density perturbation ($a_2 = \delta n_e/n_{e0}$), follows a similar exponential growth along z: inserting the expression for $\tilde{a}_1$ into Eq. (6.50) leads to:

$$\tilde{a}_2(z) = \tilde{a}_2(0)e^{\Gamma z}. \tag{6.53}$$

We can now use that result to quantify the strongly damped assumption, which we used to neglect $|v_{g2}\partial_z \tilde{a}_2|$ compared to $|v_2 \tilde{a}_2|$; since $\partial_z \tilde{a}_2 = \Gamma \tilde{a}_2$ and $v_{g2} = c_2^2 k_2/\omega_2$, we have the following condition:

$$\boxed{\frac{\Gamma}{k_2} \ll \frac{v_2}{\omega_2}\left(\frac{\omega_2}{k_2 c_2}\right)^2}. \tag{6.54}$$

The squared term on the right-hand side is ≈ 1 for an IAW since $c_2 = c_s \approx \omega_2/k_2$, and to $[3(k_2\lambda_{De})^2]^{-1}$ (typically $\gg 1$) for an EPW since $c_2 = \sqrt{3}v_{Te}$ and $\omega_2 \approx \omega_{pe}$. Physically, this means that the "strongly damped" approximation is justified as long as the damping rate of the plasma wave per oscillation period is larger than its spatial growth rate per wavelength (with an extra margin of $[3(k_2\lambda_{De})^2]^{-1}$ for an EPW).

Alternatively, recognizing that $v_{g2}\partial_z$ represents a time derivative in the frame moving at v_{g2}, or that $v_{g2}\Gamma$ represents a temporal growth rate in the frame moving at v_{g2}, the strongly damped approximation implies that this growth rate shall remain smaller than v_2

Again, we should keep in mind that this condition is distinct from the actual amount of damping of the plasma wave: being in the strongly damped regime for instabilities depends on the growth rate and thus on the pump wave intensity, whereas the strong damping regime of plasma waves (assuming Landau damping) means that the phase velocity of the wave ω_2/k_2 cannot be much larger than the thermal velocity of the particle distribution v_{Ts}, which will allow particles in the tail to interact with the wave and lead to Landau damping. It is, for example, possible to be in the strongly damped regime of instabilities while having weakly damped plasma waves with either $k\lambda_{De} \ll 1$ (EPWs) or $ZT_e/T_i \ll 1$ (IAWs).

Note that the spatial growth rate can be viewed as an imaginary perturbation to the wave-vector or refractive index as seen by wave #1, i.e., $k_1 \rightarrow k_1 + \delta k_1$ or $n_1 \rightarrow n_1 + \delta n_1$, with $\delta k_1, \delta n_1 \in \mathbb{C}$ such that

$$\Gamma = -\mathrm{Im}(\delta k_1) = -\frac{\omega_1}{c}\mathrm{Im}(\delta n_1). \tag{6.55}$$

Furthermore, since the imaginary part of the refractive index (i.e., the spatial growth rate) depends on ω_1, its real part must depend on ω_1 as well, the two being connected by the Kramers-Kronig relations [6]. The physical interpretation is that when a light wave is spatially amplified by a three-wave instability in a plasma, the amplification (i.e., imaginary perturbation to the index) must be accompanied by a variation in the real part of the index as well. In other words, the plasma refractive index is not exactly $n = \sqrt{1 - n_e/n_c}$ anymore, but has a perturbation (typically small) which is connected to the amplification rate (and therefore depends on the intensity of the pump). This has been measured experimentally in Ref. [7], and will be described in greater detail in Sect. 7.2.4.

6.4.5 Instability Dispersion Relation

As we mentioned at the beginning of this section, every three-wave coupling instability description starts from a wave equation for each of the three waves with a coupling term on the right-hand side which depends on the other two waves' physical quantities (e.g., electric field or density perturbation).

We assume again an un-depleted pump regime where the pump wave, described by $a_0 = \frac{1}{2}\tilde{a}_0 \exp[i\psi_0] + c.c.$ with $\psi_0 = \mathbf{k}_0 \cdot \mathbf{r} - \omega_0 t$, is unaffected by the growth of the daughter waves. We can therefore assume that $\tilde{a}_0$ is a constant $\in \mathbb{R}$. We will typically start from a system of coupled equations for the daughter waves of the form

$$D_1(\partial_t, \nabla)a_1 = C_1 a_0 a_2, \tag{6.56}$$

$$D_2(\partial_t, \nabla)a_2 = C_2 a_0 a_1, \tag{6.57}$$

where $D_j(\partial_t, \nabla) = \partial_t^2 + 2\nu_j\partial_t + (1 - \delta_{ja})\omega_{pe}^2 - c_j^2\nabla^2$ is the propagator for wave # $j = 1, 2$, with $c_j = c, c_s$ or $\sqrt{3}v_{Te}$ for an EMW, IAW or EPW as was already defined earlier. The damping term will be omitted through the rest of the section. Taking a Fourier transform of the coupled equations leads to:[2]

---

[2] More rigorously, we are in fact taking a Fourier transform in space and Laplace transform in time, with ω being related to the Laplace variable s via $s = i\omega$. The relations after the Fourier vs. Laplace transform look similar (we set the initial conditions at $t = 0$ to 0), except that ω is a complex variable, whose imaginary part corresponds to an instability with exponential growth

$$\hat{D}_1(\mathbf{k}, \omega)\hat{a}_1(\mathbf{k}, \omega) = \frac{1}{2}C_1\bar{a}_0\left[\hat{a}_2(\mathbf{k} - \mathbf{k}_0, \omega - \omega_0) + \hat{a}_2(\mathbf{k} + \mathbf{k}_0, \omega + \omega_0)\right], \qquad (6.58)$$

$$\hat{D}_2(\mathbf{k}, \omega)\hat{a}_2(\mathbf{k}, \omega) = \frac{1}{2}C_2\bar{a}_0\left[\hat{a}_1(\mathbf{k} - \mathbf{k}_0, \omega - \omega_0) + \hat{a}_1(\mathbf{k} + \mathbf{k}_0, \omega + \omega_0)\right]. \qquad (6.59)$$

The hat denotes Fourier transforms or operators, and the propagator in Fourier space is $\hat{D}_j(\mathbf{k}, \omega) = -\omega^2 + (1 - \delta_{ja})\omega_{pe}^2 + k^2 c_j^2$. To simplify the notations, we define:

$$\hat{D}_j(n) = \hat{D}_j(\mathbf{k} + n\mathbf{k}_0, \omega + n\omega_0), \qquad (6.60)$$

$$\hat{a}_j(n) = \hat{a}_j(\mathbf{k} + n\mathbf{k}_0, \omega + n\omega_0). \qquad (6.61)$$

The coupled equations can be expressed as

$$\hat{D}_1(0)\hat{a}_1(0) = \frac{1}{2}C_1\bar{a}_0[\hat{a}_2(-1) + \hat{a}_2(1)], \qquad (6.62)$$

$$\hat{D}_2(0)\hat{a}_2(0) = \frac{1}{2}C_2\bar{a}_0[\hat{a}_1(-1) + \hat{a}_1(1)]. \qquad (6.63)$$

Let us pause here and look at Eq. (6.62). If the frequency and wave-vector $(\omega, \mathbf{k})$ satisfy the dispersion relation $\hat{D}_1$ (at least approximately), then the coupling between wave #1 and the pump #0 gives rise to two oscillation components for wave #2 at $(\omega - \omega_0, \mathbf{k} - \mathbf{k}_0)$ and $(\omega + \omega_0, \mathbf{k} + \mathbf{k}_0)$. The former is called the Stokes component: it corresponds the phase matching condition discussed in the previous sections, i.e., a wave at the phase $\exp[+i\psi_2]$ with $\psi_2 = \mathbf{k}_2 \cdot \mathbf{r} - \omega_2 t$ such that $\psi_0 = \psi_1 + \psi_2$ (with $\omega = \omega_1, \mathbf{k} = \mathbf{k}_1$). In other words: this expresses mathematically what we had previously assumed based on intuitive physical arguments about the beat between two waves (0 and 1) resonantly exciting a third wave (EPW or IAW) at a phase $\psi_2 = \psi_0 - \psi_1$ such that $(\omega_2, \mathbf{k}_2)$ satisfies the dispersion relation for an EPW or IAW (i.e., $\hat{D}_2(-1) \approx 0$). We can also see that linearizing the dispersion relation in the vicinity of the resonant frequency and k-vector for each wave and transforming back to real space, as we briefly described in Sect. 6.4.1, leads back to coupled equations for the slowly varying envelopes with a form similar to Eqs. (6.40), if we only consider the Stokes component (cf. Problem 6.2).

However, we also notice the apparition of another modulation component at $(\omega + \omega_0, \mathbf{k} + \mathbf{k}_0)$: this is the anti-Stokes component. It will only contribute to the coupling if $(\omega + \omega_0, \mathbf{k} + \mathbf{k}_0)$ also satisfies the dispersion relation of wave #2, at least approximately (i.e., $\hat{D}_2(1) \approx 0$). In most cases, the frequency and wave-vectors of the different waves involved in the coupling will not allow to satisfy $\hat{D}_2(-1) \approx 0$ and $\hat{D}_2(1) \approx 0$ simultaneously, and only the Stokes component will contribute to the coupling. This is the "decay" regime of instability, involving three waves, as we have discussed so far in the previous sections. However, under certain conditions, both the Stokes and anti-Stokes component will contribute, and the coupling will

---

if > 0—just like it corresponded to an exponential decay when < 0 for Landau damping, cf. Sect. 1.3.4.

involve four waves instead of three. The most famous and relevant example is the filamentation instability, which we will discuss in detail in Sect. 7.3.

Going back to Eqs. (6.62)–(6.63): eliminating $\hat{a}_2$ from the system gives the relation:

$$\hat{D}_1(0)\hat{a}_1(0) = \frac{1}{4}C_1C_2\tilde{a}_0^2\left[\frac{\hat{a}_1(-2)}{\hat{D}_2(-1)} + \frac{\hat{a}_1(0)}{\hat{D}_2(-1)} + \frac{\hat{a}_1(0)}{\hat{D}_2(1)} + \frac{\hat{a}_1(2)}{\hat{D}_2(1)}\right].\qquad(6.64)$$

Note that a similar relation can be obtained connecting $\hat{a}_1(n)$ to $\hat{a}_1(n+2)$ and $\hat{a}_1(n-2)$ for any $n \in \mathbb{N}$. This is a recurrence relation which constitutes an infinite system that needs to be truncated. For most practical cases, the terms $\hat{a}_1(\pm 2)$ will be non-resonant, as it will be difficult to find situations where both $(\omega, \mathbf{k})$ and $(\omega \pm 2\omega_0, \mathbf{k} \pm 2\mathbf{k}_0)$ satisfy the dispersion relation of wave #1 simultaneously, i.e., $\hat{D}_1(0) \approx 0$ and $\hat{D}_1(\pm 2) \approx 0$. We are then left with:

$$\hat{D}_1(0) = \frac{1}{4}C_1C_2\tilde{a}_0^2\left[\frac{1}{\hat{D}_2(-1)} + \frac{1}{\hat{D}_2(1)}\right],\qquad(6.65)$$

showing the contributions from the Stokes and anti-Stokes components $\hat{D}_2(-1)$ and $\hat{D}_2(+1)$.

This is the general dispersion relation for three- or four-wave instabilities. It consists in a polynomial in ω; the imaginary components of the roots correspond to instability growth rates, and can generally be obtained with some simplifying assumptions (concrete examples will be given when we study the filamentation instability in Sect. 7.3).

Since the role of waves 1 and 2 are interchangeable in the derivation, we can obtain the equivalent relation (e.g., by eliminating $\hat{a}_1$ instead of $\hat{a}_2$ in Eqs. (6.62)–(6.63)):

$$\hat{D}_2(0) = \frac{1}{4}C_1C_2\tilde{a}_0^2\left[\frac{1}{\hat{D}_1(-1)} + \frac{1}{\hat{D}_1(1)}\right].\qquad(6.66)$$

If we restrict ourselves to the three-wave instability and neglect the anti-Stokes component, we obtain the following dispersion relation for the decay (three-wave) instability:

$$\hat{D}_2(0)\hat{D}_1(-1) = \frac{1}{4}C_1C_2\tilde{a}_0^2.\qquad(6.67)$$

Likewise, we could also have gotten to the equivalent relation:

$$\hat{D}_1(0)\hat{D}_2(-1) = \frac{1}{4}C_1C_2\tilde{a}_0^2.\qquad(6.68)$$

Since these equations are related to the coupled mode equations for the slowly varying envelopes but expressed in Fourier space (Problem 6.2), it is also possible to get a general expression for a temporal growth rate from them. Taking $\omega \approx \omega_1 + i\gamma_0$ where $\hat{D}_1(\omega_1, \mathbf{k}_1) = \hat{D}_2(\omega_1 - \omega_0, \mathbf{k}_1 - \mathbf{k}_0) = 0$ and γ_0 is the growth rate (cf. footnote 2), gives $\hat{D}_1(0) = -2i\omega_1\gamma_0$ and $\hat{D}_2(-1) = 2i\omega_2\gamma_0$ after a Taylor expansion of $\hat{D}_1$ and $\hat{D}_2$ near $\omega_1, \mathbf{k}_1$, leading to

$$\boxed{\gamma_0^2 = \frac{C_1 C_2 \tilde{a}_0^2}{16\omega_1\omega_2}}, \tag{6.69}$$

which is equivalent to Eq. (6.44) which was derived starting from the slowly varying envelope equations Eqs. (6.40). This expression provides a simple way to extract the temporal growth rate from a system of coupled wave equations of the form of Eqs. (6.56)–(6.57) simply by replacing the constants C_1 and C_2 by their appropriate expressions in the coupled equations.

6.4.6   Conservation Laws in Three-Wave Coupling; The Manley–Rowe Relations

We will see in later chapters that under the slowly varying envelope equations, the system of equations (6.40) can be re-written using the action amplitude quantities related to each of the three involved waves and defined in Sect. 1.3.5, Eqs. (1.214)–(1.216), as:

$$(\partial_t + \nu_0 + v_{g0}\partial_{z0})\tilde{\alpha}_0 = -iK\tilde{\alpha}_1\tilde{\alpha}_2, \tag{6.70}$$

$$(\partial_t + \nu_1 + v_{g1}\partial_{z1})\tilde{\alpha}_1 = -iK\tilde{\alpha}_0\tilde{\alpha}_2^*, \tag{6.71}$$

$$(\partial_t + \nu_2 + v_{g2}\partial_{z2})\tilde{\alpha}_2 = -iK\tilde{\alpha}_0\tilde{\alpha}_1^*, \tag{6.72}$$

where $K = \gamma/|\tilde{\alpha}_0(0)|$ ($K \in \mathbb{R}$) is the temporal growth rate normalized to the initial action amplitude of the pump wave $|\tilde{\alpha}_0(0)|$ before any interaction with the other waves (i.e., the initial, un-depleted value). The tilde denotes a slowly varying envelope like in the rest of the chapter (and throughout most of the book). Recall that the complex action amplitudes simply represent any of the fluid variables associated with a wave in a plasma introduced in Sect. 1.3.1, i.e., the electric field, fluid velocity perturbation, or density modulation (for EPWs and IAWs), and verify $\mathcal{A}_j = |\tilde{\alpha}_j|^2$ where $\mathcal{A}_j$ is the action density for the wave j, i.e., the density of energy quanta N_j (multiplied by Planck's constant $\hbar$), $\mathcal{A} = \hbar N$ (cf. Sect. 1.3.5).

In the following we will neglect the damping of the waves, and assume 1D propagation along a same direction z (the group velocities can be < 0). Another classic form of these equations is then:

$$(\partial_t + v_{g0}\partial_z)\tilde{\beta}_0 = -\gamma\tilde{\beta}_1\tilde{\beta}_2 \,, \tag{6.73}$$

$$(\partial_t + v_{g1}\partial_z)\tilde{\beta}_1 = \gamma\tilde{\beta}_0\tilde{\beta}_2^* \,, \tag{6.74}$$

$$(\partial_t + v_{g2}\partial_z)\tilde{\beta}_2 = \gamma\tilde{\beta}_0\tilde{\beta}_1^* \,, \tag{6.75}$$

where the variables $\tilde{\beta}_j$ are defined following

$$\tilde{\alpha}_j = |\tilde{\alpha}_0(0)|e^{i\theta_j}\tilde{\beta}_j \,, \tag{6.76}$$

with $\theta_0 = \pi/2$ and $\theta_1 = \theta_2 = 0$; the variables $\tilde{\beta}_j$ are simply another way to express the action amplitude in dimensionless units, normalizing to the initial amplitude of the pump such that

$$|\tilde{\beta}_j|^2 = \frac{\mathcal{A}_j}{\mathcal{A}_0(0)} \,. \tag{6.77}$$

We will not try to prove that these forms of the three-wave coupling equations are correct in general for any wave-coupling process: instead, we will derive the set of coupled equations specific to each instability in the later chapters, and then show that the system of equations always reduce to the system above when using action amplitudes—but we are now going to see why this *has* to be true.

Multiplying Eq. (6.73) by $\tilde{\beta}_0^*$ and adding its complex conjugate, and likewise, multiplying Eq. (6.74) by $\tilde{\beta}_1^*$ and adding the complex conjugate yields:

$$(\partial_t + v_{g0}\partial_z)\mathcal{A}_0 = -2\gamma\mathcal{A}_0(0)\text{Im}[\tilde{\beta}_0^*\tilde{\beta}_1\tilde{\beta}_2] \,, \tag{6.78}$$

$$(\partial_t + v_{g1}\partial_z)\mathcal{A}_1 = 2\gamma\mathcal{A}_0(0)\text{Im}[\tilde{\beta}_0^*\tilde{\beta}_1\tilde{\beta}_2] = -(\partial_t + v_{g0}\partial_z)\mathcal{A}_0 \,. \tag{6.79}$$

Proceeding similarly with the third equation gives the following relations:

$$(\partial_t + v_{g0}\partial_z)\mathcal{A}_0 = -(\partial_t + v_{g1}\partial_z)\mathcal{A}_1 \tag{6.80}$$

$$(\partial_t + v_{g0}\partial_z)\mathcal{A}_0 = -(\partial_t + v_{g2}\partial_z)\mathcal{A}_2 \tag{6.81}$$

$$(\partial_t + v_{g1}\partial_z)\mathcal{A}_1 = (\partial_t + v_{g2}\partial_z)\mathcal{A}_2 \,. \tag{6.82}$$

Introducing the total spatial derivative for the wave j moving at the group velocity v_{gj}, $d_z\mathcal{A}_j[t, z(t)] = (1/v_{gj})(\partial_t + v_{gj}\partial_z)\mathcal{A}_j$, allows to rewrite the equations above as

$$\boxed{d_z(v_{g0}\mathcal{A}_0 + v_{g1}\mathcal{A}_1) = d_z(v_{g0}\mathcal{A}_0 + v_{g2}\mathcal{A}_2) = d_z(v_{g1}\mathcal{A}_1 - v_{g2}\mathcal{A}_2) = 0} \,. \tag{6.83}$$

These relations are called the Manley–Rowe relations [8], and are applicable to a wide range of physics areas, in particular, nonlinear optics. They express the fact that for each energy quantum from wave 0 disappearing in the three-wave coupling, one new energy quantum appears in both wave 1 and wave 2.

Alternatively, we can also combine these relations to express the conservation of the flux of energy (in energy per unit time and area) $\mathcal{I}_j = \omega_j v_{gj} \mathcal{A}_j$:

$$
\begin{aligned}
d_z(\mathcal{I}_0 + \mathcal{I}_1 + \mathcal{I}_2) &= d_z(\omega_0 v_{g0} \mathcal{A}_0 + \omega_1 v_{g1} \mathcal{A}_1 + \omega_2 v_{g2} \mathcal{A}_2) \\
&= \omega_0(-d_z v_{g1} \mathcal{A}_1) + \omega_1 d_z v_{g1} \mathcal{A}_1 + \omega_2 d_z v_{g1} \mathcal{A}_1 \\
&= (-\omega_0 + \omega_1 + \omega_2) d_z v_{g1} \mathcal{A}_1 \\
&= 0 .
\end{aligned}
\tag{6.84}
$$

Obviously, the frequency-matching condition $\omega_0 = \omega_1 + \omega_2$ can also be understood as the conservation of energy via the waves' quanta, $\hbar\omega_0 = \hbar\omega_1 + \hbar\omega_2$: once again, this simply illustrates the idea that for each quantum from wave 0 disappearing, a new quantum appears in both wave 1 and wave 2.

This is essentially why any three-wave instability *has* to reduce to a system of equations like Eqs. (6.70)–(6.72) or (6.73)–(6.75) above, in order to satisfy energy conservation. Note that similarly, the wave-vector matching condition $\mathbf{k}_0 = \mathbf{k}_1 + \mathbf{k}_2$ expresses conservation of momentum in the three-wave coupling. For example, in the case of SBS or SRS (two EMWs coupled via an EPW or IAW), frequency- and wave-vector-matching imply that the "pump" (the laser) will deposit some of its energy and momentum not only in the scattered light wave but also in the plasma wave. This can lead to significant transfer of energy to the plasma electrons for SRS (cf. Sect. 8.4), or to momentum deposition in the bulk plasma motion in the case of SBS or CBET (cf. Sect. 5.7).

6.5 Spatial Amplification in a Non-Uniform Plasma: The Rosenbluth Gain Formula

When the plasma is non-uniform (which it always is in real life), the wave-vectors and dispersion relations of the waves that can exist in the plasma vary in space, as we have seen in Chap. 3. Therefore, the phase-matching relations for three-wave coupling can only exist in a localized region of space. It is intuitively obvious that the spatial amplification of the daughter waves will only take place in the vicinity of that phase-matching location. This section will describe and quantify the amplification level and the physical size of the amplification region, and derive the so-called Rosenbluth gain formula, after the classic 1972 paper [9].

6.5.1 Coupled Mode Equations in a Non-Uniform Plasma

Let us consider the steady-state problem only, in one spatial dimension. We therefore neglect the time variations of the plasma discussed in Sect. 3.5, and assume that we have a triplet of waves labeled 0 (the pump), 1 and 2 with perfect frequency matching, $\omega_0 = \omega_1 + \omega_2$. On the other hand, the k-matching relation is now assumed to be satisfied only at one location in space, which we take as $z = 0$. We define $\kappa(z)$ as the k-vector mismatch between the waves:

$$\kappa(z) = k_0(z) - k_1(z) - k_2(z) , \quad \kappa(0) = 0. \tag{6.85}$$

Throughout this section we will assume that the inhomogeneity is approximately linear over the growth region, so that

$$\kappa(z) \approx \kappa' z, \tag{6.86}$$

where κ' is a constant.

We saw in Sect. 3.1 that the waves can be modeled using the WKB approximation:

$$a_j(z,t) = \frac{1}{2}\tilde{a}_j(z)\exp\left[i\psi_j(0) + i\int \delta k_j(z)dz\right], \tag{6.87}$$

$$\psi_j(0) = k_j(0)z - \omega_j t, \tag{6.88}$$

$$\delta k_j(z) = k_j(z) - k_j(0). \tag{6.89}$$

The slowly varying envelope (SVE) approximation can be obtained via the same procedure as in Sect. 6.4.1. It is customary to assume that the plasma conditions do not vary significantly over the size of the amplification region, so we can assume a constant group velocity and neglect the term $\propto k'_j(z)$ in the derivation of the propagator under the SVE approximation. All that is left is a phase mismatch: defining $\psi_j(z,t) = \psi_j(0) + \int \delta k_j(z)dz$, we have

$$\psi_0 - \psi_1 - \psi_2 = i\int \kappa(z)dz. \tag{6.90}$$

The coupled mode equations then take the form:

$$
\begin{aligned}
(\partial_t + \nu_0 + v_{g0}\partial_z)\tilde{a}_0 &= iK_0\tilde{a}_1\tilde{a}_2 e^{-i\int\kappa(z)dz} , \\
(\partial_t + \nu_1 + v_{g1}\partial_z)\tilde{a}_1 &= iK_1\tilde{a}_0\tilde{a}_2^* e^{i\int\kappa(z)dz} , \\
(\partial_t + \nu_2 + v_{g2}\partial_z)\tilde{a}_2 &= iK_2\tilde{a}_0\tilde{a}_1^* e^{i\int\kappa(z)dz} .
\end{aligned}
\tag{6.91}
$$

In the following we will solve the system of coupled equations to express both the amplification of the daughter waves and the size of the amplification region, under the assumptions of negligible pump depletion (i.e., $\tilde{a}_0 = cst.$ in the equations above) and steady-state (i.e., $\partial_t = 0$).

6.5.2 Weakly Damped Waves

We first consider the case where the two daughter waves are weakly damped. This is similar to the derivation of Rosenbluth in Ref. [9]. Our set of coupled equations for the daughter waves, assuming steady-state and no pump depletion, is:

$$v_{g1}\tilde{a}_1' = i K_1 \tilde{a}_0 \tilde{a}_2^* e^{i \int \kappa(z) dz} , \qquad (6.92)$$

$$v_{g2}\tilde{a}_2' = i K_2 \tilde{a}_0 \tilde{a}_1^* e^{i \int \kappa(z) dz} , \qquad (6.93)$$

where the prime represents the spatial derivative along z and the two coefficients K_1, K_2 will generally be > 0.

Multiplying the first equation by $\tilde{a}_1^*$ and adding the complex conjugate to obtain $\partial_z |\tilde{a}_1|^2$, and proceeding likewise for $\partial_z |\tilde{a}_2|^2$ gives the Manley–Rowe conservation law (cf. Sect. 6.4.6):

$$\frac{\partial |\tilde{a}_1|^2}{\partial z} - \frac{K_1}{K_2} \frac{v_{g2}}{v_{g1}} \frac{\partial |\tilde{a}_2|^2}{\partial z} = 0 . \qquad (6.94)$$

If the two daughter waves are co-propagating ($v_{g1}, v_{g2} > 0$), then their spatial amplitude profiles will be proportional. If they are counter-propagating (e.g., $v_{g1} > 0$, $v_{g2} < 0$), then the sum of their intensities (after scaling $|\tilde{a}_2|^2$ by $K_1 v_{g2}/K_2 v_{g1}$) will be constant. Total flux conservation is not ensured here because we are neglecting the depletion of the pump.

Let us start with the simpler situation where the two daughter waves are co-propagating, with v_{g1} and v_{g2} both > 0. Eliminating $\tilde{a}_2'$ from the system of equations gives:

$$\tilde{a}_1'' - i\kappa\tilde{a}_1' - \frac{\gamma_0^2}{v_{g1}v_{g2}}\tilde{a}_1 = 0 , \qquad (6.95)$$

where $\gamma_0 = \sqrt{K_1 K_2}|\tilde{a}_0|$ is the temporal growth rate, as defined earlier in Eq. (6.44). This equation can be simplified further by introducing the variable α_1 defined as

$$\tilde{a}_1 = \alpha_1 \exp\left[\frac{i}{2} \int \kappa(z) dz\right] , \qquad (6.96)$$

leading to:

$$\alpha_1'' + \left(\frac{\kappa^2}{4} - \frac{\gamma_0^2}{v_{g1}v_{g2}} \right) \alpha_1 = 0. \tag{6.97}$$

Here we assumed that $|\kappa'| \ll \gamma_0^2/v_{g1}v_{g2}$, which as we will see below means that the spatial amplification must be $\gg 1$.

This equation is a parabolic cylinder equation (because $\kappa = \kappa' z$ where κ' is assumed constant over the amplification region), whose solutions are non-trivial (cf. Problem 6.3); however, we can immediately notice some important features:

- if $\kappa^2 > 4\gamma_0^2/v_{g1}v_{g2}$, i.e., $|z| > z_t$ where

$$z_t = \frac{2\gamma_0}{|\kappa'|\sqrt{v_{g1}v_{g2}}}, \tag{6.98}$$

then the solution for α_1 is purely oscillatory, with a spatially dependent spatial frequency that increases with z, i.e., away from the phase matching (and spatial growth) location (cf. Problem 6.4).
- if $|z| < z_t$, i.e., in the vicinity of the growth region, the term in parenthesis in Eq. (6.97) is < 0, thus we expect an exponential behavior. For $|z| \ll z_t$, neglecting the first term in the parenthesis in Eq. (6.97) gives $\alpha_1(z) \sim \exp[\gamma_0 z/\sqrt{v_{g1}v_{g2}}]$ so we can already expect the spatial amplification from $-z_t$ to z_t to have a gain exponent in intensity on the order of

$$G \approx \frac{2\gamma_0 z_t}{\sqrt{v_{g1}v_{g2}}} = \frac{2\gamma_0^2}{|\kappa'|v_{g1}v_{g2}}. \tag{6.99}$$

The behavior is reminiscent of the WKB analysis and Airy solution near the turning point that we described in Chap. 3. Therefore, we apply a WKB analysis again, and look for a solution of the type

$$\alpha_1 = A(z) \exp[i\phi(z)], \tag{6.100}$$

which after insertion into Eq. (6.97) gives

$$A'' + 2i\phi'A' + i\phi''A - \phi'^2A + \left(\frac{\kappa'^2}{4}z^2 - \frac{\gamma_0^2}{v_{g1}v_{g2}} \right) A = 0. \tag{6.101}$$

Away from the turning points (which are located at $z = \pm z_t$), the WKB approximation assumes that the function in parenthesis in Eq. (6.97) varies on shorter spatial scales than the envelope A; therefore, collecting the real terms in the equation above and neglecting A'' leads to

$$\phi' = \pm \frac{\gamma_0}{\sqrt{v_{g1} v_{g2}}} \sqrt{(z/z_t)^2 - 1}. \tag{6.102}$$

As we had intuited above, we see that for $|z| < z_t$, ϕ' is imaginary, i.e., $\alpha_1 = A \exp[i\phi]$ varies exponentially vs. z: this is the amplification region. The spatial amplification from $-z_t$ to z_t can then be approximated by integrating the phase between the two turning points,

$$|\alpha_1(z_t)| = |\alpha_1(-z_t)| \exp \left[\frac{\gamma_0}{\sqrt{v_{g1} v_{g2}}} \int_{-z_t}^{z_t} \sqrt{1 - (z/z_t)^2} dz \right]. \tag{6.103}$$

Using $\int_{-1}^{1} \sqrt{1 - x^2} dx = \pi/2$,[3] and the expression of the turning point z_t from Eq. (6.98), finally gives the amplification factor in intensity

$$|\tilde{a}_1(z_t)|^2 = |\tilde{a}_1(-z_t)|^2 e^{G_R}, \tag{6.104}$$

where G_R is the so-called Rosenbluth gain,

$$\boxed{G_R = \frac{2\pi \gamma_0^2}{|\kappa' v_{g1} v_{g2}|}}. \tag{6.105}$$

As expected, the gain is similar to our previous guess from Eq. (6.99), up to a numerical pre-factor.

The amplification length l_a in the weakly damped regime is equal to $2z_t$, i.e.,

$$\boxed{l_a = \frac{4\gamma_0}{|\kappa'| \sqrt{v_{g1} v_{g2}}}}. \tag{6.106}$$

To illustrate the process, we show in Fig. 6.3 a numerical solution of Eqs. (6.92)–(6.93) for a gain exponent $G_R = 3$ (i.e., intensity amplification ≈ 20) and $|\tilde{a}_1(0)| = 1$ (units are arbitrary). As discussed above, the behavior is oscillatory for $|z| > z_t$, and quite reminiscent of the behavior of a light wave's electric field as it approaches its turning point (cf. Sects. 3.1, 3.2, and Fig. 3.1): as we approach $\pm z_t$ (from the "outside," i.e., $|z| > z_t$), the oscillation wavelength of the envelope increases, and by virtue of flux conservation, the amplitude increases as well (similar to field swelling, like we saw in Sect. 3.1; cf. Problem 6.4).[4] The amplification region is limited to $|z| < z_t$.

---

[3] Note that $y = \sqrt{1 - x^2}$ for $x \in [-1, 1]$ is the equation of a circle of radius 1, so the integral is half the area of the unit circle, or $\pi/2$.

[4] The oscillation of the envelope on the $z < z_t$ side is not discernable on the scale of the plot but is present.

Fig. 6.3 Spatial amplification for un-damped waves in co-propagating geometry ($v_{g1}v_{g2} > 0$), for a spatial amplification gain exponent $G_R = 3$ (i.e., amplification of ≈ 20) and $|\tilde{a}_1(0)| = 1$. The second daughter wave's profile $|\tilde{a}_2(z)| \propto |\tilde{a}_1(z)|$. The amplification occurs over a length $l_a = 2z_t$ around the phase-matching location (here at $z = 0$)

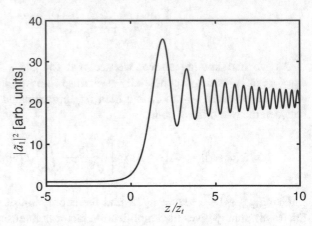

Fig. 6.4 Spatial amplification for two counter-propagating daughter waves (wave #2, in gray, propagates to the left; wave #1, in black, propagates toward $z > 0$). The spatial amplification gain exponent is $G_R = 3$

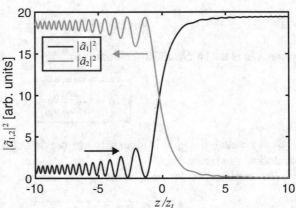

By virtue of the Manley–Rowe conservation relations and Eq. (6.94) the spatial profile of $|\tilde{a}_2|^2$ looks the same as for $|\tilde{a}_1|^2$, except for a scaling factor.

The counter-propagating case is more complicated, as was noted by Rosenbluth [9]. Indeed, in this case the term in parenthesis in Eq. (6.97) is always positive, which means that the behavior of $\tilde{a}_1$ is always oscillatory (see also [10]). The derivation of the gain requires an analytic continuation in the complex plane; the mathematical result for the intensity gain exponent ends up being identical to the one for the co-propagating case, Eq. (6.105). To illustrate the behavior, a numerical simulation for the counter-propagating geometry is shown in Fig. 6.4.

6.5.3 Strongly Damped Plasma Wave

We are now going to describe the spatial amplification in the presence of damping of the daughter waves, and more specifically when wave #2 (the plasma wave) is strongly damped (by which we mean that the damping rate is greater than the spatial amplification rate, cf. Sect. 6.4.4[5]). As we will see, the amplification factor turns out to be the same as in the un-damped case described above, Eq. (6.105) [11]; however, the size of the amplification region and the spatial profiles of the daughter waves' amplitudes are going to be different.

We start from the following system of coupled equations:

$$(v_{g1}\partial_z + v_1)\tilde{a}_1 = iK_1\tilde{a}_0\tilde{a}_2^* e^{i\int\kappa(z)dz} , \tag{6.107}$$

$$(v_{g2}\partial_z + v_2)\tilde{a}_2 = iK_2\tilde{a}_0\tilde{a}_1^* e^{i\int\kappa(z)dz} . \tag{6.108}$$

We introduce the variable α_2 defined by

$$\alpha_2 = \tilde{a}_2 e^{-i\int\kappa(z)dz} . \tag{6.109}$$

The system of coupled equations then becomes:

$$(v_{g1}\partial_z + v_1)\tilde{a}_1 = iK_1\tilde{a}_0\alpha_2^* , \tag{6.110}$$

$$(v_2 + iv_{g2}\kappa'z)\alpha_2 = iK_2\tilde{a}_0\tilde{a}_1^* , \tag{6.111}$$

where we assumed that the wave #2 was strongly damped, i.e., $v_2\alpha_2 \gg v_{g2}\alpha_2'$, and that the k-vector mismatch was approximately linear in the region of interest, $\kappa(z) \approx \kappa'z$ with κ' a constant.

Eliminating α_2 gives the following equation for $\tilde{a}_1$:

$$(v_{g1}\partial_z + v_1)\tilde{a}_1 = \frac{\gamma_0^2}{v_2 - iv_{g2}\kappa'z}\tilde{a}_1, \tag{6.112}$$

where like before, $\gamma_0^2 = K_1K_2|\tilde{a}_0|^2$ is the temporal growth rate.

Multiplying by $\tilde{a}_1^*$ and adding the complex conjugate gives the following relation for the evolution of the intensity $|\tilde{a}_1|^2$:

$$\frac{\partial|\tilde{a}_1|^2}{\partial z} = 2\frac{\gamma_0^2}{v_{g1}v_2}\left[\frac{1}{1 + (z/z_c)^2} - \frac{v_1v_2}{\gamma_0^2}\right]|\tilde{a}_1|^2 , \tag{6.113}$$

where the quantity z_c, which we can already recognize as directly connected to the length of the amplification region, is defined as

$$z_c = \frac{v_2}{|\kappa'v_{g2}|} . \tag{6.114}$$

---

[5] Per this criterion only one daughter wave can be strongly damped, otherwise neither wave can grow and there cannot be an instability.

Note that the second term in the bracket in Eq. (6.113) is the ratio of the threshold for the instability $\nu_1\nu_2$ to the squared temporal growth rate, as seen in Sect. 6.4.3. We will generally assume that the instability is well above threshold, $\nu_1\nu_2/\gamma_0^2 \ll 1$ (which is in fact equivalent to saying that the amplification of wave #1 should largely dominate over its absorption, i.e., wave #1 needs to be in the weakly damped regime; cf. footnote 5 above and Problem 6.5). We can also verify that if $\kappa' = 0$, i.e., in the homogeneous plasma limit, then $|\tilde{a}_1(z)|^2 = |\tilde{a}_1(0)|^2 \exp[2\Gamma z]$, where $\Gamma = (\gamma_0^2 - \nu_1\nu_2)/(v_{g1}\nu_2)$ is the uniform spatial growth rate, as seen in Sect. 6.4.4 (Eq. 6.48).

For $\kappa' \neq 0$, and in the limit of $\nu_1\nu_2/\gamma_0^2 \ll 1$, Eq. (6.113) can be integrated, giving

$$|\tilde{a}_1(z)|^2 = |\tilde{a}_1(z_i)|^2 \exp\left[2\frac{\gamma_0^2}{|\kappa' v_{g1} v_{g2}|} (\arctan(z/z_c) - \arctan(z_i/z_c))\right]. \quad (6.115)$$

Therefore, when integrating from an initial position $z_i < 0$ to a final position $z_f > 0$ which are both well outside the amplification region, i.e., $|z_i|, z_f \gg z_c$, we obtain

$$|\tilde{a}_1(z_f)|^2 = |\tilde{a}_1(z_i)|^2 \exp[G_R] \,, \quad G_R = \frac{2\pi\gamma_0^2}{|\kappa' v_{g1} v_{g2}|}. \quad (6.116)$$

We recover the expression of the Rosenbluth gain derived in the previous section for the case of un-damped daughter waves, Eq. (6.105). However, the size of the amplification region l_a is different, and is now defined as

$$l_a = 2z_c = 2\frac{\nu_2}{|\kappa' v_{g2}|}. \quad (6.117)$$

We see that the size of the amplification region depends on the damping of wave #2 rather than the temporal growth rate like in the weakly damped case seen earlier in Eq. (6.106).

The expression for the plasma wave amplitude follows from Eq. (6.111). Taking the expression for the spatial evolution of wave #1, Eq. (6.115), we immediately obtain

$$\frac{|\tilde{a}_2(z)|}{|\tilde{a}_2(0)|} = \frac{|\tilde{a}_1(z)|/|\tilde{a}_1(0)|}{\sqrt{1 + (z/z_c)^2}}. \quad (6.118)$$

The amplification process in the strongly damped regime is illustrated in Fig. 6.5. Unlike for the un-damped regime, the amplitudes of the daughter waves are not modulated; the amplitude of wave #2 is maximum in the vicinity of the resonance region and drops on the outside due to its damping.

The case where the damping of both waves is included, i.e., when $\nu_1\nu_2/\gamma_0^2$ is kept in Eq. (6.113), is discussed in Problem 6.5.

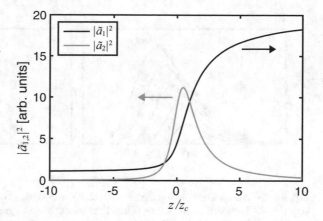

Fig. 6.5 Spatial amplification of the daughter waves a_1 and a_2 when a_2 is strongly damped (in arbitrary units, for $G_R = 3$)

6.6 Absolute vs. Convective Instabilities

6.6.1 Physical Picture

So far we have mostly focused our attention on the spatial amplification of waves through three-wave coupling. The full spatiotemporal instability analysis is more complex; in this section we provide a more complete physics picture and discuss the distinction between "absolute" and "convective" instabilities.

Let us first define these terms:

- An *absolute instability* is an instability which can grow exponentially in time[6] from a fixed location in space.
- A *convective instability* is an instability that grows exponentially *in a moving frame*, but where there is no infinite temporal growth from any location in space.

In the convective regime, a fixed location in space might see temporal growth for only a limited amount of time, followed by a decay after the amplification region, which is moving at some velocity V_c which we will define below, has passed. This is illustrated in Fig. 6.6.

We can attempt to give an intuitive physical picture for the nature of the instability. Consider a small initial disturbance for each of the daughter waves at $t = 0$, each propagating at the group velocity of its wave. If the system is unstable, each disturbance will start to grow as it propagates, taking energy off the pump (which is assumed constant and un-depleted in the following); while growing, the perturbations also extend in size. If the waves propagate in opposite directions, and if the growth is strong enough, the size increase can lead to net growth at some fixed spatial location even though the two waves keep moving away from each other: this

---

[6] In reality growth will always be limited by nonlinear and saturation effects, which are ignored in this section.

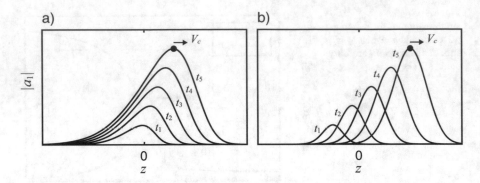

Fig. 6.6 (**a**) Absolute instability: the location of the maximum may move (like shown here), but some spatial locations (such as $z = 0$ here) see an exponential growth vs. time. (**b**) Convective instability: exponential temporal growth occurs only in the frame moving at the velocity of the maximum V_c, but a fixed point in space will never see infinite temporal growth (the instability will just "pass by" the location)

is the absolute instability. If the growth is not strong enough, or if the waves are moving in the same direction, the perturbation will simply "pass by" any location in space as illustrated in Fig. 6.6. All these points will be quantified below as we analyze the properties of the Green's function of the three-wave system below (cf. also [3]).

6.6.2   Green's Function of the Three-Wave System

Let us now go back to the 1D system of coupled equations (6.40); we assume a uniform and infinite plasma and ignore pump depletion effects—i.e., we consider the pump amplitude a_0 constant:

$$(\partial_t + v_1 + v_{g1}\partial_{z1})\tilde{a}_1 = i K_1 \tilde{a}_0 \tilde{a}_2^* \,, \tag{6.119}$$

$$(\partial_t + v_2 + v_{g2}\partial_{z2})\tilde{a}_2 = i K_2 \tilde{a}_0 \tilde{a}_1^* \,. \tag{6.120}$$

The group velocities v_{g1}, v_{g2} can be positive or negative. Combining these two equations by applying the operator from the second equation to the first leads to

$$\left[(\partial_t + v_2 + v_{g2}\partial_z)(\partial_t + v_1 + v_{g1}\partial_z) - \gamma_0^2\right]\tilde{a}_1 = 0 \,, \tag{6.121}$$

where γ_0 is the temporal growth rate, cf. Eq. (6.44).

The behavior of an initial disturbance at $t = 0$ and $z = 0$ is described by the Green's function $G(z, t)$ of the equation above, which satisfies

$$\left[(\partial_t + v_2 + v_{g2}\partial_z)(\partial_t + v_1 + v_{g1}\partial_z) - \gamma_0^2\right] G(z, t) = \delta(z)\delta(t) \,. \tag{6.122}$$

To proceed we introduce the new variables $\zeta_1 = z - v_{g1}t$ and $\zeta_2 = z - v_{g2}t$, corresponding to the spatial coordinate in the frame moving with the daughter wave #1 or #2, respectively. Changing from the variables (z, t) to (ζ_1, ζ_2), we use the transformations

$$\partial_t = -v_{g1}\partial_{\zeta_1} - v_{g2}\partial_{\zeta_2}, \tag{6.123}$$

$$\partial_z = \partial_{\zeta_1} + \partial_{\zeta_2} \tag{6.124}$$

and obtain

$$\Delta v \left[v_1 \partial_{\zeta_1} - v_2 \partial_{\zeta_2} - \Delta v \partial_{\zeta_1} \partial_{\zeta_2} + \frac{v_1 v_2 - \gamma_0^2}{\Delta v} \right] G(\zeta_1, \zeta_2) = \delta(\zeta_1)\delta(\zeta_2), \tag{6.125}$$

with $\Delta v = v_{g2} - v_{g1}$.

The equation can be further simplified by introducing the new variable $\hat{G}$ such that

$$G = \hat{G} \exp\left[\frac{v_1 \zeta_2 - v_2 \zeta_1}{\Delta v} \right], \tag{6.126}$$

leading to:

$$\left[\Delta v^2 \partial_{\zeta_1} \partial_{\zeta_2} + \gamma_0^2 \right] \hat{G} = -\delta(\zeta_1)\delta(\zeta_2). \tag{6.127}$$

The Green's function can be found by first solving the homogeneous equation with solution h,

$$\left[\partial_{\zeta_1} \partial_{\zeta_2} + \frac{\gamma_0^2}{\Delta v^2} \right] h = 0. \tag{6.128}$$

We make an ansatz and look for solutions in the form of an infinite series:

$$h = \sum_{n=0}^{\infty} \alpha_n \zeta_1^n \zeta_2^n. \tag{6.129}$$

Inserting into Eq. (6.128) and looking for the recurrence relation ensuring that the infinite sum is equal to zero, we obtain the following expression for the series coefficients α_n:

$$\alpha_n = (-1)^n \frac{[\gamma_0/\Delta v]^{2n}}{(n!)^2} \alpha_0, \tag{6.130}$$

leading to

$$h = \alpha_0 \sum_{n=0}^{\infty} \frac{(-\zeta_1 \zeta_2 \gamma_0^2 / \Delta v^2)^n}{(n!)^2} . \tag{6.131}$$

At this point we can recognize that the resulting expression matches the definition of the modified Bessel function of the first kind I_0:

$$I_0(x) = \sum_{n=0}^{\infty} \frac{1}{(n!)^2} \left(\frac{x}{2}\right)^{2n} . \tag{6.132}$$

Our solution then becomes

$$h = \alpha_0 I_0 \left[\frac{2\gamma_0}{|\Delta v|} \sqrt{-\zeta_1 \zeta_2} \right] \tag{6.133}$$

(the constant α_0 will not be relevant for the rest of the discussion). Since Eq. (6.127) is first order with respect to each of the variables ζ_1, ζ_2, the full solution for $\hat{G}$ is obtained by multiplying the homogeneous solution h by Heaviside step functions:

$$\hat{G}(\zeta_1, \zeta_2) = I_0 \left[\frac{2\gamma_0}{|\Delta v|} \sqrt{-\zeta_1 \zeta_2} \right] H(\pm \zeta_1) H(\mp \zeta_2) . \tag{6.134}$$

The two signs in the Heaviside functions need to be opposite in order to recover the minus sign in front of the delta functions in Eq. (6.127)—cf. Problem 6.6.

And finally, using the asymptotic expansion $I_0(x) \approx e^x / \sqrt{2\pi x}$ for $x \to \infty$ and switching back from $\hat{G}$ to G, we obtain the final asymptotic expression of our solution valid for $t \to \infty$:

$$G(z, t) \propto (-\zeta_1 \zeta_2)^{-1/4} \exp\left[f(\zeta_1, \zeta_2) \right] H(\pm \zeta_1) H(\mp \zeta_2) , \tag{6.135}$$

$$f(\zeta_1, \zeta_2) = \frac{2\gamma_0}{|\Delta v|} \sqrt{-\zeta_1 \zeta_2} + \frac{1}{\Delta v} (v_1 \zeta_2 - v_2 \zeta_1) . \tag{6.136}$$

We notice that the Heaviside functions ensure that $\zeta_1 \zeta_2 < 0$ and thus that the term in the square root term remains positive. As illustrated in Fig. 6.7, this means that at a fixed time t, the instability is spatially limited to the domain $[\min(v_{gj})t, \max(v_{gj})t]$, where $j \in \{1, 2\}$. Two situations emerge:

- if $v_{g1} v_{g2} > 0$: the instability domain is getting larger with time but its boundaries move in the same direction: this means that a fixed location in space can only be inside the domain for a finite amount of time, thus eliminating the possibility of absolute instability (Fig. 6.7a);
- if $v_{g1} v_{g2} < 0$: the domain extends in time in both directions, $z > 0$ and $z < 0$, so once a fixed spatial location finds itself inside the domain it will stay there forever. Absolute instability is possible.

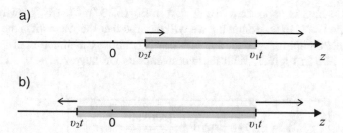

Fig. 6.7 (**a**) When $v_{g1}v_{g2} > 0$: the z-locations that can satisfy $\zeta_1\zeta_2 < 0$ are contained in the spatial domain $[\min(v_{gj})t, \max(v_{gj})t]$ (shown in gray in the figure) which is stretching in time but also moving, so that any location in space will only be inside this domain for a finite amount of time. (**b**) When $v_{g1}v_{g2} < 0$ the z-locations that can satisfy $\zeta_1\zeta_2 < 0$ are also contained in the spatial domain $[\min(v_{gj})t, \max(v_{gj})t]$—but now, the domain boundaries move in opposite directions so that once a point in space is in the domain, it will stay in forever. The situation in (**a**), i.e., $v_{g1}v_{g2} > 0$, cannot support absolute instabilities; the situation in (**b**), $v_{g1}v_{g2} < 0$, can: this defines a necessary (but not sufficient) condition for absolute instability

In summary: the condition $v_{g1}v_{g2} < 0$ constitute a necessary (but not sufficient) condition for the existence of absolute instability.

6.6.3   Convective and Absolute Growth Rates and Thresholds

To quantify the absolute vs. convective behavior of the instability, we first calculate the position of the maximum of Eq. (6.135) at a fixed time. Since the
asymptotic behavior is dominated by the exponential term, with the term $\propto$
$(-\zeta_1\zeta_2)^{-1/4}$ only contributing as a small correction, we are looking for z_m such that
$\partial_z f[\zeta_1(z,t), \zeta_2(z,t)] = 0$ where f is given by Eq. (6.136).
   We obtain:

$$z_m(t) = V_c t \;, \quad V_c = \frac{1}{2}\left[v_\Sigma \pm \frac{|\Delta v \Delta v|}{\sqrt{4\gamma_0^2 + \Delta v^2}} \right], \tag{6.137}$$

with $\Delta v = v_2 - v_1$, $v_\Sigma = v_{g1} + v_{g2}$ and $\Delta v = v_{g2} - v_{g1}$. V_c is the *convective velocity*, i.e.,
the velocity at which the location of the maximum moves, as illustrated in Fig. 6.6.
The ambiguity about the sign can be cleared by calculating the corresponding
growth rate γ_c, which is the *convective growth rate*, i.e., the growth rate in the frame
moving at V_c. Inserting the expression for z_m back into f leads to $f(z_m, t) = \gamma_c t$,
with

$$\gamma_c = \mp \mathrm{sign}(\Delta v \Delta v)\sqrt{\gamma_0^2 + \Delta v^2/4} - \frac{1}{2}(v_1 + v_2), \tag{6.138}$$

where the $\mp$ sign is opposite to the $\pm$ sign in Eq. (6.137). Clearly, growth requires $\mp\text{sign}(\Delta v \Delta v) = 1$. In the following we will assume that the wave #2 is the one with the strongest damping, $v_2 > v_1$, and (arbitrarily) choose the direction of z such that $v_{g1} > v_{g2}$ like in Fig. 6.7. With these conventions we have $\Delta v > 0$, $\Delta v < 0$ and therefore,

$$
V_c = \frac{1}{2} \left[v_{g1} + v_{g2} + \frac{|\Delta v \Delta v|}{\sqrt{4\gamma_0^2 + \Delta v^2}} \right] , \tag{6.139}
$$

$$
\gamma_c = \sqrt{\gamma_0^2 + \frac{\Delta v^2}{4}} - \frac{v_1 + v_2}{2} . \tag{6.140}
$$

Growth is only possible if the convective growth rate is positive, leading to

$$
\gamma_0 > \gamma_{th,c} = \sqrt{v_1 v_2} , \tag{6.141}
$$

as we had already established in Eq. (6.48).

It is also interesting to consider the weakly vs. strongly damped regimes. In the weakly damped limit where both v_1 and v_2 are negligible, we simply have $V_c \approx (v_{g1} + v_{g2})/2$ (i.e., the convective velocity is equal to the average of the two group velocities) and $\gamma_c \approx \gamma_0$.

In the limit of $v_2 \gg \gamma_0$, we have $\gamma_c \approx (\gamma_0^2 - \gamma_{th,c}^2)/v_2$ (thus $\gamma_c \approx \gamma_0^2/v_2$ well above threshold) and $V_c \approx v_{g1}$, i.e., the maximum follows the least damped wave. This will be a common situation for backscatter instabilities, where the damping of wave #2 (the Landau-damped plasma wave) is usually stronger than that of wave #1 (the scattered light wave). In this case the growth is spatial in nature, since the seed disturbance for wave #1 grows while propagating at v_{g1}. The spatial amplification rate is simply $\Gamma = \gamma_c/v_{g1}$ (since $d_t \tilde{a}_1 = v_{g1} d_z \tilde{a}_1$), i.e.,

$$
\Gamma = \frac{\gamma_0^2}{v_2 v_{g1}} . \tag{6.142}
$$

We recover the steady-state spatial amplification rate derived earlier in Sect. 6.4.4, Eq. (6.52). The convective amplification is illustrated in Fig. 6.8.

For the absolute instability, the growth rate can be calculated by fixing $z = 0$ in the expression for f above. We obtain $f(z = 0, t) = \gamma_{abs} t$, with

$$
\gamma_{abs} = \frac{2\gamma_0 \sqrt{|v_{g1} v_{g2}|} - v_2 |v_{g1}| - v_1 |v_{g2}|}{|v_{g1}| + |v_{g2}|} . \tag{6.143}
$$

Absolute instability requires $\gamma_{abs} > 0$, which leads to the condition

Fig. 6.8 Illustration of the spatial amplification under convective instability in the strongly damped regime, $v_2 \gg \gamma_0$, where $V_c \approx v_{g1}$

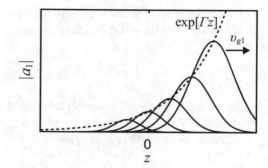

$$\boxed{\gamma_0 > \gamma_{th,abs} = \frac{1}{2}\sqrt{|v_{g1}v_{g2}|}\left(\frac{v_2}{|v_{g2}|} + \frac{v_1}{|v_{g1}|}\right).}$$ (6.144)

It is interesting to look at the ratio of the absolute to convective thresholds; we get:

$$\frac{\gamma_{th,abs}}{\gamma_{th,c}} = \frac{1}{2}\left(\sqrt{\frac{|v_{g1}|v_2}{|v_{g2}|v_1}} + \sqrt{\frac{|v_{g2}|v_1}{|v_{g1}|v_2}}\right).$$ (6.145)

First, since the term in parenthesis is of the form $x + 1/x$, which admits a minimum of 2 at $x = 1$, we see that we always have $\gamma_{abs,th} > \gamma_{th,c}$: absolute instability requires a higher coupling γ_0 than convective instability.

We can also see that when $|v_{g1}| \gg |v_{g2}|$ and $v_2 \gg v_1$, which is often the case for backscatter instabilities (since wave #1 is a light wave with velocity $\approx c$ in underdense plasma and wave #2 is a plasma wave subject to Landau damping), the ratio can become $\gg 1$, making the absolute instability harder to reach than the convective one.

6.6.4 Final Remarks

The analysis of the asymptotic behavior of the Green's function, while providing valuable physical insight into the nature of the instability, is far from complete and contains many restricting hypotheses. First, we only calculated the Green's function of the problem; to find the true solution to a physical problem we will still need to convolve the Green's function with the initial and boundary conditions.

The next important point is that our analysis assumed an infinite and uniform plasma. Accounting for the effects of finite length and inhomogeneity can change the conditions of existence of absolute instability, in particular:

- For a finite length (but uniform) plasma, it has been shown [11] that the instability can still be absolute if the length of the plasma exceeds some critical length

$$L_c = \frac{\pi}{2} \frac{\sqrt{|v_{g1}v_{g2}|}}{\gamma_0} . \tag{6.146}$$

- For an infinite plasma, Rosenbluth showed that a linear gradient in the phase mismatch $\Delta k(z) \propto z$, similar to what we described in Sect. 6.5, prevents the existence of absolute instability, even if the slope of the non-uniformity is very small [9].
- For a non-uniformity that is parabolic, $\Delta k(z) \propto z^2$, or even higher-order in power of z, absolute instability becomes possible again but with a new threshold that depends on the shape of the non-uniformity (and is higher than the equivalent absolute threshold in a uniform plasma) [9, 12–14].

Problems

6.1   List of Allowed Three-Wave Instabilities in Plasmas

Using the conditions that $\omega_0 > \omega_1, \omega_2$ and wave #2 is a plasma wave (EPW or IAW) with $\omega_{IAW} \ll \omega_{EPW} \leq \omega_{EMW}$, show that only the processes listed in Table 6.1 can exist as three-wave instabilities (i.e., list every other combination and show that it violates one of these conditions).

6.2   From the Dispersion Relation to the Coupled Equations for the Slowly Varying Envelopes

Starting from Eq. (6.62), express the frequency and wave-vectors as small variations around a pair $(\omega_1, \mathbf{k}_1)$ that exactly satisfies the dispersion relation of wave 1, i.e., $\omega = \omega_1 + \delta\omega$, $\mathbf{k} = \mathbf{k}_1 + \delta\mathbf{k}$ with $\hat{D}_1(\omega_1, \mathbf{k}_1) = 0$. Taylor-expand $\hat{D}_1(0)$ near $(\omega_1, \mathbf{k}_1)$ in Eq. (6.62), keeping only the term corresponding to the Stokes component on the right-hand side, inverse Fourier transform and show that the resulting equation is similar to the second equation from Eqs. (6.40). Proceed similarly for the third equation of Eqs. (6.40), starting from Eq. (6.63).

6.3   Rosenbluth Gain via Parabolic Cylinder Functions

- Find the change of variable transforming Eq. (6.97) into an equation of the form $f''(z) + (\frac{1}{4}z^2 - a)f(z) = 0$, which is one of the two Weber equations whose solutions constitute the parabolic cylinder functions.
- Consult the section on parabolic cylinder functions in Ref. [10]: match the different regimes of asymptotic expansions to the physical regimes discussed in Sect. 6.5.2 (co-propagating vs. counter-propagating, behavior inside vs. outside the amplification region), and recover the key results from that section.

6.4 Oscillatory Behavior Outside of the Amplification Region

- Analyze the asymptotic behavior of the scattered wave amplitude in the co-propagating case outside the amplification region using the WKB method (Sect. 3.1). Show that the spatial variation of the oscillation frequency (with an oscillation $\propto \exp[iCz^2]$ with C a constant) is accompanied by a reduction of the oscillation amplitude ($\propto 1/\sqrt{|z|}$) that is consistent with our choice of a linear variation of the wave-number mismatch, $\kappa(z) = \kappa'z$ (the variation in the spatial frequency and the amplitude of the oscillations are visible in Fig. 6.3).
- Derive an expression for the distance outside the amplification region where the modulation period becomes comparable to the wavelength (in which case the slowly varying envelope approximation breaks down).

6.5 Influence of the Absorption of Daughter Wave #1 on Spatial Amplification in the Regime of Strongly Damped Wave #2

- Considering Eq. (6.113), verify that spatial amplification is not possible when the instability is below threshold per the criterion established in Sect. 6.4.3, Eq. (6.48).
- Calculate the expression for the spatial amplification in the strongly damped limit near the instability threshold, i.e., without neglecting $\nu_1\nu_2/\gamma_0^2$. Show that the amplification gain peaks at a location

$$z_{max} = z_c\sqrt{\frac{1-\nu}{\nu}}, \tag{6.147}$$

where $\nu = \nu_1\nu_2/\gamma_0^2$ and then goes back down beyond z_{max}, eventually going back to zero.
- Show that the absorption fraction over the amplification length l_a (Eq. 6.117) is equal to $(2/\pi)G_R\nu$, and that ν is therefore a direct measure of the ratio of absorption to amplification of the wave #1 over the phase-matching region.

6.6 Green's Function for the Three-Wave Coupled Mode Equations

Show that the Green's function solution in Eq. (6.134) does indeed satisfy Eq. (6.127). Hint: you will need to show that $\partial_{\zeta_1}h|_{\zeta_2=0} = \partial_{\zeta_2}h|_{\zeta_1=0} = 0$, and use $I_0(0) = 1$ which will eliminate the constant α_0 from the expression of $\hat{G}$.

References

1. J.P. Palastro, J.G. Shaw, R.K. Follett, A. Colaïtis, D. Turnbull, A.V. Maximov, V.N. Goncharov, D.H. Froula, Phys. Plasmas **25**, 123104 (2018)
2. R.C. Davidson, *Methods in Nonlinear Plasma Theory* (Academic Press, New York, 1972)
3. D. Pesme, Fusion thermonucléaire inertielle par laser - partie 1 - volume 1: L'interaction laser-matière, Chap. 2 (Eyrolles, 1993)
4. D.W. Forslund, J.M. Kindel, E.L. Lindman, Phys. Fluids **18**, 1002 (1975)
5. J.F. Drake, P.K. Kaw, Y.C. Lee, G. Schmidt, C.S. Liu, M.N. Rosenbluth, Phys. Fluids **17**, 778 (1974)

6. R.W. Boyd, *Nonlinear Optics*, 3rd edn. (Academic Press, Orlando, FL, USA, 2008)
7. D. Turnbull, C. Goyon, G.E. Kemp, B.B. Pollock, D. Mariscal, L. Divol, J.S. Ross, S. Patankar, J.D. Moody, P. Michel, Phys. Rev. Lett. **118**, 015001 (2017)
8. J.M. Manley, H.E. Rowe, Proc. IRE **44**, 904 (1956)
9. M.N. Rosenbluth, Phys. Rev. Lett. **29**, 565 (1972)
10. M. Abramowitz, I. Stegun, *Handbook of Mathematical Functions: With Formulas, Graphs, and Mathematical Tables*, Applied Mathematics Series (Dover Publications, 1965)
11. D. Pesme, G. Laval, R. Pellat, Phys. Rev. Lett. **31**, 203 (1973)
12. G. Picard, T.W. Johnston, Phys. Fluids **28**, 859 (1985)
13. T.W. Johnston, G. Picard, J.P. Matte, V. Fuchs, M. Shoucri, Plasma Phys. Control. Fusion **27**, 473 (1985)
14. E.A. Williams, T.W. Johnston, Phys. Fluids B Plasma Phys. **1**, 188 (1989)

Chapter 7
Wave Coupling Instabilities via Ion Acoustic Waves

In this chapter we describe the three-wave coupling processes involving two light waves and an ion acoustic wave (IAW). These include stimulated Brillouin scattering (SBS) and crossed-beam energy transfer (CBET). In SBS, an intense light wave (the laser) couples to a weak light wave (from noise fluctuations) via an IAW, such that the ponderomotive force from the beat wave between the two light waves resonantly drives the IAW. CBET is similar, except that the second light wave is another laser beam; the beat between the two beams is not necessarily resonant with an IAW, and the associated amplification gains are usually much smaller than for SBS, but even a relatively small amplification gain can lead to significant energy transfer between the two beams.

We will also investigate the filamentation instability, which is technically a four-wave coupling process (three light waves and one IAW) with a smooth transition to forward SBS. This is the most common example of a situation where the anti-Stokes component during the scattering of a light wave off an IAW, which we mentioned in Sect. 6.4.5, has to be accounted for. The filamentation instability is also intimately connected to the self-focusing process, as we will describe.

7.1 Stimulated Brillouin Scattering (SBS)

Stimulated Brillouin scattering (SBS) is a three-wave instability consisting in the coupling of two EMWs (the "pump", i.e., the laser, and a scattered light wave) via an IAW. SBS can occur in any medium that can sustain acoustic waves and was first reported in crystals [1], followed by liquids [2], gases [3], and finally plasmas [4, 5].

As we shall see below, SBS in plasmas is generally most effective in the backscatter geometry, which means that it tends to reflect part of an intense light wave propagating in a plasma, leading to a loss of coupling to the target. This has been a high concern and the subject of an intense research effort since the early

© The Author(s), under exclusive license to Springer Nature Switzerland AG 2023
P. Michel, *Introduction to Laser-Plasma Interactions*, Graduate Texts in Physics,
https://doi.org/10.1007/978-3-031-23424-8_7

days of ICF; significant reflectivities were measured in experiments using $1.06\,\mu$m neodymium lasers [4–6], $10.6\,\mu$m CO_2 lasers [7–10], and later on with shorter wavelength lasers, using the second $(0.53\,\mu$m) [11], third $(0.35\,\mu$m) [12], and even fourth $(0.26\,\mu$m) [11] harmonics of a neodymium laser.

The subject is still highly relevant today, as very high reflectivities, in excess of 40%, were for example measured on preliminary experiments on the first test beamline of the National Ignition Facility laser [13]. SBS can even pose a risk of physical damage to the laser optics if the reflected light is intense enough [14].

7.1.1   Phase-Matched Coupled Mode Equations and Temporal Growth Rate

The key features of SBS in its most common form can be derived using phase-matched coupled equations as presented in Sect. 6.4. We start from the driven wave equations for an EMW and an IAW, Eqs. (6.20), (6.29). The pump laser and scattered light wave are represented by their normalized vector potentials $\mathbf{a}_0$ and $\mathbf{a}_1$ and the IAW by its density modulation δn_e. To simplify the notations, we will simply use $n_s = n_{s0} + \delta n_s$ and $\mathbf{v}_s = \mathbf{v}_{s0} + \delta\mathbf{v}_s$ for the fluid density and velocity perturbations and not track the (1) and (2) superscripts indicating first-order vs. second-order quantities [cf. Eqs. (6.22)–(6.23) and associated discussion].

For each light wave, the coupling term on the right-hand side, $\propto \delta n_e \delta\mathbf{v}_{e\perp}$, involves the IAW via δn_e and the other light wave via $\delta v_{e\perp}$ (recall that δn_e can only be associated with a longitudinal wave and not a light wave, cf. Sect. 1.3.1). Since we wish to describe the coupling of each wave to the beat wave from the other two waves, we write the following coupled wave equations for the light waves:

$$(\partial_t^2 + \omega_{pe0}^2 - c^2\nabla^2)\mathbf{a}_0 = -\omega_{pe0}^2 \frac{\delta n_e}{n_{e0}}\mathbf{a}_1 \,, \tag{7.1}$$

$$(\partial_t^2 + \omega_{pe0}^2 - c^2\nabla^2)\mathbf{a}_1 = -\omega_{pe0}^2 \frac{\delta n_e}{n_{e0}}\mathbf{a}_0 \,. \tag{7.2}$$

Now for the IAW, since we consider coupling to two EMWs, we can ignore all the terms involving a density modulation δn_e on the right-hand side of Eq. (6.30). We are left with the first two terms on the right-hand side; the ion velocity perturbation associated with the EMWs corresponds to the slow ion quiver in the light wave's electric field, $\delta v_i = (Ze/M_i\omega)E = (Zm/M_i)\delta v_e$: since the term $\propto \delta v_i$ is squared in Eq. (6.29), it is smaller than the first term on the right-hand side by a factor Zm/M_i and can be neglected.[1] We are simply left with the first term, with

---

[1] The first two terms on the right-hand side of Eq. (6.29) are the ponderomotive force on the electrons and on the ions (cf. Eqs. (5.8) and (5.19)); the ponderomotive force on the ions is smaller due to the higher ion inertia, cf. Sect. 5.1.

$\delta \mathbf{v}_e^2 = 2c^2 \mathbf{a}_0 \cdot \mathbf{a}_1$ (i.e., the ponderomotive force on the electrons from the beat wave) and obtain the following driven wave equation for the IAW, coupled to the previous two for the EMWs:

$$(\partial_t^2 - c_s^2)\frac{\delta n_e}{n_{e0}} = \frac{Zmc^2}{M_i}\nabla^2 \mathbf{a}_0 \cdot \mathbf{a}_1 \,. \tag{7.3}$$

Next we use enveloped expressions for the three waves:

$$\mathbf{a}_0 = \frac{1}{2}\tilde{\mathbf{a}}_0 e^{i\psi_0} + c.c. \,, \tag{7.4}$$

$$\mathbf{a}_1 = \frac{1}{2}\tilde{\mathbf{a}}_1 e^{i\psi_1} + c.c. \,, \tag{7.5}$$

$$\delta n_e = \frac{1}{2}\delta\tilde{n}_e e^{i\psi_2} + c.c. \,, \tag{7.6}$$

where $\psi_j = \mathbf{k}_j \cdot \mathbf{r} - \omega_j t$ (for $j \in \{0, 1, 2\}$), and we assume perfect phase matching, i.e., $\psi_0 = \psi_1 + \psi_2$. Using these expressions together with the slowly varying envelope approximation from Eq. (6.35) in Eqs. (7.1)–(7.3) above and collecting terms oscillating at the phase $\propto e^{i\psi_0}$, $\propto e^{i\psi_1}$, and $\propto e^{i\psi_2}$ lead to the three coupled mode equations for SBS for the slowly varying envelopes of the three waves:

$$(\partial_t + \mathbf{v}_{g0} \cdot \nabla)\tilde{a}_0 = -i\frac{\omega_{pe0}^2}{4\omega_0}\tilde{a}_1\frac{\delta\tilde{n}_e}{n_{e0}}\cos(\epsilon) \,, \tag{7.7}$$

$$(\partial_t + \mathbf{v}_{g1} \cdot \nabla)\tilde{a}_1 = -i\frac{\omega_{pe0}^2}{4\omega_1}\tilde{a}_0\frac{\delta\tilde{n}_e^*}{n_{e0}}\cos(\epsilon) \,, \tag{7.8}$$

$$(\partial_t + \mathbf{c}_s \cdot \nabla)\frac{\delta\tilde{n}_e}{n_{e0}} = -i\frac{Zm}{M_i}\frac{k_2^2 c^2}{4\omega_2}\tilde{a}_0\tilde{a}_1^*\cos(\epsilon) \,, \tag{7.9}$$

where ϵ is the angle between $\mathbf{a}_0$ and $\mathbf{a}_1$.

The temporal growth rate is immediately obtained from the same procedure explained in Sect. 6.4.3 (cf. Eq. (6.44)), i.e., taking $\nabla = 0$ in the equations above and solving vs. time, which gives

$$\gamma = \frac{k_2 c}{4}\frac{\omega_{pi}}{\sqrt{\omega_1\omega_2}}|\tilde{a}_0|\cos(\epsilon) \,, \tag{7.10}$$

where we used $\omega_{pi} = c_s/\lambda_{De}$ in the limit of $ZT_e \gg 3T_i$.

We immediately see that the growth is maximum when the two light waves' polarizations are aligned ($\cos(\epsilon) = 1$), which maximizes the beat wave amplitude between their electric fields, and when k_2 is maximum, which corresponds to the backscatter geometry. The corresponding maximum growth rate, for backscatter (backward SBS or "BSBS") and aligned polarizations, is then (taking $\omega_1 \approx \omega_0$ and $k_2 \approx 2k_0$)

$$\gamma_{BSBS} = \frac{\omega_0}{\sqrt{8}} \frac{|\tilde{a}_0|c}{v_{Te}} \sqrt{n_0 \frac{n_{e0}}{n_c} \frac{c_s}{c}} \,, \tag{7.11}$$

where $n_0 = \sqrt{1 - n_{e0}/n_c}$ is the plasma refractive index.

In practical units, we have

$$\gamma_{BSBS}[s^{-1}] \approx 4 \times 10^{13} \left[\sqrt{1 - n_{e0}/n_c} \frac{n_{e0}}{n_c} \sqrt{\frac{Z}{AT_{ek}}} I_{16} \right]^{1/2}, \tag{7.12}$$

with T_{ek} the electron temperature in keV and I_{16} the laser intensity in units of 10^{16} W/cm^2.

7.1.2 Spatial Amplification in Uniform and Non-uniform Plasmas

The spatial amplification rate is related to the temporal growth via Eq. (6.52), $\Gamma = \gamma_0^2/(v_{g1}v_2)$, where $v_{g1} \approx v_{g0} = k_0 c^2/\omega_0$ is the scattered light wave's group velocity, as discussed in Sect. 6.4.4. For the backscatter geometry, when the SBS growth is maximum and $k_2 \approx 2k_0$, the fluid expression for the spatial amplification rate takes the form:

$$\Gamma_{BSBS} = \frac{|\tilde{a}_0|^2}{16 k_0 \lambda_{De}^2 v_2/\omega_2} \,. \tag{7.13}$$

In practical units, we have

$$\Gamma[m^{-1}] \approx 1.5 \times 10^8 \, I_{16} \frac{\lambda_\mu}{v_2/\omega_2} \frac{n_{e0}/n_c}{T_{ek}\sqrt{1 - n_{e0}/n_c}} \,, \tag{7.14}$$

with the same definitions as above.

Since laser–plasma experiments can often involve plasmas with mixed ion species, a kinetic treatment will be strongly preferred in these cases in order to correctly account for the presence of multiple acoustic modes, as we discussed in Sect. 1.3.6. In this case we use the kinetic expression for the density modulation from a beat wave obtained in Sect. 5.4, Eq. (5.61). Combining with the coupled equations for the two EMWs derived above, Eqs. (7.7)–(7.9), taken in steady-state ($\partial_t = 0$), gives the following system of equations:

$$\partial_z \tilde{a}_0 = i \frac{\omega_{pe0}^2}{4 k_0 c^2} \frac{\delta \tilde{n}_e}{n_{e0}} \tilde{a}_1 \,, \tag{7.15}$$

$$\partial_z \tilde{a}_1 = -i \frac{\omega_{pe0}^2}{4k_0 c^2} \frac{\delta \tilde{n}_e^*}{n_{e0}} \tilde{a}_0 \,, \tag{7.16}$$

$$\frac{\delta \tilde{n}_e}{n_{e0}} = -\frac{2k_0^2 c^2}{\omega_{pe0}^2} F_\chi \tilde{a}_0 \tilde{a}_1^* \,, \tag{7.17}$$

where $F_\chi = \chi_e(1+\chi_i)/(1+\chi_e+\chi_i)$ and we assumed the pure backscatter geometry (with $k_2 = 2k_0$) with z the propagation direction of the backscatter (the laser goes toward $-z$). Inserting the expression for $\delta \tilde{n}_e$ from the last equation into the first two gives

$$\partial_z \tilde{a}_0 = -i \frac{k_0}{2} F_\chi |\tilde{a}_1|^2 \tilde{a}_0 \,, \tag{7.18}$$

$$\partial_z \tilde{a}_1 = i \frac{k_0}{2} F_\chi^* |\tilde{a}_0|^2 \tilde{a}_1 \,. \tag{7.19}$$

In the absence of pump depletion ($\tilde{a}_0$ constant), the second equation immediately gives the kinetic expression for the spatial amplification rate for BSBS:

$$\boxed{\Gamma_{\text{BSBS}} = \frac{k_0}{2} \text{Im}(F_\chi) |\tilde{a}_0|^2} \,. \tag{7.20}$$

In a single-ion species plasma, the kinetic expression above recovers the fluid expression from Eq. (7.13) (cf. Problem 7.1). However, for multi-species plasmas, the kinetic expression is the most appropriate, with the ion susceptibility being the sum of the susceptibility from each species as explained in Sect. 1.3.6.

Next we turn our attention to the effect of a plasma flow on SBS and spatial flow gradients. Since plasma flow velocities in ICF or HED experiments are typically on the order of the sound speed (at least as a rough order of magnitude) and the frequency shift between the laser and SBS scattered wave is also equal to the acoustic frequency, Doppler shifts can easily detune the SBS resonance for the IAW.

In the case of a perfectly uniform plasma with a uniform flow, we can repeat the analysis of SBS by substituting $\partial_t \to \partial_t + \mathbf{V} \cdot \nabla$ in Eq. (7.3), as was explained in Sect. 1.3.2, Eq. (1.149) [15]. Here $\mathbf{V}$ is the background plasma flow. The substitution is equivalent to replacing ω_2 by $\omega_2 - \mathbf{k}_2 \cdot \mathbf{V}$ in Fourier, i.e., in the dispersion relation, which now reads

$$\omega_2 = k_2 c_s + \mathbf{k}_2 \cdot \mathbf{V} \,. \tag{7.21}$$

We see that the most sensitive fluid quantity in the IAW dispersion relation is the flow, followed to a lesser extent by the temperature (since $c_s \propto \sqrt{T_e}$), and to an even smaller extent by the density, since $k_2 \propto \sqrt{1 - n_e/n_c}$. Furthermore, in typical laser–plasma experiments, temperature profiles tend to be relatively uniform due to heat conduction, whereas flows can have very steep gradients—like for the isothermal expansion model from Sect. 1.5. Therefore, spatial gradients in flow are the ones most prone to spatially limit the spatial amplification of SBS in realistic conditions. In the following we calculate the spatial amplification for SBS in a flow gradient, applying the Rosenbluth gain formula.

Consider a region of plasma where the flow satisfies the SBS resonance at $z = 0$ in the 1D backscatter geometry. We approximate the flow in this region as

$$\mathbf{V} = V_{\parallel}\mathbf{e}_z + V_{\perp}\mathbf{e}_{\perp}, \tag{7.22}$$

$$V_{\parallel} \approx V_0 + V'z, \tag{7.23}$$

where V_0 can be positive or negative (only the flow component along z can lead to a Doppler shift).

All other plasma parameters are considered fixed in the region of interest. Likewise, the frequencies of the three waves ω_j (j=0,1, and 2 for the laser, backscattered EMW, and IAW, respectively) are also assumed fixed, so the phase-matching condition is spatially dependent via the IAW wave-number,

$$k_2(z) = \frac{\omega_2}{c_s + V_0 + V'z} \approx \frac{\omega_2}{c_s + V_0}\left(1 - \frac{V'z}{c_s + V_0}\right), \tag{7.24}$$

where we assumed that the variation in flow over the region of interest $V'L$ (where L is the length of that region and will be specified below) remains small compared to $c_s + V_0$.

We can quantify the deviation from resonance via the phase mismatch parameter κ like in Sect. 6.5, such that

$$\kappa(z) \equiv |k_0| + |k_1| - |k_2(z)| = \frac{k_2(0)V'z}{c_s + V_0}, \tag{7.25}$$

with $k_2(0) = \omega_2/(c_s + V_0)$.

We can then directly insert this expression in the Rosenbluth gain formula, $G_r = 2\pi\gamma_0^2/|\kappa'v_{g1}v_{g2}|$ (Eq. (6.105)), with γ_0 from Eq. (7.11), $v_{g2} \approx c_s + V_0$ (as expressed in the flowing plasma frame), $v_{g1} \approx v_{g0} = k_0c^2/\omega_0$ and $\kappa' = k_2(0)V'/v_{g2}$. We obtain

$$\boxed{G_r = \frac{\pi}{4}\frac{|\tilde{a}_0^2|\omega_{pi}^2}{c_s k_2 V'}}, \tag{7.26}$$

in accordance with Ref. [16]. As discussed in Sect. 6.5, the characteristic amplification length l_a depends on whether the IAW damping is larger or smaller than the spatial amplification rate, corresponding to the strongly vs. weakly damped regime, respectively:

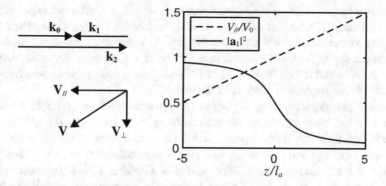

Fig. 7.1 SBS (backscatter) spatial amplification in a velocity gradient. The scattered light wave, whose intensity is represented here in arbitrary units, propagates from right to left (with a smooth, non-oscillating profile characteristic of the strong IAW damping limit, cf. Sect. 6.5). The flow matches SBS resonance conditions at $z = 0$; l_a is the characteristic amplification length

- For strong damping, i.e., $\nu_2/\omega_2 \gg \Gamma/(2k_0)$ (cf. Sect. 6.4.4):

The amplification length is $l_a = 2\nu_2/(\kappa' v_{g2})$ (cf. Eq. (6.117)). For SBS backscatter, this becomes

$$l_a = \frac{\nu_2}{k_0 V'}. \tag{7.27}$$

- For weak damping, i.e., $\nu_2/\omega_2 \ll \Gamma/(2k_0)$ (cf. Sect. 6.4.4):

The amplification length is $l_a = 4\gamma_0/|\kappa'\sqrt{v_{g1}v_{g2}}|$ (cf. Eq. (6.106)), i.e.,

$$l_a = \frac{|\tilde{a}_0|\omega_{pi}}{\sqrt{2}}\frac{\sqrt{1+M}}{k_0 V'}, \tag{7.28}$$

where $M = V_0/c_s$ is the flow Mach number.

The SBS amplification in a velocity gradient is illustrated in Fig. 7.1.

7.1.3 Dispersion Relation: Strongly vs. Weakly Coupled Regimes

It will be useful to derive the dispersion relation of the SBS instability (following the procedure described in Sect. 6.4.5) for several reasons. First, we will see that under certain conditions the growth rate can become comparable to the acoustic frequency (because that frequency is typically small): in this case the slowly varying envelope analysis presented above is inadequate, and we will obtain a different expression for the growth rate [17]. Second, we will see later on that SBS in the forward scattering geometry merges naturally with the filamentation instability; the merge occurs if we account for the anti-Stokes wave (filamentation is a four-wave process, involving the laser, IAW, and the scattered Stokes and anti-Stokes light waves), which appears naturally in the dispersion relation of the instability.

The analysis neglects pump depletion and starts with a Fourier–Laplace analysis of the coupled wave equations for the scattered light wave and the IAW derived above in Eqs. (7.2)–(7.3). We assume that the laser is a plane monochromatic wave, $\mathbf{a}_0 = \frac{1}{2}a_0 \exp[i\psi_0] + c.c.$ with $\psi_0 = \mathbf{k}_0 \cdot \mathbf{r} - \omega_0 t$, but make no assumption about the IAW or scattered light wave. We obtain after a Fourier–Laplace transform of the two wave equations (Laplace in time and Fourier in space, with the Laplace variable $s = i\omega$ and $\omega \in \mathbb{C}$):

$$\hat{D}_1(0)\hat{a}_1(0) = -\omega_{pe0}^2 \frac{a_0}{2}\left[\frac{\delta\hat{n}_e}{n_{e0}}(-1) + \frac{\delta\hat{n}_e}{n_{e0}}(+1)\right], \tag{7.29}$$

$$\hat{D}_2(0)\frac{\delta\hat{n}_e}{n_{e0}}(0) = -\frac{k^2 Z m c^2}{M_i}\frac{a_0}{2}\left[\hat{a}_1(-1) + \hat{a}_1(+1)\right], \tag{7.30}$$

where we used the same notations as in Sect. 6.4.5: $\hat{D}_1(n) = -(\omega+n\omega_0)^2 + \omega_{pe0}^2 + c^2(\mathbf{k}+n\mathbf{k}_0)^2$, $\hat{D}_2(n) = -(\omega+n\omega_0)^2 + c_s^2(\mathbf{k}+n\mathbf{k}_0)^2$, and $\hat{a}_1(n) = \hat{a}_1(\omega+n\omega_0, \mathbf{k}+n\mathbf{k}_0)$, $\delta\hat{n}_e(n) = \delta\hat{n}_e(\omega+n\omega_0, \mathbf{k}+n\mathbf{k}_0)$ with $n \in \mathbb{N}$. The hat denotes quantities in Fourier–Laplace space. We also assumed that the pump and scattered light waves' polarizations are aligned, which maximizes the beating between the electric fields and hence the coupling.

Following the same procedure as in Sect. 6.4.5 (i.e., eliminating $\hat{a}_1(0)$ and truncating the recurrence to keep only the terms of order 0 and ± 1) gives the dispersion relation:

$$\hat{D}_2(0) = \frac{k^2 a_0^2 c^2 \omega_{pe0}^2}{4}\frac{Zm}{M_i}\left[\frac{1}{\hat{D}_1(-1)} + \frac{1}{\hat{D}_1(+1)}\right] \tag{7.31}$$

$$\approx \frac{k^2 a_0^2 c^2 c_s^2}{4\lambda_{De}^2}\left[\frac{1}{\hat{D}_1(-1)} + \frac{1}{\hat{D}_1(+1)}\right], \quad ZT_e \gg T_i. \tag{7.32}$$

Expanding the expressions of $\hat{D}_1(\pm 1)$ assuming that $\omega \sim \omega_{IAW} \ll \omega_0, \omega - \omega_0$ and using the fact that $\omega_0^2 = \omega_{pe0}^2 + k_0^2 c^2$ give the simplified expressions:

$$\hat{D}_1(\pm 1) = k^2 c^2 \pm 2(\mathbf{k}\cdot\mathbf{k}_0 c^2 - \omega\omega_0). \tag{7.33}$$

Inserting into Eq. (7.32) gives the following dispersion relation:

$$(\omega^2 - k^2 c_s^2) \left[\omega - \mathbf{k} \cdot \mathbf{k}_0 \frac{c^2}{\omega_0} + \frac{k^2 c^2}{2\omega_0} \right] \left[\omega - \mathbf{k} \cdot \mathbf{k}_0 \frac{c^2}{\omega_0} - \frac{k^2 c^2}{2\omega_0} \right]$$

$$= \frac{a_0^2 k^4 c^4}{8} \frac{\omega_{pi}^2}{\omega_0^2}. \tag{7.34}$$

This dispersion relation describes the coupling between a monochromatic pump wave, an IAW, and both the Stokes and anti-Stokes scattered waves. The terms in the parenthesis, first bracket, and second bracket represent the dispersion relation of the IAW, Stokes, and anti-Stokes waves, respectively. SBS consists in the coupling with the Stokes component only; the anti-Stokes component will typically be non-resonant unless scattering is in the forward direction, when $\psi \to 0$ where ψ is the angle between the laser and Stokes waves' directions (with $\mathbf{k} \cdot \mathbf{k}_0 = k k_0 \sin(\psi/2)$) as illustrated in Fig. 7.2. Indeed, if $(\omega, \mathbf{k})$ satisfy the dispersion relation of an IAW and $(\omega_0 - \omega, \mathbf{k}_0 - \mathbf{k})$ (i.e., the Stokes component) that of an EMW, then the anti-Stokes component can only come close to satisfying an EMW dispersion relation if $\hat{D}_1(+1) \approx 0$, i.e., if $|\mathbf{k}_0 + \mathbf{k}| \approx k_0$, or $\psi \approx 0$. In this case the SBS instability transitions smoothly to the filamentation instability, which we investigate in Sect. 7.3; the transition between SBS and filamentation is discussed later in Sect. 7.3.4. In the following, we will quantify the conditions under which the Stokes component can indeed be neglected.

To investigate the different instability regimes arising from the dispersion relation Eq. (7.34), we will take various limits and solve the equation for complex $\omega = \omega_R + i\gamma$ and a real k, a positive imaginary component γ indicating the presence of an instability with temporal growth rate γ.

For SBS, we will be looking for perturbations around the frequency and wave-vector $(\omega_2, \mathbf{k}_2)$ satisfying both the IAW and Stokes EMW dispersion relations, i.e.,

$$\omega_2 = k_2 c_s, \tag{7.35}$$

$$(\omega_0 - \omega_2)^2 = \omega_{pe0}^2 + |\mathbf{k}_0 - \mathbf{k}_2|^2 c^2. \tag{7.36}$$

Assuming $\omega_2 \ll \omega_0$ and using the pump's EMW dispersion relation in Eq. (7.36) give the simplified relation $2\omega_0 \omega_2 - 2c^2 k_0 k_2 \sin(\psi/2) + k_2^2 c^2 = 0$, which after eliminating ω_2 via Eq. (7.35) gives the expression for the resonant IAW wave-

Fig. 7.2 Stokes and anti-Stokes components for SBS. The anti-Stokes is non-resonant except for forward scattering ($\theta \to \pi/2$), in which case SBS transitions to the filamentation instability

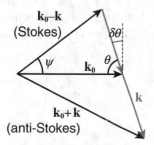

vector:

$$k_2 = 2k_0 \sin(\psi/2) - 2\frac{c_s \omega_0}{c^2}. \tag{7.37}$$

For all angles except near-forward scattering, we simply have $k_2 \approx 2k_0 \sin(\psi/2)$.

Now let us perturb the frequency around ω_2, by taking $\omega = \omega_2 + \delta\omega$, with $\delta\omega \in \mathbb{C}$. Inserting in the dispersion relation Eq. (7.34) with $k = k_2$ gives

$$\delta\omega^2 (2\omega_2 + \delta\omega) \left(\delta\omega - \frac{k_2^2 c^2}{\omega_0} \right) = \frac{a_0^2 k_2^4 c^4 \omega_{pi}^2}{8\omega_0^2}. \tag{7.38}$$

The three-wave coupling regime of SBS (as opposed to four-wave for filamentation) is defined by the condition:

$$|\delta\omega| \ll \frac{k_2^2 c^2}{\omega_0}, \tag{7.39}$$

which allows us to ignore the anti-Stokes resonance. The remaining dispersion relation is

$$\delta\omega^2 (2\omega_2 + \delta\omega) = -\frac{a_0^2 k_2^2 c^2 \omega_{pi}^2}{8\omega_0}. \tag{7.40}$$

If a complex solution exists with a positive imaginary component γ, then the system is unstable. Two limits can be taken to extract a simple solution from Eq. (7.40):

- $|\delta\omega| \ll \omega_2$: "weakly coupled" regime of SBS.

In this regime, Eq. (7.40) is readily solved, giving a purely imaginary solution $\delta\omega = i\gamma$ with

$$\gamma = \frac{a_0 k_2 c \omega_{pi}}{4\sqrt{\omega_0 \omega_2}} = \frac{a_0}{2\sqrt{2}} \omega_{pi} \sqrt{n_0 \frac{c}{c_s} \sin(\psi/2) - 1}, \tag{7.41}$$

where $n_0 = \sqrt{1 - n_{e0}/n_c}$ is the background plasma refractive index.

For all angles except forward scattering, the first term in the square root is $\gg 1$. In particular, this expression shows that the growth rate is maximized for pure backscatter ($\psi = \pi$), leading to the growth rate:

$$\boxed{\gamma_0 = \frac{a_0}{\sqrt{8}} \omega_{pi} \sqrt{n_0 \frac{c}{c_s}},} \tag{7.42}$$

which is the same as Eq. (7.11) (making use of the fact that $\omega_{pi} = c_s \lambda_{De}$ in the limit of $ZT_e \gg T_i$ as is assumed here), as expected.[2]

Now let us quantify the conditions of existence of SBS in this regime. The first assumption we have used is to neglect the anti-Stokes component, Eq. (7.39). Inserting the expression for the SBS growth rate for arbitrary geometry, Eq. (7.41), gives the following condition:

$$a_0 \ll 8\sqrt{2}\lambda_{De}c_s \frac{\omega_0}{c^2} \left(\frac{cn_0}{c_s} \sin(\psi/2) - 1 \right)^{3/2}. \tag{7.43}$$

Our second assumption concerns the definition of the weakly coupled regime of SBS, $|\delta\omega| \ll \omega_2$, which becomes

$$a_0 \ll 4\sqrt{2}\lambda_{De}c_s \frac{\omega_0}{c^2} \sqrt{\frac{cn_0}{c_s} \sin(\psi/2) - 1}. \tag{7.44}$$

Equations (7.43) and (7.44) define the conditions of existence of SBS in the weakly coupled regime. Now we turn to the other regime of SBS:

- $|\delta\omega| \gg \omega_2$: "strongly coupled" SBS.

This regime corresponds to the case where the pump is so strong that it forces the plasma response away from an IAW resonance, to become a "quasi-mode." In this case the slowly varying envelope approximation is invalid, since the envelope amplitude evolves on faster time scales than the IAW frequency. Equation (7.40) solved in this limit gives

$$\delta\omega = e^{i\pi/3} \left[\frac{a_0 k_2 c\omega_{pi}}{2\sqrt{2\omega_0}} \right]^{2/3}, \tag{7.45}$$

i.e., the plasma acquires a real frequency shift in addition to the imaginary part (the growth rate). Substituting k_2 from Eq. (7.37) gives

$$\gamma = \frac{\sqrt{3}}{2} \left[a_0 \frac{c_s^2}{\lambda_{De}c} \sqrt{\omega_0/2} \left(\frac{n_0 c}{c_s} \sin(\psi/2) - 1 \right) \right]^{2/3}. \tag{7.46}$$

For SBS in the backscatter geometry ($\psi = \pi$), this simplifies to

$$\boxed{\gamma = \frac{\sqrt{3}}{2} \left(a_0 \omega_{pi} n_0 \sqrt{\omega_0/2} \right)^{2/3}.} \tag{7.47}$$

---

[2] Our previous analysis used the slowly varying envelope approximation: under that approximation, the temporal evolution of the envelope, which is $\propto e^{\gamma t}$ for a pure temporal growth, has to remain slow compared to the oscillation of the wave $\propto e^{i\omega_2 t}$, which is equivalent to the assumption $|\delta\omega| = \gamma \ll \omega_2$ used in this section.

Let us describe the conditions of validity for SBS in the strongly coupled regime. The first one is that the anti-Stokes component be negligible, Eq. (7.39), which when inserting the expression for $\delta\omega$ from Eq. (7.45) gives

$$a_0 \ll 8\sqrt{2}\lambda_{De}\frac{\omega_0 c_s}{c^2}\left(\frac{n_0 c}{c_s}\sin(\psi/2) - 1\right)^2. \tag{7.48}$$

And the second condition is that $|\delta\omega| \gg \omega_2$, per the definition of the strongly coupled regime, which leads to

$$a_0 \gg 4\frac{\omega_0}{c^2}\lambda_{De}c_s\sqrt{\frac{n_0 c}{c_s}\sin(\psi/2) - 1}. \tag{7.49}$$

For given plasma conditions, we see that the conditions of validity for SBS in the weakly and strongly coupled regimes depend on the pump intensity and the scattering angle. Later in Sect. 7.3.4, we will complement these by the conditions of validity of the filamentation instability, which also exists in a weakly and strongly coupled regime. Together these conditions will allow us to map out the various regimes of coupling between a laser and an IAW (SBS vs. filamentation and weakly vs. strongly coupled regime) in an (a_0, ψ) parameter space.

7.1.4   Mitigation of SBS Using Mixed-Species Plasmas

As we discussed in Sect. 1.3.6, the presence of multiple ion species in a plasma can significantly impact the features of the IAWs. Several modes can exist: in two-species plasma with well-separated ion masses, we can have a "slow" mode whose phase velocity sits in between the thermal velocities of the heavy and light ions, and a "fast" mode, with a phase velocity larger than the thermal velocities of both species (in all cases the phase velocity remains much smaller than the electron thermal velocity). The mode with the least amount of Landau damping is the one that might be observable in experiments. When the fast and slow modes are not very well separated, as is for example the case with hydrocarbon plasmas, even the most weakly damped IAW mode can experience significant Landau damping due to the presence of co-existing modes.

A particularly useful application of mixed-species plasmas consists in manipulating the IAW damping in order to mitigate the growth of IAW-driven instabilities such as SBS. Introducing a small fraction of low-Z dopant in a plasma was shown to increase the damping of IAWs [18]; this idea has since been applied to mitigate SBS in laser–plasma experiments [9, 19–23]. While this section focuses on SBS mitigation via increased IAW damping, it should also be noted that increasing the IAW damping can also inhibit the decay of EPWs into IAWs via the Langmuir decay instability (LDI), which is a saturation mechanism for EPW-driven instabilities like stimulated Raman scattering (SRS) or two plasmon decay (TPD), and can therefore enhance these instabilities; LDI is discussed in detail in Sect. 10.2.1.

To illustrate how manipulating the plasma composition can help mitigate SBS, we show in Fig. 7.3 the normalized spatial amplification rate from a kinetic model $Im(F_\chi)$ [cf. Eq. (7.20); as we mentioned in Sect. 1.3.6, a kinetic model should always be used when treating multi-species plasmas]. We show the case of a carbon/hydrogen mixture (a–b) and a tantalum/oxygen mixture (c–d).

For a C/H mixture, Fig. 7.3a represents the maximum amplification rate (at the resonant IAW frequency) as a function of the fraction of light ions (here hydrogen), showing a strong drop as the fraction of hydrogen is increased. Figure 7.3b shows the amplification rate vs. IAW phase velocity (normalized to the carbon thermal velocity) for two particular hydrogen fractions: 0 (i.e., pure carbon plasma) and 4/5 (i.e., CH_4 plasma). The IAW in pure carbon is weakly damped with a narrow resonance peak, whereas for CH_4 the curve has a double-hump characteristic of the two modes of CH, as we discussed in Sect. 1.3.6, and as a result, the IAW is strongly damped and the peak amplification rate much reduced compared to pure carbon (the amplification rate for CH_4 is multiplied by 10 in the figure for better visibility).

Likewise, Fig. 7.3c shows the maximum amplification rate for tantalum/oxygen as a function of the fraction of light ion (now oxygen), and Fig. 7.3d shows the amplification rate vs. phase velocity (normalized to the thermal velocity of Ta) for $f_O = 0$ (i.e., pure Ta) and 5/7 (i.e., Ta_2O_5). Tantalum-oxide (Ta_2O_5) is what was used in Ref. [23]: there, a thin layer of Ta_2O_5 was deposited on the hohlraum wall, with the tantalum replacing gold to produce the X-ray drive on the capsule of nuclear fuel, and the oxygen being introduced to increase Landau damping of IAWs, resulting in a strong mitigation of the measured SBS in the experiments.

Other mixtures have been tested and used in ICF experiments to mitigate SBS, such as He/H (instead of pure He) [22] or Au/B (instead of pure Au) [22, 23]. One should keep in mind that the effectiveness of such mixtures depends on a variety of factors, like the electron-to-ion temperature ratio, and whether SBS amplification is limited by velocity gradients (cf. Sect. 7.1.2), in which case it becomes insensitive to the IAW damping rate.

7.2 Crossed-Beam Energy Transfer (CBET)

7.2.1 Introduction

Crossed-beam energy transfer (CBET) is the process where two laser beams overlapping in a plasma can resonantly or *near-resonantly* drive an IAW. It is very similar to SBS, except that it involves two high-amplitude, "pre-existing" waves (i.e., with externally imposed frequencies, directions, spot sizes, intensities,

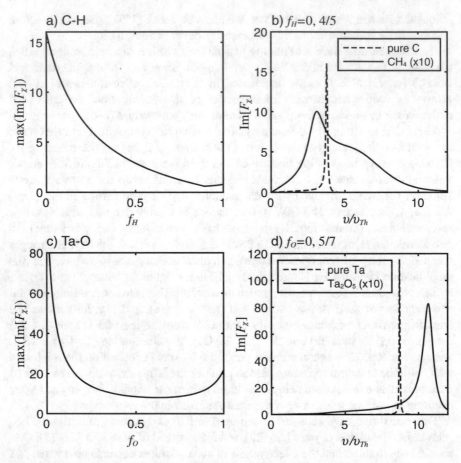

Fig. 7.3 (a) Maximum SBS amplification rate $Im[F_\chi]$ for a C/H mixture as a function of the fraction of hydrogen, for $k\lambda_{De} = 0.67$ (typical of backscatter geometry at $n_e/n_c \approx 0.05$ and $T_e = 3\,\text{keV}$) and $T_e/T_i = 2.5$. (b) Amplification rate vs. phase velocity normalized to the heavy ion (here C) thermal velocity, for two particular fractions of hydrogen: 0 (i.e., pure C) and 4/5 (i.e., CH_4); the curve for CH_4 is multiplied by 10 for better visibility. Similarly for Ta/O mixtures, (c) shows the maximum amplification rate vs. fraction of O and (d) the amplification rate vs. velocity (normalized to the Ta thermal velocity) for $f_O = 0$ and 5/7

polarizations, etc.) as opposed to one pre-defined high-amplitude wave and a low-amplitude seed whose conditions automatically match the maximum coupling conditions for SBS. By automatically matching the maximum coupling we mean that the SBS seed, which originates from broadband noise, will emerge with a frequency (shifted from the pump by the IAW frequency), direction (typically counter-propagating relative to the pump), and polarization (aligned with the pump) that ensure its maximum amplification. Therefore, in our analysis of CBET, we will have to be more general than for SBS and assume arbitrary frequency, propagation direction, polarization, etc. for the two light waves.

CBET was first demonstrated in laboratory experiments using two microwave sources in a plasma, with a frequency shift between the two sources tuned to map out the IAW resonance [24]. Its relevance to indirect-drive experiments was pointed out by Kruer more than a decade later [25], and a significant research effort began that same year [26, 27], which led to the first demonstration of the phenomenon in laser experiments by Kirkwood later that year [28] using a frequency mismatch between the beams, similar to the first experiments by Pawley using microwaves [24].

For energy transfer to occur, a frequency shift between the two beams must be present *in the plasma frame*. This can be the result of an externally imposed frequency shift between the beams or, for two beams of equal frequency, be the result of a Doppler shift due to plasma flow. Of course, both effects can exist at the same time, i.e., an externally imposed frequency shift on top of a Doppler-induced shift due to flows. Experiments by Wharton demonstrated the presence of CBET between equal-frequency beams in flowing plasmas due to Doppler shifts [29].

CBET is a crucial phenomenon in ICF experiments. For indirect-drive geometries, CBET occurs near the entrance holes of the "hohlraum" targets where all the laser beams overlap and can lead to energy exchange between groups of beams. This can in turn be utilized to control the laser energy deposition inside the hohlraum and the implosion symmetry of the capsule [30–34]. The geometry of CBET for indirect-drive ICF experiments is illustrated in Fig. 7.4a.

CBET also occurs in spherical implosion geometries like for direct-drive ICF experiments. There, the process is entirely driven by plasma flows, as illustrated in Fig. 7.4b, and goes as follows: as rays near the edges of a given beam refract away from the target, they intersect incoming rays from other beams. When the intersection occurs in the vicinity of the Mach 1 flow surface, the stationary interference pattern between the incoming and outgoing rays appears to be moving at the sound speed in the plasma frame, which can lead to a resonant excitation of IAWs and to transfer from the incoming to the outgoing beam [35]. Therefore, the process constitutes a loss of coupling to the target [36, 37].

7.2.2 Fluid Analysis

Let us first derive a fluid analysis of CBET. We start from the same coupled wave equations as for SBS, Eqs. (7.1)–(7.3), describing the general coupling of two light waves via an EMW. Since plasma flows typically play a very important role for CBET, we include the background flow in the IAW wave equation, as was introduced in Sect. 1.3.2, Eq. (1.149):

$$\left[(\partial_t + \nu + \mathbf{V} \cdot \nabla)^2 - c_s^2 \nabla^2 \right] \frac{\delta n_e}{n_{e0}} = \frac{Zmc^2}{M_i} \nabla^2 \mathbf{a}_0 \cdot \mathbf{a}_1 , \tag{7.50}$$

where $\mathbf{V}$ is the background plasma flow and ν the IAW damping.

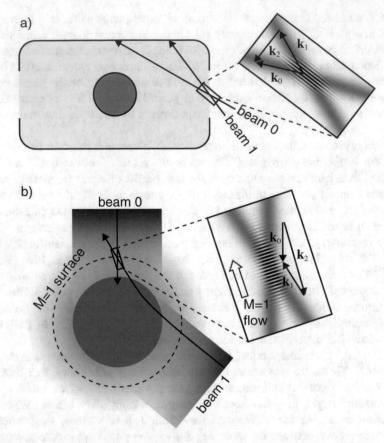

Fig. 7.4 CBET geometry in ICF experiments: (**a**) indirect-drive configuration: CBET occurs at the laser entrance holes, where all the laser beams from a given hemisphere overlap in the same volume of plasma, and (**b**) direct-drive configuration: CBET leads to energy transfer from incoming rays near the center of a beam to outgoing rays from the wings of another beam being refracted away from the target

Next we envelope each of the three waves at its main frequency and wave-number; for the driven IAW, we envelope at the beat frequency and wave-number but will keep in mind that the beat wave is not generally exactly resonant with an IAW mode:

$$\mathbf{a}_0 = \frac{1}{2}\tilde{\mathbf{a}}_0 e^{i\psi_0} + c.c., \tag{7.51}$$

$$\mathbf{a}_1 = \frac{1}{2}\tilde{\mathbf{a}}_1 e^{i\psi_1} + c.c., \tag{7.52}$$

$$\delta n_e = \frac{1}{2}\delta\tilde{n}_e e^{i\psi_2} + c.c., \tag{7.53}$$

where $\psi_j = \mathbf{k}_j \cdot \mathbf{r} - \omega_j t$ (for $j \in \{0, 1, 2\}$) and $\psi_2 = \psi_0 - \psi_1$.

Using a slowly varying envelope (SVE) approximation and assuming steady-state (i.e., $\partial_t \tilde{a}_{0,1} = 0$) lead to the following simplification for the EMW wave operators:

$$(\partial_t^2 + \omega_{pe0}^2 - c^2 \nabla^2) a_{0,1} \approx -\frac{e^{i\psi_{0,1}}}{2} 2i k_{0,1} c^2 \partial_{z_{0,1}} \tilde{a}_{0,1} + c.c., \qquad (7.54)$$

where z_0 and z_1 are the directions of beams 0 and 1, respectively.

Likewise for the IAW, we apply a slowly varying envelope approximation (cf. Sect. 6.4.1) and assume steady-state ($\partial_t \delta\tilde{n}_e = 0$) as well as the strongly damped regime (i.e., we assume the wave damps faster than it grows due to CBET, cf. Sect. 6.4.4, which is generally easily satisfied as CBET gains are often not very high as the waves are driven off-resonance), i.e., we neglect the spatial derivatives of $\delta\tilde{n}_e$ as well. Applying these assumptions simplifies the wave operator in Eq. (7.50) into

$$\left[(\partial_t + \nu + \mathbf{V} \cdot \nabla)^2 - c_s^2 \nabla^2 \right] \delta n_e \approx \frac{e^{i\psi_2}}{2} \left(-\hat{\omega}_2^2 + k_2^2 c_s^2 - 2i\nu\hat{\omega}_2 \right) \delta\tilde{n}_e + c.c., \qquad (7.55)$$

where

$$\hat{\omega}_2 = \omega_2 - \mathbf{k}_2 \cdot \mathbf{V} \qquad (7.56)$$

is the beat frequency as seen in the frame of the flowing plasma (cf. Problem 7.2). Here we note that unlike for SBS, the driven wave is not generally at resonance with an IAW mode, i.e., $\hat{\omega}_2 \neq k_2 c_s$.

Inserting these SVE operators into the driven wave equations for the EMWs and the IAW, Eqs. (7.1)–(7.2) and (7.50), leads to the system of coupled equations:

$$\partial_{z_0} \tilde{a}_0 = -i \frac{\omega_{pe0}^2}{4k_0 c^2} \frac{\delta\tilde{n}_e}{n_{e0}} \tilde{a}_1, \qquad (7.57)$$

$$\partial_{z_1} \tilde{a}_1 = -i \frac{\omega_{pe0}^2}{4k_0 c^2} \frac{\delta\tilde{n}_e^*}{n_{e0}} \tilde{a}_0, \qquad (7.58)$$

$$\left(-\hat{\omega}_2^2 + k_2^2 c_s^2 - 2i\nu\hat{\omega}_2 \right) \frac{\delta\tilde{n}_e}{n_{e0}} = -k_2^2 \frac{Zmc^2}{2M_i} \tilde{a}_0 \tilde{a}_1^*. \qquad (7.59)$$

Note that the derivatives are taken along two different spatial directions, so this system of coupled equations is really 2D. The system can actually be solved exactly (with the nonlinear interaction of each beam on the other), in 2D, as we will see in Sect. 10.1.3.

We can now eliminate $\delta\tilde{n}_e$ from the system of equations, leading to a system of two coupled equations for the two light waves:

$$\partial_{z_0} \tilde{a}_0 = \omega_{pi}^2 \frac{k_2^2}{8k_0} |\tilde{a}_1|^2 \frac{-2\nu\hat{\omega}_2 + i(k_2^2 c_s^2 - \hat{\omega}_2^2)}{(k_2^2 c_s^2 - \hat{\omega}_2^2)^2 + (2\nu\hat{\omega}_2)^2} \tilde{a}_0 , \tag{7.60}$$

$$\partial_{z_1} \tilde{a}_1 = \omega_{pi}^2 \frac{k_2^2}{8k_0} |\tilde{a}_0|^2 \frac{2\nu\hat{\omega}_2 + i(k_2^2 c_s^2 - \hat{\omega}_2^2)}{(k_2^2 c_s^2 - \hat{\omega}_2^2)^2 + (2\nu\hat{\omega}_2)^2} \tilde{a}_1 , \tag{7.61}$$

where we assumed $k_0 \approx k_1$. We also assumed that the two light waves were s-polarized, i.e., with their polarizations aligned and normal to the plane of incidence.[3] In general, the two beams will not necessarily have this simple polarization geometry; the analysis then requires to account for arbitrary polarization arrangements and will be presented in detail in Sect. 7.2.5.

The spatial amplification gains for each beam (ignoring the depletion of the other beam) $\Gamma_{0,1}$ are then obtained by taking the real part of the coupling term on the right-hand side of these equations:

$$\boxed{\Gamma_{0,1} = \mp \omega_{pi}^2 \frac{k_2^2}{8k_0} |\tilde{a}_{1,0}|^2 \frac{2\nu\hat{\omega}_2}{(k_2^2 c_s^2 - \hat{\omega}_2^2)^2 + (2\nu\hat{\omega}_2)^2}} , \tag{7.62}$$

where Γ_0 (respectively, Γ_1) is the amplification rate of beam 0 (respectively, beam 1) with a minus sign (respectively, plus sign) on the right-hand side. The direction of the transfer depends on the sign of $\hat{\omega}_2$: for $\hat{\omega}_2 > 0$, i.e., $\omega_0 > \omega_1 + \mathbf{k}_2 \cdot \mathbf{V}$ ($\omega_0 > \omega_1$ in the frame of the flowing plasma), we have $\Gamma_0 < 0$ and $\Gamma_1 > 0$: transfer goes from beam 0 (acting as a pump) to beam 1—and vice versa, transfer goes $1 \to 0$ when $\hat{\omega}_2 < 0$.

The Doppler shift effect (term $\mathbf{k}_2 \cdot \mathbf{V}$) can be very significant in experiments because plasma flows are typically comparable to the sound speed c_s (at least within the same order of magnitude). The effect is very sensitive to the interaction geometry, as only the component of the flow aligned with the beat wave (i.e., $\parallel \mathbf{k}_2$) matters. In particular, if the component of the plasma flow along the direction of the beat wave is equal to c_s, then resonant CBET will occur for two equal-frequency lasers $\omega_0 = \omega_1$, as was demonstrated experimentally by Wharton [29].

We can easily verify that when the beat wave is exactly at the IAW resonance, i.e., $\hat{\omega}_2 = k_2 c_s$, and in the absence of flow, i.e., $\hat{\omega}_2 = \omega_2 = k_2 c_s$, the amplification rate Γ_1 exactly recovers the expression we derived for SBS earlier, Eq. (7.13).

Note that besides the spatial amplification, the imaginary part of the coupling coefficient on the right-hand sides of Eqs. (7.60)–(7.61) produces a phase shift of the waves' envelope. This is equivalent to a small change in the real part of the plasma refractive index (i.e., a change in the magnitude of each wave's wave-vector $k_{0,1} = n_{0,1}\omega_{0,1}/c$), as will be discussed in more detail in the next section.

---

[3] The plane of incidence is defined as the plane containing $\mathbf{k}_0$, $\mathbf{k}_1$, and $\mathbf{k}_0 - \mathbf{k}_1$.

7.2.3 Kinetic Analysis

As mentioned earlier, a kinetic analysis of CBET will often be preferable since it is not necessary to specify a damping explicitly and the resulting expression is valid for multi-species plasmas.

The beginning of the derivation follows the fluid analysis; we envelope the fields for the three waves following

$$\mathbf{a}_0 = \frac{1}{2}\tilde{\mathbf{a}}_0 e^{i\psi_0} + c.c., \tag{7.63}$$

$$\mathbf{a}_1 = \frac{1}{2}\tilde{\mathbf{a}}_1 e^{i\psi_1} + c.c., \tag{7.64}$$

$$\delta n_e = \frac{1}{2}\delta\tilde{n}_e e^{i\psi_2} + c.c., \tag{7.65}$$

where $\psi_j = \mathbf{k}_j \cdot \mathbf{r} - \omega_j t$ (for $j \in \{0, 1, 2\}$) and $\psi_2 = \psi_0 - \psi_1$.

Under the same approximations used in our fluid analysis, i.e., steady-state, slowly varying envelope and s-polarized lasers, we arrive at the same equations relating the EMW envelopes to the driven density perturbation, Eqs. (7.57)–(7.58). However, we now use the kinetic expression for the density perturbation from the beat wave's ponderomotive force as given in Sect. 5.4, Eq. (5.61):

$$\frac{\delta\tilde{n}_e}{n_{e0}} = \frac{k_2^2 c^2}{2\omega_{p0}^2} F_\chi \tilde{a}_0 \tilde{a}_1^*, \tag{7.66}$$

$$F_\chi \equiv \frac{\chi_e(1 + \chi_i)}{1 + \chi_e + \chi_i}. \tag{7.67}$$

The susceptibilities are evaluated at $(\hat{\omega}_2, \mathbf{k}_2)$, where $\hat{\omega}_2 = \omega_2 - \mathbf{k}_2 \cdot \mathbf{V}$ is the beat frequency in the frame of the flowing plasma.

Inserting back into Eqs. (7.57)–(7.58) leads to

$$\frac{\partial\tilde{a}_0}{\partial z_0} = i\frac{k_2^2}{8k_0} F_\chi |\tilde{a}_1|^2 \tilde{a}_0, \tag{7.68}$$

$$\frac{\partial\tilde{a}_1}{\partial z_1} = i\frac{k_2^2}{8k_1} F_\chi^* |\tilde{a}_0|^2 \tilde{a}_1. \tag{7.69}$$

Taking the real part of the coupling coefficient on the right-hand sides of these equations, we obtain the spatial amplification rate for each beam (ignoring the amplitude variation of the other beam—the nonlinear analysis including pump depletion will be presented in Sect. 10.1):

$$\boxed{\Gamma_{0,1} = \mp\frac{k_2^2}{8k_{0,1}}\text{Im}[F_\chi]|\tilde{a}_{1,0}|^2,} \tag{7.70}$$

where like for the fluid expression in Eq. (7.62), Γ_0 (respectively, Γ_1) is the amplification rate of beam 0 (respectively, beam 1) with a minus sign (respectively, plus sign) on the right-hand side. The energy transfer can go from beam 0 to 1 or vice versa depending on the sign of $\hat{\omega}_2$. Since the wavelength separations involved with CBET are typically a very small fraction of the laser wavelength, we can usually use k_0 in the denominator for both Γ_0 and Γ_1.

For a single-species plasma, expanding the kinetic expression of the amplification rate near the resonances recovers the fluid expression as expected. Near an IAW resonance, ignoring the plasma flow for simplicity, the beat wave phase velocity satisfies $v_{Ti} \ll \omega_2/k_2 \ll v_{Te}$, therefore the expressions for the susceptibilities simplify to $\chi_e \approx 1/(k_2\lambda_{De})^2$ and $\chi_i \approx -\omega_{pi}^2/\omega_2^2$. Introducing a phenomenological damping term ν for the IAWs (typically Landau damping) via the substitution $\omega_2 \to \omega_2 + i\nu$ with $\nu/\omega_2 \ll 1$ leads to

$$\chi_i \approx -\frac{\omega_{pi}^2}{\omega_2^2}\left(1 - 2i\frac{\nu}{\omega_2}\right). \tag{7.71}$$

Furthermore, near the IAW resonance, we also have, by definition, $1 + \chi_e + \mathrm{Re}[\chi_i] \approx 0$, and hence

$$F_\chi \approx \chi_e \frac{1 + \chi_i}{i\,\mathrm{Im}[\chi_i]}. \tag{7.72}$$

We obtain

$$\mathrm{Im}[\chi_i] = \frac{\omega_2}{2\nu_2 k_2^2 \lambda_{De}^2}, \tag{7.73}$$

which inserted into the spatial amplification rate Eq. (7.70) leads to the following fluid approximation for the fluid CBET amplification rate:

$$\Gamma_1 = \frac{|\tilde{a}_0|^2}{16 k_0 \lambda_{De}^2 \nu/\omega_2}, \tag{7.74}$$

i.e., we recover Eq. (7.62) in the limit of $\hat{\omega}_2 = k_2 c_s$ (IAW resonance).

7.2.4 Electric Field Dephasing and Variation of the Refractive Index in Off-resonance CBET

For the kinetic analysis like for the fluid analysis, in addition to the variation in amplitude, there is also a variation in the phase of the wave's electric field, from the real part of F_χ (or the imaginary part of the coupling coefficients on the right-hand sides of Eqs. (7.60)–(7.61)). Assuming that the plasma is uniform for simplicity, Eq. (7.69) has for solution

$$\tilde{a}_1(z_1) = \tilde{a}_1(0) \exp\left[\frac{k_2^2}{8k_0}|\tilde{a}_0|^2 \left(\text{Im}[F_\chi] + i\text{Re}[F_\chi]\right) z_1\right]. \tag{7.75}$$

Since $\tilde{a}_1$ denotes the slowly varying amplitude of the full field $a_1 = \frac{1}{2}\tilde{a}_1 \exp[i\psi_1] +$ c.c., the expression for $\tilde{a}_1$ means that the wave-number k_1 is effectively substituted by $k_1 + \delta k_1$, with $\delta k_1 = \delta k_{1R} + i\delta k_{1I}$ a complex perturbation of the wave-number defined by

$$\delta k_{1R} = \frac{k_2^2}{8k_0}|\tilde{a}_0|^2\text{Re}[F_\chi], \tag{7.76}$$

$$\delta k_{1I} = -\frac{k_2^2}{8k_0}|\tilde{a}_0|^2\text{Im}[F_\chi] = -\Gamma_1. \tag{7.77}$$

Equivalently, this means that the plasma refractive index $n^{(1)} = k_1 c/\omega_1$ experienced by the beam #1 is modified by the presence of beam 0 (and vice versa) and becomes

$$n^{(1)} = n_0 + \delta n_R^{(1)} + i\delta n_I^{(1)}, \tag{7.78}$$

with

$$n_0 = \sqrt{1 - n_{e0}/n_c}, \tag{7.79}$$

$$\delta n_R^{(1)} = \frac{c}{\omega_1}\delta k_{1R} = n_0\frac{k_2^2}{8k_0^2}|\tilde{a}_0|^2\text{Re}[F_\chi], \tag{7.80}$$

$$\delta n_I^{(1)} = \frac{c}{\omega_1}\delta k_{1I} = -n_0\frac{k_2^2}{8k_0^2}|\tilde{a}_0|^2\text{Im}[F_\chi]. \tag{7.81}$$

The superscript $^{(1)}$ is meant to highlight the fact that these index variations are for beam 1 only: likewise, beam 0 also experiences a complex index perturbation $\delta n^{(0)} \propto |\tilde{a}_1|^2$ due to the presence of beam 1. Similar expressions for the real and imaginary parts of the refractive index can immediately be obtained from the fluid analysis as well (cf. Problem 7.3).

The real and imaginary parts of the refractive index are connected via the Kramers–Kronig relations [38]. This is a well-known fact in optics: any variation in the amplitude of a light wave vs. its frequency (e.g., absorption near an atomic transition) is always accompanied by a variation of the refractive index—and vice versa.

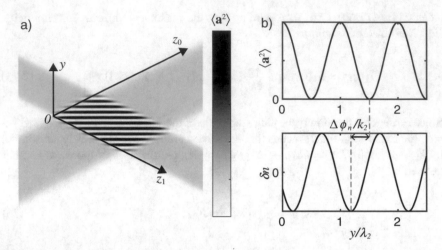

Fig. 7.5 CBET geometry: (**a**) illustration of the overlap between two beams 0 and 1; the overlap region shows their beat wave intensity pattern. (**b**) Beat wave intensity pattern $\langle a^2 \rangle$ vs. y, the direction of $\mathbf{k}_2 = \mathbf{k}_0 - \mathbf{k}_1$, and refractive index modulation δn vs. y. The two patterns are generally shifted by an amount $\Delta \phi_n$: this dephasing determines whether energy exchange can occur or not

We shall also note that this variation in the phase of the wave's electric field is connected to a dephasing $\Delta \phi_n$ between the beat wave's intensity pattern and the resulting index modulation, as represented in Fig. 7.5. From the definition of $\delta \tilde{n}_e$ above (Eq. (7.66)), defining ϕ_0, ϕ_1, and ϕ_n as the phases of $\tilde{a}_0$, $\tilde{a}_1$, and $\delta \tilde{n}_e$, respectively (such that $\tilde{a}_0 = |\tilde{a}_0| \exp[i\phi_0]$, etc.), we obtain

$$\Delta \phi_n \equiv \phi_n - (\phi_0 - \phi_1) = \mathrm{Arg}[F_\chi]. \tag{7.82}$$

This dephasing, which corresponds to a lag in the medium response to an imposed driver (here the ponderomotive force), dictates whether the interaction can lead to energy transfer or not: as can be immediately seen from Eqs. (7.68) and (7.69) above, if $\Delta \phi_n = 0$ (i.e., $F_\chi \in \mathbb{R}$), then no energy transfer can occur; this typically happens when the two beams have the same frequency (in the flowing plasma frame). In this case, while no energy transfer can occur, there is however a dephasing of the two waves' electric fields, i.e., each beam will experience a change in the real refractive index of the plasma due to the other beam (and proportional to the other beam's intensity). This situation is illustrated in Fig. 7.6.

On the other hand, if $\Delta \phi_n = \pm \pi/2$ (i.e., F_χ purely imaginary), energy transfer is maximized. This corresponds to the situation when the beat wave is exactly resonant with an IAW; in this case, no dephasing of the electric fields occurs, i.e., each beam only experiences the "background" plasma index $n_0 = \sqrt{1 - n_{e0}/n_c}$.

This dephasing or change in refractive index will play an important role when we investigate polarization effects of CBET in Sect. 7.2.5 (as we will see, the change

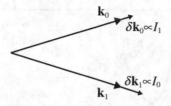

Fig. 7.6 Interaction between two equal-frequency beams in a plasma: no energy transfer occurs, but each beam experiences a change in the plasma refractive index—i.e., an increase in its wavenumber—proportional to the other beam's intensity (in this illustration, we would have $I_0 > I_1$)

of refractive index for one beam only occurs along the direction aligned with the other beam's electric field, i.e., the plasma becomes birefringent), as well as for the filamentation instability, where it is key to understand the process in terms of four-wave phase matching (cf. Sect. 7.3.5). In particular, it will be interesting to express the refractive index variation for $\hat{\omega}_2 = 0$, $\delta n_R(\omega_2 = 0)$, which can be derived using $Z'(0) = -2$ in the definition of the susceptibilities (assuming a Maxwellian background of plasma electrons). As $\chi_i(0) \gg \chi_e(0)$ and $F_\chi \approx \chi_e \approx 1/(k_2\lambda_{De})^2$ per our initial assumption that $ZT_e \gg T_i$, we obtain

$$\frac{\delta n_R(\hat{\omega}_2 = 0)}{n_0} = \frac{|\tilde{a}_0|^2}{8k_0^2\lambda_{De}^2}. \tag{7.83}$$

Figure 7.7 summarizes these results (with $\mathbf{V} = 0$ for simplicity). The amplitude of the density modulation, which is proportional to $|F_\chi|$ (Fig. 7.7a), is maximum at the IAW resonances, $\omega_2 = \pm k_2 c_s$ (i.e., when the phase velocity of the beat wave is equal to the plasma sound speed). The spatial amplification rate Γ_1 (for wave 1) is also maximum at the resonances (Fig. 7.7b); it is positive at $\omega_2 = k_2 c_s$, meaning wave 1 gains energy from wave 0 (consistent with $\omega_2 = \omega_0 - \omega_1 > 0$), and negative at $\omega_2 = -k_2 c_s$, corresponding to exponential decay of wave 1 as it transfers energy to wave 0. Figure 7.7c shows the variation of the plasma refractive index $\delta n_R \propto \text{Re}[F_\chi]$ experienced by wave 1. Note that $\delta n_R = 0$ at the resonances. Finally, Fig. 7.7d shows the dephasing $\Delta\phi_n = \text{Arg}[F_\chi]$; the dephasing is 0 at $\omega_2 = 0$, i.e., the beat wave intensity pattern and refractive index modulation (Fig. 7.5b) are aligned (in which case no energy transfer occurs, $\Gamma = 0$), and $\pm\pi/2$ at the resonances.

7.2.5 Polarization Effects

In experiments involving CBET, the polarization arrangement of the interacting beams will often be arbitrary, as opposed to the idealized situation assumed above

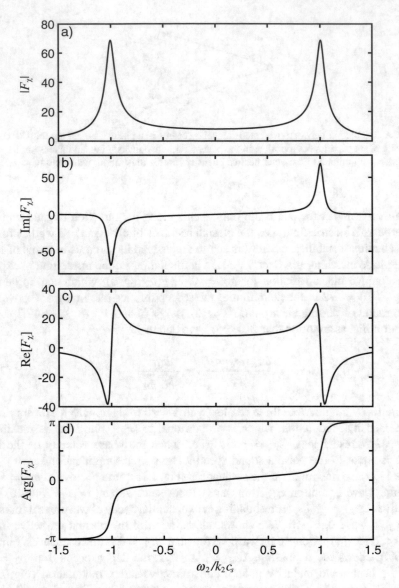

Fig. 7.7 Main features of wave-mixing (/CBET) in plasma, showing (**a**) $|F_\chi| \propto |\delta\tilde{n}_e|$, the amplitude of the electron density modulation, (**b**) $\mathrm{Im}[F_\chi] \propto \Gamma$, the spatial amplification rate, (**c**) $\mathrm{Re}[F_\chi] \propto \delta n_R$, the (real) variation in refractive index, and (**d**) $\mathrm{Arg}[F_\chi] = \Delta\phi_n$, the dephasing between the beat wave intensity pattern and the resulting density (or the refractive index) modulation. The parameters used here were $k_2\lambda_{De} = 1/3$, $ZT_e/T_i = 9$, and $Z = A$

Fig. 7.8 Geometry of CBET for arbitrary polarizations for the two beams

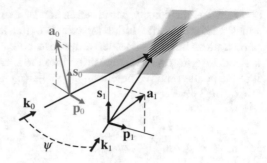

of two linearly polarized beams, both s-polarized (i.e., orthogonal to the plane of incidence). This is an important aspect because the interaction between beams can rotate their polarization (or change from linear to elliptical) even in the absence of energy transfer [39–41].

To describe the polarization of the beams, we need to track the projection of the electric field onto the two transverse directions of the beam, perpendicular to the propagation direction. Let us define the unit vectors $(\mathbf{p}_j, \mathbf{s}_j)$ for each beam j ($=0,1$), such that $\mathbf{p}_j$ represents the p-polarized component with respect to the plane of incidence (containing $\mathbf{k}_0$, $\mathbf{k}_1$, and $\mathbf{k}_0 - \mathbf{k}_1$) and $\mathbf{s}_0 = \mathbf{s}_1$ is along the s-component, orthogonal to the plane of incidence (cf. Fig. 7.8).

The vector potential of each beam is now decomposed into its s- and p-components:

$$\mathbf{a}_j = \frac{1}{2}(\tilde{a}_{jp}\mathbf{p}_j + \tilde{a}_{js}\mathbf{s}_j)e^{i\psi_j} + c.c. \tag{7.84}$$

We now have to deal with the s- and p-components of each wave, i.e., four coupled quantities. The description can be greatly simplified by introducing the Jones vectors of the light waves in their respective polarization bases $(\mathbf{p}_j, \mathbf{s}_j)$:

$$|a_j\rangle = \begin{pmatrix} \tilde{a}_{jp} \\ \tilde{a}_{js} \end{pmatrix}. \tag{7.85}$$

The complementary "bra" notation will be used to denote the conjugate transpose of a "ket", i.e.,

$$\langle a_j| = (\tilde{a}_{jp}^* \ \ \tilde{a}_{js}^*), \tag{7.86}$$

such that

$$\langle a_j|a_j\rangle = |\tilde{a}_j|^2, \tag{7.87}$$

$$|a_j\rangle \langle a_j| = \begin{pmatrix} |\tilde{a}_{jp}|^2 & \tilde{a}_{jp}\tilde{a}_{js}^* \\ \tilde{a}_{jp}^*\tilde{a}_{js} & |\tilde{a}_{js}|^2 \end{pmatrix}. \tag{7.88}$$

Likewise, the density perturbation can be expressed as a function of the beams' Jones vectors. In the following we use the kinetic description of the density perturbation, but the same derivation can be carried out with the fluid expression instead. Inserting the decomposition into the s- and p-components of the two beams into the ponderomotive potential, $\varphi_p = -\langle(\mathbf{a}_0 + \mathbf{a}_1)^2\rangle/2$, and collecting the terms $\propto \exp[i\psi_2]$ give

$$\frac{\delta\tilde{n}_e}{n_{e0}} = -\frac{k_2^2 c^2}{2\omega_{pe0}^2} F_\chi (\tilde{a}_{0p}\mathbf{p}_0 + \tilde{a}_{0s}\mathbf{s}_0) \cdot (\tilde{a}_{1p}^*\mathbf{p}_1 + \tilde{a}_{1s}^*\mathbf{s}_1) \tag{7.89}$$

$$= -\frac{k_2^2 c^2}{2\omega_{pe0}^2} F_\chi (\tilde{a}_{0p}\tilde{a}_{1p}^* \cos(\psi) + \tilde{a}_{0s}\tilde{a}_{1s}^*) \tag{7.90}$$

$$= -\frac{k_2^2 c^2}{2\omega_{pe0}^2} F_\chi \langle a_1| \Pi |a_0\rangle, \tag{7.91}$$

where

$$\Pi = \begin{pmatrix} \cos(\psi) & 0 \\ 0 & 1 \end{pmatrix} \tag{7.92}$$

is a projection matrix from $(\mathbf{p}_0, \mathbf{s}_0)$ onto $(\mathbf{p}_1, \mathbf{s}_1)$ ($\psi = \arccos(\mathbf{p}_0 \cdot \mathbf{p}_1)$ is the crossing angle between the two beams).

The coupled equations for arbitrary polarizations,

$$\partial_{z0}(\tilde{a}_{0p}\mathbf{p}_0 + \tilde{a}_{0s}\mathbf{s}_0) = -i\frac{\omega_{pe0}^2}{4k_0 c^2} F_\chi \frac{\delta\tilde{n}_e}{n_{e0}} (\tilde{a}_{1p}\mathbf{p}_1 + \tilde{a}_{1s}\mathbf{s}_1) \tag{7.93}$$

$$\partial_{z1}(\tilde{a}_{1p}\mathbf{p}_1 + \tilde{a}_{1s}\mathbf{s}_1) = -i\frac{\omega_{pe0}^2}{4k_0 c^2} F_\chi^* \frac{\delta\tilde{n}_e^*}{n_{e0}} (\tilde{a}_{0p}\mathbf{p}_0 + \tilde{a}_{0s}\mathbf{s}_0), \tag{7.94}$$

can be considerably simplified by using the Jones vectors' notations: multiplying the equations above by $\mathbf{p}_j$ and $\mathbf{s}_j$, we obtain the following system of coupled equations describing CBET (and more generally wave-mixing in plasmas) under arbitrary polarizations:

$$\partial_{z0} |a_0\rangle = i\frac{k_2^2}{8k_0} F_\chi \Pi |a_1\rangle \langle a_1| \Pi |a_0\rangle, \tag{7.95}$$

$$\partial_{z1} |a_1\rangle = i\frac{k_2^2}{8k_0} F_\chi^* \Pi |a_0\rangle \langle a_0| \Pi |a_1\rangle. \tag{7.96}$$

We can easily verify that for s-polarized waves, we recover Eqs. (7.68) and (7.69).

Nonlinear solutions (i.e., including variations in both beams) cannot be expressed in general, except for special geometries like when the crossing angle is either very

Fig. 7.9 Definition of the
CBET geometry in the
polarization plane $(\mathbf{p}_1, \mathbf{s}_1)$

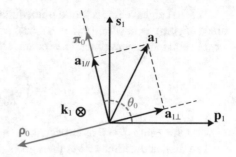

small or $\pi/2$ [40, 41]; however, it is very instructive to solve this system in the un-depleted pump limit, where $\tilde{a}_0$ is assumed uniform in space. Equation (7.96) then takes the form:

$$\partial_{z_1} |a_1\rangle = i \frac{k_2^2}{8k_0} F_\chi^* M_0 |a_1\rangle,\tag{7.97}$$

$$M_0 \equiv |\pi_0\rangle \langle \pi_0| = \begin{pmatrix} |\tilde{a}_{0p}|^2 \cos^2(\psi) & \tilde{a}_{0p}\tilde{a}_{0s}^* \cos(\psi) \\ \tilde{a}_{0p}^* \tilde{a}_{0s} \cos(\psi) & |\tilde{a}_{0s}|^2 \end{pmatrix},\tag{7.98}$$

where

$$|\pi_0\rangle = \Pi |a_0\rangle = \begin{pmatrix} \tilde{a}_{0p} \cos(\psi) \\ \tilde{a}_{0s} \end{pmatrix}\tag{7.99}$$

is the projection of $\mathbf{a}_0$ on the polarization basis of beam 1, $(\mathbf{p}_1, \mathbf{s}_1)$ (cf. Fig. 7.9).

M_0 is independent of z_1 based on our un-depleted pump assumption. Assuming uniform plasma conditions, we have the solution

$$|a_1(z)\rangle = \exp\left[i \frac{k_2^2}{8k_0} F_\chi^* M_0 z \right] |a_1(0)\rangle,\tag{7.100}$$

where the exponential notation refers to the matrix exponential operator, defined as

$$\exp[X] = \sum_{k=0}^{\infty} \frac{X^k}{k!}.\tag{7.101}$$

To express the matrix exponential, we first need to diagonalize the matrix. The eigenvalues of M_0, found by solving $\det(M_0 - \lambda I) = 0$, where I is the 2×2 identity matrix, are 0 and $|\tilde{a}_{0p}|^2 \cos^2(\psi) + |\tilde{a}_{0s}|^2 = \langle \pi_0|\pi_0\rangle = |\pi_0|^2$.

The first eigenvector is found immediately from the definition of $M_0 = |\pi_0\rangle \langle\pi_0|$: since $M_0 |\pi_0\rangle = |\pi_0\rangle \langle\pi_0|\pi_0\rangle = \langle\pi_0|\pi_0\rangle |\pi_0\rangle$, and $\langle\pi_0|\pi_0\rangle$ being an eigenvalue, then $|\pi_0\rangle$ itself is an eigenvector. The second eigenvector $|\rho_0\rangle$, such that $M_0 |\rho_0\rangle = 0$, is

$$|\rho_0\rangle = \begin{pmatrix} -\tilde{a}_{0s}^* \\ \tilde{a}_{0p}^* \cos(\psi) \end{pmatrix}. \tag{7.102}$$

We have $\langle\pi_0|\rho_0\rangle = 0$ as expected.

The diagonalization of M_0 then gives $M_0 = PDP^{-1}$, where P and D are the column eigenvectors matrix and the eigenvalues diagonal matrix, respectively:

$$P = \begin{pmatrix} \tilde{a}_{0p} \cos(\psi) & -\tilde{a}_{0s}^* \\ \tilde{a}_{0s} & \tilde{a}_{0p}^* \cos(\psi) \end{pmatrix}, \quad D = \begin{pmatrix} |\pi_0|^2 & 0 \\ 0 & 0 \end{pmatrix}. \tag{7.103}$$

Since $M_0^k = PD^k P^{-1}$, the matrix exponential can be expressed as

$$\exp\left[i\frac{k_2^2}{8k_0}F_\chi^* M_0 z\right] = \sum_{k=0}^\infty \frac{1}{k!}\left(i\frac{k_2^2}{8k_0}F_\chi^* z PDP^{-1}\right)^k \tag{7.104}$$

$$= \sum_{k=0}^\infty \frac{1}{k!}P\left(i\frac{k_2^2}{8k_0}F_\chi^* z D\right)^k P^{-1} \tag{7.105}$$

$$= P\begin{pmatrix} \exp\left[i\frac{k_2^2}{8k_0}F_\chi^* z|\pi_0|^2\right] & 0 \\ 0 & 1 \end{pmatrix}P^{-1}. \tag{7.106}$$

This solution takes a more transparent form if we assume that $\mathbf{a}_0$ is linearly polarized. Taking $\tilde{a}_{0p}$ and $\tilde{a}_{0s}$ as real quantities (without loss of generality), then $|\pi_0\rangle$ can be expressed as

$$|\pi_0\rangle = |\pi_0|\begin{pmatrix} \cos(\theta_0) \\ \sin(\theta_0) \end{pmatrix}, \tag{7.107}$$

where θ_0 is the angle between $\mathbf{p}_1$ and π_0 (cf. Fig. 7.9). Likewise,

$$|\rho_0\rangle = |\pi_0|\begin{pmatrix} -\sin(\theta_0) \\ \cos(\theta_0) \end{pmatrix}, \tag{7.108}$$

so P can be expressed in a more visual way (and easily inverted), as $P = |\pi_0|R(\theta_0)$, $P^{-1} = R(-\theta_0)/|\pi_0|$ with $R(\theta_0)$ the rotation matrix in the basis $(\mathbf{p}_1, \mathbf{s}_1)$:

$$R(\theta_0) = \begin{pmatrix} \cos(\theta_0) & -\sin(\theta_0) \\ \sin(\theta_0) & \cos(\theta_0) \end{pmatrix}. \tag{7.109}$$

Inserting into Eq. (7.100) gives the following solution:

$$|a_1(z)\rangle = R(\theta_0) \begin{pmatrix} \exp\left[i\frac{k_2^2}{8k_0}F_\chi^* z|\pi_0|^2\right] & 0 \\ 0 & 1 \end{pmatrix} R(-\theta_0)\,|a_1(0)\rangle. \qquad (7.110)$$

The physical interpretation of this solution is clear (see Fig. 7.9): since the two rotations back and forth correspond to a change of basis from $(\mathbf{p}_1, \mathbf{s}_1)$ to (π_0, ρ_0) and then back, the component of the polarization vector $|a_1\rangle$ aligned with π_0, $\mathbf{a}_{1/\!/}$ in the figure, is multiplied by $\exp[ik_2^2/(8k_0)F_\chi^* z|\pi_0|^2]$, while the component orthogonal to π_0 remains unchanged (multiplied by 1). This simply means that only the component of $\mathbf{a}_1$ aligned with the pump's electric field interacts with it. The interaction can lead to a variation in amplitude or in phase, as we saw in Sect. 7.2.4 (Fig. 7.7b, c). The dephasing, which is always present except if the beat wave is exactly at resonance, means that the pump makes the plasma birefringent for beam #1, which experiences a modified refractive index along π_0 (the variation being generally complex, i.e., dephasing and amplitude variation) but only sees the background plasma index along the direction orthogonal to π_0.

If both beams are s-polarized, the results from the previous section are recovered. These polarization effects of CBET have been verified experimentally [42, 43] and can impact the estimates of CBET in ICF experiments or lead to systematic implosion asymmetries for direct-drive geometry if the polarization arrangement of the beams is not symmetric [44, 45].

To conclude this section, let us consider the situation of random polarizations. This will be typical of many laser–plasma experiments with high-energy lasers, where beams are equipped with polarization smoothing schemes (cf. Sect. 9.2). Since $\partial_{z_1}|\tilde{a}_1|^2 = \langle a_1|a_1\rangle' = \langle a_1'|a_1\rangle + \langle a_1|a_1'\rangle$ (where the prime denotes ∂_{z_1}), by multiplying Eq. (7.97) by $\langle a_1|$ on the left side and adding the complex conjugate, we obtain

$$\partial_{z_1}|\tilde{a}_1|^2 = -\frac{k_2^2}{4k_0}\mathrm{Im}\left[F_\chi^*\langle a_1|M_0|a_1\rangle\right]. \qquad (7.111)$$

We first assume that the pump is randomly polarized but make no assumption on beam 1. Taking $\tilde{a}_{0p} = |\tilde{a}_{0p}|$ and $\tilde{a}_{0s} = |\tilde{a}_{0s}|\exp[i\delta\varphi]$ (without loss of generality, since only the phase difference between $\tilde{a}_{0p}$ and $\tilde{a}_{0s}$ is relevant), we can write

$$\tilde{a}_{0p} = |\tilde{a}_0|\cos(\theta_0)\,, \qquad (7.112)$$

$$\tilde{a}_{0s} = |\tilde{a}_0|\sin(\theta_0)e^{i\delta\varphi} \qquad (7.113)$$

such that $\cos(\theta_0) \equiv |\tilde{a}_{0p}|/|\tilde{a}_0|$ and $\sin(\theta_0) = |\tilde{a}_{0s}|/|\tilde{a}_0|$. As the pump's polarization is random, we assume that θ_0 and $\delta\varphi$ are random variables uniformly distributed in $[0, 2\pi]$ and uncorrelated to one another, i.e., $\langle\cos(\theta_0)\rangle_{\mathrm{PS}} = \langle\sin(\theta_0)\rangle_{\mathrm{PS}} =$

$\langle \cos(\delta\varphi) \rangle_{\mathrm{PS}} = 0$. Here the brackets denote a statistical average over polarization states, not to be confused with the bra–ket notation of the Jones vectors.

We now wish to take the statistical average of Eq. (7.111); the average is only applied to M_0 and can be applied to each matrix element since the matrix operations are linear; we get

$$\langle M_0 \rangle_{\mathrm{PS}} = \frac{|\tilde{a}_0|^2}{2} \begin{pmatrix} \cos^2(\psi) & 0 \\ 0 & 1 \end{pmatrix}. \tag{7.114}$$

Therefore, Eq. (7.111) becomes

$$\partial_{z_1} |\tilde{a}_1|^2 = \frac{k_2^2}{8k_0} \mathrm{Im}[F_\chi] |\tilde{a}_0|^2 \left(\cos^2(\psi) |\tilde{a}_{1p}|^2 + |\tilde{a}_{1s}|^2 \right). \tag{7.115}$$

The p- and s-components of beam 1 may evolve differently depending on the crossing angle and polarization arrangements.

If both beams are randomly polarized, then $\langle |\tilde{a}_{1p}|^2 \rangle_{\mathrm{PS}} = \langle |\tilde{a}_{1s}|^2 \rangle_{\mathrm{PS}} = |\tilde{a}_1|^2/2$, and the equation above can be solved, giving

$$|\tilde{a}_1(z)| = |\tilde{a}_1(0)| \exp\left[\Gamma_{1\mathrm{PS}} z \right], \tag{7.116}$$

$$\Gamma_{1\mathrm{PS}} = \frac{k_2^2}{32k_0} \mathrm{Im}[F_\chi] \left(1 + \cos^2(\psi) \right) |\tilde{a}_0|^2 = \Gamma_1 \frac{1 + \cos^2(\psi)}{4}, \tag{7.117}$$

where Γ_1 is the amplification rate for two linearly s-polarized beams from Eq. (7.70) and $\Gamma_{1\mathrm{PS}}$ is the amplification rate with random polarization (i.e., polarization smoothing, cf. Sect. 9.2). This indicates that the amplification rate is reduced by a factor varying between 2 (for small crossing angles, $\cos^2(\psi) \ll 1$) and 4 (for $\psi = \pi/2$) compared to the s-polarized amplification rate.

7.3 The Filamentation Instability

7.3.1 Introduction

The filamentation instability is a process which leads to the nonlinear breakup of a "large" beam (large enough that it can be locally approximated as a plane wave; we will quantify this condition later) into small filaments. The physical process is illustrated in Fig. 7.10: as the beam interacts with density fluctuations from noise (or, equivalently, as it develops random noise in its phase front), it undergoes small-angle scattering off mostly transverse density fluctuations with a broad spectrum of wave-numbers. For a given fluctuation wave-vector $\mathbf{k}_2$, the laser at $\mathbf{k}_0$ can scatter into either $\mathbf{k}_1 = \mathbf{k}_0 + \mathbf{k}_2$ or $\mathbf{k}_{-1} = \mathbf{k}_0 - \mathbf{k}_2$ (since the fluctuation is assumed steady-

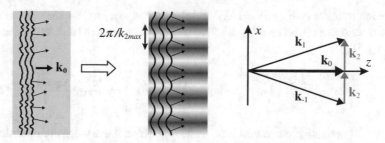

Fig. 7.10 Filamentation process: random fluctuations in the plasma density lead to a noisy phase front of the laser; a particular mode of density fluctuation, transverse to the propagation direction and at a wave-vector $\mathbf{k}_{2max}$ (to be defined later), maximizes the instability growth, leading to a transverse density modulation and in turn to a breakup of the laser into filaments separated by $2\pi / k_{2max}$

state, it has no associated direction, so $\mathbf{k}_2$ and $-\mathbf{k}_2$ are equivalent since the mode is not propagating).

As we will see, there exists a particular mode at k_{2max} for which the instability growth is maximum; it means that the density is going to develop a modulation at k_{2max}, leading to a beam breakup due to refraction (and eventually self-focusing) into the low density troughs of the modulation pattern (cf. Fig. 7.10 or Fig. 7.12a for an actual simulation result). In other words, the beam breaks into filaments whose typical size (or separation distance) is $2\pi / k_{2max}$.

The filamentation instability was first described in the general framework of nonlinear optics (i.e., in the presence of an intensity-dependent refractive index, the ponderomotive force in plasma being one example among many) by Bespalov [46] and Chiao [47] (see also Ref. [38] for a summarized derivation based on the same approach). Here we treat the process similarly to SBS via the instability dispersion relation (cf. Sect. 7.1.3) but keep the anti-Stokes scattered wave. As we will see, the two approaches lead to the same results, which can be physically interpreted in terms of phase matching of the four-wave mixing process, as we discuss in Sect. 7.3.5.

Filamentation in plasma was investigated experimentally in ICF-relevant conditions by imposing a periodic intensity modulation on the laser beam profile chosen close to the optimum modulation period k_{2max} mentioned above and measuring the associated plasma density modulation [48]. However, most experimental work was done with optical smoothing techniques (e.g., random phase plates), which will be discussed in more detail in Sect. 9.5.1.

7.3.2 Dispersion Relation and Growth Rate

Filamentation can also be treated as a special case of coupling between a pump wave (the laser, modeled as a plane wave), an IAW, and both the Stokes and anti-Stokes scattered EMW's. The analysis is the same as in Sect. 7.1.3, up to Eq. (7.34), the dispersion relation coupling the IAW and the Stokes and anti-Stokes waves, which can be rewritten as (with a_0 the peak field amplitude without a tilde, consistent with Sect. 7.1.3)

$$(\omega^2 - k^2 c_s^2)\left[\left(\omega - \mathbf{k} \cdot \mathbf{k}_0 \frac{c^2}{\omega_0}\right)^2 - \frac{k^4 c^4}{4\omega_0^2}\right] = \frac{a_0^2 k^4 c^4 \omega_{pi}^2}{8\omega_0^2}. \qquad (7.118)$$

Since filamentation corresponds to scattering off stationary density modulations, we set $\omega = 0$ in the equation above. Therefore,

$$1 - 4\frac{(\mathbf{k} \cdot \mathbf{k}_0)^2}{k^4} = \frac{a_0^2}{2k^2 \lambda_{De}^2}. \qquad (7.119)$$

Furthermore, we also assume that the Stokes and anti-Stokes scattered modes contribute nearly equally to the process (which is what distinguishes filamentation, a four-wave-mixing process, from SBS, a three-wave process). Going back to our discussion in Sect. 7.1.3 and to Eq. (7.33), we see that the two EMW modes contribute equally in the limit of $\mathbf{k} \cdot \mathbf{k}_0 \to 0$ (assuming $\omega = 0$), such that the two waves' dispersion relations become similar, $D_1(+1) \approx D_1(-1)$. This is the limit of forward scattering (cf. Fig. 7.2).

Next we look for a solution for the scattered EMWs that is spatially amplified, i.e., $\mathbf{k} = \mathbf{k}_2 - i\Gamma \mathbf{e}_z$ with $\mathbf{k}_2 \cdot \mathbf{k}_0 \approx 0$ (such that the Stokes and anti-Stokes modes are both $\propto \exp[i(\mathbf{k} \pm \mathbf{k}_0) \cdot \mathbf{r}] \propto \exp[\Gamma z]$). Inserting into Eq. (7.119) above (together with $\omega = 0$ and $\mathbf{k} \cdot \mathbf{k}_0 = -i\Gamma k_0$) leads to

$$\boxed{\Gamma = \frac{k_2^2}{2k_0}\sqrt{\frac{a_0^2}{2k_2^2 \lambda_{De}^2} - 1}}. \qquad (7.120)$$

The instability can only develop if $\Gamma > 0$, i.e., if

$$k_2 < \frac{a_0}{\sqrt{2}\lambda_{De}}. \qquad (7.121)$$

We can also see from solving Eq. (7.120) that the growth is maximum for a particular mode k_2 given by

$$\boxed{k_{2max} = \frac{a_0}{2\lambda_{De}}}, \qquad (7.122)$$

with the peak amplification rate at k_{2max} given by

$$\Gamma_{max} = \frac{a_0^2}{8k_0 \lambda_{De}^2} \qquad (7.123)$$

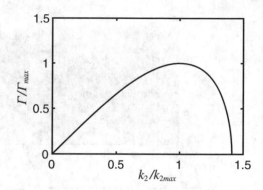

Fig. 7.11 Filamentation spatial amplification rate Γ normalized to Γ_{max}, as a function of the density modulation wave-number k_2 normalized to the value k_{2max} maximizing the amplification (cf. definitions in the text)

$$= k_0 \frac{\delta n}{n_0}, \tag{7.124}$$

where $\delta n/n_0 = a_0^2/(8k_0^2\lambda_{De}^2)$ is the variation in refractive index experienced by the scattered light waves (Stokes and anti-Stokes) due to the presence of the pump, as we saw in Sect. 7.2.5 Eq. (7.83).

The expression for the growth rate can then be recast as

$$\Gamma = \Gamma_{max} \sqrt{\left(\frac{k_2}{k_{2max}}\right)^2 \left[2 - \left(\frac{k_2}{k_{2max}}\right)^2\right]}. \tag{7.125}$$

The growth rate is represented in Fig. 7.11.

The density modulation wave-number of maximum growth k_{2max} is a key result because it provides the typical filament size δx_{fil} that will emerge from the background noise and that the beam will break up into,

$$\delta x_{fil} \approx \frac{2\pi}{k_{2max}} = \frac{4\pi\lambda_{De}}{a_0}. \tag{7.126}$$

The existence of a particular mode for the density modulation that maximizes the filamentation instability can be physically explained in terms of phase-matching conditions for the four-wave-mixing process, as will be discussed in Sect. 7.3.5.

To illustrate the process, let us visualize a 2D simulation showing the early stage of the filamentation instability. In Fig. 7.12, we simulated the nonlinear propagation of a Gaussian beam with a transverse profile of $1/e$ width $w_0 = 10\delta x_{fil}$ with a nonlinearity $\delta n/n_0 = 4 \times 10^{-4}$ and a broadband noise added to the beam in Fourier space at a level of 2×10^{-4} (in amplitude).

Figure 7.12a shows the filaments developing near the end of the simulation box; a lineout of the transverse intensity profile at the beginning vs. at the end of the simulation box is represented in Fig. 7.12b, showing the filaments having reached approximately twice the initial laser intensity. The spacing between filaments is

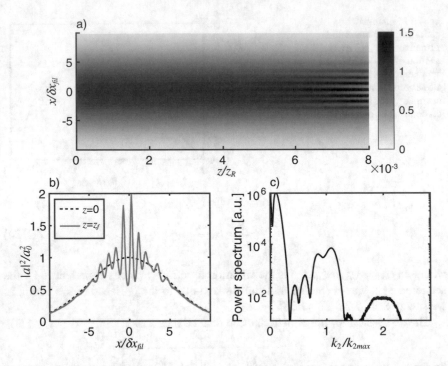

Fig. 7.12 Simulation of the filamentation process (conditions in the text): (**a**) intensity profile, (**b**) intensity profile, lineouts at the beginning and end of the simulation box, and (**c**) Fourier spectrum at the end of the simulation box, showing the growth of modes at k_{2max} and its harmonics (from diffraction into higher-orders)

approximately δx_{fil} as expected. Figure 7.12c shows the spectrum at the end of the simulation box: it shows the main mode at k_{2max} emerging from the background noise, as well as harmonics at $2k_{2max}$, corresponding to the transition to the Raman-Nath regime of diffraction off the density modulation (i.e., diffraction into higher orders, cf. Ref. [38]). Things rapidly turn highly nonlinear if one lets the laser propagate for a longer distance as the filaments start to self-focus: indeed, as we explain below in Sect. 7.3.6, the size of the filaments that emerge from the filamentation instability is such that the power contained in a filament coincides with the critical power for self-focusing. In other words, the development of filaments from the filamentation instability will always be rapidly followed by self-focusing of these filaments and the evolution toward highly nonlinear beam breakup.

7.3.3 Weakly vs. Strongly Coupled Regimes of Filamentation

More generally, one can also look for complex solutions for ω in Eq. (7.118). If such solutions exist with an imaginary component $\gamma > 0$, then they correspond to an instability. In this section we do not assume that the density modulation is stationary anymore, i.e., $\omega \neq 0$: the modulation can be a traveling wave with a finite frequency. The zero-frequency case assumed in the previous section corresponds to the maximum growth of the instability; when growth is not maximum, a finite frequency component exists (we will see in Sect. 7.3.5 that the zero-frequency case corresponds to four-wave phase-matching conditions between the laser, density modulation, and the Stokes and anti-Stokes scattered light waves). We will see in the next section that there is a smooth transition between the "pure" filamentation instability with $\omega = 0$ and the forward SBS instability with $\omega \neq 0$.

Solutions for the temporal growth rate of the filamentation instability can be found in two regimes, similar to SBS:

- $|\omega| \ll kc_s$: "weakly coupled" regime

Solving the dispersion relation Eq. (7.118) for ω in this limit leads to

$$\omega = \mathbf{k} \cdot \mathbf{k}_0 \frac{c^2}{\omega_0} + \sqrt{\frac{k^4 c^4}{4\omega_0^2} - \frac{a_0^2 k^2 c^4}{8\omega_0^2 \lambda_{De}^2}}. \tag{7.127}$$

Unstable solutions exist if $\gamma = \text{Im}[\omega] > 0$, i.e.,

$$a_0 > \sqrt{2} k \lambda_{De}. \tag{7.128}$$

When this condition is satisfied, the temporal growth rate can be expressed as

$$\boxed{\gamma = \frac{k^2 c^2}{2\omega_0} \sqrt{\frac{a_0^2}{2k^2 \lambda_{De}^2} - 1}} \tag{7.129}$$

We see that the temporal growth rate is related to the spatial amplification rate Γ from the previous section, Eq. (7.120) via

$$\gamma = \Gamma v_{g1}, \tag{7.130}$$

where $v_{g1} = k_0 c^2 / \omega_0$ is the scattered light wave's group velocity in the small angle limit: the spatial amplification rate corresponds to the temporal growth rate expressed in the frame moving at the scattered wave's group velocity.

We also note that in general, the real frequency of the density modulation is not zero but $\omega_R = \mathbf{k} \cdot \mathbf{k}_0 c^2 / \omega_0$ such that $\omega = \omega_R + i\gamma$. The real frequency ω_R is only equal to zero when $\mathbf{k} \cdot \mathbf{k}_0 = 0$, i.e., for the conditions that maximize the growth rate due to perfect phase matching as we will see in Sect. 7.3.5.

The condition of validity for filamentation in the weakly coupled regime is $|\omega| \ll kc_s$, i.e., $\omega_R \ll kc_s$ and $\gamma \ll kc_s$. The first condition leads to

$$\sin(\psi/2) \ll \frac{c_s}{n_0 c}, \tag{7.131}$$

with ψ the angle between the laser and Stokes waves (cf. Fig. 7.2) and $n_0 = k_0 c / \omega_0 = \sqrt{1 - n_{e0}/n_c}$ the background refractive index, while the second condition, $\gamma \ll k c_s$, taken for the maximum growth rate γ_0 at $k = k_{max} = a_0/2\lambda_{De}$ leads to

$$a_0 \ll 4 k_0 \lambda_{De} \frac{c_s}{c n_0}. \tag{7.132}$$

- $|\omega| \gg k c_s$, $\mathbf{k} \cdot \mathbf{k}_0 c^2/\omega_0$: "strongly coupled" regime

  Under these assumptions, the dispersion relation Eq. (7.118) becomes

$$\omega^2 \left(\omega^2 - \frac{k^4 c^4}{4 \omega_0^2} \right) = a_0^2 k^4 c^4 \frac{\omega_{pi}^2}{8 \omega_0^2}. \tag{7.133}$$

This equation admits purely imaginary solutions $\omega = i\gamma$ such that

$$\gamma^2 = \frac{k^4 c^4}{8 \omega_0^2} \left[\sqrt{1 + 8 a_0^2 \omega_{pi}^2 \frac{\omega_0^2}{k^4 c^4}} - 1 \right]. \tag{7.134}$$

For $k \to 0$, or more specifically $k^2 c^2 \ll 2\sqrt{2} a_0 \omega_0 \omega_{pi}$, we have

$$\gamma \approx k c [a_0 \omega_{pi}/(\sqrt{8}\omega_0)]^{1/2}. \tag{7.135}$$

In the other limit, $k^2 c^2 \gg 2\sqrt{2} a_0 \omega_0 \omega_{pi}$, the growth rate has the asymptotic limit $\gamma \to \gamma_0$ (the maximum growth rate for the strongly coupled filamentation instability) with

$$\boxed{\gamma_0 = \frac{a_0 \omega_{pi}}{\sqrt{2}}.} \tag{7.136}$$

This maximum growth rate for the strongly coupled regime is approximately reached for $k \geq k_{max}$, where

$$k_{max}^2 = \frac{\sqrt{8}}{c^2} a_0 \omega_0 \omega_{pi}. \tag{7.137}$$

The filamentation growth rate for the strongly coupled regime is represented in Fig. 7.13.

The first condition of validity for filamentation in the strongly coupled regime is $\gamma \gg k c_s$, which for $\gamma \approx \gamma_0$ (and hence $k \geq k_{max}$) implies $\gamma_0 \gg k_{max} c_s$, i.e.,

$$a_0 \gg 4\sqrt{2} \frac{k_0 \lambda_{De}}{n_0} \frac{c_s}{c}. \tag{7.138}$$

Fig. 7.13 Temporal growth
rate of the filamentation
instability in the strongly
coupled regime

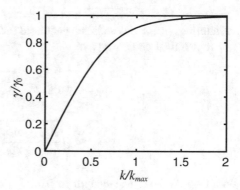

The second validity condition is $\gamma \gg \mathbf{k} \cdot \mathbf{k}_0 c^2 / \omega_0$, which leads to

$$a_0 \gg 16\sqrt{2} \frac{k_0 \lambda_{De}}{n_0} \frac{c_s}{c} \left[\frac{n_0 c}{2 c_s} \sin(\psi/2) \right]^2 . \tag{7.139}$$

These conditions will allow us to map out the conditions for SBS vs. filamentation in the weakly vs. strongly coupled regimes, as discussed in the next section below.

7.3.4   Transition from Forward SBS to Filamentation

We now have the conditions of validity for the weakly and strongly coupled regimes of SBS and filamentation: they are Eqs. (7.43), (7.44) (SBS weakly coupled), (7.48), (7.49) (SBS strongly coupled), (7.131), (7.132) (filamentation weakly coupled), and (7.138), (7.139) (filamentation strongly coupled). A rapid inspection of these conditions show that they are complementary (up to numerical factors of order $\sim O(1)$ which do not matter since these are all strong inequalities) and for fixed plasma conditions only depend on the pump amplitude a_0 and the angle θ between the laser and IAW, as shown in Fig. 7.2. The (a_0, θ) parameter space can be divided into four regions corresponding to these four instabilities [49].

Following Pesme [49],[4] we introduce the dimensionless quantities:

$$\epsilon_0 = \frac{a_0 c^2}{\sqrt{8}\lambda_{De}\omega_0 c_s} , \tag{7.140}$$

$$\kappa_0 = \frac{n_0 c}{c_s} \cos(\theta) \approx \frac{n_0 c}{c_s} \delta\theta , \tag{7.141}$$

---

[4] Our definitions differ slightly from Ref. [49].

where $n_0 = \sqrt{1 - n_{e0}/n_c}$ and $\delta\theta = \pi/2 - \theta$, with $\delta\theta \ll 1$ for near-forward scattering geometries as is relevant here (cf. Fig. 7.2).

In practical units, we have

$$\epsilon_0 \approx 29.3 \frac{\lambda_\mu}{T_{ek}} \sqrt{I_{16} \frac{n_{e0}}{n_c} \frac{A}{Z}} , \tag{7.142}$$

$$\kappa_0 \approx \sqrt{\frac{A(1 - n_{e0}/n_c)}{Z T_{ek}}} \delta\theta \; [\text{mrad}] , \tag{7.143}$$

with λ_μ the laser wavelength in μm, T_{ek} the electron temperature in keV, I_{16} the intensity in 10^{16} W/cm^2, and Z and A the ionization level and atomic number.

The conditions of validity for SBS and filamentation in the weakly and strongly coupled regimes, from Eqs. (7.43), (7.44), (7.48), (7.49), (7.131), (7.132), (7.138), and (7.139), then take the simple form:

- SBS weakly coupled: $\epsilon_0 \ll 2\sqrt{\kappa_0 - 1}$, $\epsilon_0 \ll 4(\kappa_0 - 1)^{3/2}$ (with $\kappa_0 > 1$).
- SBS strongly coupled: $\sqrt{2\kappa_0 - 2} \ll \epsilon_0 \ll 4(\kappa_0 - 1)^2$.
- Filamentation weakly coupled: $\kappa_0 \ll 1$, $\epsilon_0 \ll \sqrt{2}$.
- Filamentation strongly coupled: $\epsilon_0 \gg 2$, $\epsilon_0 \gg 2\kappa_0^2$.

Plotting these inequalities divides the (ϵ_0, κ_0) (i.e., $(a_0, \delta\theta)$) parameter space into the four regions of existence of these instabilities, as shown in Fig. 7.14. These four instabilities are simply strong limits of the same dispersion relation, Eq. (7.118), describing the coupling between an IAW, one or two scattered EMWs (Stokes and anti-Stokes), and the pump laser. The limits between them (the dashed lines on the plot) are smooth transitions.

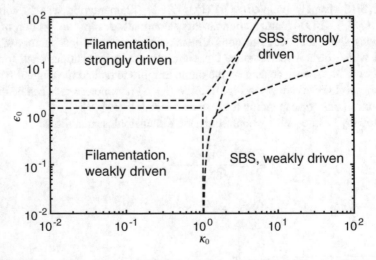

Fig. 7.14 Regions of the existence of the SBS and filamentation instabilities, as a function of the normalized laser intensity and scattering angle ϵ_0, κ_0 (cf. definitions in the text)

When interpreting this plot, the approximation $\cos(\theta) \approx \sin(\psi/2)$ used in our description of SBS is not valid anymore. In other words, we cannot directly relate the angle θ to the scattering angle ψ, due to the variation in the scattered light's wave-number due to its interaction with the pump (as discussed in Sect. 7.2.4; see also Fig. 7.6), which starts to play a role at very small scattering angles. This should become more clear as we interpret the filamentation instability in terms of resonant four-wave phase matching in the next section below.

7.3.5 Physical Interpretation of the Filamentation Instability in Terms of Four-Wave-Mixing Phase-Matching Condition

We saw in Sect. 7.2.4 that when two light waves with the same frequency interact in a plasma, each wave experiences a refractive index that is slightly modified from the background index $n_0 = \sqrt{1 - n_{e0}/n_c}$ by an amount δn which is proportional to the intensity of the other wave (cf. Fig. 7.6), as given by Eq. (7.83).

For filamentation, the laser couples to random electromagnetic fluctuations as it propagates in the plasma; this noise source corresponds to a continuous background of EMWs. As a given mode $\mathbf{k}_1$ from the background noise with the same frequency as the laser interacts with the pump at $\mathbf{k}_0$ (with $k_0 = k_1$ since the frequencies are equal), this particular mode will experience an increase in its wave-number $\delta k_1 = (\omega_0/c)\delta n_1$ via the increase in the (nonlinear) index δn_1 given by Eq. (7.83), i.e.,

$$\frac{\delta k_1}{k_0} = \frac{\delta n_1}{n_0} = \frac{a_0^2}{8k_0^2\lambda_{De}^2} . \tag{7.144}$$

We assume that the pump is much more intense than the background noise and therefore neglect the increase in the pump's wave-number ($\delta k_0 \propto a_1^2$).

The "true" beat wave between the noise and the pump will thus be at $\mathbf{k}_2 = \mathbf{k}_0 - (\mathbf{k}_1 + \delta\mathbf{k}_1)$. The peak gain for filamentation occurs when the beat wave is exactly perpendicular to the laser propagation direction, with $\mathbf{k}_2 = \mathbf{k}_{2max} \perp \mathbf{k}_0$. In this case, even though the coupling between the laser and the mode at $\mathbf{k}_1$ cannot lead to energy transfer (cf. Sect. 7.2.4), the density modulation it creates will be reinforced by another mode from the noise also at the same frequency ω_0 and at $\mathbf{k}_{-1}$, subject to the same wave-number increase $\delta k_{-1} = \delta k_1$ and whose beating with $\mathbf{k}_0$ will exactly coincide with $\mathbf{k}_{2max}$ [cf. Fig. 7.15].[5] The two modes reinforce each other's index modulation, and the system becomes unstable, with the amplitude of both modes and the modulation at $\mathbf{k}_{2max}$ growing exponentially along z. The modulation acts like a plasma grating and will lead to a modulation of the beam intensity and

---

[5] Since the EMWs all have the same frequency, the beat waves between $\mathbf{k}_0$ and $\mathbf{k}_1$ or $\mathbf{k}_0$ and $\mathbf{k}_{-1}$ are stationary, and the direction of $\mathbf{k}_2$ does not matter.

Fig. 7.15 Four-wave phase matching for the filamentation instability: the beatings between $(\mathbf{k}_1 + \delta\mathbf{k}_1, \mathbf{k}_0)$ and $(\mathbf{k}_{-1} + \delta\mathbf{k}_{-1}, \mathbf{k}_0)$ produce the same beat wave (i.e., density modulation) at k_{2max} when $\delta k_1 = \delta k_{-1} \propto a_0^2$ is such that $\mathbf{k}_0 \cdot \mathbf{k}_{2max} = 0$

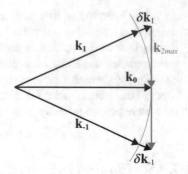

its breakup into filaments. The filaments are then prone to self-focusing and other nonlinear phenomena.

The mode $\mathbf{k}_{2max}$ can simply be estimated by geometry: since $k_{2max}^2 = (k_1 + \delta k_1)^2 - k_0^2 \approx 2k_0\delta k_1$ for $\delta k_1 \ll k_0$, we immediately get

$$k_{2max} = k_0\sqrt{2\frac{\delta n_1}{n_0}} = \frac{a_0}{2\lambda_{De}}. \qquad (7.145)$$

We exactly recover the result from Eq. (7.122).

7.3.6 Connection Between Self-focusing and Filamentation

We have seen in Sect. 7.3.2 and in the discussion above that a spatially uniform laser beam may break up into filaments whose size $\delta x_{fil} = 2\pi/k_{2max}$ is given by Eq. (7.126). Conversely, a laser beam whose transverse dimension is smaller than δx_{fil} cannot filament; since $\delta x_{fil} \propto 1/a_0$, this sets a condition on the laser intensity as well as its size.

To clarify the meaning of this condition, we can define the "critical" laser spot size w_c such that $w_c = \delta x_{fil} = \lambda_0/\sqrt{2n_0\delta n}$, where $\delta n = n_2 I_0$ is the refractive index perturbation. The corresponding beam power is $P_c \approx I_0\pi(w_c/2)^2$, which after substituting for w_c becomes

$$P_c \approx \frac{\pi\lambda_0^2}{8\pi n_0 n_2}. \qquad (7.146)$$

We recognize the critical power for self-focusing derived in Sect. 5.6, Eqs. (5.70) or (5.92) (these expressions being accurate up to a numerical factor $\sim O(1)$). The filament size δx_{fil} that maximizes the filamentation growth rate, Eq. (7.126), also happens to contain the critical power for self-focusing within a filament. Consequently, a laser beam whose transverse size exceeds δx_{fil} is necessarily above the critical power for self-focusing. Or, equivalently, when a laser beam exceeds the critical power for self-focusing, meaning that at least a few filaments can develop within the beam envelope, then each filament will be at the critical power for self-focusing. The filamentation instability will therefore trigger self-focusing of the individual filaments; the breakup into filaments only represents the initial phase of a typically highly nonlinear evolution of the beam propagation.

The evolution toward whole beam self-focusing vs. filamentation depends on several factors, such as the level of noise in the beam phase front (which is related to the level of background fluctuations in the plasma), the initial conditions (i.e., a beam close to the threshold will self-focus since it is not quite large enough to filament), and the use of optical smoothing by continuous or random phase plates, which will be discussed in Chap. 9 (filamentation of optically-smoothed beams is the subject of Sect. 9.5.1).

7.3.7 Thermal Filamentation

Thermal filamentation arises when the laser energy deposition via collisional heating locally heats the plasma electrons, which in turn move toward the colder surrounding regions via heat conduction [50–52]. The combination of heating and thermal conduction plays a similar role as the ponderomotive force in pushing electrons away from the high-intensity regions and setting up a space charge potential that the ions will respond to as they also get pulled toward the regions of lower intensity.

To treat thermal filamentation, we need to go back to the derivation of the IAW wave equation in the presence of a laser driver (Sect. 5.2) but include variations in electron density instead of the isothermal hypothesis used so far. Our derivation will largely follow Schmitt's [52]. Going back to the equation of motion for the electron fluid, Eq. (1.70), we now allow for variations in T_e in the pressure $p_e = n_e T_e$; therefore, the (last) pressure term takes the form:

$$\frac{\nabla p_e}{n_e m} = v_{Te}^2 \left(\frac{\nabla n_e}{n_e} + \frac{\nabla T_e}{T_e} \right). \tag{7.147}$$

Proceeding like in Sect. 5.2, we now linearize both the electron density and temperature, $n_e = n_{e0} + \delta n_e$, $T_e = T_{e0} + \delta T_e$, where the perturbations correspond to second order quantities ($\propto E^2$) with the notations from Sect. 5.2. Equation (5.24) then becomes, with the normalization $\varphi = (e/mc^2)\Phi$,

$$\nabla(\varphi + \varphi_p) = \frac{v_{Te}^2}{c^2} \left(\frac{\nabla \delta n_e}{n_{e0}} + \frac{\nabla \delta T_e}{T_{e0}} \right). \tag{7.148}$$

The rest of the derivation is the same as in Sect. 5.2: we linearize the ion fluid equations, eliminate the electrostatic potential φ using the expression above, and obtain the following wave equation:

$$(\partial_t^2 - c_s^2 \nabla^2) \frac{\delta n_e}{n_{e0}} = -\frac{Zmc^2}{M_i} \nabla^2 \varphi_p + \frac{Z}{M_i} \nabla^2 \delta T_e. \tag{7.149}$$

Next we must relate the temperature variation δT_e to the absorption and conduction. The heat transport equation is

$$\frac{3}{2} n_{e0} \frac{\partial T_e}{\partial t} = \frac{1}{2} mc^2 a_0^2 \nu_{ei} n_{e0} + \nabla \cdot (\kappa_e \nabla T_e). \tag{7.150}$$

The heating rate from collisional absorption was derived in Sect. 4.3; κ_e is the electron conductivity. Assuming thermal equilibrium (i.e., setting $\partial_t T_e = 0$) and linearizing the temperature like above lead to

$$\nabla^2 \delta T_e = -\frac{mc^2 a_0^2 \nu_{ei} n_{e0}}{2\kappa_e}, \tag{7.151}$$

where κ_e is taken at equilibrium (n_{e0}, T_{e0}). Inserting into the IAW wave equation above with $\varphi_p = -\langle \mathbf{a}^2 \rangle / 2$ gives

$$(\partial_t^2 - c_s^2 \nabla^2) \frac{\delta n_e}{n_{e0}} = \frac{Zmc^2}{2M_i} \left(\nabla^2 \langle \mathbf{a}^2 \rangle - 2 \frac{\nu_{ei} n_{e0}}{\kappa_e} \langle \mathbf{a}^2 \rangle \right). \tag{7.152}$$

Next we proceed like for the ponderomotively driven filamentation (or SBS, cf. Sect. 7.1.3) and decompose the field as $\mathbf{a} = \mathbf{a}_0 + \mathbf{a}_1$ with $\mathbf{a}_0 = \mathbf{e}_0 a_0 \cos(\psi_0)$ and $\mathbf{a}_1$ the scattered wave and look for the IAW response to the beat of $\mathbf{a}_0$ and $\mathbf{a}_1$. Taking a Fourier–Laplace transform gives the IAW dispersion relation in the presence of both the ponderomotive and thermal (heating) effects from the laser:

$$(\omega^2 - k^2 c_s^2) \frac{\delta \hat{n}_e}{n_{e0}} = -\frac{Zmc^2}{2M_i} k^2 a_0 \left(1 + 2 \frac{\nu_{ei} n_{e0}}{k^2 \kappa_e} \right) [\hat{a}_1(-1) + \hat{a}_1(+1)], \tag{7.153}$$

where like in the previous sections $\hat{a}_1(n) = \hat{a}_1(\omega + n\omega_0, \mathbf{k} + n\mathbf{k}_0)$. The second term in the parenthesis on the right-hand side is the perturbation from collisional heating and conduction; the 1 is from the ponderomotive force (cf. Eq. (7.30)), and we see that thermal effects are only dominant if

$$2 \frac{\nu_{ei} n_{e0}}{k^2 \kappa_e} \geq 1. \tag{7.154}$$

If $2\nu_{ei} n_{e0}/(k^2 \kappa_e) \ll 1$, then only the ponderomotive force is left and the dispersion relation reduces to Eq. (7.30). Focusing on thermal effects for now, i.e., assuming $2\nu_{ei} n_{e0}/(k^2 \kappa_e) \gg 1$, we can derive the general dispersion relation of the coupling of the pump laser to an IAW via the Stokes and anti-Stokes scattered EMWs from thermal effects only. We obtain

Fig. 7.16 Spatial growth rate of the thermal filamentation instability

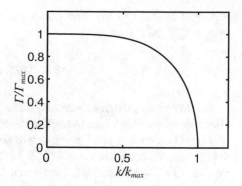

$$(\omega^2 - k^2 c_s^2)\left[\left(\omega - \mathbf{k} \cdot \mathbf{k}_0 \frac{c^2}{\omega_0}\right)^2 - \frac{k^4 c^4}{4\omega_0^2}\right] = \frac{a_0^2 k^2 c^4 \omega_{pi}^2 \nu_{ei} n_{e0}}{4\omega_0^2 \kappa_e}. \tag{7.155}$$

This is similar to Eqs. (7.34) and (7.118) but for a thermal drive instead of ponderomotive; we can then describe the filamentation instability following the same procedure as in Sect. 7.3.2. Taking the stationary limit of $\omega = 0$ and looking for solutions of the form $\mathbf{k} = \mathbf{k}_2 - i l \, \mathbf{e}_z$ with $\mathbf{k}_2 \cdot \mathbf{k}_0 = 0$ in the weakly coupled regime, we obtain the following expression for the spatial growth rate:

$$\Gamma = \Gamma_{max}\sqrt{1 - (k/k_{max})^4}, \tag{7.156}$$

with the maximum growth rate

$$\boxed{\Gamma_{max} = \frac{a_0}{2k_0\lambda_{De}}\sqrt{\frac{\nu_{ei}n_{e0}}{\kappa_e}}} \tag{7.157}$$

and the characteristic wave-number

$$\boxed{k_{max}^4 = \frac{a_0^2 \nu_{ei} n_{e0}}{\kappa_e \lambda_{De}^2}}. \tag{7.158}$$

The spatial growth rate is represented in Fig. 7.16. The dependence on k is functionally different from the ponderomotive instability (Fig. 7.11): while the ponderomotive filamentation favors a particular wavelength (δx_{fil}, Eq. (7.126)), here the growth rate is maximum for perturbations with infinite wavelengths ($k \rightarrow 0$). Thermal filamentation will therefore tend to develop on larger spatial scales compared to its ponderomotive counterpart (for numerical illustrations, see Ref. [52]).

To compare the relative importance of thermal vs. ponderomotive filamentation, we take the ratio of the thermal to ponderomotive maximum spatial growth rates,

Eqs. (7.157) and (7.123). The condition for thermal filamentation to dominate, $\Gamma_{max}^{th}/\Gamma_{max}^{pond} \gg 1$, leads to

$$a_0^2 \ll 16\lambda_{De}^2 \frac{v_{ei}n_{e0}}{\kappa_e}. \tag{7.159}$$

In other words, the laser intensity cannot be too high (otherwise ponderomotive effects dominate) and the absorption needs to be significant (otherwise no heating occurs). On the other hand, too much absorption will prevent any instability from developing; the condition $\kappa_c \ll \Gamma_{max}^{th}$, i.e., the absorption rate (derived in Sect. 4.1.2, Eq. (4.24)) being smaller than the filamentation growth rate, leads to

$$a_0^2 \gg 4\frac{v_e^2}{c^4}\frac{v_{ei}\kappa_e}{n_{e0}}. \tag{7.160}$$

The thermal filamentation instability can therefore only exist and dominate over the ponderomotive instability in the parameter space defined by these last two inequalities. Combining these two inequalities together, we see that the existence of such a parameter space where the thermal filamentation can exist puts the following requirement on the heat conduction:

$$\kappa_e \ll 2\frac{n_{e0}c^2\omega_0}{\omega_{pe}^2}. \tag{7.161}$$

We shall keep in mind that these conditions are necessary but not sufficient for thermal filamentation to develop.

In general, non-local heat transport effects will also play a role and lead to a growth rate which recovers both the ponderomotive and thermal limits [53]. Berger also showed that in addition to heat conduction, Landau damping can smooth out the small-scale temperature perturbations associated with thermal filamentation and limit the contribution of thermal filamentation compared to ponderomotive [54].

Problems

7.1 SBS Spatial Amplification: From Kinetic to Fluid
Starting from the kinetic expression of the SBS spatial amplification rate, Eq. (7.20), with the electron and ion susceptibilities defined in Sect. 1.3.3, introduce a phenomenological damping in the IAW frequency, $\omega_2 \rightarrow \omega_2 + i\nu_2$, and express the electron and ion susceptibilities in the IAW limit with the damping present (the damping only appears in the ion susceptibility). Insert these expressions in Eq. (7.20) and verify that you recover the fluid expression from Eq. (7.13).

Fig. 7.17 Illustration of the
Doppler shift of the light
waves as seen from the
flowing plasma

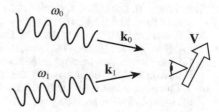

7.2 Doppler Shift from Flow in CBET (or SBS)

Recall that due to the Doppler effect, the frequency of a light wave of frequency ω_j actually "seen" by an observer in the frame of the flowing plasma is

$$\omega_j^{plasma} = \omega_j \sqrt{\frac{1-\beta_j}{1+\beta_j}} \approx \omega_j (1-\beta_j), \qquad (7.162)$$

where β_j is the velocity of the receiver relative to the source normalized to $v_{\varphi,j} = c/n_0$, the phase velocity of the light wave in the plasma (cf. Fig. 7.17).

Show that the beat wave frequency between two crossing light waves $\omega_2 = \omega_0 - \omega_1$ seen from the frame of the flowing plasma is then

$$\hat{\omega}_2 = \omega_2^{plasma} = \omega_0^{plasma} - \omega_1^{plasma} \approx \omega_2 - \mathbf{k}_2 \cdot \mathbf{V}, \qquad (7.163)$$

where it was assumed that $\omega_0 \approx \omega_1$.

7.3 Fluid Expressions of the Real and Imaginary Components of the Refractive Index Perturbations from CBET

Write down the fluid expressions of the real and imaginary perturbations of n, the plasma refractive index (with $n = n_0 + \delta n_R + i\delta n_I$; start from Eqs. (7.60)–(7.61)).

References

1. R.Y. Chiao, C.H. Townes, B.P. Stoicheff, Phys. Rev. Lett. **12**, 592 (1964)
2. E. Garmire, C.H. Townes, Appl. Phys. Lett. **5**, 84 (1964)
3. E.E. Hagenlocker, W.G. Rado, Appl. Phys. Lett. **7**, 236 (1965)
4. C. Yamanaka, T. Yamanaka, T. Sasaki, J. Mizui, H.B. Kang, Phys. Rev. Lett. **32**, 1038 (1974)
5. B.H. Ripin, J.M. McMahon, E.A. McLean, W.M. Manheimer, J.A. Stamper, Phys. Rev. Lett. **33**, 634 (1974)
6. D.W. Phillion, W.L. Kruer, V.C. Rupert, Phys. Rev. Lett. **39**, 1529 (1977)
7. A.A. Offenberger, M.R. Cervenan, A.M. Yam, A.W. Pasternak, J. Appl. Phys. **47**, 1451 (1976)
8. A. Ng, L. Pitt, D. Salzmann, A.A. Offenberger, Phys. Rev. Lett. **42**, 307 (1979)
9. C.E. Clayton, C. Joshi, A. Yasuda, F.F. Chen, Phys. Fluids **24**, 2312 (1981)
10. H.A. Baldis, C.J. Walsh, Phys. Fluids **26**, 3426 (1983)
11. C. Labaune, E. Fabre, A. Michard, F. Briand, Phys. Rev. A **32**, 577 (1985)

12. K. Tanaka, B. Boswell, R.S. Craxton, L.M. Goldman, F. Guglielmi, W. Seka, R.W. Short, J.M. Soures, Phys. Fluids **28**, 2910 (1985)
13. S.H. Glenzer, D.H. Froula, L. Divol, M. Dorr, R.L. Berger, S. Dixit, B.A. Hammel, C. Haynam, J.A. Hittinger, J.P. Holder, O.S. Jones, D.H. Kalantar, O.L. Landen, A.B. Langdon, S. Langer, B.J. MacGowan, A.J. Mackinnon, N. Meezan, E.I. Moses, C. Niemann, C.H. Still, L.J. Suter, R.J. Wallace, E.A. Williams, B.K.F. Young, Nat. Phys. **3**, 716 (2007)
14. T. Chapman, P. Michel, J.-M.G. Di Nicola, R.L. Berger, P.K. Whitman, J.D. Moody, K.R. Manes, M.L. Spaeth, M.A. Belyaev, C.A. Thomas, B.J. MacGowan, J. Appl. Phys. **125**, 033101 (2019)
15. R.L. Berger, Phys. Fluids **27**, 1796 (1984)
16. G.R. Mitchel, T.W. Johnston, H. Pépin, Phys. Fluids **26**, 2292 (1983)
17. D.W. Forslund, J.M. Kindel, E.L. Lindman, Phys. Fluids **18**, 1002 (1975)
18. I. Alexeff, W.D. Jones, D. Montgomery, Phys. Rev. Lett. **19**, 422 (1967)
19. H.X. Vu, J.M. Wallace, B. Bezzerides, Phys. Plasmas **1**, 3542 (1994)
20. P. Neumayer, R.L. Berger, L. Divol, D.H. Froula, R.A. London, B.J. MacGowan, N.B. Meezan, J.S. Ross, C. Sorce, L.J. Suter, S.H. Glenzer, Phys. Rev. Lett. **100**, 105001 (2008)
21. P. Neumayer, R.L. Berger, D. Callahan, L. Divol, D.H. Froula, R.A. London, B.J. MacGowan, N.B. Meezan, P.A. Michel, J.S. Ross, C. Sorce, K. Widmann, L.J. Suter, S.H. Glenzer, Phys. Plasmas **15**, 056307 (2008)
22. N.B. Meezan, L.J. Atherton, D.A. Callahan, E.L. Dewald, S. Dixit, E.G. Dzenitis, M.J. Edwards, C.A. Haynam, D.E. Hinkel, O.S. Jones, O. Landen, R.A. London, P.A. Michel, J.D. Moody, J.L. Milovich, M.B. Schneider, C.A. Thomas, R.P.J. Town, A.L. Warrick, S.V. Weber, K. Widmann, S.H. Glenzer, L.J. Suter, B.J. MacGowan, J.L. Kline, G.A. Kyrala, A. Nikroo, Phys. Plasmas **17**, 056304 (2010)
23. J.E. Ralph, A. Kemp, N.B. Meezan, R.L. Berger, D. Strozzi, B.J. MacGowan, O. Landen, N. Lemos, M. Belyaev, M. Biener, D.A. Callahan, T. Chapman, L. Divol, D.E. Hinkel, J. Moody, A. Nikroo, O. Jones, S. Schiaffino, M. Stadermann, P. Michel, Phys. Plasmas **28**, 072704 (2021)
24. C.J. Pawley, H.E. Huey, N.C. Luhmann, Phys. Rev. Lett. **49**, 877 (1982)
25. W.L. Kruer, S.C. Wilks, B.B. Afeyan, R.K. Kirkwood, Phys. Plasmas **3**, 382 (1996)
26. C.J. McKinstrie, J.S. Li, R.E. Giacone, H. X. Vu, Phys. Plasmas **3**, 2686 (1996)
27. V.V. Eliseev, W. Rozmus, V.T. Tikhonchuk, C.E. Capjack, Phys. Plasmas **3**, 2215 (1996)
28. R.K. Kirkwood, B.B. Afeyan, W.L. Kruer, B.J. MacGowan, J.D. Moody, D.S. Montgomery, D.M. Pennington, T.L. Weiland, S.C. Wilks, Phys. Rev. Lett. **76**, 2065 (1996)
29. K.B. Wharton, R.K. Kirkwood, S.H. Glenzer, K.G. Estabrook, B.B. Afeyan, B.I. Cohen, J.D. Moody, C. Joshi, Phys. Rev. Lett. **81**, 2248 (1998)
30. E.A. Williams, B.I. Cohen, L. Divol, M.R. Dorr, J.A. Hittinger, D.E. Hinkel, A.B. Langdon, R.K. Kirkwood, D.H. Froula, S.H. Glenzer, Phys. Plasmas **11**, 231 (2004)
31. P. Michel, L. Divol, E.A. Williams, S. Weber, C.A. Thomas, D.A. Callahan, S.W. Haan, J.D. Salmonson, S. Dixit, D.E. Hinkel, M.J. Edwards, B.J. MacGowan, J.D. Lindl, S.H. Glenzer, L.J. Suter, Phys. Rev. Lett. **102**, 025004 (2009)
32. S.H. Glenzer, B.J. MacGowan, P. Michel, N.B. Meezan, L.J. Suter, S.N. Dixit, J.L. Kline, G.A. Kyrala, D.K. Bradley, D.A. Callahan, E.L. Dewald, L. Divol, E. Dzenitis, M.J. Edwards, A.V. Hamza, C.A. Haynam, D.E. Hinkel, D.H. Kalantar, J.D. Kilkenny, O.L. Landen, J.D. Lindl, S. LePape, J.D. Moody, A. Nikroo, T. Parham, M.B. Schneider, R.P.J. Town, P. Wegner, K. Widmann, P. Whitman, B.K.F. Young, B.V. Wonterghem, L.J. Atherton, E.I. Moses, Science **327**, 1228 (2010)
33. P. Michel, S.H. Glenzer, L. Divol, D.K. Bradley, D. Callahan, S. Dixit, S. Glenn, D. Hinkel, R.K. Kirkwood, J.L. Kline, W.L. Kruer, G.A. Kyrala, S. Le Pape, N.B. Meezan, R. Town, K. Widmann, E.A. Williams, B.J. MacGowan, J. Lindl, L.J. Suter, Phys. Plasmas **17**, 056305 (2010)
34. J.D. Moody, P. Michel, L. Divol, R.L. Berger, E. Bond, D.K. Bradley, D.A. Callahan, E.L. Dewald, S. Dixit, M.J. Edwards, S. Glenn, A. Hamza, C. Haynam, D.E. Hinkel, N. Izumi, O. Jones, J.D. Kilkenny, R.K. Kirkwood, J.L. Kline, W.L. Kruer, G.A. Kyrala, O.L. Landen,

S. LePape, J.D. Lindl, B.J. MacGowan, N.B. Meezan, A. Nikroo, M.D. Rosen, M.B. Schneider, D.J. Strozzi, L.J. Suter, C.A. Thomas, R.P.J. Town, K. Widmann, E.A. Williams, L.J. Atherton, S.H. Glenzer, E.I. Moses, Nat. Phys. **8**, 344 (2012)

35. C.J. Randall, J.R. Albritton, J.J. Thomson, Phys. Fluids **24**, 1474 (1981)
36. I.V. Igumenshchev, D.H. Edgell, V.N. Goncharov, J.A. Delettrez, A.V. Maximov, J.F. Myatt, W. Seka, A. Shvydky, S. Skupsky, C. Stoeckl, Phys. Plasmas **17**, 122708 (2010)
37. I.V. Igumenshchev, W. Seka, D.H. Edgell, D.T. Michel, D.H. Froula, V.N. Goncharov, R.S. Craxton, L. Divol, R. Epstein, R. Follett, J.H. Kelly, T.Z. Kosc, A.V. Maximov, R.L. McCrory, D.D. Meyerhofer, P. Michel, J.F. Myatt, T.C. Sangster, A. Shvydky, S. Skupsky, C. Stoeckl, Phys. Plasmas **19**, 056314 (2012)
38. R.W. Boyd, *Nonlinear Optics*, 3rd edn. (Academic Press, Orlando, 2008)
39. P. Michel, L. Divol, D. Turnbull, J.D. Moody, Phys. Rev. Lett. **113**, 205001 (2014)
40. P. Michel, E. Kur, M. Lazarow, T. Chapman, L. Divol, J.S. Wurtele, Phys. Rev. X **10**, 021039 (2020)
41. E. Kur, M. Lazarow, J.S. Wurtele, P. Michel, Opt. Express **29**, 1162 (2021)
42. D. Turnbull, P. Michel, T. Chapman, E. Tubman, B.B. Pollock, C.Y. Chen, C. Goyon, J.S. Ross, L. Divol, N. Woolsey, J.D. Moody, Phys. Rev. Lett. **116**, 205001 (2016)
43. D. Turnbull, C. Goyon, G.E. Kemp, B.B. Pollock, D. Mariscal, L. Divol, J.S. Ross, S. Patankar, J.D. Moody, P. Michel, Phys. Rev. Lett. **118**, 015001 (2017)
44. D. Turnbull, A. Colaitis, R.K. Follett, J.P. Palastro, D.H. Froula, P. Michel, C. Goyon, T. Chapman, L. Divol, G.E. Kemp, D. Mariscal, S. Patankar, B.B. Pollock, J.S. Ross, J.D. Moody, E.R. Tubman, N.C. Woolsey, Plasma Phys. Control. Fusion **60**, 054017 (2018)
45. D.H. Edgell, P.B. Radha, J. Katz, A. Shvydky, D. Turnbull, D.H. Froula, Phys. Rev. Lett. **127**, 075001 (2021)
46. V. Bespalov, V. Talanov, Sov. Phys. JETP **3**, 307 (1966)
47. R.Y. Chiao, P.L. Kelley, E. Garmire, Phys. Rev. Lett. **17**, 1158 (1966)
48. P.E. Young, H.A. Baldis, R.P. Drake, E.M. Campbell, K.G. Estabrook, Phys. Rev. Lett. **61**, 2336 (1988)
49. D. Pesme, Fusion thermonucléaire inertielle par laser - partie 1 - volume 1: L'interaction laser-matière, (Eyrolles, 1993) Chap. 2
50. F.W. Perkins, E.J. Valeo, Phys. Rev. Lett. **32**, 1234 (1974)
51. A.J. Palmer, Phys. Fluids **14**, 2714 (1971)
52. A.J. Schmitt, Phys. Fluids **31**, 3079 (1988)
53. E.M. Epperlein, Phys. Rev. Lett. **65**, 2145 (1990)
54. R.L. Berger, B.F. Lasinski, T.B. Kaiser, E.A. Williams, A.B. Langdon, B.I. Cohen, Phys. Fluids B: Plasma Phys. **5**, 2243 (1993)

Chapter 8
Wave Coupling Instabilities via Electron Plasma Waves

In this chapter we investigate in more detail the instabilities involving the coupling of a laser and an EPW. We are concerned with two main mechanisms: coupling mediated by another light wave, i.e., EMW → EMW + EPW, or coupling via another EPW, i.e., EMW → EPW + EPW. The former instability is the stimulated Raman scattering process (SRS), and the latter is the two plasmon decay (TPD) instability. These processes are very important for ICF experiments, as they can degrade the laser to target coupling (as we will see SRS is most effective in the backscatter direction, in which case some of the incident laser energy is "reflected" back out of the target instead of depositing its energy) and produce anomalous absorption at the quarter critical density for TPD. We will also describe the ion acoustic decay (IAD) instability, whereby a light wave decays into an EPW and an IAW; this process can only occur in the vicinity of the critical density due to the frequency-matching conditions.

Another important aspect is that these processes can excite high-amplitude EPWs, which are prone to trapping and accelerating electrons to high energies (Sect. 2.2). These suprathermal electrons are a major concern for ICF, as they can preheat the target and reduce its compressibility. In the following we will describe the SRS and TPD mechanisms, as well as the associated generation of suprathermal electrons and their measurement via the hard x-ray bremsstrahlung emission they generate during their collisions with plasma ions.

8.1 Stimulated Raman Scattering (SRS)

Stimulated Raman scattering (SRS) is the coupling of an intense light wave (the laser) to a scattered light wave via an EPW, such that the beat intensity pattern between the two light waves, which is modulated at $(\omega_0 - \omega_1, \mathbf{k}_0 - \mathbf{k}_1)$, satisfies the EPW dispersion relation and can thus resonantly drive an EPW.

© The Author(s), under exclusive license to Springer Nature Switzerland AG 2023   269
P. Michel, *Introduction to Laser-Plasma Interactions*, Graduate Texts in Physics,
https://doi.org/10.1007/978-3-031-23424-8_8

SRS has been observed in many laser–plasma experiments, beginning with Nd-glass lasers at $\approx 1.05\,\mu\text{m}$ wavelength [1–4] and then with frequency-tripled lasers which allowed for better propagation in the context of ICF and HED experiments [5–7]. Many of the features of SRS, concerning the scattered light or the generation of suprathermal electrons due to the excitation of high-amplitude EPWs, were consistent with particle-in-cell simulations [8–11]. Significant amounts of SRS have been measured in more recent experiments at the NIF laser, in conditions relevant to both indirect-drive [12, 13] and direct-drive [14] geometries.

8.1.1 Generalities: Spectrum of Raman Scattered Light vs. Plasma Conditions

As already mentioned in Sect. 6.2, since both daughter waves can only propagate if their frequencies are larger than the plasma frequency, $\omega_2, \omega_1 \geq \omega_{pe}$, the frequency matching condition $\omega_0 = \omega_1 + \omega_2$ implies that $\omega_0 \geq 2\omega_{pe}$ (here the subscripts 0, 1, and 2 refer to the laser, scattered light wave, and EPW, respectively). In other words, SRS can only exist for plasma densities below quarter-critical for the laser, $n_e/n_c \leq 1/4$, where n_c is the critical density for the laser frequency.

In the low density limit, $n_e/n_c \ll 1$, we have $\omega_{pe} \ll \omega_0$, and therefore, $\omega_1 \approx \omega_0$ and $k_1 \approx k_0$. On the other hand, in the high density limit of SRS, near $n_c/4$, we have $\omega_0 \approx 2\omega_{pe}$, and therefore, $\omega_1 \approx \omega_{pe} \approx \omega_0/2$. The range of possible wavelengths for the SRS scattered light is therefore

$$\lambda_1 \in [\lambda_0, 2\lambda_0]. \tag{8.1}$$

Furthermore, near the quarter-critical density region, we also have $k_1 \to 0$ for the scattered light wave. This means that SRS scattered light originating near $n_c/4$ will be near its critical density, i.e., near its turning point at normal incidence. Following the arguments developed for ray tracing in Sect. 3.4, this means that the SRS scattered light will propagate from that point directly following the density gradient, toward regions of lower densities (cf. Fig. 8.1). This also means that the description from this section, where all three waves are simply treated as plane waves (i.e., we make a WKB assumption for a local region of space, neglecting swelling effects), will fail when the density where SRS originates gets too close to $n_c/4$, as WKB becomes invalid near a light wave's turning point (cf. Sect. 3.2). SRS near the quarter-critical density requires a separate treatment; the quarter-critical region can also be the subject of the two plasmon decay instability, which we will study in a later section.

It is also instructive to represent the range of scattered wavelengths (and hence the range of existence) of SRS on an (n_e, T_e) map. This is obtained via $\omega_1/\omega_0 = 1 - \omega_2/\omega_0$ (due to frequency matching), which after inserting the EPW frequency from its dispersion relation $\omega_2^2 = \omega_{pe}^2(1 + 3k_2^2\lambda_{De}^2)$ gives

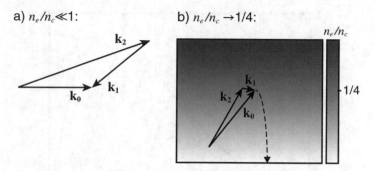

Fig. 8.1 SRS geometry for **(a)** low density limit, $n_e/n_c \ll 1$, where $\omega_1 \to \omega_0$, and **(b)** high density limit, $n_e/n_c \to 1/4$, where the scattered light wave originates near its normal incidence-turning point and propagates down the density gradient along the dashed path

$$\frac{\omega_1}{\omega_0} = 1 - \sqrt{\frac{n_e}{n_c}(1 + 3k_2^2\lambda_{De}^2)}. \qquad (8.2)$$

This is shown in Fig. 8.2 for the backscatter geometry; the solid line represents the Landau cutoff at $k\lambda_{De} = 0.3$ (meaning that SRS is not expected to occur on the left side of this curve, where Landau damping is too strong to allow the EPW to grow substantially). The contour lines represent the ratio of the SRS scattered light wavelength to the laser wavelength, λ_1/λ_0. As mentioned above, the scattered light wavelength becomes close to λ_0 and $2\lambda_0$ near $n_e \to 0$ and $n_e \to n_c/4$, respectively; the range of temperatures where SRS can exist without significant Landau damping of the EPW becomes narrower and limited to cold temperatures as the density becomes smaller.

This partly explains why most SRS experiments in hot plasmas typically see a cutoff at short wavelengths in the scattered light spectrum [7], as the strong Landau damping prevents SRS at the corresponding low densities if the temperature is high enough; the other reason for the cutoff is the sharp drop in SRS growth rate at densities below a few percents, as we show in the next section.

8.1.2 Coupled Mode Equations: Temporal Growth Rate

To derive the SRS coupled mode equations and temporal growth rate, we start from the nonlinear driven waves equations described in Sect. 6.3. Each light wave ("pump" laser and "daughter" scattered light wave) is described by Eq. (6.20), and the EPW is described via Eq. (6.24). We describe the EMWs via their normalized vector potentials $\mathbf{a}_{0,1} = c\delta\mathbf{v}_{e0,1}$ and the EPW via its density perturbation $\delta n_e/n_{e0}$ with $n_{e0} = n_e - \delta n_e$ the background electron density (and $\omega_{pe0}^2 = n_{e0}e^2/m\varepsilon_0$). For each wave, the nonlinear coupling terms on the right-hand side correspond to the beating of the other two waves; therefore, we obtain

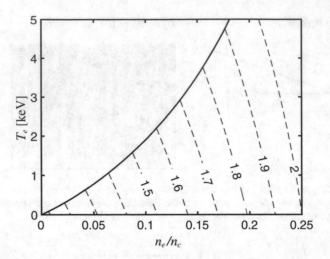

Fig. 8.2 Contour plot of λ_1/λ_0 (SRS scattered light wavelength normalized to the laser wavelength) for SRS backscatter. The solid line is the $k\lambda_{De} = 0.3$ contour: the region on the left of this curve has $k\lambda_{De} \geq 0.3$, and SRS is not expected to develop there as Landau damping of the EPW becomes too strong

$$(\partial_t^2 + \omega_{pe0}^2 - c^2\nabla^2)\mathbf{a}_0 = -\omega_{pe0}^2 \frac{\delta n_e}{n_{e0}}\mathbf{a}_1 , \tag{8.3}$$

$$(\partial_t^2 + \omega_{pe0}^2 - c^2\nabla^2)\mathbf{a}_1 = -\omega_{pe0}^2 \frac{\delta n_e}{n_{e0}}\mathbf{a}_0 , \tag{8.4}$$

$$(\partial_t^2 + \omega_{pe0}^2 - 3v_{Te}^2\nabla^2)\frac{\delta n_e}{n_{e0}} = c^2\nabla^2(\mathbf{a}_1 \cdot \mathbf{a}_0) . \tag{8.5}$$

For the EPW, we ignored the second and third terms on the right-hand side of Eq. (6.24), which both involve a density perturbation δn_e which can only exist for a longitudinal wave (EPW or IAW, cf. Sect. 1.3.1)—whereas for SRS, the EPW couples to two transverse EMWs. The squared velocity perturbation (the first term on the right-hand side, which we recognize as the ponderomotive force) was then expanded following $\nabla^2\delta v_e^2 = \nabla^2(\delta\mathbf{v}_{e0} + \delta\mathbf{v}_{e1})^2 \approx 2c^2\nabla^2\mathbf{a}_0 \cdot \mathbf{a}_1$.

Next we envelope the fields for the three waves at their frequencies and wave-vectors, i.e.,

$$\mathbf{a}_0(\mathbf{r}, t) = \frac{1}{2}\tilde{\mathbf{a}}_0(\mathbf{r}, t)e^{i\psi_0} + c.c. , \quad \psi_0 = \mathbf{k}_0 \cdot \mathbf{r} - \omega_0 t , \tag{8.6}$$

$$\mathbf{a}_1(\mathbf{r}, t) = \frac{1}{2}\tilde{\mathbf{a}}_1(\mathbf{r}, t)e^{i\psi_1} + c.c. , \quad \psi_1 = \mathbf{k}_1 \cdot \mathbf{r} - \omega_1 t , \tag{8.7}$$

$$\delta n_e(\mathbf{r}, t) = \frac{1}{2}\delta\tilde{n}_e(\mathbf{r}, t)e^{i\psi_2} + c.c. , \quad \psi_2 = \mathbf{k}_2 \cdot \mathbf{r} - \omega_2 t . \tag{8.8}$$

We can then apply the slowly varying envelope approximation from Eq. (6.35) to Eqs. (8.3)–(8.5) and collect the terms $\propto e^{i\psi_0}$, $\propto e^{i\psi_1}$ and $\propto e^{i\psi_2}$ with $\psi_0 = \psi_1 + \psi_2$ (i.e., assuming perfect phase matching) to arrive at the SRS coupled mode equations for the slowly varying envelopes:

$$(\partial_t + \mathbf{v}_{g0} \cdot \nabla)\tilde{a}_0 = -i\frac{\omega_{pe0}^2}{4\omega_0}\frac{\delta\tilde{n}_e}{n_{e0}}\tilde{a}_1 \cos(\epsilon), \qquad (8.9)$$

$$(\partial_t + \mathbf{v}_{g1} \cdot \nabla)\tilde{a}_1 = -i\frac{\omega_{pe0}^2}{4\omega_1}\frac{\delta\tilde{n}_e^*}{n_{e0}}\tilde{a}_0 \cos(\epsilon), \qquad (8.10)$$

$$(\partial_t + \mathbf{v}_{g2} \cdot \nabla)\frac{\delta\tilde{n}_e}{n_{e0}} = -i\frac{k_2^2 c^2}{4\omega_2}\tilde{a}_0\tilde{a}_1^* \cos(\epsilon), \qquad (8.11)$$

where ϵ is the angle between the two light waves' electric fields.

Setting the spatial derivatives to zero in Eqs. (8.10) and (8.11) and solving these two coupled equations while neglecting pump depletion (i.e., assuming $\tilde{a}_0$ constant), following the procedure introduced in Sect. 6.4.3, lead to the SRS temporal growth rate:

$$\gamma = \frac{k_2 c}{4}\frac{\omega_{pe0}}{\sqrt{\omega_2\omega_1}}|\tilde{a}_0|\cos(\epsilon). \qquad (8.12)$$

Notice that the expression is the same as the one we obtained for SBS in Sect. 7.1, Eq. (7.10), with the substitution $\omega_{pi} \rightarrow \omega_{pe}$.

The maximum growth rate is obtained when the polarizations of the two light waves are parallel ($\epsilon = 0$), as expected, since this maximizes the amplitude of the beat wave between their electric fields; this has also been confirmed experimentally, with measurements showing SRS strongly polarized along the same direction as the laser [7]. The coupled mode equations (8.9)–(8.11) reduce to the form of Eqs. (6.73)–(6.75) when using normalized action amplitudes for each of the three waves (cf. Problem 8.1).

Taking $\omega_1 = \omega_0 - \omega_2$ and $\omega_2 \approx \omega_{pe0}$, with $\epsilon = 0$, the maximum SRS growth rate expression simplifies to

$$\boxed{\gamma = \frac{k_2|\tilde{a}_0|c}{4}\sqrt{\frac{\omega_{pe0}}{\omega_1}}.} \qquad (8.13)$$

We notice that like for SBS, it is the backscatter geometry, which maximizes k_2, that has the highest growth rate. In this geometry, we have $k_2 = k_0 + k_1$ (where $k_j = |\mathbf{k}_j| > 0$); using the EPW and EMW dispersion relations $\omega_2^2 = \omega_{pe0}^2 + 3k_2^2 v_{Te}^2$ and $\omega_1^2 = \omega_{pe0}^2 + k_1^2 c^2$ and solving for k_2 as a function of $n_{e0}/n_c = \omega_{pe0}^2/\omega_0^2$ give

$$k_2 = \frac{\omega_0}{c}\left[\sqrt{1 - n_{e0}/n_c} + \sqrt{1 - 2\sqrt{n_{e0}/n_c}}\right]. \qquad (8.14)$$

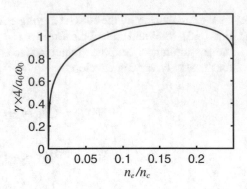

Fig. 8.3 SRS growth rate for the backscatter geometry, normalized to $\omega_0 a_0/4$, i.e., function $f(n_{e0}/n_c)$ in the text

When inserted back into the growth rate expression above, we obtain the following expression for the backscatter SRS ("BSRS") growth rate:

$$\gamma_{\text{BSRS}} = \frac{\omega_0 |\tilde{a}_0|}{4} f(n_{e0}/n_c), \tag{8.15}$$

where the function f is defined as

$$f(n_{e0}/n_c) = \sqrt{\eta(1+\eta)} + \sqrt{\frac{\eta(1-2\eta)}{1-\eta}}, \quad \eta = \sqrt{\frac{n_{e0}}{n_c}}. \tag{8.16}$$

In practical units, we have

$$\gamma_{\text{BSRS}}[\text{s}^{-1}] \approx 4 \times 10^{13} \sqrt{I_{16}} f(n_{e0}/n_c), \tag{8.17}$$

where I_{16} is the intensity in units of 10^{16} W/cm^2.

The backscatter growth rate normalized to $\omega_0 |\tilde{a}_0|/4$ (i.e., the function $f(n_{e0}/n_c)$) is represented in Fig. 8.3. For most of the range of electron densities, $n_{e0}/n_c \in [0.02–0.25]$, f is approximately equal to 1 (to $\approx 20\%$ error), and the growth rate can be approximated as $\gamma_{\text{BSRS}} \approx |\tilde{a}_0|\omega_0/4$.

8.1.3   Spatial Amplification in Uniform and Non-uniform Plasma

The spatial amplification rate in a uniform plasma is directly obtained from the temporal growth rate, using $\Gamma = \gamma^2/(v_{g1}v_2)$ from Sect. 6.4.4, Eq. (6.52). Here $v_{g1} = k_1 c^2/\omega_1$ is the scattered light wave's group velocity, v_2 is the EPW damping rate, and γ is the temporal growth rate from Eq. (8.13). We obtain

$$\boxed{\Gamma = \frac{k_2^2 |a_0|^2}{16 k_1 v_2/\omega_2}}. \tag{8.18}$$

In practical units, we have

$$\Gamma[\mathrm{m}^{-1}] \approx 1.2 \times 10^4 I_{16} \frac{\lambda_\mu}{v/\omega_{pe}} g(n_{e0}/n_c), \qquad (8.19)$$

with λ_μ the laser wavelength in μm, I_{16} the intensity in units of 10^{16} W/cm^2 and where we defined the function g which takes the following form for the backscatter geometry only:

$$g(n_{e0}/n_c) = \frac{1}{4} \frac{\left[\sqrt{1-\eta^2} + \sqrt{1-2\eta}\right]^2}{\sqrt{1-2\eta}} , \quad \eta = \sqrt{n_{e0}/n_c} \qquad (8.20)$$

(in the following, n_e will denote the background density).

For most of the range of relevant densities, between 0 and $\approx 0.22 n_c$, we have $g \approx 1$ to within 20%. However, g and therefore the spatial amplification rate become infinite at $n_c/4$; our WKB analysis also becomes invalid there since the scattered light is at its turning point. The scattered light wave can also be significantly absorbed near $n_c/4$, which raises the instability threshold (cf. Sect. 6.4.3, Eq. (6.48)).

The kinetic expression for the CBET amplification rate from Eq. (7.70), which described the general situation of two EMWs coupled via the plasma density perturbation driven by their beat wave (making no assumption on whether this drive is resonant with a plasma mode, i.e., EPW or IAW or non-resonant), is valid for SRS as well:

$$\Gamma = \frac{k_2^2}{8k_1} \mathrm{Im}[F_\chi] |\tilde{a}_0|^2 , \qquad (8.21)$$

with $F_\chi = \chi_e/(1 + \chi_e)$ (cf. Problem 8.2). This expression automatically includes the EPW Landau damping and can in principle be applied to non-Maxwellian distribution functions.

Next we investigate the finite spatial amplification of SRS in a non-uniform plasma, following the Rosenbluth gain as introduced in Sect. 6.5. Since SRS is most sensitive to the plasma density (due to $\omega_2 \propto \sqrt{n_e}$) and density gradients are typically easier to establish than temperature gradients in HED plasmas as heat conduction tends to rapidly smooth out temperature non-uniformities, we look at the case of a variation in density.

We express the wave-vector spatial mismatch assuming 1D backscatter geometry with the pump and scattered light wave propagating toward $z > 0$ and $z < 0$, respectively (cf. Fig. 8.4):

$$\kappa(z) = k_0(z) + k_1(z) - k_2(z), \qquad (8.22)$$

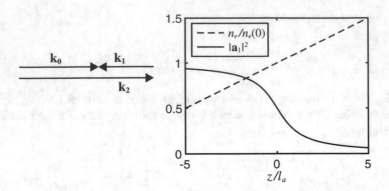

Fig. 8.4 SRS (in 1D backscatter geometry) amplification through a density gradient, with phase matching at $z = 0$ and the scattered light wave (whose intensity is shown here in arbitrary units) propagating toward $z < 0$

where $k_j = |\mathbf{k}_j|$, $j = 0, 1, 2$. The Taylor expansion of the electron density near $z = 0$, considered the location of phase matching ($\kappa(0) = 0$), is

$$n_e(z) \approx n_e(0) \left(1 + \frac{z}{L_n} \right), \tag{8.23}$$

where $L_n = n_e(0)/\partial_z n_e$.

The expressions for the wave-numbers, $k_{0,1}(z) = (\omega_{0,1}/c)\sqrt{1 - \omega_{pe}^2(z)/\omega_{0,1}^2}$, $\sqrt{3}v_{Te}k_2(z) = \omega_2\sqrt{1 - \omega_{pe}^2(z)/\omega_2^2}$, where $\omega_{pe}^2(z) = \omega_{pe}^2(0)(1 + z/L_n)$, give the following expression for the wave-vector mismatch, assuming that the variations of density remain small over the distance considered here, i.e., $z \ll L_n$:

$$\kappa(z) = \frac{\omega_{pe}^2(0)z}{2L_n} \left(-\frac{1}{k_0c^2} - \frac{1}{k_1c^2} + \frac{1}{3k_2v_{Te}^2} \right) \approx \frac{1}{6k_2\lambda_{De}^2 L_n}z, \tag{8.24}$$

since $v_{Te} \ll c$.

Inserting the resulting $\kappa' = \partial_z\kappa(z)$ coefficient in the Rosenbluth gain formula, $G_r = 2\pi\gamma^2/|\kappa' v_{g1}v_{g2}|$ [Eq. (6.105), such that $|\tilde{a}_1|^2$ gets amplified by e^{G_r} after propagating through the resonance region], gives

$$\boxed{G_r = \frac{\pi a_0^2 k_2^2 L_n}{4k_1}}. \tag{8.25}$$

The characteristic amplification length depends on the EPW damping:

- Weakly damped limit, $\Gamma/k_2 \ll (v_2/\omega_2)/(3k_2^2\lambda_{De}^2)$:

The amplification length is $l_a = 4\gamma_0/|\kappa'\sqrt{v_{g1}v_{g2}}|$ (cf. Eq. (6.106)), which for SRS becomes

$$l_a = 2\sqrt{3}a_0 k_2 \lambda_{De} L_n \sqrt{\frac{k_2}{k_1}}.$$ (8.26)

- Strongly damped limit, $\Gamma/k_2 \gg (v_2/\omega_2)/(3k_2^2\lambda_{De}^2)$:

The amplification length is $l_a = 2v_2/(\kappa' v_{g2})$ (cf. Eq. (6.117)), i.e., for SRS,

$$l_a = 4\frac{v_2}{\omega_{pe}}L_n.$$ (8.27)

8.1.4 Scattering Geometry: Forward- vs. Side- vs. Back-Scatter

Stimulated Raman scattering can occur in various geometries; the different configurations have different characteristics and behaviors that can be measured in experiments and observed in simulations. Let us first discuss forward- and backward-scattering geometries, which are usually the most relevant because this is where the scattered light is aligned with the propagation axis of the pump laser and can interact with it for the longest (without leaving its envelope sideways). Next we will discuss Raman side-scatter, which specifically refers to the geometry where the pump laser is incident at some angle in a density gradient and the scattered light originates perpendicular to the direction of the density gradient (so that it will follow a fixed-density plasma until refraction deflects it toward lower densities).

8.1.4.1 Back- vs. Forward-Scatter

The geometry of forward vs. backward scattering for SRS is illustrated in Fig. 8.5.

Fig. 8.5 SRS geometry for forward vs. backward scattering

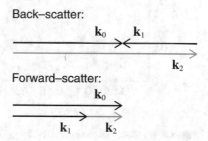

As mentioned above, these two geometries are significant because the scattered light wave propagates parallel to the pump laser and can presumably interact with it over longer distances, without leaving its envelope. However there are significant differences between the forward and backward geometries: as we can see from Fig. 8.5, the EPW driven by forward scatter will have a smaller wave-vector than the one driven by backscatter. This means that on the one hand, the temporal growth rate γ will always be higher for backscatter, per Eq. (8.13), but on the other hand the spatial amplification rate Γ in the strongly damped limit will typically be higher for forward SRS because a smaller wave-vector for the EPW will lead to less Landau damping (since a smaller $k\lambda_{De} \approx v_{Te}/v_\phi$ means a higher v_ϕ/v_{Te} ratio, which means less electrons near the wave's phase velocity and thus less Landau damping), which increases Γ per Eq. (8.18).

To clarify this, let us first quantify the differences in wave-vector and $k\lambda_{De}$. For a given density n_e, the dispersion relations for the two light waves, $\omega_j^2 = \omega_{pe}^2 + k_j^2 c^2$, $j = 0, 1$ give the following expressions for the EPW wave-vector k_2 in the forward and backward geometries:

$$\text{Forward: } |k_2| = |k_0| - |k_1| = \frac{\omega_0}{c}\left[\sqrt{1 - n_e/n_c} - \sqrt{1 - 2\sqrt{n_e/n_c}}\right], \quad (8.28)$$

$$\text{Backward: } |k_2| = |k_0| + |k_1| = \frac{\omega_0}{c}\left[\sqrt{1 - n_e/n_c} + \sqrt{1 - 2\sqrt{n_e/n_c}}\right]. \quad (8.29)$$

The EPW wave-number k_2 (normalized to the laser wave-number in vacuum, ω_0/c) is represented in Fig. 8.6a for forward- and backward-scattering geometries. Note that for $n_e/n_c \to 1/4$, we have $k_1 \to 0$ and therefore $k_2 \approx k_0 \approx (\sqrt{3}/2)\omega_0/c$ for both geometries.

To visualize the impact on the EPW damping, we plot $k_2\lambda_{De}$ vs. n_e/n_c in Fig. 8.6b. As we know from Sect. 1.3.4, EPWs will be strongly damped and unlikely

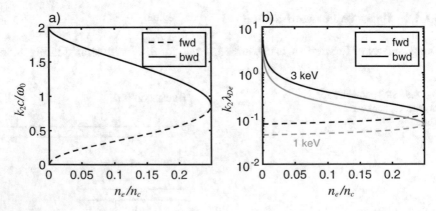

Fig. 8.6 (a) Wave-vector k_2 and (b) $k_2\lambda_{De}$ (for T_e=1 and 3 keV) as a function of n_e/n_c for forward vs. backward scattering geometries

to be generated for $k_2\lambda_{De} \geq 0.3 - 0.4$. At low densities, the expression of k_2 for the forward SRS geometry in Eq. (8.28) reaches the asymptotic limit $k_2 \approx (\omega_0/c)\sqrt{n_e/n_c}$, and therefore we have $k_2\lambda_{De} \to v_{Te}/c$ (and not 0, even though $k_2 \to 0$) for $n_e/n_c \to 0$, as can be easily verified by the asymptotic limits at both temperatures in Fig. 8.6b.

At these values of $k_2\lambda_{De}$, Landau damping essentially goes to zero and largely overcomes the reduction of k_2 in the expression for the amplification rate Γ, which goes to infinity. Forward SRS can indeed be observed in simulations with uniform plasma densities before the onset of backscatter.

However, we should remember that even when Landau damping goes to zero, some collisional damping of the EPWs will always be present, preventing v_2 from going toward 0 (cf. Sect. 4.5).

But more importantly, forward SRS is strongly reduced by density gradients. When damping becomes small, the size of the amplification region where convective amplification occurs, which is proportional to the damping rate per Eq. (8.27), becomes very small as well,[1] and the amplification is limited to the Rosenbluth gain or $\exp[G_r]$ in intensity where G_r was derived in Eq. (8.25). We know that when the amplification is limited by gradients, in the Rosenbluth regime, then the amplification does not depend on damping anymore, and the exponential gain G_r is proportional to k_2^2, largely favoring backscatter over forward scatter. The strong sensitivity of forward SRS to density gradients has been observed in early SRS simulations [11].

8.1.4.2 Side-Scatter

As we just discussed, SRS can be detuned by density gradients, which degrade the phase-matching condition and prevent amplification. When a laser is incident on a density gradient at some incidence angle, it is then natural to expect that an SRS scattered light wave seeded at some density will preferentially try to follow that density profile, which will maintain phase matching and maximize the amplification. Side-scatter SRS refers to this specific geometry, where the scattered light wave emerges perpendicular to the density gradient, as shown in Fig. 8.7. This geometry can also lead to an absolute instability at densities below $n_c/4$ [15–18].

Raman side-scatter is characterized by $\mathbf{k}_1 \cdot \nabla n_e = 0$, where $\mathbf{k}_1$ is the wave-vector of the scattered light where it originates. We can in principle have two distinct geometries, where $\mathbf{k}_0 \cdot \mathbf{k}_1 < 0$ (Fig. 8.7a) or $\mathbf{k}_0 \cdot \mathbf{k}_1 > 0$ (Fig. 8.7b) [18]. The former is the most likely to be seen, as it has the highest EPW wave-number k_2 and therefore

---

[1] Keep in mind that the "strongly damped" approximation for convective amplification which justifies using Eq. (8.27) is unrelated to the actual, "absolute" value of the damping, but instead means that the ratio between damping and amplification per unit propagation distance must be $\gg 1$, as explained in Sect. 6.4.4.

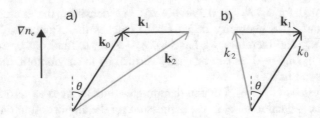

Fig. 8.7 Geometry for Raman side-scattering in a density gradient, where the scattered light originates perpendicular to the density gradient, $\mathbf{k}_1 \cdot \nabla n_e = 0$. Two arrangements are possible, with $\mathbf{k}_0 \cdot \mathbf{k}_1 < 0$ (**a**) or $\mathbf{k}_0 \cdot \mathbf{k}_1 > 0$ (**b**). The former (**a**) has the highest temporal growth rate and convective growth

the highest temporal growth rate and spatial amplification rate (in a "Rosenbluth" sense), as we will see below.

The relation $\mathbf{k}_1 \cdot \nabla n_e = 0$ also means that the scattered light originates at (or near) its turning point in the density gradient, as we have seen in Sect. 3.3. This geometry is the one where the scattered light follows the longest propagation distance while staying resonant (i.e., satisfying the frequency- and wave-vector-matching conditions), before leaving the resonance region due to refraction, as illustrated in Fig. 8.8a. It can thus be the dominant geometry for SRS over backscatter, as long as the profile remains approximately 1D and the laser envelope in the transverse direction (e.g., x in the figure) is large enough to allow it.

The scattered light refracts toward the regions of lower densities and eventually leaves the amplification region. This leads to a very distinctive feature of Raman side-scatter for experiments involving a near-1D density profile (like the one generated by a planar target expanding in vacuum), which is that there is a unique relation between the wavelength of the scattered light λ_1 and the angle θ_{1v} at which it exits the plasma.

This is illustrated in Fig. 8.8a: for a 1D density profile along z, we know from Sect. 3.3 that the scattered light originating at a given density n_{em} (and electron frequency ω_{pem}) where phase matching occurs is given by

$$\sin(\theta_{1v}) = \sqrt{1 - \frac{\omega_{pem}^2}{\omega_1^2}}, \qquad (8.30)$$

where ω_1 is the frequency of the scattered light wave.

Neglecting the temperature corrections, i.e., simply assuming that the frequency of the EPW $\omega_2 = \omega_0 - \omega_1 \approx \omega_{pem}$, leads to the following relationship between the scattering angle with respect to the density gradient and the frequency of the scattered light:

$$\boxed{\theta_{1v} = \arcsin\sqrt{2\frac{\omega_0}{\omega_1} - \frac{\omega_0^2}{\omega_1^2}}.} \qquad (8.31)$$

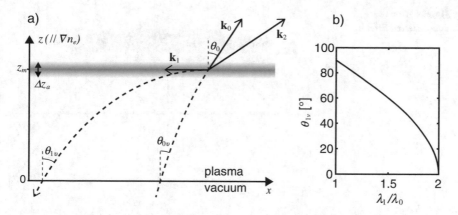

Fig. 8.8 (a) Geometry of SRS side-scatter convective amplification. The density gradient is along z, and the resonance (ω- and **k**-matching) occurs at z_m. Amplification occurs over a width Δz_a along z (shaded region), before the scattered light wave refracts toward lower density regions. (b) Relation between the observed scattering angle (with respect to z, assuming a 1D density profile and neglecting temperature corrections) and the wavelength of the scattered light λ_1 (normalized to the laser wavelength λ_0)

This means that in experiments involving a nearly 1D density profile, side-scatter SRS is distinguishable by the fact that the scattered light wavelength should depend on the observation angle according to the relation above, as represented in Fig. 8.8b. Note that the limit of $\lambda_1/\lambda_0 \rightarrow 1$ corresponds to $n_e/n_c \rightarrow 0$, for which we do indeed expect to have $\theta_{1v} \rightarrow \pi/2$ (grazing incidence), whereas $\lambda_1/\lambda_0 \rightarrow 2$ corresponds to $n_e/n_c \rightarrow 1/4$, when the laser is at near-normal incidence. This expression is independent of the laser incidence angle: in reality the laser incidence comes in through thermal corrections, which we neglected by assuming $\omega_2 \approx \omega_{pem}$ and which typically remain a relatively small effect.[2]

The angularly dependent scattered light wavelength signature was observed in several experiments involving planar targets [14, 19, 20].

A convective amplification factor can be estimated for side-scatter [20]. Going back to our general presentation of convective amplification in a gradient in Sect. 6.5, we recall that the expression for the spatial amplification rate $\Gamma(z)$ in amplitude in a gradient along z with phase matching at $z = 0$ under the strongly damped approximation is of the form, per Eq. (6.113),

$$\Gamma(z) = \frac{\Gamma_0}{1 + (z/z_c)^2},$$ (8.32)

---

[2] Thermal corrections involve the laser incidence angle θ_{0v} (cf. Fig. 8.8a) via the dependence on k_2 in the EPW dispersion relation, k_2 itself depending on k_1 and $k_0(n_{em}, \theta_{0v})$.

Fig. 8.9 Ray trajectory of
side-scattered SRS light

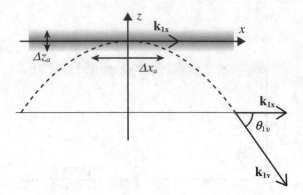

where $\Gamma_0 = \gamma_0^2/(v_{g1}v_2)$ is the spatial amplification rate in a uniform plasma, and $z_c = v_2/|\kappa' v_{g2}|$. The total gain exponent for side-scatter can be estimated by integrating this amplification rate along the path integral of a scattered light ray. We choose a coordinate system centered at the turning point of the scattered light, as shown in Fig. 8.9. The ray-tracing equations, Eqs. (3.55)-(3.56), allow us to express the ray coordinates parametrically, as a function of time (see Problem 3.1). For simplicity we choose a linear density profile, of the form

$$n_e(z) = n_{em}\left(1 + \frac{z}{L}\right),\tag{8.33}$$

where n_{em} is the density at the SRS turning point. We will see later that it is merely sufficient to have an approximately linear profile within the longitudinal ($\propto z$) amplification length Δz_a, which we will define below.

The ray trajectory, with the time origin picked as the time at which the ray is at the turning point (0,0), follows a parabola:

$$x(t) = c\sin(\theta_{1v})t,\tag{8.34}$$

$$z(t) = -\frac{c^2}{4L}\frac{\omega_{pem}^2}{\omega_1^2}t^2 = -\frac{\omega_{pem}^2}{4L\sin^2(\theta_{1v})\omega_1^2}x(t)^2.\tag{8.35}$$

The side-scatter gain can be estimated by integrating $\Gamma(z)$ along the path of the side-scattered light C,

$$G = \int_C \frac{\Gamma_0}{1 + (z/z_c)^2}ds.\tag{8.36}$$

The elementary arc length ds can be estimated using our parametrization against time as $v_{g1} = ds/dt$. We will assume that the value of v_{g1} remains approximately constant inside the amplification region (the gray region in Fig. 8.9), as the most important spatial dependence comes through the resonance function in the integrand. We then have

$$G = \Gamma_0 v_{g1} \int_{-\infty}^{\infty} \frac{dt}{1 + z^2(t)/z_c^2}. \tag{8.37}$$

Rigorously, the integration is to be taken between two times t_i and t_f (or locations in space $\mathbf{r}_i = \mathbf{r}(t_i)$ and $\mathbf{r}_f = \mathbf{r}(t_f)$) that are sufficiently far away from the resonance region near the turning point, i.e., $|\mathbf{r}_{i,f} \cdot \mathbf{e}_z| \gg \Delta z_a$, where Δz_a is derived below.

Putting all the pieces together with what we already derived in the previous section, i.e., using Γ_0 from Eq. (8.18), $z_c = l_a/2$ with l_a from Eq. (8.27), and noticing that $\kappa' v_{g2} \approx \omega_{pem}/2L$ from Eq. (8.24), we finally obtain the side-scatter convective gain:

$$G = G_r \sqrt{\frac{\omega_1^2 - \omega_{pem}^2}{\omega_{pem}^2}} \sqrt{\frac{\omega_{pem}}{v_2}}, \tag{8.38}$$

where we used the definite integral formula:

$$\int_{-\infty}^{\infty} \frac{dx}{a^n + x^n} = \frac{2\pi a^{1-n}}{n \sin(\pi/n)}. \tag{8.39}$$

The physical size of the resonance region can be estimated by taking the FWHM of the resonance function in Eq. (8.37), which is obtained for $z(t) = z_c$, leading to

$$\Delta z_a = z_c = 2L \frac{v_2}{\omega_{pem}}, \tag{8.40}$$

$$\Delta x_a = 4L \sqrt{\frac{\omega_1^2 - \omega_{pem}^2}{\omega_{pem}^2}} \sqrt{\frac{2v_2}{\omega_{pem}}}, \tag{8.41}$$

where we used Eq. (8.35) to derive Δx_a.

These results show that the convective gain for side-scatter is typically larger than for backscatter, at least in this gradient-dominated regime. Other factors need to be considered as well, such as the transverse size of the laser (which must remain larger than the transverse size of the amplification region Δx_a) and the deviations from 1D for the density profile. Note that as we saw in Sect. 3.2, the ray-tracing (i.e., WKB) approach we took here does not accurately describe the electric field at the turning point; however, it can be shown that under most typical conditions, the Airy skin depth (defined in Eq. (3.28)) will be much smaller than the longitudinal size of the resonance region, and numerical integrations without the WKB approximation have indeed confirmed the validity of these results for typical ICF conditions [20].

8.2   Two Plasmon Decay (TPD)

The two plasmon decay (TPD) instability consists in the coupling of a "pump" light wave to an EPW via another EPW. Like for other three-wave instabilities (cf. Sect. 6.1), resonant coupling occurs when the beat intensity pattern between the light wave and the first EPW's electric fields satisfies the dispersion relation for the second EPW.

Because the two EPW frequencies ω_1 and ω_2 are close to ω_{pe} (as long as $k\lambda_{De} \leq 0.3$, as required for the Landau damping not to be too strong), the frequency-matching condition $\omega_0 = \omega_1 + \omega_2$ implies that $\omega_0 \approx 2\omega_{pe}$, i.e., $n_e \approx n_c/4$. In other words, the TPD instability is limited to the vicinity of the quarter-critical density.

Another important feature is that since the ponderomotive force driving the coupling originates from the beat between a transverse light wave and a longitudinal EPW, the electric fields cannot be orthogonal for TPD to occur, i.e., we need $\mathbf{E}_0 \cdot \mathbf{E}_1 \neq 0$. Since $\mathbf{k}_0 \perp \mathbf{E}_0$ and $\mathbf{k}_1 /\!/ \mathbf{E}_1$, this means that unlike for SRS and SBS where this is often the primary configuration, the daughter waves' wave-vectors with TPD cannot be aligned with the laser's. It also means that the EPWs will be primarily in the plane $(\mathbf{k}_0, \mathbf{E}_0)$ and that TPD cannot occur in the plane $(\mathbf{k}_0, \mathbf{B}_0)$ (cf. Fig. 8.10). This feature has been confirmed experimentally, via direct measurements of the suprathermal electrons generated by the high-amplitude EPWs from TPD which were found to lie primarily in the plane $(\mathbf{k}_0, \mathbf{E}_0)$ [21].

8.2.1   Coupled Mode Equations and Temporal Growth Rate

Like for every three-wave instability, the derivation of the coupled mode equations begins with the second-order wave equations derived in Sect. 6.3. Here, since TPD involves an EMW and two EPWs, we need the wave equations for EPWs and EMWs. We will describe the EMW (the pump, i.e., the laser) via the normalized vector potential $\mathbf{a}_0$ or electron quiver velocity $\delta\mathbf{v}_0 = \mathbf{a}_0 c$ and each of the two EPWs by either their velocity perturbation $\delta\mathbf{v}_{1,2}$ or their density perturbation $\delta n_{e1,2} = n_{e0}\delta v_{1,2}/v_{\phi1,2}$ (cf. Eq. (1.124)).

Starting with the EPWs, Eq. (6.24), and assuming planes waves,[3] we consider the coupling of each wave to the beat-wave from the other two waves, i.e.,

$$\left[\partial_t^2 + \omega_{pe0}^2 - 3v_{Te}^2\nabla^2\right]\frac{\delta n_{e1}}{n_{e0}} = \nabla^2(\delta\mathbf{v}_0 \cdot \delta\mathbf{v}_2) - \partial_t\nabla\left(\frac{n_{e2}}{n_{e0}}\delta\mathbf{v}_0\right), \quad (8.42)$$

$$\left[\partial_t^2 + \omega_{pe0}^2 - 3v_{Te}^2\nabla^2\right]\frac{\delta n_{e2}}{n_{e0}} = \nabla^2(\delta\mathbf{v}_0 \cdot \delta\mathbf{v}_1) - \partial_t\nabla\left(\frac{n_{e1}}{n_{e0}}\delta\mathbf{v}_0\right), \quad (8.43)$$

---

[3] We assume plane waves "in a WKB sense," i.e., considering a region of plasma where the waves can be enveloped at a given frequency and wave-number.

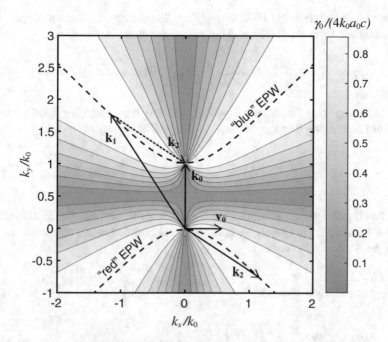

Fig. 8.10 TPD geometry, shown here for EPWs in the $(\mathbf{E}_0, \mathbf{k}_0)$ plane (the most prominent situation). The color scale represents the growth rate (normalized to $4k_0v_0$, where $v_0 = ca_0$), and the dashed lines are the two hyperbolas of maximum growth rate ($\gamma_0 = k_0v_0/4$). The upper hyperbola corresponds to the higher frequency ("blue") EPW (with the higher wave-number), whereas the lower hyperbola corresponds to the lower frequency ("red") EPW

where we took $\delta v_e^2 = (\delta\mathbf{v}_0 + \delta\mathbf{v}_2)^2 \approx 2\delta\mathbf{v}_0 \cdot \delta\mathbf{v}_2$ in the first equation's right-hand side and $\delta v_e^2 \approx 2\delta\mathbf{v}_0 \cdot \delta\mathbf{v}_1$ in the second's. The third term on the right-hand side of Eq. (6.24) ($\propto \delta n_e^2$) does not intervene since each EPW is coupled to another EPW and an EMW, and the latter does not have an associated density perturbation (cf. Sect. 1.3.1).

Likewise, the driven EMW equation, Eq. (6.20), takes the form:

$$\left[\partial_t^2 + \omega_{pe0}^2 - c^2\nabla^2\right]\delta\mathbf{v}_0 = -\omega_{pe0}^2\left(\frac{\delta n_{e1}}{n_{e0}}\delta\mathbf{v}_2 + \frac{\delta n_{e2}}{n_{e0}}\delta\mathbf{v}_1\right). \qquad (8.44)$$

Next we envelope each of the three waves at its frequency- and wave-vector:

$$\mathbf{a}_0(\mathbf{r}, t) = \frac{1}{2}\tilde{\mathbf{a}}_0(\mathbf{r}, t)e^{i\psi_0} + c.c. \,, \quad \psi_0 = \mathbf{k}_0 \cdot \mathbf{r} - \omega_0 t \,, \qquad (8.45)$$

$$\delta n_{e1}(\mathbf{r}, t) = \frac{1}{2}\delta\tilde{n}_{e1}(\mathbf{r}, t)e^{i\psi_1} + c.c. \,, \quad \psi_1 = \mathbf{k}_1 \cdot \mathbf{r} - \omega_1 t \,, \qquad (8.46)$$

$$\delta n_{e2}(\mathbf{r}, t) = \frac{1}{2}\delta\tilde{n}_{e2}(\mathbf{r}, t)e^{i\psi_2} + c.c. \,, \quad \psi_2 = \mathbf{k}_2 \cdot \mathbf{r} - \omega_2 t \,, \qquad (8.47)$$

with the phase-matching conditions $\omega_0 = \omega_1 + \omega_2$, $\mathbf{k}_0 = \mathbf{k}_1 + \mathbf{k}_2$.

Inserting these into Eq. (8.42), applying the slowly varying envelope (SVE) approximation from Eq. (6.35), and collecting the terms $\propto e^{i\psi_1}$ give

$$(\partial_t + \mathbf{v}_{g1} \cdot \nabla)\frac{\delta\tilde{n}_{e1}}{n_{e0}} = -\frac{i}{4\omega_1}\left(k_1^2\delta\tilde{\mathbf{v}}_0 \cdot \delta\tilde{\mathbf{v}}_2^* + \omega_1\frac{\delta\tilde{n}_{e2}^*}{n_{e0}}\mathbf{k}_1 \cdot \delta\tilde{\mathbf{v}}_0\right). \qquad (8.48)$$

Substituting $\delta\tilde{\mathbf{v}}_{e2}$ for $\delta\tilde{n}_{e2}/n_{e0}$ via Eq. (1.124) (i.e., from the fluid continuity equation) and noting that $\mathbf{k}_2 \cdot \delta\tilde{\mathbf{v}}_0 = (\mathbf{k}_0 - \mathbf{k}_1) \cdot \delta\tilde{\mathbf{v}}_0 = -\mathbf{k}_1 \cdot \delta\tilde{\mathbf{v}}_0$ yield

$$(\partial_t + \mathbf{v}_{g1} \cdot \nabla)\frac{\delta\tilde{n}_{e1}}{n_{e0}} = -\frac{i}{4}\left(\frac{k_1^2\omega_2}{k_2^2\omega_1} - 1\right)\frac{\delta\tilde{n}_{e2}^*}{n_{e0}}\mathbf{k}_2 \cdot \delta\tilde{\mathbf{v}}_0. \qquad (8.49)$$

Following the same procedure for the other EPW and for the EMW as well, starting then from Eq. (8.44), leads to the three coupled mode equations for TPD under the SVE approximation:

$$(\partial_t + \mathbf{v}_{g0} \cdot \nabla)\tilde{a}_0 = -\frac{i}{4}\frac{\omega_{pe}^2}{c\omega_0}\frac{\omega_2}{k_2^2}\left(\frac{k_2^2\omega_1}{k_1^2\omega_2} - 1\right)(\mathbf{k}_1 \cdot \mathbf{e}_0)\frac{\delta\tilde{n}_{e1}}{n_{e0}}\frac{\delta\tilde{n}_{e2}}{n_{e0}}, \qquad (8.50)$$

$$(\partial_t + \mathbf{v}_{g1} \cdot \nabla)\frac{\delta\tilde{n}_{e1}}{n_{e0}} = -\frac{ic}{4}\left(\frac{k_1^2\omega_2}{k_2^2\omega_1} - 1\right)(\mathbf{k}_2 \cdot \mathbf{e}_0)\frac{\delta\tilde{n}_{e2}^*}{n_{e0}}\tilde{a}_0, \qquad (8.51)$$

$$(\partial_t + \mathbf{v}_{g2} \cdot \nabla)\frac{\delta\tilde{n}_{e2}}{n_{e0}} = -\frac{ic}{4}\left(\frac{k_2^2\omega_1}{k_1^2\omega_2} - 1\right)(\mathbf{k}_1 \cdot \mathbf{e}_0)\frac{\delta\tilde{n}_{e1}^*}{n_{e0}}\tilde{a}_0, \qquad (8.52)$$

where $\mathbf{e}_0$ is a unit vector along the laser polarization direction, i.e., $\mathbf{a}_0 = a_0\mathbf{e}_0$.

The temporal growth rate is obtained from these equations by setting the spatial derivatives to zero and solving in time assuming the pump wave remains constant—or, more immediately, by taking the product of the coupling terms on the right-hand sides of the equations for the daughter waves, Eqs. (8.51)–(8.52), as described in Sect. 6.4.3, Eq. (6.44).

We obtain

$$\gamma = \frac{|(\mathbf{k}_2 \cdot \mathbf{e}_0)(k_1^2 - k_2^2)|}{4k_1k_2}c|\tilde{a}_0|. \qquad (8.53)$$

We can once again verify that the coupled equations, Eqs. (8.50)–(8.52), reduce to the form of Eqs. (6.73)–(6.75) if we use normalized action amplitudes for the three waves (cf. Problem 8.3).

Since we know that TPD occurs predominantly in the plane containing $\mathbf{e}_0$ and $\mathbf{k}_0$, as we discussed above, we now consider a 2D geometry with x and y chosen along the directions of $\mathbf{e}_0$ and $\mathbf{k}_0$, respectively (cf. Fig. 8.10). The region of maximum growth in the (k_x, k_y) plane, where $\mathbf{k} = k_x\mathbf{e}_x + k_y\mathbf{e}_y$ (since waves 1 and 2 are interchangeable in Eq. (8.53) we drop the 1 and 2 indices: the analysis below applies to either of the two EPWs), can be found by taking the derivative of the growth rate with respect to k_y and setting it equal to zero, $\partial \gamma / \partial k_y = 0$. We obtain

$$k_y (k_y - k_0) = k_x^2, \tag{8.54}$$

which leads to the following relation describing the location of the maximum growth rate in the (k_x, k_y) plane:

$$k_y = \frac{1}{2} \left[k_0 \pm \sqrt{k_0^2 + 4k_x^2} \right]. \tag{8.55}$$

This relation describes a pair of hyperbolas, which are represented in Fig. 8.10. Each hyperbola corresponds to one of the two EPWs, with one EPW propagating "forward" with the laser ($\mathbf{k}_1 \cdot \mathbf{k}_0 > 0$) and the other going backward ($\mathbf{k}_2 \cdot \mathbf{k}_0 < 0$). If we insert Eq. (8.55) back into the expression for the growth rate Eq. (8.53), we obtain the maximum growth rate $\gamma_0 = \max(\gamma)$ for TPD, which is constant along the hyperbola:

$$\boxed{\gamma_0 = \frac{k_0 c a_0}{4}}. \tag{8.56}$$

In practical units, we have:

$$\gamma_0 [\text{s}^{-1}] \approx 3.5 \times 10^{13} \sqrt{I_{16}}, \tag{8.57}$$

with I_{16} the laser intensity in units of 10^{16} W/cm^2. Notice that the growth rate is very similar to the one for SRS.

The growth rate is constant all along the hyperbolas. Numerical simulations of TPD routinely show these hyperbolas when Fourier analyzing the electric field of the EPWs [22, 23]; usually there will be a cutoff around $k_{1,2}\lambda_{De} \approx 0.3$, beyond which EPWs are strongly Landau-damped and cannot sustainably grow.

We can see from the EPW dispersion relation that the two EPW frequencies are slightly shifted from ω_{pe} when the waves have a finite wave-number and that the shift increases with the magnitude of k. Since $\omega_0 = \omega_1 + \omega_2 \geq \omega_{pe}$, the resonant density for TPD will move further away from $n_c/4$ as the EPWs' wave-numbers increase, i.e., as we move further away from the vertex on the hyperbolas. In fact, the electron density can be directly mapped out onto the hyperbolas [24]. Considering a pair of daughter EPWs with wave-vectors $\mathbf{k}_1, \mathbf{k}_2$, with $\mathbf{k}_0 = \mathbf{k}_1 + \mathbf{k}_2$, we define $k_x = k_{1x}$ and $k_y = k_{1y}$, so that $k_{2x} = -k_{1x}$ and $k_{2y} = k_0 - k_{1y}$, with the geometry of Fig. 8.10. From the dispersion relations of the EPW and EMW and the phase-matching condition, we can find the following relation for the EPW's wave-number (cf. Problem 8.4):

$$\left(\frac{k_y}{k_0} - \frac{1}{2} \right)^2 + \left(\frac{k_x}{k_0} \right)^2 = \frac{c^2}{12 v_{Te}^2} \frac{1 - 4n_e/n_c}{1 - n_e/n_c} - \frac{1}{4}. \tag{8.58}$$

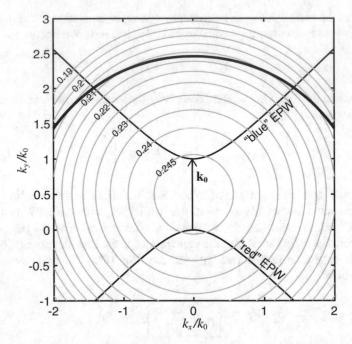

Fig. 8.11 Density mapping onto the TPD hyperbolas shown here for $T_e = 2$ keV. The gray circles are the locations of the TPD k-vectors for electron densities shown in the figure, normalized to n_c (the circles are drawn by increment of $n_e/n_c = 0.005$). The thick line is the Landau cutoff for $k\lambda_{De} = 0.3$

This means that the TPD EPWs are located on a circle centered at $k_x = 0$, $k_y = k_0/2$ in k-space, where the circle radius depends on n_e and T_e. Assuming a fixed temperature, we can therefore map out the density onto the hyperbola, as represented in Fig. 8.11 for $T_e = 2$ keV. On that figure, we also represented the Landau cutoff, chosen here at $k\lambda_{De} = 0.3$, beyond which EPWs are too strongly damped and TPD ceases to exist. For the example shown here, the larger EPW wave-numbers by the Landau cutoff correspond to TPD developing at $n_e \approx 0.2n_c$, whereas TPD developing near the vertex of the hyperbola (i.e., in which case one of the EPWs has $\mathbf{k}_1 \approx \mathbf{k}_0$ and the other $\mathbf{k}_2 \approx 0$, as will be discussed below) occurs near $n_e \approx 0.245n_c$.

It is also useful to note that the pair of EPWs generated by TPD have different wave-numbers and therefore different frequencies. This will be a crucial point to understand the scattered light signatures of TPD, which we will discuss in Sect. 8.2.4. We first express the two EPW frequencies via a simple Taylor expansion assuming that $k_{1,2}\lambda_{De} \ll 1$:

$$\omega_{1,2} \approx \omega_{pe} + \frac{3}{2}\frac{k_{1,2}^2 v_{Te}^2}{\omega_{pe}}. \tag{8.59}$$

The upper hyperbola in Figs. 8.10 and 8.11 corresponds to the larger k-vector and thus the higher frequency EPW (i.e., on the "blue" side of the average wavelength), whereas the lower hyperbola corresponds to the lower frequency ("red") EPW, as indicated in the figures.

The two frequencies are equally shifted around the central frequency $\omega_0/2$ (since $|\omega_1 - \omega_0/2| = |(\omega_0 - \omega_2) - \omega_0/2| = |\omega_2 - \omega_0/2|$) by $\Delta\omega$, such that $\omega_{1,2} = \omega_0/2 \pm \Delta\omega$, with

$$\Delta\omega = \frac{3}{4}\frac{v_{Te}^2}{\omega_{pe}}\left|k_1^2 - k_2^2\right| . \tag{8.60}$$

For EPWs on the hyperbolas with $k_y(k_y - k_0) = k_x^2$, we obtain

$$\Delta\omega = \frac{3}{2}\frac{k_0^2 v_{Te}^2}{\omega_{pe}}\left|\frac{k_{y1}}{k_0} - \frac{1}{2}\right|, \tag{8.61}$$

or

$$\boxed{\frac{\Delta\omega}{\omega_0} \approx \frac{9}{4}\frac{v_{Te}^2}{c^2}\left|\frac{\mathbf{k}_{1,2}\cdot\mathbf{k}_0}{k_0^2} - \frac{1}{2}\right|}, \tag{8.62}$$

where we assumed $n_e/n_c \approx 1/4$ and used the fact that $|k_{y1}/k_0 - 1/2| = |k_{y2}/k_0 - 1/2|$.

In practical units,

$$\frac{\Delta\omega}{\omega_0} \approx 4.4 \times 10^{-3} T_{ek}\left|\frac{\mathbf{k}_{1,2}\cdot\mathbf{k}_0}{k_0^2} - \frac{1}{2}\right|, \tag{8.63}$$

where T_{ek} represents the temperature in keV [25].

We already mentioned that the TPD instability does not allow the three wave-vectors to be co-aligned (as it would prevent the electric field of either EPW from interacting and beating with the pump, as $\mathbf{E}_0 \perp \mathbf{k}_0$, whereas $\mathbf{E}_{1,2} \parallel \mathbf{k}_{1,2}$). This corresponds to the dark contour with $\gamma_0 = 0$ along $k_x = 0$ in Fig. 8.10. However, one can see that there is an (asymptotic) exception located at $k_x = 0, k_y = k_0$, where the hyperbola intersects the k_y axis. There, the growth rate stays equal to γ_0, like everywhere else on the hyperbola. This is the case where $\mathbf{k}_1 \rightarrow \mathbf{k}_0$: plasma wave #2 has a vanishing wave-vector, $k_2 \rightarrow 0$ while also becoming nearly orthogonal to $\mathbf{k}_0$ and $\mathbf{k}_1$. Our analysis, developed using the WKB approximation, is therefore invalid at this location. Because the wave-vector becomes almost zero, the dispersion relation of this EPW becomes undistinguishable from that of a light wave at its turning point, as both satisfy $k \approx 0$ and thus $\omega \approx \omega_{pe}$. This means that at $n_c/4$, TPD and SRS become strongly coupled, since the beating of waves 0 and 1 can resonantly excite both an EPW (TPD) and an EMW (SRS). Both instabilities can

Fig. 8.12 TPD geometry in a
density gradient (with the
hyperbola shown in dashed
gray)

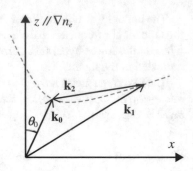

become absolute in this small k limit [16–18, 26–28]. This situation was described
under a common mathematical framework using variational methods in Ref. [29].

8.2.2   Spatial Amplification in a Density Gradient

Two plasmon decay occurs near the quarter-critical density, which is typically
located in an expanding plasma with density gradients. The convective spatial
amplification of TPD in a linear density gradient was derived in Ref. [26].
Proceeding like before with SBS and SRS, we assume that the gradient near $n_c/4$ is
approximately linear and in the z-direction,

$$\omega_{pe}^2(z) \approx \omega_{pe0}^2 \left(1 + \frac{z}{L} \right), \tag{8.64}$$

with $\omega_{pe0} \approx \omega_0/2$.

We directly apply the Rosenbluth gain formula, $G_r = 2\pi\gamma^2/|\kappa' v_{g1z}v_{g2z}|$ (cf.
Sect. 6.5), where $\kappa' = \partial_z[k_{0z}(z) - k_{1z}(z) - k_{2z}(z)]$ and v_{gjz} is the group velocity
of the EPW #j ($j = 1, 2$) along z (the Rosenbluth formula is valid for arbitrary
directions of the waves, as long as we choose the components along the gradient in
the formula).

We know from our earlier analysis of linear propagation of EMWs and EPWs at
oblique incidence in density gradients (Sect. 3.3) that the x-component of the wave-
vector of all three waves is independent of z. Here x is the direction transverse to z
such that the interaction occurs in the plane (x, z), as shown in Fig. 8.12; we assume
that the laser electric field is also in the plane (x, z), for the reasons mentioned
earlier.

Therefore, the wave-numbers' z-components can be expressed as

$$k_{0z}^2(z) = \frac{\omega_0^2}{c^2} - k_{0x}^2 - \frac{\omega_{pe}^2(z)}{c^2} \tag{8.65}$$

$$k_{1,2z}^2(z) = \frac{\omega_{1,2}^2}{3v_{Te}^2} - k_{1,2x}^2 - \frac{\omega_{pe}^2(z)}{3v_{Te}^2}. \tag{8.66}$$

It follows that

$$\partial_z k_{0z} = -\frac{\omega_{pe0}^2}{2Lc^2k_{0z}}, \tag{8.67}$$

$$\partial_z k_{1,2z} = -\frac{\omega_{pe0}^2}{6Lv_{Te}^2 k_{1,2z}}, \tag{8.68}$$

and since $v_{Te} \ll c$, we have $|\partial_z k_{1,2z}| \gg |\partial_z k_{0z}|$, like we already saw for SRS.

Inserting back into the Rosenbluth gain formula with $\kappa' \approx -\partial_z[k_{1z} + k_{2z}]$, and assuming that $\omega_1 \approx \omega_2 \approx \omega_{pe0} \approx \omega_0/2$, we finally obtain

$$\boxed{G_r = \frac{\pi}{12} \frac{k_0 L}{\cos(\theta_0)} \frac{a_0^2 c^2}{v_{Te}^2}}, \tag{8.69}$$

where θ_0 is the angle between $\mathbf{k}_0$ and $\mathbf{e}_z$ at $n_c/4$ (cf. Fig. 8.12). We also used $k_{1z} + k_{2z} \approx k_{0z}$ and took the growth rate on the TPD hyperbola, $\gamma_0 = k_0 a_0 c/4$. The gain exponent is for the amplitude of the waves, consistent with the rest of the book (the gain exponent in intensity is simply $2G_r$). This can be expressed in practical units as

$$G_r \approx 6.2 L_\mu \lambda_\mu \frac{I_{16}}{\cos(\theta_0) T_{ek}}, \tag{8.70}$$

with L_μ, λ_μ the gradient scale-length and laser wavelength both in μm, I_{16} the laser intensity in W/cm^2, and T_{ek} the electron temperature in keV.

Interestingly, the amplification remains the same for any pair of EPWs regardless of their location on the hyperbola, which means that a broad spectrum of EPWs will typically be excited, up to the Landau cutoff. The amplification is sensitive to the angle θ_0 between the light wave and the density gradient at $n_c/4$ and becomes larger for higher incidence angles.

We should also note that the TPD instability can become absolute. The absolute process was investigated in Refs. [26–28, 30], and it was found that the instability admits a threshold of the form

$$\frac{a_0^2 c^2}{v_{Te}^2} k_0 L \geq K, \tag{8.71}$$

where K is typically between 2 and 4, depending on the value of v_{os}/v_{Te} [27]. Note that the general scaling for the absolute threshold is the same as for the spatial amplification gain from Eq. (8.69) above, being roughly equivalent to $4G_r \geq 1$. It has been found in experiments that this expression constitutes a good metric for the onset of TPD. Simulations of TPD have shown that both absolute and convective instabilities typically develop together. The absolute mode develops primarily near $n_c/4$ and $\mathbf{k}_1 \rightarrow \mathbf{k}_0$, $k_2 \rightarrow 0$, whereas the convective mode tends to develop at slightly lower densities, due to the finite thermal corrections to both EPWs.

8.2.3 Multi-beam Process

It was also shown in simulations [23, 31–34] and confirmed by experiments [35, 36] that TPD is prone to multi-beam couplings, where multiple laser beams arranged in a cone (as is typical in ICF experiments) commonly drive a shared EPW. This is illustrated in Fig. 8.13 for the case of two laser beams with wave-vectors $\mathbf{k}_{0a}$ and $\mathbf{k}_{0b}$: as we can see from Fig. 8.13a, the two hyperbolas from each laser beam will cross at a particular point, such that an EPW located at that point (with wave-vector $\mathbf{k}_1$ in the figure) will be simultaneously driven by both lasers. This is a primarily convective instability. In Fig. 8.13b, we show the same arrangement but for absolute instability: now each laser beam drives an EPW $\mathbf{k}_{2a,b}$ with $k_{2a,b} \rightarrow 0$ (it is non-dispersive), so that EPW #2 is essentially the same for both beams and can be driven simultaneously by both laser beams.

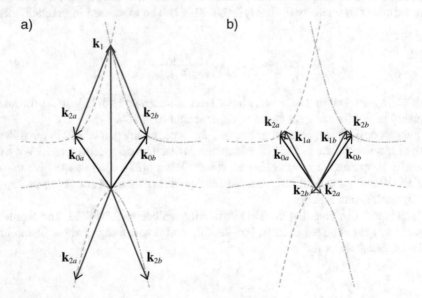

Fig. 8.13 Multi-beam TPD, shown here for two laser beams with wave-vectors $\mathbf{k}_{0a}$, $\mathbf{k}_{0b}$: (**a**) convective instability, with a shared EPW $\mathbf{k}_1$ located at the intersection of the two beams' hyperbolas (shown here as dashed and dash-dotted lines for beam a and b, respectively), and (**b**) absolute instability, with the $k_{2a} = k_{2b} \approx 0$, so that EPW #2 is the same for both laser beams and is being driven by both

Growth rates and spatial amplification gains can in principle be easily calculated for multi-beam processes involving a shared daughter wave; some attention will need to be given to the polarization arrangement of the beams, especially with polarization smoothing schemes that are routinely used in high-power lasers (cf. Sect. 9.2). But in general, the main result is that the key formulas for the TPD instability (growth rate, gain, threshold) are valid for symmetric multi-beams arrangements (with beams arranged in a cone and the shared EPW along the cone axis) if we simply replace the single-beam intensity by the (incoherent) sum of the intensities. This has also proven to match experimental observables in multiple experiments [35–39].

8.2.4 Experimental Signatures of TPD from Half-Harmonic ($\omega_0/2$ and $3\omega_0/2$) Emission

The most revealing signature of TPD is the emission of half-harmonics of the laser frequency, at $\omega_0/2$ and $3\omega_0/2$. This emission has been observed in simulations [10] and experiments [1–3, 5–7]. The exact origin of the emission and its interpretation have been somewhat controversial; however, some of the signatures are now fairly well understood and constitute a powerful tool not only as an indicator that TPD is present but also as a plasma temperature diagnostic.

The $3\omega_0/2$ emission is a result of Thomson up-scattering of the light wave (either the "main" incident laser or a combination of multiple beams in multi-beam experiments) off the EPWs generated by TPD. The frequency- and wave-vector of the scattered light (ω_s, $\mathbf{k}_s$) must satisfy the phase-matching condition:

$$\omega_{s1,2} = \omega_0 + \omega_{1,2} \tag{8.72}$$

$$\mathbf{k}_{s1,2} = \mathbf{k}_0 + \mathbf{k}_{1,2}, \tag{8.73}$$

where $\omega_{1,2}$ and $\mathbf{k}_{1,2}$ are the frequency and wave-vector of the two EPWs driven by TPD. Since the two EPWs are separated in frequency by $\pm\Delta\omega$ around $\omega_0/2$, such that $\omega_{1,2} = \omega_0/2 \pm \Delta\omega$ (cf. Eq. (8.62)), the scattered light emission for a given pair of EPWs will also consist of two lines symmetrically shifted around $3\omega_0/2$ as long as the k-matching condition can also be satisfied:

$$\omega_s = \frac{3}{2}\omega_0 \pm \Delta\omega. \tag{8.74}$$

Since the frequency shift between the two EPWs is directly related to the electron temperature, from Eq. (8.62), it is tempting to try and infer the temperature from the measured frequency splitting in the $3\omega_0/2$ spectrum. However, a broad spectrum of EPWs can generally be excited by TPD, with wave-vectors ranging from ≈ 0 to the Landau cutoff, meaning that k_{y1}/k_0 can range from 1 to ≈ 2 for a typical Landau cutoff of 0.25–0.3, yielding a variation by a factor of ≈ 3 in $|k_{y1}/k_0 - 1/2|$

and therefore in the estimated temperature measurement. It is therefore difficult in general to extract quantitative information from the $3\omega_0/2$ emission besides the basic signature of the presence of TPD.

However, there are particular geometries where the spectrum can be directly related to the temperature. One example is the simplest situation of a single laser beam at normal incidence onto a 1D density profile, as can be created by using planar targets in laboratory experiments. In this case, we can see by geometry that the Thomson up-shift is only possible with a "red" EPW (lower hyperbola). Indeed, in the vicinity of $n_c/4$, we have $k_0 \approx (\sqrt{3}/2)\omega_0/c$ and $k_s \approx \sqrt{2}\omega_0/c$ (assuming $\omega_s \approx 3\omega_0/2$), and therefore,

$$k_s = \sqrt{\frac{8}{3}}k_0. \tag{8.75}$$

The only way to fulfill the condition $\mathbf{k}_s = \mathbf{k}_0 + \mathbf{k}_{1,2}$ is then with a "red" EPW (i.e., $\mathbf{k}_2$ in Fig. 8.14), as it is not possible to satisfy that condition and have $k_s = \sqrt{8/3}k_0$ with a blue EPW $\mathbf{k}_1$. The resulting scattered wave is almost orthogonal to the density gradient (angle of $\approx 83°$) and, after propagating and refracting toward lower densities, leaves the plasma at an angle of $69°$ as shown in Fig. 8.14 (cf. Problem 8.5).

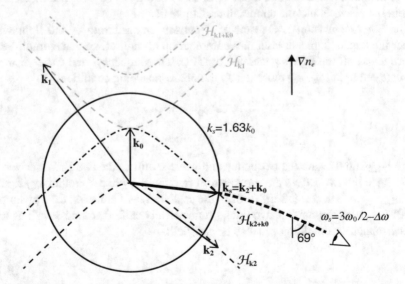

Fig. 8.14 Thomson up-scattering of the laser beam off the "red" EPW driven by TPD. $\mathcal{H}_{k1}$, $\mathcal{H}_{k2}$ are the hyperbolas for the "blue" and "red" EPW from TPD, respectively; $\mathcal{H}_{k1+k0}$ and $\mathcal{H}_{k2+k0}$ are the locations of $\mathbf{k}_1 + \mathbf{k}_0$ and $\mathbf{k}_2 + \mathbf{k}_0$. The circle of radius $k_s = \sqrt{8/3}k_0$ represents the possible locations of the up-scattered light wave wave-vector $\mathbf{k}_s$. Since $\mathbf{k}_s = \mathbf{k}_0 + \mathbf{k}_{1,2}$, the only possible locations for $\mathbf{k}_s$ are where the circle intersects $\mathcal{H}_{k0+k2}$ (the circle does not intersect $\mathcal{H}_{k0+k1}$, and therefore the up-scattering process cannot occur with a blue EPW). The scattered light is red-shifted by $\Delta\omega$ from $3\omega_0/2$ and exits the plasma at $69°$ from ∇n_e as indicated (assuming normal laser incidence, $\mathbf{k}_0 // \nabla n_e$)

As a consequence, if the experiment allows the collection of scattered light at the 69° exit angle indicated in the figure, then the collected light will have a frequency $\omega_s = 3\omega_0/2 - \Delta\omega$, and the measurement of $\Delta\omega$ can be used to infer the temperature at the quarter-critical density following Eq. (8.62), i.e.,

$$T_{ek} \approx 0.13 \frac{\Delta\lambda_{3/2}[\text{nm}]}{\lambda_\mu},$$ (8.76)

where T_{ek} is the temperature in keV and λ_μ the laser wavelength in μm (the measured wavelength shift is expressed in nm). This geometry was for example used in Ref. [40], where the temperature inferred from the red-shifted $3\omega_0/2$ satellite was found to be in good agreement with hydrodynamic simulations.

However, if multiple beams are present and/or the density profile is not 1D, which is often the case in ICF or HED experiments, then scattering can occur from multiple EPWs and the $3\omega_0/2$ spectrum usually shows a pair of satellites whose interpretation becomes a lot more complicated and depends on the precise geometry of the experiment.

Another important signature of TPD is the emission of light at $\omega_0/2$. This signal has been attributed to the small k limit of TPD, where $k_1 \approx k_0$ ("blue" EPW) and $k_2 \approx 0$ ("red" EPW). In this limit the beat wave pattern between k_0 and k_1 can not only excite the TPD EPW at k_2 but also an EMW at $k_{\omega_0/2} \approx k_2$ (cf. Fig. 8.15) since k_2 can satisfy both the EPW and EMW dispersion relations in the small k-limit (with $\omega \approx \omega_{pe}$).

In some cases, this emission can also be attributed to SRS at $n_e \approx n_c/4$, where the instability can become absolute. Measurements of the polarization of the $\omega_0/2$ emission usually reveal that the light is unpolarized, indicating its origin is rather from TPD [1, 7]. However, experiments have shown that the $\omega_0/2$ emission can become polarized along the laser's polarization direction if the laser intensity is strong enough, past a threshold that was interpreted as the threshold for absolute SRS instability at $n_c/4$ [5]. In the following we will discuss the more common situation where the light is unpolarized and attributed to TPD rather than absolute SRS.

The $\omega_0/2$ light wave is generated at its turning point and will therefore propagate directly down the density gradient, with a frequency $\omega_2 = \omega_0/2 - \Delta\omega$ (since it originates on the "red" hyperbola), where $\Delta\omega$ is the frequency shift between the two EPWs driven by TPD given by Eq. (8.62). In this case, since $\mathbf{k}_1 \to \mathbf{k}_0$ and $k_2 \to 0$, we have $|\mathbf{k}_{1,2} \cdot \mathbf{k}_0/k_0^2 - 1/2| = 1/2$, and therefore the temperature at $n_c/4$ can be directly inferred from the measured wavelength shift via Eq. (8.62), leading to [10]:

$$\boxed{\frac{T_e}{mc^2} = \frac{8}{9} \frac{\Delta\omega_{1/2}}{\omega_0}},$$ (8.77)

where $\Delta\omega_{1/2} = \omega_0/2 - \omega_2$ is the frequency shift between the measured frequency and $\omega_0/2$.

Fig. 8.15 Signature of TPD at $\omega_0/2 - \Delta\omega$, with $\Delta\omega$ given by Eq. (8.62) with $|\mathbf{k}_{1,2} \cdot \mathbf{k}_0/k_0^2 - 1/2| = 1/2$. The light can be collected in a narrow cone angle directly along the density gradient, and the splitting $\Delta\omega$ from $\omega_0/2$ provides a measurement of T_e near $n_c/4$ via Eq. (8.77)

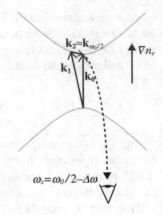

In practical units, we have

$$T_e[\text{keV}] \approx 0.45 \frac{\Delta\lambda_{1/2}[\text{nm}]}{\lambda_\mu}, \tag{8.78}$$

with the same notations as above.

This feature constitutes a diagnostic of the electron temperature in the vicinity of $n_c/4$ [2, 7]. The feature is visible only in a narrow cone angle around the normal to the density gradient, as predicted by the refraction of the $\omega_0/2$ light (cf. Fig. 8.15). It is largely insensitive to the incidence angle of the laser, as the $\omega_0/2$ light will propagate down the density gradient regardless. It has also been correlated to the absolute multi-beam TPD process discussed earlier in direct-drive ICF geometry [39].

Note that both $3\omega_0/2$ and $\omega_0/2$ spectra often feature blue-shifted components as well; discussions on the origin and correct interpretation of these features are not fully settled yet. But generally speaking, they do correspond to the "blue" EPWs from TPD as the splitting is comparable to the red-shift from $\omega_0/2$ and $3\omega_0/2$ discussed above, but toward the "blue" (i.e., $+\Delta\omega$). Inverse resonance absorption has been invoked to explain the $\omega_0/2 + \Delta\omega$ feature (whereby EPW #1 propagates up the density gradient up to its turning point and can generate an EMW via mode conversion, i.e., the opposite process of the resonance absorption discussed in Sect. 4.8). Broader features on both the blue and red sides for either ω_0 or $3\omega_0/2$ are attributed to a variety of processes, such as Thomson down-scattering of the laser off EPWs after these EPWs have propagated and refracted enough to allow proper phase matching [25] or scattering of the incident light wave off secondary EPWs resulting from the decay of the primary TPD EPWs into secondary EPWs via IAWs in the Langmuir decay instability [41]—just to cite a few. Note that the $\omega_0/2$ signal could also be due to the electromagnetic decay instability (EDI) [42], which will be described in Sect. 10.2.3.

8.3 Ion Acoustic Decay (IAD) Instability

The ion acoustic decay (IAD) instability, sometimes referred to as parametric decay instability, is the decay of an EMW (wave #0) into an EPW (wave #1) and an IAW (wave #2)[4] [43–45]. The frequency-matching condition, $\omega_0 = \omega_1 + \omega_2$, with $\omega_2 \ll \omega_0, \omega_1$, implies that $\omega_0 \approx \omega_1 \approx \omega_{pe}$ (since wave #1 is an EPW). Therefore, this instability can only occur in the vicinity of the critical density for the laser wavelength.

Another consequence of $\omega_0 \approx \omega_1$ is obtained from the EMW and EPW dispersion relations, $\omega_0^2 = \omega_{pe}^2 + k_0^2 c^2$ and $\omega_1^2 = \omega_{pe}^2 + 3k_1^2 v_{Te}^2$, which yields

$$k_0 \approx \sqrt{3}\frac{v_{Te}}{c}k_1 \ll k_1 . \tag{8.79}$$

This relation can help define the region of existence of IAD: since $k_0 = (\omega_0/c)\sqrt{1 - n_e/n_c}$, we have

$$k_1\lambda_{De} \approx \frac{1}{\sqrt{3}} \frac{\sqrt{1 - n_e/n_c}}{\sqrt{n_e/n_c}} , \tag{8.80}$$

from which we see that the Landau cutoff for the daughter EPW, $k_1\lambda_{De} \approx 0.3$, corresponds to $n_e/n_c \approx 0.8$. In other words, IAD is limited to densities in the range $[0.8 - 1]n_c$.

The wave-vector matching of the IAD instability is shown in Fig. 8.16. Since $\mathbf{E}_0 \perp \mathbf{k}_0$ whereas $\mathbf{E}_1 \parallel \mathbf{k}_1$, waves 0 and 1 cannot be co-aligned; we will see below that the geometry where $\mathbf{E}_0 \parallel \mathbf{k}_1$ (as shown in Fig. 8.16), i.e., where $\mathbf{k}_0 \cdot \mathbf{k}_1 = 0$ and the daughter waves are in the plane perpendicular to $\mathbf{B}_0$, maximizes the instability growth rate.

The coupled mode equations for the instability are derived starting from Eqs. (6.21), (6.25), and (6.30) for the EMW, EPW, and IAW, respectively. We describe the pump EMW via its normalized vector potential $\mathbf{a}_0 = a_0\mathbf{e}_0$ (with $\mathbf{e}_0$ a unit vector) and the EPW and IAW via their electron density modulations δn_1 and δn_2. For the pump EMW, Eq. (6.21) becomes

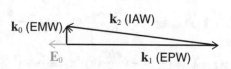

Fig. 8.16 Geometry of the ion acoustic decay (IAD) instability, in the maximum growth configuration ($\mathbf{k}_0 \cdot \mathbf{k}_1 = 0$, with $\mathbf{E}_0$ aligned with $\mathbf{k}_1$)

---

[4] This instability is arbitrarily placed in the EPW chapter but could as well have been in the previous chapter on primary instabilities via IAWs.

$$(\partial_t^2 + \omega_{pe0}^2 - c^2\nabla^2)\mathbf{a}_0 = -\omega_{pe0}^2 \frac{c_s}{c}(\mathbf{e}_0 \cdot \mathbf{e}_2)\frac{\delta n_1}{n_{e0}}\frac{\delta n_2}{n_{e0}}, \tag{8.81}$$

where $\mathbf{e}_2$ is a unit vector along $\mathbf{k}_2$ (likewise $\mathbf{e}_1$ will be a unit vector along $\mathbf{k}_1$).

Next, for the EPW, we obtain

$$(\partial_t^2 + \omega_{pe0}^2 - 3v_{Te}^2\nabla^2)\frac{\delta n_1}{n_{e0}} = cc_s(\mathbf{e}_0 \cdot \mathbf{e}_2)\nabla^2\left(a_0\frac{\delta n_2}{n_{e0}}\right) - c\partial_t\nabla\cdot\left(\mathbf{a}_0\frac{\delta n_2}{n_{e0}}\right). \tag{8.82}$$

And, for the IAW, we have

$$(\partial_t^2 - c_s^2\nabla^2)\frac{\delta n_2}{n_{e0}} = \frac{Zm}{M_i}\frac{\omega_1 c}{k_1}(\mathbf{e}_0 \cdot \mathbf{e}_1)\nabla^2\left(a_0\frac{\delta n_1}{n_{e0}}\right) \tag{8.83}$$

(since the IAW is coupled to the EMW and the EPW, neither of which involves ion motion, all the terms containing an ion density or velocity perturbation on the right-hand side of Eq. (6.30) must be ignored, as well as the last term $\propto (\delta n_e/n_{e0})^2$ since the EMW does not have a density modulation associated with it).

Next we envelope the three quantities,

$$a_0 = \frac{1}{2}\tilde{a}_0 e^{i\psi_0} + c.c., \tag{8.84}$$

$$\frac{\delta n_1}{n_{e0}} = \frac{1}{2}\frac{\delta\tilde{n}_1}{n_{e0}}e^{i\psi_1} + c.c., \tag{8.85}$$

$$\frac{\delta n_2}{n_{e0}} = \frac{1}{2}\frac{\delta\tilde{n}_2}{n_{e0}}e^{i\psi_2} + c.c., \tag{8.86}$$

with $\psi_j = \mathbf{k}_j \cdot \mathbf{r} - \omega_j t$ for $j \in \{0, 1, 2\}$. Using the slowly-varying envelope approximation for the waves, Eq. (6.35), assuming perfect phase matching ($\psi_0 = \psi_1 + \psi_2$), and collecting terms $\propto e^{i\psi_0}$, $e^{i\psi_1}$, and $e^{i\psi_2}$ in Eqs. (8.81), (8.82), and (8.83), respectively, lead to the three slowly varying coupled mode equations for IAD:

$$(\partial_t + v_{g0}\partial_{z_0})\tilde{a}_0 = -i\frac{\omega_{pe0}^2}{4\omega_0}\frac{c_s}{c}(\mathbf{e}_0 \cdot \mathbf{e}_2)\frac{\delta\tilde{n}_1}{n_{e0}}\frac{\delta\tilde{n}_2}{n_{e0}}, \tag{8.87}$$

$$(\partial_t + v_{g1}\partial_{z_1})\frac{\delta\tilde{n}_1}{n_{e0}} = -i\frac{k_1 c}{4}(\mathbf{e}_0 \cdot \mathbf{e}_1)\tilde{a}_0\frac{\delta\tilde{n}_2^*}{n_{e0}}, \tag{8.88}$$

$$(\partial_t + v_{g2}\partial_{z_2})\frac{\delta\tilde{n}_2}{n_{e0}} = -i\frac{\omega_{pe0}}{4}\frac{c}{c_s}\frac{Zm}{M_i}(\mathbf{e}_0 \cdot \mathbf{e}_1)\tilde{a}_0\frac{\delta\tilde{n}_1^*}{n_{e0}}. \tag{8.89}$$

(we can verify during the development that the first term on the right-hand side of Eq. (8.82) is negligible compared to the second).

The growth rate is obtained by expressing the coefficients K_1 and K_2 from Eqs. (6.40) after identification with Eqs. (8.87)–(8.89) and applying Eq. (6.44), leading to

$$\gamma = \frac{|\tilde{a}_0|c}{4} \sqrt{\frac{Zm}{M_i}} \sqrt{\frac{k_1 \omega_{pe0}}{c_s}} (\mathbf{e}_0 \cdot \mathbf{e}_1),$$ (8.90)

in accordance with Ref. [45].

The maximum growth rate γ_0 is obtained for $\mathbf{e}_0 \cdot \mathbf{e}_1 = 1$, i.e., when $\mathbf{k}_0 \cdot \mathbf{k}_1 = 0$ as shown in Fig. 8.16, leading to

$$\boxed{\gamma_0 = \frac{|\tilde{a}_0|c}{4} \sqrt{\frac{Zm}{M_i}} \sqrt{\frac{k_1 \omega_{pe0}}{c_s}}.}$$ (8.91)

Note that since the instability occurs near $n_e = n_c$, the WKB approximation used here will not be valid for regions within an Airy skin depth of the critical density— see Sect. 3.2 and Ref. [45].

The presence of IAD has been inferred in early laser–plasma experiments from measurements of red-shifts in the second harmonic of the laser frequency in the back- or side-scatter geometry [1, 46, 47]; the observed scattered light, slightly red-shifted from $2\omega_0$, was attributed to wave mixing between pairs of IAD-generated EPWs [48]. Later on, short-wavelength optical Thomson scattering allowed more comprehensive measurements and less controversial interpretations by directly measuring the EPWs generated by IAD in the maximum growth geometry shown in Fig. 8.16 [49].

8.4 Suprathermal Electrons from Driven EPWs

8.4.1 General Considerations

As we saw in Sect. 2.2, electrons can get trapped in EPWs, leading to an acceleration at and above the wave's phase velocity $v_\phi = \omega/k$ (the corresponding increase in energy from the plasma electrons being balanced by the energy decrease of the EPW via Landau damping).

Early particle-in-cell simulations of laser–plasma instabilities involving EPWs indicated the formation of a population of suprathermal or "hot" electrons at energies beyond $mv_\phi^2/2$ and being well characterized by a Maxwellian distribution with temperature $T_{hot} = \frac{1}{2}mv_\phi^2$. Such distributions have been observed in computer simulations for most LPI processes involving strongly driven EPWs, such as SRS (backward or forward), TPD, or resonance absorption [11, 50–52]. Few attempts were made to understand the detailed physics process behind the formation of such a "thermalized" tail of hot electrons, as opposed to a more localized distribution in the vicinity of v_ϕ and cutting off near $v_\phi + v_{trap}$ according to the simple 1D picture from Sect. 2.2. The explanation advanced by Bezzerides in Ref. [52], for the specific case of resonance absorption, emphasized the role of the broad spectrum of fields (due to the nonlinear behavior of the waves), leading to stochastic heating of the tail up to velocities significantly beyond v_ϕ. This picture was later confirmed (at least qualitatively) through detailed investigations of TPD [23, 31, 32, 53].

But more importantly, these distributions were confirmed experimentally, and laser–plasma experiments involving strongly driven EPWs systematically show electron spectra with Maxwellian distributions whose temperatures are roughly consistent with the simple $T_{hot} = m v_\phi^2 / 2$ scaling [4, 54–58]. The energy and temperature of hot electrons are often measured indirectly, via measurements of the bremsstrahlung emission from collisions between these electrons and plasma ions. This emission is typically in the hard x-ray range, as the radiation temperature is equal to the hot electron temperature, i.e., typically tens of keV under ICF or HED conditions (cf. the next section).

Hot electrons were identified as a potential issue for ICF very early on due to their capability to preheat the cold fuel and reduce its compressibility [59]. Understanding and mitigating hot electrons has been the subject of a major effort in the ICF community throughout the following decades (and still active at the time of writing) for both direct- and indirect-drive ICF (cf. Refs. [60–63] and the references therein). The issue is particularly important for direct-drive geometry, where hot electrons are generated in close proximity to the nuclear fuel.

Since the two main mechanisms generating high-amplitude EPWs in current typical HED and ICF conditions are SRS and TPD,[5] it is interesting to show the expected range of hot electron temperatures that are accessible by each process. For both instabilities, the hot electron temperature can be expressed as a function of $k\lambda_{De}$ using the EPW dispersion relation, leading to

$$\frac{T_{hot}}{T_e} = \frac{1}{2}\frac{\omega_2^2}{k_2^2 v_{Te}^2} = \frac{1}{2}\frac{1}{(k_2\lambda_{De})^2} + \frac{3}{2},\tag{8.92}$$

where ω_2 and k_2 are the EPW frequency- and wave-number.

For SRS, we can express T_{hot}/T_e via the expressions for k_2 in the forward- and backward-scattering geometries from Eqs. (8.28) and (8.29). The curves of T_{hot}/T_e as a function of n_e/n_c for $T_e = 2\,\text{keV}$ are shown in Fig. 8.17a; the gray region between the forward and backward curves represents the range of possible hot electron temperatures. In particular, we have the following limits:

---

[5] Resonance absorption is not believed to be a major contributor in today's experiments as long plasma scale-lengths make it difficult for the laser to reach the vicinity of the critical density due to collisional absorption.

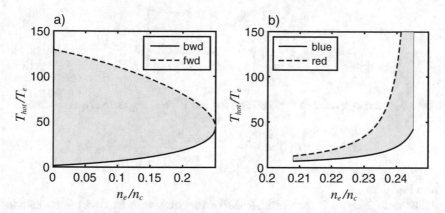

Fig. 8.17 Typical ranges of hot electron temperatures T_{hot} (gray areas), relative to the background electron temperature T_e, for (**a**) SRS, and (**b**) TPD, as a function of electron density and for a temperature of 2 keV. SRS is delimited by the scattering geometry, between back- and forward-scatter (with most of the scattering usually expected for backscatter, i.e., in the lower range of the domain along the solid line). TPD is bracketed between the "blue" and the "red" plasmons, with a Landau cutoff at $k\lambda_{De} = 0.3$ (corresponding to the density cutoff slightly below $0.21 n_c$). The red plasmon can have $k \to 0$ as the density approaches $n_c/4$, which can theoretically lead to infinite phase velocities and acceleration energies. The highest density ($\approx 0.245 n_c$) corresponds to the limit of $k \to 0$ for the red plasmon

- For $n_e/n_c \approx 0$,

$$\frac{T_{hot}}{T_e} \approx \frac{3}{2} \quad \text{(backward)} \tag{8.93}$$

$$\approx \frac{1}{2} \frac{c^2}{v_{Te}^2} \quad \text{(forward)} . \tag{8.94}$$

- For $n_e/n_c \approx 1/4$,

$$\frac{T_{hot}}{T_e} \approx \frac{c^2}{6 v_{Te}^2} \quad \text{(backward and forward)} . \tag{8.95}$$

We should keep in mind that in most situations, backward SRS is believed to be dominant, i.e., the expected hot electron temperature will mostly be confined to the lower temperature side of the gray region, by the solid line.

Likewise for TPD, we can also express the EPW wave-number as a function of the density (at a given temperature); combining Eqs. (8.55) and (8.58) to eliminate k_y, we obtain

$$\frac{k_x}{k_0} = \pm \sqrt{\frac{\rho(n_e)^2}{2} - \frac{1}{8}} , \tag{8.96}$$

$$\frac{k_{yB}}{k_0} = \frac{1}{2}\left[1 + \sqrt{1 + 4k_x^2/k_0^2}\right], \tag{8.97}$$

$$\frac{k_{yR}}{k_0} = \frac{1}{2}\left[1 - \sqrt{1 + 4k_x^2/k_0^2}\right], \tag{8.98}$$

where k_{yB} and k_{yR} refer to the blue and the red EPWs, respectively, and where

$$\rho^2(n_e) = \frac{c^2}{12v_{Te}^2}\frac{1 - 4n_e/n_c}{1 - n_e/n_c} - \frac{1}{4} \tag{8.99}$$

from Eq. (8.58).

The ratio T_{hot}/T_e is then estimated like for SRS, with $k_2^2 = k_x^2 + k_y^2$ inserted in Eq. (8.92). The corresponding plots are shown in Fig. 8.17b, with the curves cut off at $k\lambda_{De} = 0.3$ on the lower density side (just below $0.21n_c$, as was already observed in Fig. 8.11 plotted for the same parameters). As we discussed in the previous section on TPD, the red EPW can in principle have arbitrarily small wave-numbers, meaning arbitrarily large phase velocity (and hot electron temperatures).

For SRS, if scattered light measurements are available, it is possible to infer the fraction of energy deposited in the plasma waves via the Manley–Rowe relations (cf. Sect. 6.4.6). We know from these relations (Eq. (6.83)) that the action density of the scattered light wave and EPW must be the same, i.e., $\mathcal{A}_s = \mathcal{A}_{EPW}$; therefore, the ratio of the waves' energies to frequencies must be identical as well, i.e., $\mathcal{E}_s/\omega_s = \mathcal{E}_{EPW}/\omega_{EPW}$ or, using $\omega_{EPW} = \omega_0 - \omega_s$,

$$\boxed{\mathcal{E}_{EPW} = \left(\frac{\lambda_s}{\lambda_0} - 1\right)\mathcal{E}_s}. \tag{8.100}$$

For example, if one measures 20% reflectivity from SRS in an experiment with an SRS signal at 550 nm for a 351 nm laser, the EPW energy represents about 57% of the scattered light's energy, meaning that on top of the 20% deposited in the (measured) backscattered light, the laser also lost an extra 11% in the EPW.

In case the SRS backscattered light has a finite spectrum, the formula above still applies, if one replaces λ_s by the power-averaged scattered light wavelength (cf. Problem 8.6).

An important point to keep in mind when trying to assess the generation of hot electrons from EPW instabilities like SRS or TPD is the fact that trapping and hot electron generation is associated with Landau damping of the waves. However, for the instability to grow and be above threshold (Sect. 6.4.3, and Eq. (6.48)), Landau damping cannot be too important; a typical limit of 0.25 to 0.3 is usually used as a cutoff, consistent with experiments [39]. Therefore, while it may tempting to assume that the laser energy converted to EPW energy via SRS or TPD is entirely converted to suprathermal electrons, one should keep in mind that these EPWs may also be absorbed by collisional absorption (cf. Sect. 4.5) rather than Landau damping.

This was pointed out in Ref. [39] and fully explored by Turnbull et al. for direct-drive conditions, where it was found that the measured energy in hot electrons (measured via their bremsstrahlung emission, cf. the next section) could not be directly related to the absorbed laser energy via TPD because collisional absorption was the primary absorption mechanism for these EPWs [64].

8.4.2 Measurement of Suprathermal Electrons via Bremsstrahlung Emission

Early experiments on SRS and TPD did offer direct measurements of the accelerated electrons via magnetic electron spectrometers [7, 21], which allowed to validate some of the important features of the electron generation such as the primary excitation in the plane of polarization of the laser for TPD [21]. However, the arrangement of ICF experiments, particularly with the indirect-drive geometry where the laser–plasma interaction is confined inside a high-Z hohlraum, does not always allow for a direct measurement of these electrons. One has to rely on indirect signatures via the bremsstrahlung radiation emitted during the collisions of suprathermal electrons with ions.

In this section we present the physics argument behind the measurement of suprathermal electrons generated by laser–plasma instabilities. These "hot" electrons are measured via their bremsstrahlung emission, i.e., the cumulative radiation from each electron due to its acceleration (i.e., deflection) as it passes near an ion. The emission at a given frequency ω is due to the contributions of electrons at energies $\geq \hbar\omega$ (the maximum photon energy being the electron's own energy); radiation at higher frequencies will have fewer contributors among the fast electrons, so the bremsstrahlung radiation spectrum is related to the distribution of electrons.

Brueckner [65, 66] proposed a way to infer the fast electron energy and distribution from the observed bremsstrahlung radiation spectrum, which, with later improvements by other authors (described below), constitutes the basis for quantifying these electrons in ICF and HED experiments.

The situation is illustrated in Fig. 8.18. An electron propagates through a plasma and collides with plasma particles. In each collision with an ion, its deflection will lead to the emission of radiation due to the acceleration in the electric field of the ion.[6] The radiation is emitted as a very short impulse in each collision, whose duration can be roughly approximated as

$$\tau_c \approx \frac{b}{v_0}, \tag{8.101}$$

---

[6] Electron–electron collisions do not lead to any significant emission because the two radiation waves emitted by each electron interfere destructively [67].

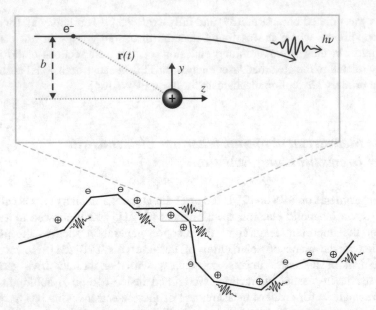

Fig. 8.18 Bremsstrahlung emission of a fast electron as it stops in a plasma: the electron loses energy with each collision (whether from another electron, represented as $\ominus$ in the figure, or an ion, $\oplus$) and emits radiation with each ion collision. The emission stops once the electron has lost too much energy to keep radiating

where b and v_0 are the impact parameter and initial velocity of the electron, respectively (this is the same approximation we already used to derive the collisional absorption in Sect. 4.1, particularly when discussing the Coulomb logarithm, Sect. 4.4). Therefore the spectrum of radiation extends up to $\omega \sim 1/\tau_c$, and as is well known, faster electrons will tend to produce radiation at higher frequencies. Suprathermal electrons from LPI will therefore emit a radiation spectrum at higher frequencies than the background plasma (typically in the hard x-ray domain, as we will see).

The key idea behind the reconstruction of the hot electron properties from their bremsstrahlung emission is that each fast electron will emit high-frequency radiation as it keeps colliding with the plasma, up to the point where it loses enough energy to bring it back to the "bulk" of the background electron distribution, at which point its emission will not be distinguishable from the bulk's anymore. The following derivation, which follows Brueckner's [65], consists in calculating the total radiated energy by the suprathermal electrons in all directions, until they "stop" due to collisions.

The two key formulas we will need from classical electrodynamics are the bremsstrahlung radiation emitted by an electron (in energy per unit frequency and unit path length) and the energy loss per unit path length due to collisions. Both are available in the literature (e.g., Jackson [68]), however for reasons that will be clear later on we are going to re-derive the bremsstrahlung formula based on the semi-heuristic approach from Hutchinson [67].

We start from the Larmor formula, which gives the total power radiated by an electron undergoing an acceleration $\dot{v}$:

$$P_r = \frac{2}{3} \frac{e^2}{4\pi \varepsilon_0 c} \left(\frac{\dot{v}}{c}\right)^2 . \tag{8.102}$$

For the case of an electron–ion collision, the radiation occurs during a very short time $\tau_c \sim b/v_0$, as the electron passes by the ion. Since the deflection is typically small (in general, and a fortiori for the suprathermal electrons we are considering here), the distance of the closest approach is approximately equal to the impact parameter $\sim b$, and the strongest acceleration occurs as the electron passes by the ion at a distance $\sim b$; the acceleration is primarily directed toward y (with the notations from Fig. 8.18) and is obtained from the Coulomb force $\mathbf{F}_{max} = m\dot{\mathbf{v}}_{max}$, i.e., $\dot{v}_{max} = -Ze^2/(4\pi \varepsilon_0 b^2 m)$. The maximum radiated power as the electron passes by the ion is then

$$P_{r,max} \approx \frac{2}{3} \frac{(Ze)^2 e^4}{(4\pi \varepsilon_0)^3} \frac{1}{m^2 c^3 b^4} , \tag{8.103}$$

and the total radiated energy is roughly approximated as $\epsilon_r \approx P_{r,max} \tau_c$, i.e.,

$$\epsilon_r = \frac{2}{3} \frac{(Ze)^2 e^4}{(4\pi \varepsilon_0)^3} \frac{1}{m^2 c^3 b^3 v_0} , \tag{8.104}$$

where v_0 is the initial electron velocity as it approaches the ion. The radiation is much stronger for heavy ions ($\epsilon_r \propto Z^2$). Now to obtain a spectral distribution, we consider the fact that the emission is similar to an impulse of duration τ_c; therefore, its spectrum extends up to $\omega \sim 1/\tau_c = v_0/b$. In the very simplified picture where the spectrum is nearly flat up to v_0/b, the spectral distribution is then given by

$$\frac{d\epsilon_r}{d\omega} \approx \epsilon_r \tau_c = \frac{2}{3} \frac{(Ze)^2 e^4}{(4\pi \varepsilon_0)^3} \frac{1}{m^2 c^3 b^2 v_0^2} \quad (\omega \ll \frac{v_0}{b}), \tag{8.105}$$

$$\approx 0 \quad (\omega \gg \frac{v_0}{b}). \tag{8.106}$$

For the last step we want to integrate over all the ions the electron is interacting with per unit propagation length dl. Proceeding like in Sect. 1.4, the number of ions in an annular volume of thickness db at a transverse distance b from the electron and with longitudinal length dl is $\delta N_i = n_i \delta V$, where n_i is the ion density and $\delta V = dl \, 2\pi b \, db$ is the annular element volume (cf. Fig. 1.16). Therefore, the total radiated energy for a single electron with initial velocity v_0 as it propagates over an infinitesimal distance dl is obtained by integrating the radiated energy from each ion over b:

$$\frac{d^2\epsilon_r}{d\omega dl} = \frac{2}{3} \frac{(Ze)^2 e^4}{(4\pi\varepsilon_0)^3} \frac{2\pi n_i}{m^2 c^3 v_0^2} \int_{b_{min}}^{b_{max}} \frac{db}{b} . \tag{8.107}$$

Note that the derivation is very similar to the calculation of the collision frequency in Sect. 1.4 or the collisional absorption in Sect. 4.1; in particular, we are dealing with yet another integral over impact parameters similar to the one providing the Coulomb logarithm. Like in Sect. 4.4, the minimum impact parameter is usually chosen to be the de Broglie wavelength, $b_{min} = \hbar/(2mv_0)$. For the maximum impact parameter, since the spectrum extends up to $\omega \approx v_0/b$, we want to integrate up to $b_{max} = v_0/\omega$. Integrating $1/b$ between these two limits yields

$$\frac{d^2\epsilon_r}{d\omega dl} = \frac{4\pi}{3} \frac{(Ze)^2 e^4}{(4\pi\varepsilon_0)^3} \frac{n_i}{m^2 c^3 v_0^2} \ln(\Lambda_r) , \tag{8.108}$$

with the logarithmic term (resembling the Coulomb logarithm we encountered before and commonly referred to as the "Gaunt factor") defined as

$$\Lambda_r = \frac{b_{max}}{b_{min}} = \frac{2mv_0^2}{\hbar\omega} . \tag{8.109}$$

The velocity term in the logarithm is then corrected to account for the (small) energy loss from the electron due to radiation; we simply take the average between the initial velocity v_0 and the velocity with the radiation loss subtracted, $v_0' = \sqrt{2(\epsilon_0 - \hbar\omega)/m}$, where $\epsilon_0 = \frac{1}{2}mv_0^2$. Substituting v_0 by $(v_0 + v_0')/2$ in Eq. (8.109) and replacing the factor $4\pi/3$ in Eq. (8.108) by a more accurate factor $16/3$,[7] we obtain the so-called Bethe and Heitler formula of bremsstrahlung radiation, which was obtained via a more rigorous quantum-mechanical calculation:

$$\frac{d^2\epsilon_r}{d\omega dl} = \frac{16}{3} \frac{(Ze)^2 e^4}{(4\pi\varepsilon_0)^3} \frac{n_i}{m^2 c^3 v_0^2} \ln(\Lambda_r) , \tag{8.110}$$

$$\Lambda_r = \frac{\left(\sqrt{\epsilon_0} + \sqrt{\epsilon_0 - \hbar\omega}\right)^2}{\hbar\omega} . \tag{8.111}$$

Now that we have the expression for the radiated energy per unit path length and frequency for a single electron, we want to sum up the total radiation from this electron due to all its collisions, until it has lost too much energy to radiate anymore. The second formula we need is the stopping power, i.e., the energy loss of the electron per unit length due to collisions [67, 68]:

---

[7] The error lies in the rough approximations used to estimate the radiated energy; for a more accurate derivation, cf. Ref. [69].

$$\frac{d\epsilon}{dl} = -4\pi Z \frac{n_i e^4}{m v_0^2} \ln(\Lambda_c),$$

(8.112)

where the logarithmic term originates once again from an integral over impact parameters and is given by

$$\Lambda_c = \frac{2\epsilon_0}{\hbar\omega}.$$

(8.113)

As the electron keeps colliding and losing energy, the radiated energy from Eq. (8.110) needs to be adjusted for the fact that v_0 (and ϵ_0) keep decreasing. To calculate the total radiated energy per unit frequency, we must first change variables from dl (unit path length) to ϵ:

$$\frac{d^2\epsilon_r}{d\omega dl} = \frac{d\epsilon}{dl} \frac{d}{d\epsilon} \frac{d\epsilon_r}{d\omega},$$

(8.114)

from which we obtain that $d^2\epsilon_r/d\epsilon d\omega$ is the ratio of $d^2\epsilon_r/dl d\omega$ to $d\epsilon/dl$. The total radiated energy per unit $d\omega$ is then obtained by integrating over the electron energies, from its initial energy ϵ_0 to a minimal energy $\epsilon_{min} = \hbar\omega$ (since the electron cannot radiate more energy than its own kinetic energy):

$$\frac{d\epsilon_r}{d\omega} = \int_{\hbar\omega}^{\epsilon_0} d\epsilon \frac{d^2\epsilon_r/d\omega dl}{d\epsilon/dl}$$

$$= \frac{4}{3\pi} \frac{Ze^2}{(4\pi\varepsilon_0)^2} \frac{1}{mc^3} \int_{\hbar\omega}^{\epsilon_0} \frac{\ln(\Lambda_r)}{\ln(\Lambda_c)} d\epsilon.$$

(8.115)

Finally, if the suprathermal electrons belong to a distribution $f_{hot}(\epsilon_0)$ (which is what we are ultimately looking to characterize), the measured emission spectrum $d\mathcal{E}_r/d\omega$ (in energy per unit frequency) is related to the distribution via

$$\frac{d\mathcal{E}_r}{d\omega} = \int_{\hbar\omega}^{\infty} f_{hot}(\epsilon_0) \frac{d\epsilon_r(\epsilon_0)}{d\omega} d\epsilon_0.$$

(8.116)

The left-hand side of this equation is what we measure in an experiment ($\mathcal{E}_r$ designates the radiation energy from the entire distribution of hot electrons, as opposed to ϵ_r which designates a single electron), and our goal is to extract $f_{hot}(\epsilon_0)$. The difficulty to invert this equation comes from the dependence of the Gaunt factors Λ_r, Λ_c on the energy ϵ. However, recall that our initial expression for the bremsstrahlung Gaunt factor as given by Eq. (8.109) is essentially the same as Λ_c except for a numerical factor: therefore, it is reasonable as a first approximation to take the ratio $\ln(\Lambda_r)/\ln(\Lambda_c)$ as a constant, approximately equal to 1. In this case we obtain after performing the integration in Eq. (8.115)

$$\frac{d\epsilon_r}{d\omega} = \frac{4}{3\pi} \frac{Ze^2}{(4\pi\varepsilon_0)^2} \frac{1}{mc^3} (\epsilon_0 - \hbar\omega) . \tag{8.117}$$

We can now insert this expression in Eq. (8.116) and take derivatives with respect to ω to extract $f_{hot}(\epsilon_0)$. We use the Leibniz integral rule:

$$\frac{d}{dx} \left[\int_{a(x)}^{b(x)} f(x,t)dt \right] = f[x,b(x)]b'(x) - f[x,a(x)]a'(x)$$

$$+ \int_{a(x)}^{b(x)} \partial_x f(x,t)dt . \tag{8.118}$$

We obtain the following derivatives:

$$\frac{d^2\mathcal{E}_r}{d\omega^2} = \int_{\hbar\omega}^{\infty} f_{hot}(\epsilon_0) \frac{d^2\epsilon_r}{d\omega^2} d\epsilon_0 , \tag{8.119}$$

$$\frac{d^3\mathcal{E}_r}{d\omega^3} = -\hbar f_{hot}(\hbar\omega) \frac{d^2\epsilon_r}{d\omega^2} \bigg|_{\epsilon_0=\hbar\omega} . \tag{8.120}$$

This last equation gives us the distribution function f_{hot} as a function of the measurable quantity $d^3\mathcal{E}_r/d\omega^3$; after inserting the expression for $d^2\epsilon_r/d\omega^2$ by taking the derivative of Eq. (8.117), we finally obtain our main result:

$$\boxed{f_{hot}(\hbar\omega) = \frac{3\pi}{4} \hbar mc^3 \frac{(4\pi\varepsilon_0)^2}{Ze^2} \frac{d^3\mathcal{E}_r}{d(\hbar\omega)^3} .} \tag{8.121}$$

Note that this result can be refined by keeping the full expression for the Gaunt factor in the bremsstrahlung emission formula with energy dependence, Eq. (8.111) [70–72]. However, the result cited here, from the original work by Brueckner [65], gives a reasonable agreement with the exact solution and has a much simpler format. The only required operation is to take the second derivative of $d\mathcal{E}_r/d(\hbar\omega)$, the measured radiation power spectrum.

The total energy in the hot electron population can then be estimated via

$$E_{hot} = \int_0^{\infty} f_{hot}(\epsilon)\epsilon d\epsilon . \tag{8.122}$$

Using an integration by parts with the boundary condition $d^2\mathcal{E}_r/d\epsilon^2 \to 0$ for $\epsilon \to \infty$ gives

$$\boxed{E_{hot} = \frac{3\pi}{4} \hbar mc^3 \frac{4\pi\varepsilon_0}{e^2} \frac{1}{Z} \frac{d\mathcal{E}_r}{d(\hbar\omega)} \bigg|_0 .} \tag{8.123}$$

Fig. 8.19 Illustration of a typical bremsstrahlung hard x-ray spectrum measurement, representing the (made up) data points and exponential fit of x-ray radiation spectrum as a function of the x-ray energy for $T_{hot} = 50$ keV, $E_{hot} = 42$ J, and assuming hot electrons are stopped in gold ($Z = 79$)

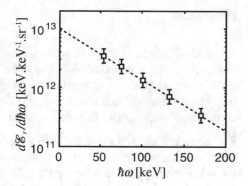

In practical units, we have

$$E_{hot}\,[\mathrm{J}] \approx 2.65 \times 10^{-11}\, \frac{1}{Z}\, \left.\frac{d\mathcal{E}_r}{d(\hbar\omega)}\right|_0 \tag{8.124}$$

(note that this is for the total bremsstrahlung emission in 4π and would therefore need to be corrected to account for the solid angle of the x-ray detection system).

This estimate is within a factor of two of the one used in Refs. [58, 73].[8] As emphasized in Ref. [58], such estimates of the hot electron energy should not be considered accurate to better than a factor of 2 anyway. A more rigorous analysis that takes the energy dependence in the Gaunt factors into account is presented in [72].

Practically, since the population of hot electrons generated by LPI will typically have a Maxwellian spectrum (cf. the previous section), the temperature (i.e., linear slope on a log scale) of the bremsstrahlung x-rays will be the same as the hot electron temperature, as seen from Eq. (8.121). The hot electron energy is simply estimated via Eq. (8.124) where $d\mathcal{E}_r/d(\hbar\omega)|_0$ is taken as the continuation of the fit to the measured x-ray spectrum at $\hbar\omega = 0$, as illustrated in Fig. 8.19. In this example, the hot electron temperature is 50 keV, and the hot electron energy can be reconstructed by inserting the value of the exponential fit (dashed curve) at $\hbar\omega = 0$ (i.e., 10^{13}) in Eq. (8.124) (accounting for the 4π solid angle correction). The hot electron energy in this example would be ≈ 42 J, assuming that the electrons stop in gold ($Z = 79$).

These measurements are usually taken using hard x-ray spectrometers such as the FFLEX (filter fluorescent x-ray diagnostic) system [73–75]. We should keep in mind that many simplifying assumptions were taken to arrive at Eq. (8.124): the radiation is assumed to be generated by electrons that have fully stopped in the plasma (by "fully stopped" we mean that they have reached background thermal velocities), the plasma is assumed to be transparent to the x-ray radiation, the radiation is assumed to be isotropic, etc. Each of these assumptions should be carefully checked (and if necessary corrected) for any given experimental situation.

---

[8] Note that the equation on p. 3222 in Ref. [58] is for the value of $d\mathcal{E}_r/d(\hbar\omega)$ at $\hbar\omega = T_{hot}$, i.e., $d\mathcal{E}_r/d(\hbar\omega)|_{T_{hot}} = e^{-1} d\mathcal{E}_r/d(\hbar\omega)|_0$.

Problems

8.1 SRS Coupled Mode Equations Using Normalized Action Amplitudes

Going from the normalized field variables a and $\delta n_e/n_{e0}$ for the EMW and the EMWs, to complex action amplitudes as defined in Eqs. (1.214), (1.215), and then to normalized action amplitudes using Eq. (6.76), verify that the SRS coupled mode equations, Eqs. (8.9)–(8.11), reduce to the form of Eqs. (6.73)–(6.75).

8.2 Kinetic Expression for the SRS Amplification Rate

Start from the steady-state plasma response to the ponderomotive drive imposed by the beat intensity pattern between two light waves, as derived in Sect. 7.2, Eq. (7.70) (remember that no assumption was made in that expression regarding the driver's frequency and wave-number and their proximity to a plasma resonance, i.e., EPW or IAW). Neglect the ion susceptibility, introduce an EPW damping by replacing ω_2 by $\omega_2 + i\nu_2$ in the electron susceptibility, and show that by taking the fluid limit of F_χ one recovers the fluid expression for the SRS spatial amplification rate, Eq. (8.18).

8.3 TPD Coupled Mode Equations Using Normalized Action Amplitudes

Going from the normalized field variables a and $\delta n_e/n_{e0}$ for the EMW and the EPWs respectively, to complex action amplitudes as defined in Eqs. (1.214), (1.215), and then to normalized action amplitudes using Eq. (6.76), verify that the TPD coupled mode equations Eqs. (8.50)–(8.52) reduce to the form of Eqs. (6.73)–(6.75).

8.4 Density Mapping onto the TPD Hyperbola, Eq. (8.58)

Demonstrate Eq. (8.58), using the EPW dispersion relation for the light wave $\omega_0^2 = \omega_{pe}^2 + k_0^2 c^2$ and the daughter EPWs, $\omega_{1,2}^2 = \omega_{pe}^2 + 3v_{Te}^2 k_{1,2}^2$, and the phase-matching relations, $\omega_0 = \omega_1 + \omega_2$, $\mathbf{k}_0 = \mathbf{k}_1 + \mathbf{k}_2$.

Hint: show that

$$(\omega_1^2 - \omega_2^2)^2 + \omega_0^4 = 2\omega_0^2(\omega_1^2 + \omega_2^2)^2 ,$$

then substitute the frequencies $\omega_{0,1,2}^2$ via the dispersion relations, and use $k_x = k_{1x}$, $k_y = k_{1y}$, so that $k_{2x} = -k_{1x}$ and $k_{2y} = k_0 - k_{1y}$, as mentioned in the text. You will neglect terms of order $O(v_{Te}^4/c^4)$.

8.5 $3\omega_0/2$ Signature for a Single Laser at Normal Incidence onto a 1D Density Profile

- Verify the numbers in the text and in Fig. 8.14 regarding the geometry of the Thomson up-shift scattered wave $\mathbf{k}_s$ scattering off a TPD EPW (in particular the exit angle of the $3\omega_0/2$ scattered light).
- Check that the temperature inferred from the frequency shift in Eq. (8.76) is consistent with the results from Ref. [40] (Fig. 6a and inferred $T_e \approx 1.3\,\mathrm{keV}$).

8.6 Fraction of Laser Energy Going into EPWs for SRS with a Broad Scattered Light Spectrum

Show that the fraction of energy going into EPWs in the SRS process when the measured scattered light has a finite frequency spectrum is given by $\mathcal{E}_{EPW} = (\bar{\lambda}_s/\lambda_0 - 1)\mathcal{E}_s$, where $\bar{\lambda}_s$ is the power-averaged scattered light wavelength,

$$\bar{\lambda}_s = \frac{\int \lambda_s P_s(\lambda_s) d\lambda_s}{\int P_s(\lambda_s) d\lambda_s}. \qquad (8.125)$$

References

1. J.L. Bobin, M. Decroisette, B. Meyer, Y. Vitel, Phys. Rev. Lett. **30**, 594 (1973)
2. J. Elazar, W.T. Toner, E.R. Wooding, Plasma Phys. **23**, 813 (1981)
3. D.W. Phillion, E.M. Campbell, K.G. Estabrook, G. E. Phillips, F. Ze, Phys. Rev. Lett. **49**, 1405 (1982)
4. D.W. Phillion, D.L. Banner, E.M. Campbell, R.E. Turner, K.G. Estabrook, Phys. Fluids **25**, 1434 (1982)
5. K. Tanaka, L.M. Goldman, W. Seka, M.C. Richardson, J.M. Soures, E.A. Williams, Phys. Rev. Lett. **48**, 1179 (1982)
6. W. Seka, E.A. Williams, R.S. Craxton, L.M. Goldman, R.W. Short, K. Tanaka, Phys. Fluids **27**, 2181 (1984)
7. H. Figueroa, C. Joshi, H. Azechi, N.A. Ebrahim, K. Estabrook, Phys. Fluids **27**, 1887 (1984)
8. D.W. Forslund, J.M. Kindel, E.L. Lindman, Phys. Rev. Lett. **30**, 739 (1973)
9. D. Biskamp, H. Welter, Phys. Rev. Lett. **34**, 312 (1975)
10. W.L. Kruer, K. Estabrook, B.F. Lasinski, A.B. Langdon, Phys. Fluids **23**, 1326 (1980). References 305
11. K. Estabrook, W.L. Kruer, B.F. Lasinski, Phys. Rev. Lett. **45**, 1399 (1980)
12. N.B. Meezan, L.J. Atherton, D.A. Callahan, E.L. Dewald, S. Dixit, E.G. Dzenitis, M.J. Edwards, C. A. Haynam, D.E. Hinkel, O.S. Jones, O. Landen, R.A. London, P.A. Michel, J.D. Moody, J.L. Milovich, M.B. Schneider, C.A. Thomas, R.P.J. Town, A.L. Warrick, S.V. Weber, K. Widmann, S. H. Glenzer, L.J. Suter, B.J. MacGowan, J.L. Kline, G.A. Kyrala, A. Nikroo, Phys. Plasmas **17**, 056304 (2010)
13. D.E. Hinkel, M.D. Rosen, E.A. Williams, A.B. Langdon, C.H. Still, D.A. Callahan, J.D. Moody, P. A. Michel, R.P.J. Town, R.A. London, S.H. Langer, Phys. Plasmas **18**, 056312 (2011)
14. M.J. Rosenberg, A.A. Solodov, J.F. Myatt, W. Seka, P. Michel, M. Hohenberger, R.W. Short, R. Epstein, S.P. Regan, E.M. Campbell, T. Chapman, C. Goyon, J.E. Ralph, M.A. Barrios, J.D. Moody, J.W. Bates, Phys. Rev. Lett. **120**, 055001 (2018)
15. H.H. Klein, W.M. Manheimer, E. Ott, Phys. Rev. Lett. **31**, 1187 (1973)
16. C.S. Liu, M.N. Rosenbluth, R.B. White, Phys. Fluids **17**, 1211 (1974)
17. M.A. Mostrom, A.N. Kaufman, Phys. Rev. Lett. **42**, 644 (1979)
18. B.B. Afeyan, E.A. Williams, Phys. Fluids **28**, 3397 (1985)
19. S. Depierreux, C. Neuville, C. Baccou, V. Tassin, M. Casanova, P.-E. Masson-Laborde, N. Borisenko, A. Orekhov, A. Colaitis, A. Debayle, G. Duchateau, A. Heron, S. Huller, P. Loiseau, P. Nicolaï, D. Pesme, C. Riconda, G. Tran, R. Bahr, J. Katz, C. Stoeckl, W. Seka, V. Tikhonchuk, C. Labaune, Phys. Rev. Lett. **117**, 235002 (2016)
20. P. Michel, M.J. Rosenberg, W. Seka, A.A. Solodov, R.W. Short, T. Chapman, C. Goyon, N. Lemos, M. Hohenberger, J.D. Moody, S.P. Regan, J.F. Myatt, Phys. Rev. E **99**, 033203 (2019)
21. N.A. Ebrahim, H.A. Baldis, C. Joshi, R. Benesch, Phys. Rev. Lett. **45**, 1179 (1980)

22. R. Yan, A.V. Maximov, C. Ren, F.S. Tsung, Phys. Rev. Lett. **103**, 175002 (2009)
23. H.X. Vu, D.F. DuBois, D.A. Russell, J.F. Myatt, Phys. Plasmas **17**, 072701 (2010)
24. J. Meyer, Y. Zhu, Phys. Rev. Lett. **71**, 2915 (1993)
25. W. Seka, B.B. Afeyan, R. Boni, L.M. Goldman, R.W. Short, K. Tanaka, T.W. Johnston, Phys. Fluids **28**, 2570 (1985)
26. C.S. Liu, M.N. Rosenbluth, Phys. Fluids **19**, 967 (1976)
27. A. Simon, R.W. Short, E.A. Williams, T. Dewandre, Phys. Fluids **26**, 3107 (1983)
28. R. Yan, A.V. Maximov, C. Ren, Phys. Plasmas **17**, 052701 (2010)
29. B.B. Afeyan, E.A. Williams, Phys. Rev. Lett. **75**, 4218 (1995)
30. M.N. Rosenbluth, Phys. Rev. Lett. **29**, 565 (1972)
31. H.X. Vu, D.F. DuBois, D.A. Russell, J.F. Myatt, Phys. Plasmas **19**, 102708 (2012)
32. J.F. Myatt, J. Zhang, J.A. Delettrez, A.V. Maximov, R.W. Short, W. Seka, D.H. Edgell, D.F. DuBois, D.A. Russell, H.X. Vu, Phys. Plasmas **19**, 022707 (2012)
33. J. Zhang, J.F. Myatt, R.W. Short, A.V. Maximov, H.X. Vu, D.F. DuBois, D.A. Russell, Phys. Rev. Lett. **113**, 105001 (2014)
34. J.F. Myatt, J. Zhang, R.W. Short, A.V. Maximov, W. Seka, D.H. Froula, D.H. Edgell, D.T. Michel, I.V. Igumenshchev, D.E. Hinkel, P. Michel, J.D. Moody, Phys. Plasmas **21** (2014)
35. D.T. Michel, A.V. Maximov, R.W. Short, S.X. Hu, J.F. Myatt, W. Seka, A.A. Solodov, B. Yaakobi, D.H. Froula, Phys. Rev. Lett. **109**, 155007 (2012)
36. D.T. Michel, A.V. Maximov, R.W. Short, J.A. Delettrez, D. Edgell, S.X. Hu, I.V. Igumenshchev, J.F. Myatt, A.A. Solodov, C. Stoeckl, B. Yaakobi, D.H. Froula, Phys. Plasmas **20** (2013)
37. C. Stoeckl, R.E. Bahr, B. Yaakobi, W. Seka, S.P. Regan, R.S. Craxton, J.A. Delettrez, R.W. Short, J. Myatt, A.V. Maximov, H. Baldis, Phys. Rev. Lett. **90**, 235002 (2003)
38. S.P. Regan, N.B. Meezan, L.J. Suter, D.J. Strozzi, W.L. Kruer, D. Meeker, S.H. Glenzer, W. Seka, C. Stoeckl, V.Y. Glebov, T.C. Sangster, D.D. Meyerhofer, R.L. McCrory, E.A. Williams, O.S. Jones, D.A. Callahan, M.D. Rosen, O.L. Landen, C. Sorce, B.J. MacGowan, Phys. Plasmas **17** (2010)
39. W. Seka, J.F. Myatt, R.W. Short, D.H. Froula, J. Katz, V.N. Goncharov, I.V. Igumenshchev, Phys. Rev. Lett. **112**, 145001 (2014)
40. W. Seka, D.H. Edgell, J.F. Myatt, A.V. Maximov, R.W. Short, V.N. Goncharov, H.A. Baldis, Phys. Plasmas **16**, 052701 (2009)
41. D.A. Russell, D.F. DuBois, Phys. Rev. Lett. **86**, 428 (2001)
42. K.L. Baker, K.G. Estabrook, R.P. Drake, B.B. Afeyan, Phys. Rev. Lett. **86**, 3787 (2001)
43. D.F. DuBois, M.V. Goldman, Phys. Rev. Lett. **14**, 544 (1965)
44. F.W. Perkins, J. Flick, Phys. Fluids **14**, 2012 (1971)
45. R.P. Drake, M.V. Goldman, J.S. DeGroot, Phys. Plasmas **1**, 2448 (1994)
46. C. Garban, E. Fabre, C. Stenz, C. Popovics, J. Virmont, F. Amiranoff, J. Phys. Lett. **39**, 165 (1978)
47. C.J. Walsh, H.A. Baldis, R.G. Evans, Phys. Fluids **25**, 2326 (1982)
48. K. Tanaka, W. Seka, L.M. Goldman, M.C. Richardson, R.W. Short, J.M. Soures, E.A. Williams, Phys. Fluids **27**, 2187 (1984)
49. K. Mizuno, R. Bahr, B.S. Bauer, R.S. Craxton, J.S. DeGroot, R.P. Drake, W. Seka, B. Sleaford, Phys. Rev. Lett. **73**, 2704 (1994)
50. K. Estabrook, W.L. Kruer, Phys. Rev. Lett. **40**, 42 (1978)
51. K. Estabrook, W.L. Kruer, Phys. Fluids **26**, 1892 (1983)
52. B. Bezzerides, S.J. Gitomer, D.W. Forslund, Phys. Rev. Lett. **44**, 651 (1980)
53. J.F. Myatt, H.X. Vu, D.F. DuBois, D.A. Russell, J. Zhang, R.W. Short, A.V. Maximov, Phys. Plasmas **20**, 052705 (2013)
54. C. Joshi, T. Tajima, J.M. Dawson, H.A. Baldis, N.A. Ebrahim, Phys. Rev. Lett. **47**, 1285 (1981)
55. R.G. Berger, R.D. Brooks, Z.A. Pietrzyk, Phys. Fluids **26**, 354 (1983)
56. D.M. Villeneuve, R.L. Keck, B.B. Afeyan, W. Seka, E.A. Williams, Phys. Fluids **27**, 721 (1984)
57. R.P. Drake, R.E. Turner, B.F. Lasinski, K.G. Estabrook, E.M. Campbell, C.L. Wang, D.W. Phillion, E.A. Williams, W.L. Kruer, Phys. Rev. Lett. **53**, 1739 (1984)

58. R.P. Drake, R.E. Turner, B.F. Lasinski, E.A. Williams, K. Estabrook, W.L. Kruer, E.M. Campbell, T.W. Johnston, Phys. Rev. A **40**, 3219 (1989)
59. J. Lindl, Nucl. Fusion **14**, 511 (1974)
60. R.S. Craxton, K.S. Anderson, T.R. Boehly, V.N. Goncharov, D.R. Harding, J.P. Knauer, R.L. McCrory, P.W. McKenty, D.D. Meyerhofer, J.F. Myatt, A.J. Schmitt, J.D. Sethian, R.W. Short, S. Skupsky, W. Theobald, W.L. Kruer, K. Tanaka, R. Betti, T.J.B. Collins, J.A. Delettrez, S.X. Hu, J.A. Marozas, A.V. Maximov, D.T. Michel, P.B. Radha, S.P. Regan, T.C. Sangster, W. Seka, A.A. Solodov, J.M. Soures, C. Stoeckl, J.D. Zuegel, Phys. Plasmas **22**, 110501 (2015)
61. E. Campbell, V. Goncharov, T. Sangster, S. Regan, P. Radha, R. Betti, J. Myatt, D. Froula, M. Rosenberg, I. Igumenshchev, W. Seka, A. Solodov, A. Maximov, J. Marozas, T. Collins, D. Turnbull, F. Marshall, A. Shvydky, J. Knauer, R. McCrory, A. Sefkow, M. Hohenberger, P. Michel, T. Chapman, L. Masse, C. Goyon, S. Ross, J. Bates, M. Karasik, J. Oh, J. Weaver, A. Schmitt, K. Obenschain, S. Obenschain, S. Reyes, B. VanWonterghem, Matter Radiation Extremes **2**, 37 (2017)
62. J. Lindl, Phys. Plasmas **2**, 3933 (1995)
63. J. Lindl, O. Landen, J. Edwards, E. Moses, Phys. Plasmas **21**, 020501 (2014)
64. D. Turnbull, A.V. Maximov, D.H. Edgell, W. Seka, R.K. Follett, J.P. Palastro, D. Cao, V.N. Goncharov, C. Stoeckl, D.H. Froula, Phys. Rev. Lett. **124**, 185001 (2020)
65. K.A. Brueckner, Phys. Rev. Lett. **36**, 677 (1976)
66. K.A. Brueckner, Phys. Rev. Lett. **37**, 1247 (1976)
67. I. Hutchinson, J. Freidberg, *22.611J—Introduction to Plasma Physics I* (Massachusetts Institute of Technology: MIT OpenCourseWare, 2003). https://ocw.mit.edu
68. J.D. Jackson, *Classical Electrodynamics*, 3rd ed. (Wiley, 1999)
69. G.B. Rybicki, A.P. Lightman, *Radiative Processes in Astrophysics* (Wiley, New York, 1985)
70. D.B. Henderson, M.A. Stroscio, Phys. Rev. Lett. **37**, 1244 (1976)
71. H. Brysk, Phys. Rev. Lett. **37**, 1242 (1976)
72. C.A. Thomas, Phys. Rev. E **81**, 036413 (2010)
73. M. Hohenberger, F. Albert, N.E. Palmer, J.J. Lee, T. Dppner, L. Divol, E.L. Dewald, B. Bachmann, A.G. MacPhee, G. LaCaille, D.K. Bradley, C. Stoeckl, Rev. Sci. Instrum. **85**, 11D501 (2014)
74. J.W. McDonald, R.L. Kauffman, J.R. Celeste, M.A. Rhodes, F.D. Lee, L.J. Suter, A.P. Lee, J.M. Foster, G. Slark, Rev. Sci. Instrum. **75**, 3753 (2004)
75. E.L. Dewald, C. Thomas, S. Hunter, L. Divol, N. Meezan, S.H. Glenzer, L.J. Suter, E. Bond, J.L. Kline, J. Celeste, D. Bradley, P. Bell, R.L. Kauffman, J. Kilkenny, O.L. Landen, Rev. Sci. Instrum. **81**, 10D938 (2010)

Chapter 9
Optical Smoothing of High-Power Lasers and Implications for Laser–Plasma Instabilities

The goal of the optical smoothing techniques described in this chapter, which are implemented on most high-energy laser facilities around the world, is to manipulate the laser phase front in the near-field (i.e., at the lens) in order to produce laser focal spots that have a controlled and well-known spatial profile. The phase plates used for optical smoothing introduce two different spatial scales within the focal spot: a large scale (compared to the laser wavelength) associated with the "average" intensity profile of the focal spot, which can be precisely shaped and controlled, and a small scale, on the order of the wavelength, which is characterized by "speckles" resulting from random interferences within the beam. The shape, statistical intensity distribution, and lifetime of these speckles have well-known statistical properties and are crucial contributors to the behavior of some of the major laser–plasma instabilities such as backscatter and filamentation. Speckles are ubiquitous to laser physics, precisely because of the coherent nature of laser light, and a vast body of literature has been devoted to the subject—for example, see [1–3] and the references therein. In fact, the speckle mitigation techniques used in high-power lasers for ICF and HED applications like polarization smoothing and smoothing by spectral dispersion, described in this chapter, follow the same original ideas developed for imaging applications [3].

In this chapter, we introduce the main optical smoothing techniques deployed at existing laser facilities, describe the essential features (e.g., size and statistical distribution) of the speckles resulting from these techniques, and will introduce the key physics effects of optical smoothing on laser–plasma instabilities.

9.1   Random (or Continuous) Phase Plates (RPP/CPP)

9.1.1   General Concept and Expression of the Resulting Electric Field

Random phase plates (RPPs), initially proposed in Ref. [4], are transmissive optical elements that sit in the near-field of the laser (i.e., near the focusing lens). An RPP contains a large number of small elements of known identical size (for example, squares of side δX_{RPP} as illustrated in Fig. 9.1) that each introduce a random dephasing with respect to each other. The dephasing is typically a random selection of either 0 or π, introduced by etching a random set of the RPP elements by a fixed depth d selected so that the propagation at the laser wavelength introduces a dephasing of π, i.e., $k_0 d = \pi$ or $2dn = \lambda_0$, where n is the refractive index of the plate and $k_0 = n\omega_0/c$. The dephasing from any RPP element must have the value 0 or π with equal probability, i.e., there must be on average as many etched elements as non-etched on the plate. We will see below that the exact same results are obtained if the RPP elements have a size that is entirely random on scales larger than d, so that the dephasing of the RPP elements has a uniform probability distribution in $[-\pi, \pi]$.

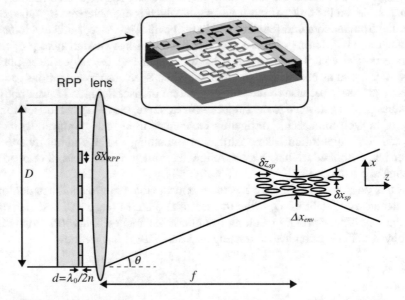

Fig. 9.1 Illustration of the RPP concept: the small elements introduce a random, equiprobable phase shift of 0 or π in the electric field via a difference in thickness $d = \lambda_0/2n$, where n is the refractive index of the RPP; the resulting focal spot consists in a large-scale envelope and many small intensity hot spots or "speckles" within, due to the random interference of the diffraction from each RPP element. The inset shows a 3D representation of an RPP (not to scale) with 16×16 elements

As we will see below, the diffraction of a laser beam with size D at the lens through the small RPP elements creates a focal spot with small intensity hot spots or "speckles" of typical size δx_{sp}, δz_{sp} within a larger-scale envelope of size Δx_{env} (cf. Fig. 9.1). The idea behind it is that if the size of the RPP elements is much smaller than the scale-length of the phase front aberrations, so that the phase through any given element is approximately constant, then the electric field after propagation through the RPP has a phase front dominated by the random phases from the RPP. The aberrations do not contribute anymore, as they simply add some random shifts to the RPP elements that already randomly break up the beam's coherence anyway (cf. Problem 9.1). The resulting focal spot will be the interference pattern between all the RPP elements, and unlike with random scale aberration, the statistical properties of its intensity pattern as well as the size of the speckles and the larger scale envelope are all well known and dictated by the RPP properties.

To describe the properties of the focal spot, let us first express the electric field near best focus, assuming vacuum propagation.[1] We follow the convention introduced in Sect. 1.1.2, whereby the electric field in the near-field flipped along x and along y corresponds to the Fourier transform of the field at best focus (the swapping $x \rightarrow -x$ and $y \rightarrow -y$ will not affect any of the results presented here). Furthermore, we will assume that the intensity profile at the lens is a "top-hat," i.e., nearly constant within a given envelope (typically a square or a disk); this is very typical in high-energy lasers, in an effort to optimize the use of the full surface of the optical elements up to the damage threshold in order to reduce the facility footprint. For example, if the beam profile is square, a typical electric field at the lens *without the RPP* is

$$\hat{E}(k_x, k_y, z, t) = \frac{1}{2}\hat{E}_0 \text{rect}\left(\frac{k_x}{\Delta k}\right)\text{rect}\left(\frac{k_y}{\Delta k}\right)e^{i\psi_0} + c.c., \tag{9.1}$$

where the rect function is defined as rect(x)= 1 if $|x| \leq 1/2$ and 0 otherwise, Δk is the size of the square beam in Fourier coordinates and $\psi_0 = k_0 z - \omega_0 t$, and the hat denotes quantities defined in the near-field. The Fourier coordinates can be simply related to the spatial coordinates in the near-field (i.e., at the lens); in particular, the full apertures of the entire beam and of a single RPP element in k-space, Δk and δk, are related to their physical sizes via

$$\Delta k = \frac{k_0}{\sqrt{f_\#^2 + 1/4}} \approx \frac{k_0}{f_\#}, \, f_\# \gg 1, \tag{9.2}$$

$$\delta k = \frac{2k_0}{\sqrt{1 + (2f/\delta X_{RPP})^2}} \approx k_0\frac{\delta X_{RPP}}{f}, \, f \gg d, \tag{9.3}$$

---

[1] Results for a uniform plasma are easily derived; however, because RPP typically produces large focal spots, it is quite common to have varying plasma density across an RPP focal spot. The resulting refraction can significantly impact the spatial frequencies of the light and therefore the speckle characteristics, as shown in Sect. 9.1.5.

where $f_\# = f/D$ is the beam's f-number with D the size of the beam at the lens and f the focal length, and δX_{RPP} is the size of an RPP element, as illustrated in Fig. 9.1.

Now with the RPP present, assuming that the RPP elements are also squares of size δk, the near-field electric field can be written as the sum of the fields from each RPP element:

$$
\hat{E}(k_x, k_y, z, t) = \frac{e^{i\psi_0}}{2} \sum_{j=1}^{N} \hat{E}_0 \text{rect}\left(\frac{k_x - k_{xj}}{\delta k}\right) \text{rect}\left(\frac{k_y - k_{yj}}{\delta k}\right) e^{i\phi_j} + c.c.,
$$

(9.4)

where (k_{xj}, k_{yj}) is the location of the center of the j-th RPP element and N is the number of RPP elements (we assume that all the RPP elements are illuminated by the laser, so that $N = \Delta k^2/\delta k^2$). Each element imparts a phase shift ϕ_j which is randomly sampled in $\{0, \pi\}$, corresponding to an element with a physical depth of 0 or $\lambda_0/2n$.

The field at best focus after the lens is obtained by performing an inverse Fourier transform; using $\mathcal{F}^{-1}[\text{rect}(k/\alpha)] = (\alpha/\sqrt{2\pi})\text{sinc}(\alpha x/2)$ with $\text{sinc}(x) = \sin(x)/x$, we obtain

$$
E(x, y, z, t) = \frac{e^{i\psi_0}}{2} \frac{\hat{E}_0 \delta k^2}{2\pi} \text{sinc}\left(\frac{\delta k x}{2}\right) \text{sinc}\left(\frac{\delta k y}{2}\right) \sum_{j=1}^{N} e^{i(\mathbf{k}_{\perp j} \cdot \mathbf{r} + \phi_j)} + c.c..
$$

(9.5)

At this point, it is instructive to derive an expression for the average intensity, where the average (as denoted by angular brackets in the following) is taken over many random realizations of the RPP pattern and represents a statistical ensemble-average (not to be confused with the same notation representing a time average in earlier chapters). This will also allow us to remove the term $\hat{E}_0$ in the expression above and replace it by a more meaningful quantity for describing the far-field. Denoting $\mathcal{E}$ the envelope of the field after separating the carrier oscillation at ψ_0, i.e.,

$$
E(\mathbf{r}, t) = \frac{1}{2}\mathcal{E}(\mathbf{r}, t)e^{i\psi_0} + c.c.,
$$

(9.6)

the average intensity is given by

$$
\langle I \rangle = \frac{c\varepsilon_0 \langle |\mathcal{E}|^2 \rangle}{2}
$$

$$
= \frac{c\varepsilon_0}{2} \frac{\hat{E}_0^2 \delta k^4}{(2\pi)^2} \text{sinc}^2\left(\frac{\delta k x}{2}\right) \text{sinc}^2\left(\frac{\delta k y}{2}\right) \left\langle \sum_{j=1}^{N} \sum_{l=1}^{N} e^{i(\mathbf{k}_{\perp j} - \mathbf{k}_{\perp l}) \cdot \mathbf{r} + i(\phi_j - \phi_l)} \right\rangle
$$

(9.7)

(the 1/2 in the definition in the first equality includes the time-average over the laser oscillation period).

Since the phases are random, we have for any RPP element j, $\langle \exp[i\phi_j] \rangle = 0$ (note that this is true whether the distribution is equiprobable within $\{0, \pi\}$ or uniformly distributed in $[-\pi, \pi]$, as we will see in Sect. 9.4), and therefore, for two RPP elements j and l, $\langle \exp[i(\phi_j - \phi_l)] \rangle = \delta_{jl}$, where δ_{jl} is the Kronecker delta. We obtain the following expression for $\langle I \rangle$, making use of $N = \Delta k^2 / \delta k^2$:

$$\langle I(x, y, z = 0, t) \rangle = \frac{c\varepsilon_0}{2} \frac{\hat{E}_0^2(t) \Delta k^2 \delta k^2}{(2\pi)^2} \text{sinc}^2\left(\frac{\delta k x}{2}\right) \text{sinc}^2\left(\frac{\delta k y}{2}\right), \quad (9.8)$$

where $z = 0$ represents the best focus position.

This result illustrates that the average intensity of the focal spot has a spatial profile that is the inverse Fourier transform of one individual RPP element. This is the large-scale envelope described earlier and illustrated in Fig. 9.1. A slowly varying time dependence (compared to ω_0) can be present as well. It is convenient to renormalize the electric field by introducing E_0 as the average field amplitude near the center of the focal spot envelope, for $x, y \ll 1/\delta k$ (such that the sinc^2 functions are both ≈ 1):

$$E_0 = \sqrt{\langle |\mathcal{E}|^2 (x = 0, y = 0) \rangle} = \frac{\hat{E}_0}{2\pi} \Delta k \delta k, \quad (9.9)$$

such that

$$\langle I \rangle = \frac{\varepsilon_0 c E_0^2}{2}. \quad (9.10)$$

This leads to our final expression for the RPP field at best focus, for a square intensity profile in the near-field:

$$\mathcal{E} = \frac{E_0}{\sqrt{N}} \text{sinc}\left(\frac{\delta k x}{2}\right) \text{sinc}\left(\frac{\delta k y}{2}\right) \sum_{j=1}^{N} e^{i(\mathbf{k}_{\perp j} \cdot \mathbf{r} + \phi_j)}$$

$$= \frac{E_0}{\sqrt{N}} \text{sinc}\left(\frac{\pi x}{\Delta x_{env}}\right) \text{sinc}\left(\frac{\pi y}{\Delta y_{env}}\right) \sum_{j=1}^{N} e^{i(\mathbf{k}_{\perp j} \cdot \mathbf{r} + \phi_j)}, \quad (9.11)$$

with $\delta k = k_0 \delta X_{\text{RPP}} / f$ per Eq. (9.3).

In this expression, the terms before the sum represent the large-scale envelope of the beam, of FWHM

$$\boxed{\Delta x_{env} = \frac{\lambda_0 f}{\delta X_{\text{RPP}}}} \quad (9.12)$$

(assuming square RPP elements and therefore $\Delta x_{env} = \Delta y_{env}$).

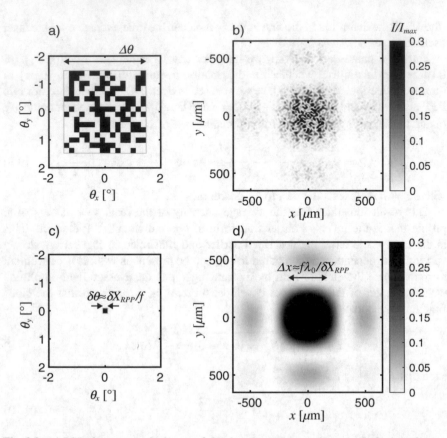

Fig. 9.2 (**a**) RPP phase pattern in the near-field; the square elements impart a dephasing of either 0 or π (black or white). The beam has a wavelength of 1.06 μm and an f-number of $f_\# = 20$ (i.e., $\Delta\theta \approx 3°$), and the RPP contains 16×16 elements. (**b**) Resulting focal spot intensity at best focus; the color scale is clamped at $0.3 I_{max}$ for better visibility. (**c**) and (**d**) represent the illumination of a single RPP element (i.e., a beam with $f_\# \approx 20 \times 16 = 320$) and the resulting focal spot, showing the characteristic diffraction pattern from a square aperture with a $sinc^2$ intensity profile, whose zeros along x or y are separated from the center by $\Delta x \approx \lambda_0 f / \delta X_{RPP}$

The sum in the expression of the field represents the random interference pattern of all the RPP elements, which is at the origin of the speckles. The term $1/\sqrt{N}$ informs us on the characteristic scaling of the electric field in the presence of speckles as a function of N (the number of RPP elements), as is usually done in the literature on statistical optics and speckles [1, 2]. It simply means that near the center of the focal spot, the characteristic electric field is $\sim E_0$, with the term in the sum (the sum of random phasors) scaling like $\sim \sqrt{N}$ as expected from a random walk process.

A typical RPP phase pattern and its corresponding focal spot profile are shown in Fig. 9.2a–b, for square RPP elements and a square beam intensity profile at the lens.

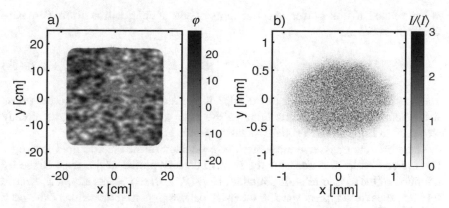

Fig. 9.3 Example of a CPP used on one of the NIF laser beams: (**a**) phase profile of the CPP in the near-field and (**b**) intensity profile at best focus, normalized to the average intensity $\langle I \rangle$ and clamped to $3\langle I \rangle$ for better visibility

Note that a significant amount of energy is contained in the side "lobes," which is generally an undesirable feature of RPPs for applications in ICF or HED. This issue was initially mitigated by using hexagonal shapes for the RPP instead of squares, which reduces the energy in the side lobes [5]. Nowadays, so-called kinoform or continuous phase plates (KPPs / CPPs) [6, 7] are being manufactured for high-power lasers, with smooth variations between phase elements instead of discrete, allowing to shape the focal spot to arbitrary profiles, the most commonly used being elliptical flat-top with minimal energy in the "wings" of the focal spot. This is illustrated in Fig. 9.3, which shows the phase profile of a CPP used on the NIF laser together with the resulting focal spot. The phase profile is continuous and not discrete like for an RPP and results in a focal spot with an elliptical shape (the optimum to enter through the circular opening of a hohlraum entrance hole at an angle) and a nearly flat-top intensity profile with virtually no wings (the speckles are too small to be resolved on the image).

9.1.2   Speckle Size

Let us now describe the characteristic size of the speckles. The autocorrelation function gives the typical electric field from a speckle; it is essentially a measure of the correlation of the field with itself as a function of space (and time, if we want to characterize the speckles' life time) [2, 3, 8]. The autocorrelation function for the electric field envelope $\mathcal{E}$ as defined above is given by

$$C_{\mathcal{E}\mathcal{E}^*}(x, y, z) = \iiint_{-\infty}^{\infty} \mathcal{E}(x + x', y + y', z + z')\mathcal{E}^*(x', y', z')dx'dy'dz' \quad (9.13)$$

and is related to the power spectral density via a Fourier transform (Wiener–Khinchin theorem),

$$C_{\mathcal{E}\mathcal{E}^*}(x, y, z) = \mathcal{F}^{-1}\left[|\hat{\mathcal{E}}(k_x, k_y, k_z)|^2\right]. \tag{9.14}$$

Since the only contribution of the RPP in the near-field is to change the phase of the field, its intensity profile remains the same as without the RPP, as already expressed in Eq. (9.1) for a square intensity profile at the lens.

Therefore, the transverse autocorrelation function of the electric field is simply the inverse Fourier transform of the beam intensity profile at the lens. Since by definition of the autocorrelation function, $C_{\mathcal{E}\mathcal{E}^*}(x, y, 0)$ represents a typical electric field in a speckle, a typical speckle intensity pattern (ignoring the value of the peak intensity, which is a random variable as we will discuss later—here we are only looking at the spatial scale) is simply obtained by taking the squared modulus of $C_{\mathcal{E}\mathcal{E}^*}$,

$$I_{sp}(x, y) \propto |C_{\mathcal{E}\mathcal{E}^*}(x, y, 0)|^2 = \left|\mathcal{F}^{-1}\left[|\mathcal{E}(k_x, k_y)|^2\right]\right|^2. \tag{9.15}$$

Let us give a couple of examples for the most common laser intensity profiles at the lens in high-power lasers:

- *Square intensity profile at the lens:* This is for example the case of the NIF laser at Lawrence Livermore National Laboratory[2] and also what we used in Fig. 9.2 and in the discussion above. With $|\hat{\mathcal{E}}|^2 = \hat{E}_0\mathrm{rect}(k_x/\Delta k)\mathrm{rect}(k_y/\Delta k)$, we obtain by the inverse Fourier transform

$$I_{sp,sq}(x, y) \propto \mathrm{sinc}^2\left(\frac{\Delta k x}{2}\right)\mathrm{sinc}^2\left(\frac{\Delta k y}{2}\right)$$

$$= \mathrm{sinc}^2\left(\frac{\pi x}{f_\# \lambda_0}\right)\mathrm{sinc}^2\left(\frac{\pi y}{f_\# \lambda_0}\right), \tag{9.16}$$

where we used again $\Delta k \approx k_0/f_\#$. The full width at half maximum (FWHM) of the $\mathrm{sinc}(x)^2$ function is ≈ 2.8, and therefore the speckle FWHM for a square beam profile at the lens is

$$\delta x = \delta y \approx \frac{2.8}{\pi} f_\# \lambda_0 \sim f_\# \lambda_0. \tag{9.17}$$

- *Circular ("flat-top") intensity profile at the lens:* This is the case of the Omega laser at the University of Rochester's Laboratory for Laser Energetics. Using the well-known formula for the Fourier transform of a disk (or the equally well-

---

[2] NIF beams are square to make them more easily stackable and reduce the facility footprint.

known diffraction pattern through a circular aperture [9]), we obtain the speckle transverse intensity profile:

$$I_{sp,circ}(x, y) = \left| \frac{J_1(\Delta k r/2)}{\Delta k r/2} \right|^2 \tag{9.18}$$

$$= \left| \frac{J_1(\pi r/f_\# \lambda_0)}{\pi r/f_\# \lambda_0} \right|^2, \tag{9.19}$$

where $r = [x^2 + y^2]^{1/2}$, $\Delta k \approx k_0/f_\#$ is the diameter of the beam at the lens in k-space and J_1 is the Bessel function of the first kind. The FWHM of $|J_1(x)/x|^2$ being ≈ 3.2, we obtain that the speckle FWHM for a circular beam profile at the lens is

$$\delta r \approx \frac{3.2}{\pi} f_\# \lambda_0 \sim f_\# \lambda_0. \tag{9.20}$$

As we can see, a square beam profile at the lens creates speckles of approximately the same size as those from a circular beam profile at the lens (with its diameter equal to the square side).

In summary, RPPs produce intensity patterns with two distinct spatial scales:

- A large-scale envelope of size $\Delta x_{env} \approx \lambda_0 f/\delta X_{RPP}$, whose intensity profile is approximately equal to the focal spot produced by a single RPP element
- Small-scale speckles of transverse size

$$\boxed{\delta r_{sp} \approx \lambda_0 f_\#}, \tag{9.21}$$

whose intensity profile is approximately equal to the focal spot of the full laser beam without RPP and without aberrations (i.e., assuming a perfectly flat phase front).

Note that the ratio of the focal spot envelope to the speckle size is approximately equal to the ratio of the size of the beam in the near-field to the size of an RPP element, $\Delta x_{env}/\delta x \approx D/\delta X_{RPP}$: therefore, the number of speckles in the focal spot is approximately equal to the number of RPP elements illuminated by the beam in the near-field.

The speckle length along the propagation direction z can simply be estimated by taking the Rayleigh length with transverse size $f_\# \lambda_0$, i.e., $\delta z_{sp} = \pi \delta x^2/\lambda_0$ or

$$\boxed{\delta z_{sp} \approx \pi f_\#^2 \lambda_0}. \tag{9.22}$$

A zoomed-in image of an RPP focal spot is shown in Fig. 9.4a, with the corresponding characteristic speckle intensity profile from Eq. (9.16) in Fig. 9.4b.

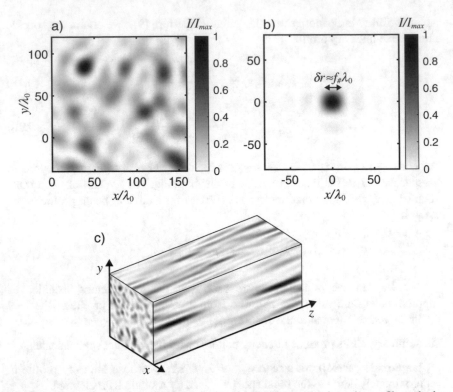

Fig. 9.4 (**a**) Focal spot of an RPP beam (zoomed in) with a square intensity profile at the lens and square RPP elements, with $f_\#=20$. (**b**) Characteristic intensity profile of a single speckle from Eq. (9.16) (same spatial scale as (**a**)), from the Fourier transform of the laser intensity profile at the lens with a planar phase front. (**c**) 3D surface plot of a speckle pattern in 3D

Here we used $f_\# = 20$ and a square laser beam intensity profile at the lens and an RPP with square elements. As we can see, the characteristic speckle intensity profile is indeed very representative of the different speckles seen in the actual RPP focal spot. A surface plot of the speckles in 3D for the same beam is shown in Fig. 9.4c.

Note that the results obtained here for the speckle size, obtained simply based on an intuitive understanding of the autocorrelation function, can be derived more rigorously by estimating the intensity profile near a local intensity maximum [8, 10]. The derivation is a lot more involved, but the results are the same.

9.1.3   Statistics of the Intensity Distribution

Let us now investigate the statistical properties of the intensity of the focal spot of an RPP beam. For simplicity, we will only consider the region near the center of the beam, $|x|, |y| \ll \lambda_0 f / \delta X_{\mathrm{RPP}}$ so that we do not have to carry the large-scale envelope and can simply express the envelope of the field $\mathcal{E}$ (where the full electric field is $E = \text{Re}[\mathcal{E}\cos(\psi_0)]$) from Eq. (9.11) as

$$\mathcal{E} = \frac{E_0}{\sqrt{N}} \sum_{j=1}^{N} e^{i(\mathbf{k}_{\perp j} \cdot \mathbf{r} + \phi_j)} . \tag{9.23}$$

The phases ϕ_j are independent and satisfy $\langle \exp[i\phi_j] \rangle = 0$ and $\langle \exp[i(\phi_j - \phi_l)] \rangle = \delta_{jl}$, where δ_{jl} is the Kronecker delta and the angular brackets represent a statistical average over many RPP realizations. This is true for RPP elements that are either uniformly distributed in $[-\pi, \pi]$ or sampled with equiprobability in $\{0, \pi\}$.

The quantity of interest to us is the intensity, $I \propto |\mathcal{E}|^2 = \mathcal{E}_R^2 + \mathcal{E}_I^2$, where $\mathcal{E}_R$ and $\mathcal{E}_I$ are the real and imaginary parts of $\mathcal{E}$,

$$\mathcal{E}_R = \frac{E_0}{\sqrt{N}} \sum_{j=1}^{N} \cos[\mathbf{k}_{\perp j} \cdot \mathbf{r} + \phi_j] , \tag{9.24}$$

$$\mathcal{E}_I = \frac{E_0}{\sqrt{N}} \sum_{j=1}^{N} \sin[\mathbf{k}_{\perp j} \cdot \mathbf{r} + \phi_j] . \tag{9.25}$$

The statistical properties of the intensity can be inferred from the properties of $\mathcal{E}_R$ and $\mathcal{E}_I$. First, given the phases distributions, we know that

$$\langle \mathcal{E}_R \rangle = \langle \mathcal{E}_I \rangle = 0 . \tag{9.26}$$

The variance, defined for a random variable X as $\sigma_X^2 = \langle X^2 \rangle - \langle X \rangle^2$, can easily be derived as well: for the real part, we have

$$\sigma_R^2 \equiv \langle \mathcal{E}_R^2 \rangle = \frac{E_0^2}{N} \sum_{j=1}^{N} \sum_{l=1}^{N} \langle \cos(\mathbf{k}_{\perp j} \cdot \mathbf{r} + \phi_j) \cos(\mathbf{k}_{\perp l} \cdot \mathbf{r} + \phi_l) \rangle . \tag{9.27}$$

Note that the terms $\mathbf{k}_{\perp j} \cdot \mathbf{r}$ do not intervene in this derivation, since they are just added to a phase that is already random anyway. They will be dropped in most of the following. Since the phases are independent, we have

$$\langle \cos(\phi_j) \cos(\phi_l) \rangle = \begin{cases} \langle \cos^2(\phi_j) \rangle = \frac{1}{2} , & j = l \\ \langle \cos(\phi_j) \rangle \langle \cos(\phi_l) \rangle = 0 , & j \neq l . \end{cases} \tag{9.28}$$

Likewise, we also have $\langle \sin(\phi_j) \sin(\phi_l) \rangle = \delta_{jl}/2$, and the variances for the real and imaginary components of the envelope are

$$\langle \mathcal{E}_R^2 \rangle = \langle \mathcal{E}_I^2 \rangle = \frac{E_0^2}{2} \equiv \sigma^2 . \tag{9.29}$$

We will also need the cross-correlation of $\mathcal{E}_R$ and $\mathcal{E}_I$, defined as $\langle \mathcal{E}_R \mathcal{E}_I \rangle$, i.e.,

$$\langle \mathcal{E}_R \mathcal{E}_I \rangle = \frac{E_0^2}{N} \sum_{j=1}^{N} \sum_{l=1}^{N} \langle \cos(\phi_j) \sin(\phi_l) \rangle = 0, \tag{9.30}$$

where we used the fact that $\langle \cos(\phi_j) \sin(\phi_l) \rangle = 0$ if $j \neq l$ and $\langle \sin(2\phi_j)/2 \rangle = 0$ if $j = l$. We conclude that $\mathcal{E}_R$ and $\mathcal{E}_I$ are uncorrelated.

We now invoke the central limit theorem (CLT) of probability, which states that the sum of a large number N of random variables sampled from a distribution of overall mean and variance μ and σ^2 becomes a normal (or Gaussian) distribution with mean and variance μ and σ^2 as $N \to \infty$, *regardless* of the initial distribution of the individual variables.

Therefore, the CLT tells us that for a large number of RPP elements, the variables $\mathcal{E}_R$ and $\mathcal{E}_I$ both have Gaussian distributions with mean 0 and variance $\sigma^2 = E_0^2/2$ and are uncorrelated.

Furthermore, we can use a theorem from the probability theory which states that if two random variables with Gaussian distributions are uncorrelated, then they are also independent (for example, cf. Ref. [11], Theorem 7.1[3]). Therefore, $\mathcal{E}_R$ and $\mathcal{E}_I$ are not only uncorrelated but also independent.

And finally, we can now use the definition of the chi-squared distribution with n degrees of freedom as the sum of the squares of n independent standard Gaussian (i.e., $\sigma^2 = 1$) random variables, with a probability density function (pdf)

$$p(x) = \frac{1}{2^{n/2}\Gamma(n/2)} x^{n/2-1} e^{-x/2}, \tag{9.31}$$

where Γ is the Gamma function. Since $\mathcal{E}_R/\sigma$ and $\mathcal{E}_I/\sigma$ follow standard Gaussian distributions, we obtain that $|\mathcal{E}|^2/\sigma^2$ follows a χ^2 distribution with two degrees of freedom ($n = 2$), i.e.,

$$p\left(\frac{|\mathcal{E}|^2}{\sigma^2}\right) = \frac{1}{2} \exp\left[-\frac{|\mathcal{E}|^2}{2\sigma^2}\right]. \tag{9.32}$$

The pdf for the intensity $I = |\mathcal{E}|^2 \varepsilon_0 c/2$ is obtained by renormalizing the expression above[4], leading to

---

[3] This can be easily demonstrated by taking the bivariate probability distribution function $f(x, y)$ of a pair of random variables (X, Y) and setting the correlation between the two variables to zero; one immediately obtains that $f(x, y) = f(x)f(y)$, proving the independence of X and Y.

[4] From the definition of the pdf $p(x)$ for a realization x of a random variable X, we know that $p(x)dx$ is the probability that $X \in [x, x + dx]$. Therefore, for any constant α, $p(\alpha x)d(\alpha x) = p(\alpha x)\alpha dx = P(\alpha X \in [\alpha x, \alpha x + \alpha dx]) = P(X \in [x, x + dx]) = p(x)dx$, and therefore, $p(\alpha x) = p(x)/\alpha$.

$$p(I) = \frac{1}{2\tilde{\sigma}^2} \exp\left[-\frac{I}{2\tilde{\sigma}^2}\right], \tag{9.33}$$

with $\tilde{\sigma}^2 = \sigma^2 \varepsilon_0 c/2 = E_0^2 \varepsilon_0 c/4$. The average intensity is then

$$\langle I \rangle = \int_0^\infty -p(I) I dI = 2\tilde{\sigma}^2 = \frac{\varepsilon_0 c E_0^2}{2}, \tag{9.34}$$

in accordance with our earlier derivation in Eq. (9.10). We can then rewrite the intensity probability distribution in the more convenient form:

$$\boxed{p(I) = \frac{1}{\langle I \rangle} e^{-I/\langle I \rangle}}. \tag{9.35}$$

The variance of the intensity is then

$$\sigma_I^2 = \langle I^2 \rangle - \langle I \rangle^2 = \int_0^\infty p(I) I^2 dI - \langle I \rangle^2 = \langle I \rangle^2, \tag{9.36}$$

and the contrast, defined as the ratio of variance to average for the intensity, is 100%,

$$C_I \equiv \frac{\sigma_I}{\langle I \rangle} = 1. \tag{9.37}$$

These are the important results of speckle statistics. Since by definition of the pdf, $p(I)dI$ is the probability that the intensity is in the small interval $[I, I + dI]$, we obtain the probability that the intensity exceeds some value I_t:

$$P(I \geq I_t) = \int_{I_t}^\infty p(I) dI = e^{-I_t/\langle I \rangle}. \tag{9.38}$$

If the number of RPP elements is large enough, then an average over many RPP realizations is equivalent to an average in space; for example, the formula above tells us that in an RPP field, about 2% of the focal spot area will contain intensities above four times the average intensity ($e^{-4} \approx 0.018$). It is not to be confused with the fraction of *power* above a certain intensity—or as is more often used, the fraction of power above n times the average intensity or FOPAI(n). The FOPAI is a more relevant metric than the fraction of surface above average and is given by

$$\boxed{\text{FOPAI}(n) = \frac{1}{\langle I \rangle} \int_{n\langle I \rangle}^\infty I p(I) dI = (1+n) e^{-n}}. \tag{9.39}$$

The fraction of power in the focal spot that is above four times $\langle I \rangle$ is now $(1+4)e^{-4} \approx 9\%$, as opposed to the $\sim 2\%$ for the fraction of *area* above $4\langle I \rangle$ mentioned above.

These results are independent of the details of the laser and RPP (beam's $f_{\#}$, spatial intensity profile at the lens, size of the RPP elements, RPP vs. CPP, etc.) and constitute standard results of random Gaussian fields theory that are also applicable to speckles encountered in imaging applications [2]. They remain valid as long as the number of RPP elements is large enough.

Another useful result that can be inferred is the typical value of the most intense speckle in a given focal spot. Since backscatter tends to get seeded in the most intense speckles and that the pdf of the intensity distribution theoretically extends to infinity but with an exponentially decaying probability, we would like to know what is the maximum intensity we can realistically expect.

A theoretical study of this problem is presented in Ref. [12]; here we merely give a simple justification for the main result, which is that the maximum intensity is on the order of $\langle I \rangle \ln(N)$, where N is the number of speckles in the focal spot (or the number of RPP elements illuminated by the laser in the near-field).

We know from Eq. (9.38) that the fraction of surface in the spot where the intensity exceeds a value I_t is $\exp[-I_t/\langle I \rangle]$. Said differently, for a surface S within the focal spot, the surface where $I \geq I_t$ is $s_t = S \exp[-I_t/\langle I \rangle]$. For very large values of I_t that surface becomes infinitely small; since the smallest surface of relevance is that of a single speckle, $s_{t,min} = s_{sp} \approx \pi(\lambda_0 f_{\#})^2$, the maximum speckle intensity $I_{sp,max}$ should satisfy $s_{sp} = S \exp[-I_{sp,max}/\langle I \rangle]$. Taking S as the focal spot area, with $S/s_{sp} \approx N$, the number of RPP elements (or the number of speckles within the focal spot), we immediately obtain

$$\boxed{I_{sp,max} \approx \langle I \rangle \ln(N)}. \tag{9.40}$$

This result can apply for a limited surface within the focal spot as well, with $I_{sp,max} = \langle I \rangle \ln(S/s_{sp})$ the typical maximum intensity within the surface S, or to the 3D focal volume, where $I_{sp,max} = \ln(V_{foc}/v_{sp})$ is the typical maximum intensity in the focal spot volume V_{foc} with $v_{sp} \approx \pi^2 \lambda_0^3 f_{\#}^4$ the speckles volume. Ref. [12] also provides the pdf of the maximum intensity, from which we can infer the probability of a rare event where the most intense speckle is larger than $I_{sp,max}$.

Note that this method cannot be applied to count the number of speckles in a given area or volume because for intensities below $I_{sp,max}$, i.e., fractional areas larger than a speckle, we cannot easily know the contribution from multiple speckles at various intensities above I_t. We must then use more advanced statistical methods for characterizing local maxima, which we discuss in the next section.

9.1.4 Statistics of Speckles as Local Maxima of Intensity

So far we have evaluated the fraction of surface and the fraction of power above a certain intensity in an RPP field. For some applications, it will be useful to quantify the actual *number of speckles* above a given intensity. The analysis involves more sophisticated statistical methods on the study of local maxima of random fields. The problem was first investigated by Rose and Dubois in Ref. [10] and later refined by Garnier [8]. In the following, we will re-derive the expression from Garnier for a 2D transverse geometry, i.e., counting speckles in a 2D (x, y) surface at a fixed location along z, the laser propagation direction (close enough to best focus to ignore envelope diffraction effects). The extension to 3D (i.e., counting speckles in a 3D volume) is more mathematically involved but follows the same logic, and we will simply mention the key results at the end of the section.

The starting point is Adler's result from the statistical theory of random fields [13], which considers a vector X of dimension d of *real* independent random variables X_j, j=1:d, such that each variable X_j follows a normal (i.e., Gaussian) statistical distribution with mean 0 and variance σ^2. Defining M_I^X as the number of local maxima of X^2 in a given surface S whose value is $\geq I$, Adler's theorem (slightly expanded in Refs. [8, 14]) states that the mean value $\langle M_I^X \rangle$ is given by

$$\langle M_I^X \rangle = \frac{2S\sqrt{|\det \Lambda|}}{(2\pi)^{3/2}\sigma^2} \left(\frac{I}{\sigma^2}\right)^{1/2} e^{-I/2\sigma^2}. \tag{9.41}$$

Here Λ is a 2D matrix whose elements Λ_{ij} for $j = 1, 2$ are given by

$$\Lambda_{ij} = \left\langle \frac{\partial X}{\partial x_i}(\mathbf{x}) \frac{\partial X}{\partial x_j}(\mathbf{x}) \right\rangle = -\left. \frac{\partial^2 C_{X^2}}{\partial x_i \partial x_j} \right|_{\mathbf{r}=0}, \tag{9.42}$$

where $x_1 = x$ and $x_2 = y$, and $C_{X^2} = \langle X(\mathbf{r})X(0) \rangle$ (we will come back to this later).

These results are only applicable to real variables. In our case, the electric field is complex, but the quantity of interest is the intensity, $I \propto |\mathcal{E}|^2 = \mathcal{E}_R^2 + \mathcal{E}_I^2$, where $\mathcal{E}_R$ and $\mathcal{E}_I$ are the real and imaginary parts of the electric field envelope $\mathcal{E}$ as defined earlier. We already saw that $\mathcal{E}_R$ and $\mathcal{E}_I$ are independent and each follows a normal distribution with mean 0 and variance $\sigma^2 = \langle I \rangle /2$. The expressions above on the number of local maxima are therefore applicable to $\mathcal{E}_R$ and $\mathcal{E}_I$ separately; however, finding the number of local maxima of $\mathcal{E}_R^2 + \mathcal{E}_I^2$ requires some subtlety.

To help guide the argument, we show in Fig. 9.5 a 2D histogram of local maxima from a simulated RPP field with $N = 1024^2 \approx 10^6$ speckles. The number of counts on the histogram is the number of local maxima in the corresponding 2D bin whose intensity $|\mathcal{E}|^2$ is greater than $5\langle I \rangle$. Each bin is localized by its position in $(\mathcal{E}_R^2, \mathcal{E}_I^2)$ space. Our goal is to estimate the *average* total number of counts on that histogram, i.e., the total number of speckles whose intensity exceeds $5\langle I \rangle$ (or more generally, $n\langle I \rangle$).

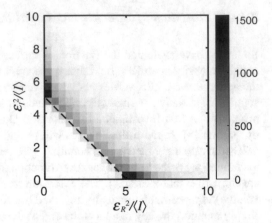

Fig. 9.5 Histogram of local intensity maxima in an $N = 10^6$ speckles RPP field whose intensity is above $5\langle I \rangle$. The dashed line represents $\mathcal{E}_R^2 + \mathcal{E}_I^2 = 5\langle I \rangle$

This figure shows that the concentration of speckles is highest in two places: along the $\mathcal{E}_R^2$ axis (near the cutoff at $5\langle I \rangle$) and along the $\mathcal{E}_I^2$ axis. This is intuitively clear: since $\mathcal{E}_R$ and $\mathcal{E}_I$ are statistically independent, a local maximum of $\mathcal{E}_R^2$ has no special property regarding $\mathcal{E}_I^2$ (and vice versa): therefore, a local intensity maximum is most likely due to a local maximum of $\mathcal{E}_R^2$, *or* $\mathcal{E}_I^2$, but it is unlikely that both are at (or near) a local maximum at the same spatial location.

Based on that idea, the first derivation by Rose & Dubois consisted in counting the local maxima of $\mathcal{E}_R^2$ (neglecting the comparatively small contributions of $\mathcal{E}_I^2$ at these locations) and the local maxima of $\mathcal{E}_I^2$ (ignoring $\mathcal{E}_R^2$ at these locations)—and summing up the two. Mathematically, this translates to

$$\left\langle M_I^{|\mathcal{E}|} \right\rangle \approx \left\langle M_I^{\mathcal{E}_R} \right\rangle + \left\langle M_I^{\mathcal{E}_I} \right\rangle, \tag{9.43}$$

i.e., the average number of speckles whose intensity $|\mathcal{E}|^2 \geq I$ is the sum of the speckles where $\mathcal{E}_R^2 \geq I$ and those where $\mathcal{E}_I^2 \geq I$.

This is visually represented in Fig. 9.6a: speckles are only counted in the gray areas, corresponding to $\mathcal{E}_R^2 \geq I$ and $\mathcal{E}_I^2 \geq I$. We notice a couple of issues:

- Speckles with both $\mathcal{E}_R^2 \geq I$ and $\mathcal{E}_I^2 \geq I$ (zone "4" in the figure) are double-counted. This is not a real concern, since these are rare events where both the real and imaginary parts (which are independent) happen to have a local maximum at the same location in space. Our previous example in Fig. 9.5 confirms this, with virtually no speckles in that area.
- Speckles with $\mathcal{E}_R^2 \leq I$ and $\mathcal{E}_I^2 \leq I$ but $\mathcal{E}_R^2 + \mathcal{E}_I^2 \geq I$ (zone 3 on Fig. 9.6a) are not counted. These are the speckles whose real component $\mathcal{E}_R^2$ is not quite above I but whose imaginary part $\mathcal{E}_I^2$ is high enough to bring the intensity $\mathcal{E}_R^2 + \mathcal{E}_I^2$ above I. This is more significant, as can be seen from our example in Fig. 9.5, showing a fair number of speckles in that area (mostly concentrated along the dashed line $\mathcal{E}_R^2 + \mathcal{E}_I^2 = I$).

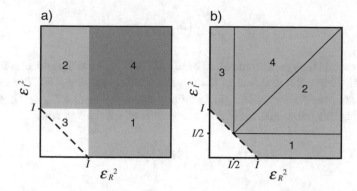

Fig. 9.6 Visualization of the approximations used to compute the number of local intensity maxima in Rose and Dubois [10] (**a**) and in Garnier [8] (**b**). Speckles are only counted in the gray areas in both cases and double-counted in the dark gray area marked "4" in (**a**)

The correction brought by Garnier consists in counting the small, but not negligible, contribution of $\mathcal{E}_I^2$ to the local maxima of $\mathcal{E}_R^2$ (and vice versa). As we already mentioned, at a local maximum of $\mathcal{E}_R^2$, the (independent) imaginary part $\mathcal{E}_I^2$ "does not know" that $\mathcal{E}_R^2$ is maximum, so that small contribution is simply represented by the distribution function of $\mathcal{E}_I^2$. From the definition of the chi-squared distribution mentioned earlier in Eq. (9.31), we know that since $\mathcal{E}_R$ is normally distributed, $\mathcal{E}_R^2$ follows a chi-squared distribution with one degree of freedom,[5]

$$p(\mathcal{E}_I^2) = \frac{1}{\sqrt{2\pi}\sigma\mathcal{E}_I} \exp\left[-\frac{\mathcal{E}_I^2}{2\sigma^2}\right].$$ (9.44)

Next we note that $M_{I-\mathcal{E}_I^2}^{\mathcal{E}_R}$ is the number of local maxima of $\mathcal{E}_R^2$ such that $\mathcal{E}_R^2 + \mathcal{E}_I^2 \geq I$: by integrating over the distribution of $\mathcal{E}_I^2$, we can then obtain the local maxima of $\mathcal{E}_R^2$ while accounting for the contribution of $\mathcal{E}_I$, i.e.,

$$\left\langle M_I^{|\mathcal{E}|}\right\rangle_R = \int_0^{I_m} \left\langle M_{I-\mathcal{E}_I^2}^{\mathcal{E}_R}\right\rangle p(\mathcal{E}_I^2) d\mathcal{E}_I^2,$$ (9.45)

where the subscript R indicates that we are counting the local maxima of $\mathcal{E}_R^2$ (and correcting by accounting for $\mathcal{E}_I$). If we integrate for values of $\mathcal{E}_I^2$ up to $I_m \to \infty$, we are obviously going to interfere with the counting of local maxima of $\mathcal{E}_I^2$. To prevent double-counting, we stop the integration at $I_m = u/2$: this is the area labelled "1" in Fig. 9.6b. The average number of speckles in zone 1 is thus

---

[5] $\mathcal{E}_I/\sigma$ follows a normal distribution of mean 0 and variance 1, and therefore the pdf $p(\mathcal{E}_I^2/\sigma^2)$ is given by Eq. (9.31) and $p(\mathcal{E}_I^2) = p(\mathcal{E}_I^2/\sigma^2)/\sigma^2$, per footnote 4.

$$\left\langle M_I^{|\mathcal{E}|} \right\rangle_1 = \int_0^{I/2} \left\langle M_{I-\mathcal{E}_I^2}^{\mathcal{E}_R} \right\rangle p(\mathcal{E}_I^2) d\mathcal{E}_I^2 . \tag{9.46}$$

Next we add the speckles in zone 2 of Fig. 9.6b.[6] These are the speckles for whom $\mathcal{E}_R^2 \geq \mathcal{E}_I^2$ (below the diagonal in the figure), with $\mathcal{E}_I^2$ varying from $I/2$ (since we stopped the previous count at $I/2$) to infinity (since we ensure $\mathcal{E}_R^2 \geq \mathcal{E}_I^2$ anyway). Translating this into math, we obtain the number of speckles in zone 2:

$$\left\langle M_I^{|\mathcal{E}|} \right\rangle_2 = \int_{I/2}^{\infty} \left\langle M_{\mathcal{E}_I^2}^{\mathcal{E}_R} \right\rangle p(\mathcal{E}_I^2) d\mathcal{E}_I^2 . \tag{9.47}$$

By symmetry, we see that zones 1 and 3 in Fig. 9.6b contain an equal number of speckles and so do zones 2 and 4. The total number of speckles whose intensity is above I is finally estimated as

$$\left\langle M_I^{|\mathcal{E}|} \right\rangle = 2 \int_0^{I/2} \left\langle M_{I-\mathcal{E}_I^2}^{\mathcal{E}_R} \right\rangle p(\mathcal{E}_I^2) d\mathcal{E}_I^2 + 2 \int_{I/2}^{\infty} \left\langle M_{\mathcal{E}_I^2}^{\mathcal{E}_R} \right\rangle p(\mathcal{E}_I^2) d\mathcal{E}_I^2 . \tag{9.48}$$

Compared to Rose and Dubois [10], this method captures the speckles in zone 3 of 9.6a. Simulations have shown that it improves the predictions on the number of speckles, especially at high intensities ($\geq 5 \langle I \rangle$).

Proceeding with the calculation, using the expressions for $p(\mathcal{E}_I^2)$ and $\left\langle M_I^{\mathcal{E}_R} \right\rangle$ from Eqs. (9.44) and (9.41), we finally obtain

$$\left\langle M_I^{|\mathcal{E}|} \right\rangle = \frac{4S\sqrt{\det\Lambda}}{\pi^2 \langle I \rangle} \left[\frac{I}{\langle I \rangle} \left(\frac{1}{2} + \frac{\pi}{4} \right) + \frac{1}{2} \right] e^{-I/\langle I \rangle} , \tag{9.49}$$

where we used the integration formula

$$\int \sqrt{\frac{a}{x} - 1} \, dx = x\sqrt{\frac{a}{x} - 1} - a \arctan\sqrt{\frac{a}{x} - 1} + K \tag{9.50}$$

with K an integration constant.

We should keep in mind that these expressions are only valid for intensities sufficiently above $\langle I \rangle$ and are incorrect for $I < \langle I \rangle$ (the initial expression from Eq. (9.41) is only valid up to a term that is small compared to $e^{-I/2\sigma^2}$). Practically, very good agreement is obtained for $I \geq \langle I \rangle$, which is more than enough since we will only be interested in speckles significantly above the average intensity for laser–plasma instabilities anyway.

---

[6] This step is not explicitly mentioned in Ref. [8]; however, it is required in order to arrive at the correct final result.

We must still obtain expressions for Λ, which as we can see from our derivation so far applies to either $\mathcal{E}_R$ or $\mathcal{E}_I$. From Eq. (9.42), using the Wiener–Khinchin theorem, we have

$$C_{\mathcal{E}_R^2} = \mathcal{F}^{-1} \left[|\hat{\mathcal{E}}_R(\mathbf{k}_\perp)|^2 \right] . \tag{9.51}$$

Assuming a square beam envelope in the near-field, as we did earlier, gives after the inverse Fourier transform:

$$C_{\mathcal{E}_R^2} = \langle \mathcal{E}_R^2 \rangle \, \text{sinc}\left(\frac{\pi x}{\rho_{sp}} \right) \text{sinc}\left(\frac{\pi y}{\rho_{sp}} \right) , \tag{9.52}$$

where $\langle \mathcal{E}_R^2 \rangle = \langle I \rangle /2$ and $\rho_{sp} = f_\# \lambda_0$ (cf. Sects. 9.1.2, 9.1.3).
  Since

$$\Lambda = - \begin{pmatrix} \partial_x^2 & \partial_x \partial_y \\ \partial_x \partial_y & \partial_y^2 \end{pmatrix} C_{\mathcal{E}_R^2} , \tag{9.53}$$

using the relations

$$\partial_x \text{sinc}\left(\frac{\pi x}{\rho_{sp}} \right) \bigg|_{x=0} = 0 , \tag{9.54}$$

$$\partial_x^2 \text{sinc}\left(\frac{\pi x}{\rho_{sp}} \right) \bigg|_{x=0} = -\frac{1}{3} \left(\frac{\pi}{\rho_{sp}} \right)^2 \tag{9.55}$$

leads to the expression for the matrix determinant,

$$\sqrt{\det \Lambda} = \frac{\pi^2 \langle I \rangle}{6 \rho_{sp}^2} . \tag{9.56}$$

We finally obtain our final result giving the number of speckles whose intensity is above I in a surface S at or near the focal plane:

$$\boxed{ \left\langle M_I^{|\mathcal{E}|} \right\rangle_{2D} = \frac{2S}{3 \rho_{sp}^2} \left[\frac{I}{\langle I \rangle} \left(\frac{1}{2} + \frac{\pi}{4} \right) + \frac{1}{2} \right] e^{-I/\langle I \rangle} . } \tag{9.57}$$

As mentioned at the beginning of this section, the 3D calculation is significantly more difficult; the main reason for that is that the propagation along z introduces a dependence between $\mathcal{E}_R$ and $\mathcal{E}_I$: the two variables are still uncorrelated, but they are not independent anymore like for the 2D case [8]. We will simply quote the main result obtained by Garnier for the specific case of an RPP with a square intensity profile in the near-field, giving the number of speckles with intensity above I in a volume V:

$$\left\langle M_I^{|\mathcal{E}|} \right\rangle_{3D} = \frac{\pi^{3/2}\sqrt{5}V}{27\rho_{sp}^2 z_{sp}} \left[\left(\frac{I}{\langle I \rangle} \right)^{3/2} - \frac{3}{10} \left(\frac{I}{\langle I \rangle} \right)^{1/2} \right] e^{-I/\langle I \rangle}, \qquad (9.58)$$

where $z_{sp} = \pi \lambda_0 f_\#^2$ is the speckle length as defined in Sect. 9.1.2.

9.1.5   Impact of Refraction on the Speckle Pattern

It will be important to keep in mind that the results presented so far were all derived for propagation in vacuum; in laboratory experiments, refraction effects in density gradients can significantly impact the speckle pattern. To illustrate this, we show results from a 3D paraxial propagation simulation of an RPP beam with a square intensity envelope in the near-field in a density gradient. The geometry is shown in Fig. 9.7: the beam comes from the left in the figure and refracts off a 1D density gradient along y, of the form $n_e(y)/n_c = (y/L_n)^2$ for $y > 0$ and 0 for $y \le 0$, with $L_n = 5000\lambda_0$.

The general behavior corresponds to what we described in Ch. 3; we notice that an average turning point for the beam is reached near $z/\lambda_0 \approx 7 \times 10^4$. There, the average intensity increases as the speckles have a "caustic" (physically, the beam overlaps with itself due to refraction).

Figure 9.8a shows the transverse Fourier spectrum of the beam at the end (right-end side) of the simulation box; the dashed square represents the power spectrum at the entrance of the box (with it is shifted because the beam is propagating at an angle with respect to the simulation box z-axis). Because of refraction, the power spectrum

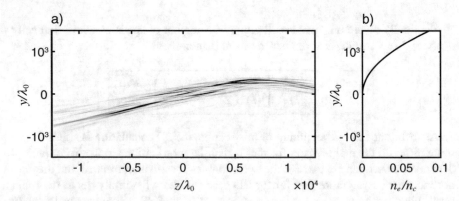

Fig. 9.7 (a) 3D simulation of the refraction of an RPP beam, showing an (y, z) slice of the intensity at $x = 0$. (b) Electron density profile used for the simulation (1D, along y only)

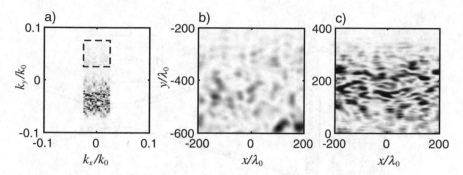

Fig. 9.8 (**a**) Fourier spectrum (for the transverse spatial coordinates x and y) of the beam at the exit of the simulation box, $z/\lambda_0 \approx 1.25 \times 10^4$. The dashed square marks the location of the power spectrum at the entrance of the simulation box (square beam in the near-field, offset from the axis in order to propagate at an angle). (**b**) and (**c**) show zoom-ins of the transverse intensity profile at the entrance (**b**) and exit (**c**) of the box (with the same color scale); (**c**) clearly shows an increase in average intensity as well as narrower speckles along y

(in spatial frequency) gets "smeared" along k_y; most of the energy ultimately shifts to $k_y < 0$, approximately at the opposite from the initial Fourier location of the lens with respect to $k_y = 0$ (which means that the light has been "reflected" off the density gradient).

The smear along k_y is equivalent to having a wider beam at the lens along that direction, which per our earlier discussions should lead to *narrower* speckles along y. This is indeed the case, as we can see in Fig. 9.8c, which shows a portion of the beam in the transverse directions at the end of the simulation box: the speckles are visibly "squeezed" along y, compared to their initial aspect at the entrance of the box as shown in Fig. 9.8b. The average intensity increase is also very noticeable (Subfigures b and c are shown on the same color scale).

9.2 Polarization Smoothing (PS)

9.2.1 General Concept

As we saw in the previous section, speckles are due to the interference of the elementary diffraction patterns from the small RPP elements in the near-field. The conundrum is that speckles are intrinsic to the coherent nature of lasers. The only way to reduce the contrast of speckles is to add different patterns *in intensity*, as opposed to adding electric fields (which will still result in interferences and speckles).

One way to do so is to superpose two independent speckle patterns that have orthogonal polarizations, so that their electric fields cannot interfere [15]; this is also known as "polarization diversity" in statistical optics [2]. This is illustrated in

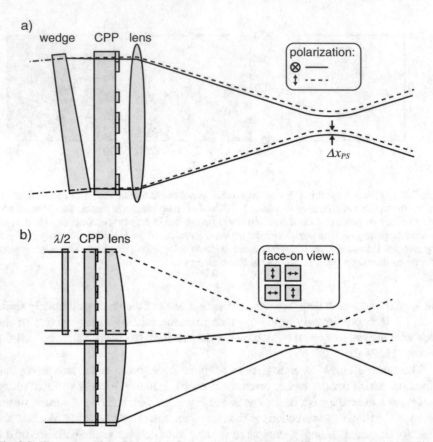

Fig. 9.9 Illustration of the PS concept from two arrangements found at the Omega and NIF laser facilities: (**a**) (Omega) a birefringent wedge is introduced in the path of the beam, splitting it into two orthogonal polarization propagating at a small angle from each other. The resulting focal spot is the incoherent sum of the same speckle intensity pattern shifted in space by a small amount Δx_{PS}. (**b**) (NIF) beams are grouped by four in "quads"; all the beams initially have the same polarization, but two beams out of the four pass through a half-waveplate ($\lambda/2$ in the figure) before the lens, rotating their polarization by 90°. The four beams are focused on the same location; the resulting focal spot is the incoherent sum (in intensity) of the two orthogonal polarization components

Fig. 9.9, for the two arrangements found at the Omega laser facility [16] and the NIF [17, 18].

In the first case (Omega laser, Fig. 9.9a), a wedge of birefringent material (typically KDP) is introduced in the path of the beam. The crystal is oriented in such a way that the direction of (linear) polarization of the laser electric field bisects the ordinary and extraordinary axes of the crystal, so that the light emerging out of the crystal is split between two orthogonal polarizations, propagating at a slight angle $\delta\theta$ from each other. The angular separation is related to the wedge angle ψ_W and its birefringence Δn via $\delta\theta = \psi_W \Delta n$. The two orthogonal (i.e., non-interacting) polarizations produce two focal spots shifted from each other by a small amount $\Delta x_{PS} \approx f_\# \delta\theta$, where $f_\#$ is the beam's f-number.

As a result, the focal spot will consist in two identical speckle patterns shifted by Δx_{PS}. If Δx_{PS} is larger than a speckle size δr_{sp}, i.e., the transverse correlation length, then the two patterns are uncorrelated. On the other hand, we also want to avoid too much spread for the overall focal spot from both polarizations, to avoid impacting the size of the focal spot designed by the CPP (as well as the loss of PS in the wings of the spot, where the two spots do not overlap), so ideally,

$$\delta r_{sp} < \Delta x_{PS} \ll \Delta x_{env}, \tag{9.59}$$

where Δx_{env} is the size of the large-scale focal spot envelope as defined previously.

Another approach is shown in Fig. 9.9b, as used on the NIF laser. There, square beams are grouped by four in quadruplets or "quads"; each beam within a quad is focused by a wedged focusing lens, which is essentially similar to a large lens split into four, so that the four beams' focal spots exactly overlap. Each beam is equipped with a CPP that is uncorrelated with the others'; all beams are initially linearly polarized in the same direction, but two of the four beams go through a half-waveplate which rotates their polarization by 90° as illustrated in the figure. As a result, the focal spot, which is the sum of the four beams' contributions, is the superposition of the two uncorrelated intensity patterns from the two orthogonal polarization directions. The polarization arrangement can either be a "checkerboard" as illustrated in the inset in Fig. 9.9b, i.e., $\begin{pmatrix} \updownarrow & \leftrightarrow \\ \leftrightarrow & \updownarrow \end{pmatrix}$, or an "up-down" configuration, $\begin{pmatrix} \leftrightarrow & \leftrightarrow \\ \updownarrow & \updownarrow \end{pmatrix}$. NIF has both arrangements present on different cones of beam.

9.2.2 Intensity Statistics and Reduction of the Speckle Contrast Using PS

To evaluate the statistics of the speckle intensity distribution with PS, we consider the laser electric field with components along both orthogonal transverse directions x and y:

$$\mathbf{E} = \frac{E_0}{\sqrt{2}} \frac{e^{i\psi_0}}{2} \mathbf{e}_x \sum_{j=1}^{N} e^{i(\mathbf{k}_j \cdot \mathbf{r} + \phi_j)} + \frac{E_0}{\sqrt{2}} \frac{e^{i\psi_0}}{2} \mathbf{e}_y \sum_{j=1}^{N} e^{i(\mathbf{k}_j \cdot \mathbf{r} + \varphi_j)} + c.c. \tag{9.60}$$

$$= \mathcal{E}_x \frac{e^{i\psi_0}}{2} \mathbf{e}_x + \mathcal{E}_y \frac{e^{i\psi_0}}{2} \mathbf{e}_y + c.c., \tag{9.61}$$

where the two components are distributed following independent RPP patterns. The normalization to $1/\sqrt{2}$ for E_0 is to account for the splitting into two polarizations. We proceed like in the previous section on RPPs with now four random variables instead of two: $\mathcal{E}_{xR}$ and $\mathcal{E}_{xI}$ are the real and imaginary parts of $\mathcal{E}_x$, and $\mathcal{E}_{yR}$ and $\mathcal{E}_{yI}$ are the real and imaginary parts of $\mathcal{E}_y$. The total intensity is

$$I = \frac{\varepsilon_0 c}{2} \left(|\mathcal{E}_{xR}|^2 + |\mathcal{E}_{xI}|^2 + |\mathcal{E}_{yR}|^2 + |\mathcal{E}_{yI}|^2 \right) . \tag{9.62}$$

Following the same arguments as before, we know that $|\mathcal{E}_{xR}|^2$, $|\mathcal{E}_{xI}|^2$, $|\mathcal{E}_{yR}|^2$ and $|\mathcal{E}_{yI}|^2$ are all following Gaussian distributions of mean 0 and variance

$$\sigma^2 = \frac{E_0^2}{4} . \tag{9.63}$$

Therefore, $(|\mathcal{E}_{xR}|^2 + |\mathcal{E}_{xI}|^2 + |\mathcal{E}_{yR}|^2 + |\mathcal{E}_{yI}|^2)/\sigma^2 = 2I/(\varepsilon_0 c \sigma^2)$ follows a chi-squared distribution with $n = 4$ degrees of freedom, with a probability density function (pdf) given by Eq. (9.31). Normalizing to obtain the pdf of I instead of $2I/(\varepsilon_0 c \sigma^2)$, we finally obtain

$$\boxed{ p(I) = \frac{4I}{\langle I \rangle^2} e^{-2I/\langle I \rangle} }, \tag{9.64}$$

with $\langle I \rangle = \varepsilon_0 c E_0^2/2$, and we can again easily verify that $\langle I \rangle = \int_0^\infty I p(I) dI$, as expected.

The variance $\sigma_I^2 = \langle I^2 \rangle - \langle I \rangle^2$ can easily be estimated as well, leading to

$$\sigma_I^2 = \frac{\langle I \rangle^2}{2} , \tag{9.65}$$

giving for the contrast:

$$C_I = \frac{\sigma_I}{\langle I \rangle} = \frac{1}{\sqrt{2}} . \tag{9.66}$$

This shows that the contrast of the speckle intensity pattern produced via PS is reduced by a factor $\sqrt{2}$ compared to a single linear polarization, where it was 100%, Eq. (9.37). The impact on the distribution of high-intensity speckles is significant as well: the fraction of power above n times the average intensity or FOPAI(n), defined previously in Eq. (9.39), now takes the form

$$\boxed{ \text{FOPAI}(n) = \frac{1}{\langle I \rangle} \int_{n\langle I \rangle}^\infty I p(I) dI = (2n^2 + 2n + 1)e^{-2n} }. \tag{9.67}$$

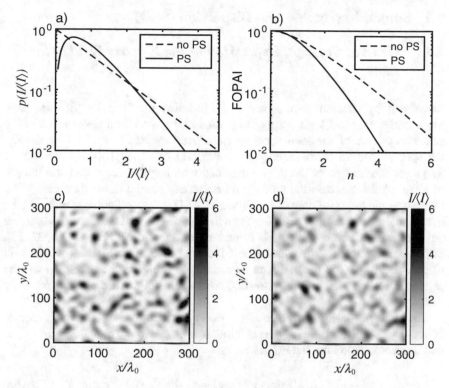

Fig. 9.10 (a) Probability density function of intensity (normalized to the average intensity $\langle I \rangle$), and (**b**) fraction of power above n times the average intensity, FOPAI(n), with and without PS, and speckle intensity pattern generated without (**c**) and with (**d**) PS

We saw earlier that for an RPP beam without PS, the fraction of power above $4\langle I \rangle$, FOPAI(4), is equal to $\approx$9% per Eq. (9.39). When PS is introduced, this fraction drops to 1.4%.

The distribution function $p(I)$ and the FOPAI with and without PS are shown in Fig. 9.10a–b. These clearly show the strong drop in the fraction of intense speckles resulting from the use of PS, with the difference from a regular RPP spot becoming more prominent at higher speckle intensities. Figure 9.10c–d shows a speckle pattern without and with PS showing the reduction in contrast when PS is introduced.

The contrast reduction from PS helps mitigate the imprint of small-scale perturbations from the speckles, which can otherwise seed hydrodynamic instabilities [16, 19, 20]. It also helps mitigate laser–plasma instabilities and provides a significant intensity margin increase to stay below threshold, as will be discussed in Sect. 9.5.

9.3 Smoothing by Spectral Dispersion (SSD)

9.3.1 General Concept, Expression of the Electric Field at the Lens

Smoothing by spectral dispersion (SSD), invented by Skupsky [21] and now implemented on most high-energy lasers around the world, mitigates some of the nefarious effects of the laser speckles introduced by RPPs or CPPs by making the speckle pattern move rapidly as a function of time. This is achieved by frequency modulation of the laser, combined with angular dispersion and the use of phase plates. Because the modulation frequency can be faster than the typical hydrodynamic response time of an ICF target, SSD is very effective at reducing the hydrodynamic instabilities resulting from the imprint of the laser speckles onto the target surface, which is particularly important for direct-drive ICF [19, 20, 22, 23]. SSD is also effective at reducing self-focusing and filamentation of lasers in plasmas, and in some cases, it can help mitigate backscatter as well, as will be explained in Sect. 9.5. In the following, we closely follow the original paper by Skupsky.

The concept is summarized in Fig. 9.11. A monochromatic laser with electric field $\hat{E} = \hat{E}_0 \cos(\mathbf{k}_0 \cdot \mathbf{r} - \omega_0 t)$ passes through an electro-optic modulator (EOM) so that its electric field after the EOM becomes

$$\hat{E} = \frac{1}{2}\hat{E}_0 \exp\left[i\left(\mathbf{k}_0 \cdot \mathbf{r} - \omega_0 t - \delta_m \sin(\omega_m t)\right)\right] + c.c., \tag{9.68}$$

where δ_m and $\omega_m = 2\pi \nu_m$ are the EOM modulation depth and frequency, respectively. The hat notation refers to the electric field before and up to the focusing lens, as opposed to the field in the plasma (near best focus).

The instantaneous frequency $\omega(t)$ after the EOM is the time derivative of the phase, $\omega(t) = -\partial\psi/\partial t$, where ψ is the term in the outer parenthesis in Eq. (9.68) (cf. Sect. 3.4), and becomes sinusoidally modulated at the frequency ω_m and with a peak-to-peak amplitude $\Delta\omega = 2\delta_m\omega_m$ (cf. Fig. 9.12a):

$$\omega(t) = \omega_0 + \omega_m\delta_m \cos(\omega_m t). \tag{9.69}$$

Alternatively, using the Jacobi–Anger expansion,

$$e^{ix\sin(\theta)} = \sum_{n=-\infty}^{\infty} J_n(x)e^{in\theta}, \tag{9.70}$$

the electric field can also be written as a sum of oscillations centered around ω_0 and separated by harmonics of the modulation frequency, weighted by the argument of the Bessel function of the first kind (cf. Fig. 9.12b):

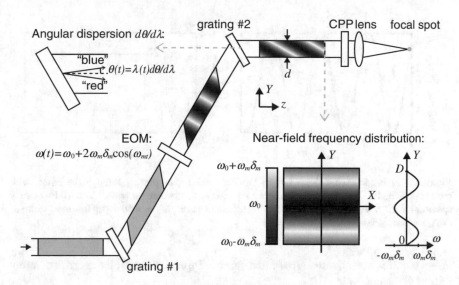

Fig. 9.11 Smoothing by spectral dispersion (SSD): a monochromatic laser beam at ω_0 passes through an electro-optic modulator (EOM) with modulation frequency ω_m and modulation depth δ_m. The light after the EOM has an instantaneous frequency $\omega(t) = \omega_0 + \omega_m \delta_m \cos(\omega_m t)$; the time variation of $\omega(t)$ is equivalent to a spatial variation along the beam's propagation direction, as shown by the gray scale gradient in the figure. Grating #2 skews the pulse front so that the transverse profile of the beam at a given time has a range of frequencies distributed along the SSD direction (Y in the figure), also varying in time. The angular dispersion imparted by grating #2 means that the different frequency components in the focal spot will be spatially shifted from each other along Y, which causes the speckle pattern to move as a function of time (periodically, at ω_m). Grating #1 is used to pre-compensate for the pulse front tilt from grating #2

$$\hat{E} = \frac{1}{2}\hat{E}_0 e^{i\mathbf{k}_0 \cdot \mathbf{r}} \sum_{j=-\infty}^{\infty} J_j(\delta_m) e^{i(\omega_0 + j\omega_m)t} + c.c. \qquad (9.71)$$

This expression, which assumes an infinite signal in time,[7] means that the spectrum of the modulated beam is composed of harmonics separated by ω_m and centered around ω_0, up to a width on the order of the peak-to-peak amplitude of the frequency oscillations $\Delta\omega_{SSD} = 2\delta_m\omega_m$, as shown in Fig. 9.12b. The most energetically significant harmonics extend up to $\pm\delta_m$, meaning that $\approx\text{round}(2\delta_m) + 1$ harmonics contribute to the bandwidth with the function "round" representing the rounding to the nearest integer (e.g., in Fig. 9.12, only ≈ 7 harmonics significantly contribute). The bandwidth introduced by SSD usually remains very small relative to the laser frequency, $\omega_m\delta_m \ll \omega_0$, since the frequency tripling used on high-power

---

[7] Practically, the expression is valid if averaged over many SSD cycles, i.e., for pulse durations $\gg 1/\nu_m$. For the effect of finite pulse durations, cf. Problem 9.2.

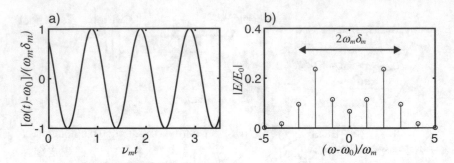

Fig. 9.12 (**a**) Instantaneous frequency of the laser beam after passing through the EOM, with modulation depth δ_m and frequency ω_m (as a function of $\nu_m t$, where $\nu_m = \omega_m/2\pi$). (**b**) Frequency spectrum of the laser after the EOM, showing harmonics at $\pm j\omega_m$ around ω_0 up to a total width of $\approx \Delta\omega = 2\delta_m\omega_m$ (shown here for $\delta_m = 3$)

lasers requires a relatively narrowband source. Typically, ν_m is on the order of a few tens of GHz and δ_m is ≤ 10.

The light is then sent to a diffraction grating ("grating 2" in Fig. 9.11), which skews the pulse front of the beam, and therefore, the spatial orientation of the frequency modulation as well, so that the transverse direction across the beam at a fixed time and propagation distance will contain a range of frequencies (cf. Fig. 9.11). The expression for the electric field at the output of the grating is easily derived by using the expression from Eq. (9.71), which essentially treats the beam as a sum of monochromatic waves of frequencies $\omega_0 \pm j\omega_m$. Taking z as the propagation direction after the grating for the fundamental frequency ω_0 (Fig. 9.11), the angle θ_j from the laser propagation axis z for the j-th harmonic is related to the angular dispersion of the grating $d\theta/d\lambda$ by

$$\theta_j = \Delta\lambda_j \frac{d\theta}{d\lambda} = -\frac{j\omega_m}{\omega_0 t}\lambda_0 \frac{d\theta}{d\lambda}, \tag{9.72}$$

where we used $\Delta\omega_j/\omega_0 \approx -\Delta\lambda_j/\lambda_0$ with $\Delta\omega_j = j\omega_m$. The field after the grating can be expressed as

$$\hat{E} = \frac{1}{2}\hat{E}_0 \sum_j J_j(\delta_m)e^{i(\mathbf{k}_j\cdot\mathbf{r}-\omega_0 t - j\omega_m t)} + c.c., \tag{9.73}$$

where $\mathbf{k}_j = k_j[\cos(\theta_j)\mathbf{e}_z + \sin(\theta_j)\mathbf{e}_Y]$ and $k_j = (\omega_0 + j\omega_m)/c$; (X, Y) represent the transverse dimensions before and up to the lens with Y the direction of the angular dispersion from the grating (we ignore the various turns that the beam might take from the grating to the lens). Developing up to first-order terms in $j\omega_m/\omega_0 \ll 1$ leads to $\mathbf{k}_j \cdot \mathbf{r} \approx k_0 z + (j\omega_m/c)z + k_0\theta_j Y$, and therefore,

$$\hat{E} = \frac{1}{2}\hat{E}_0 e^{i\psi_0} \sum_j J_j(\delta_m) \exp\left[ij\omega_m\left(\frac{z}{c} - t - \frac{\lambda_0}{c}\frac{d\theta}{d\lambda}Y\right)\right] + c.c., \quad (9.74)$$

with $\psi_0 = k_0 z - \omega_0 t$.

The expression can then be inverted back using Eq. (9.70), leading to

$$\hat{E} = \frac{1}{2}\hat{E}_0 e^{i\psi_0} \exp\left[i\delta_m \sin\left(\omega_m(z/c - t) - \alpha Y\right)\right] + c.c., \quad (9.75)$$

with

$$\alpha = 2\pi\frac{\omega_m}{\omega_0}\frac{d\theta}{d\lambda}. \quad (9.76)$$

The instantaneous frequency after the grating is now

$$\omega(z, Y, t) = \omega_0 - \omega_m\delta_m\cos\left[\omega_m(z/c - t) - \alpha Y\right]. \quad (9.77)$$

This means that at a fixed time and z location, the instantaneous frequency varies across Y, as illustrated in Fig. 9.11. The "number of color cycles" N_c is the number of times the full EOM frequency cycle is present across the beam at a given time and z. In other words, we can write the instantaneous frequency as

$$\omega(z, Y, t) = \omega_0 - \omega_m\delta_m\cos\left[\omega_m(z/c - t) - 2\pi N_c\frac{Y}{D}\right], \quad (9.78)$$

where D is the beam dimension along Y, and therefore,

$$\boxed{N_c = \frac{\alpha D}{2\pi} = D\frac{d\theta}{d\lambda}\frac{\omega_m}{\omega_0}}. \quad (9.79)$$

For example, we have $N_c \approx 1.5$ in the illustration in Fig. 9.11.

The original design includes a first grating ("grating 1" in Fig. 9.11) used to pre-compensate the pulse front tilt from grating #2 and keep the pulse front perpendicular to the propagation direction. This grating is not always used in practice (like on the NIF laser, where the EOM sits in the most upstream section where the laser is propagated in optical fibers [24]): the shear delay $t_d = N_c/v_m$ is often small compared to the duration of the laser pulse and may be left uncompensated.

Note that frequency tripling, which is often used in high-energy lasers, triples the fundamental frequency as well as the modulation depth δ_m. Another important consideration is the fact that the quantity $D(d\theta/d\lambda)$ has to be conserved throughout the propagation, as the beam divergence must scale like $1/D$.

As a result from the tilt from the angular dispersion, each location in Y at the lens will result in a focal spot envelope shifted by a small distance δy in the far-field, depending on the tilt $\theta(Y, t, z) = [\lambda(Y, t, z) - \lambda_0]d\theta/d\lambda$. Since an RPP (or CPP) is typically present at the lens when using SSD, a CPP element located at a fixed Y will therefore see a time-varying tilt $\theta(Y, t)$, which will produce an oscillating spatial shift of the focal spot envelope $\delta y(t) \approx F\theta(t)$, where F is the focal length of the lens. Since each CPP element will produce such an oscillating envelope, but with an independent random dephasing and a shift in the oscillation phase depending on their Y-location, the resulting speckle pattern at the focal spot is randomly "blinking." The recurrence time is $1/\nu_m$, typically on the order of tens of ps, which is faster than the typical hydrodynamic response time in ICF experiments. SSD is therefore very effective at mitigating imprint of speckles onto a target surface, which could otherwise seed hydrodynamic instabilities.

In the following section, we quantify these effects by deriving the electric field at best focus in the presence of SSD and an RPP.

9.3.2   Electric Field at Best Focus with an RPP and SSD

The expression for the electric field above is valid as the beam arrives at the lens. In the following, we denote the electric field envelope in the near-field (i.e., at the lens) and far-field (best focus) as $\hat{\mathcal{E}}$ and $\mathcal{E}$, respectively, such that the electric field is given by $\hat{E} = \text{Re}[\hat{\mathcal{E}}e^{i\psi_0}]$ and $E = \text{Re}[\mathcal{E}e^{i\psi_0}]$ (like in our description of RPP, the hat notation refers to near-field quantities).

The field envelope in the presence of SSD only is obtained from Eq. (9.75):

$$\hat{\mathcal{E}}_{\text{SSD}} = \hat{E}_0 \exp\left[-i\delta_m \sin(\omega_m t + \alpha f k_y/k_0)\right] . \tag{9.80}$$

The spatial coordinate Y in the near-field was converted to a Fourier coordinate k_y via $k_y/k_0 \approx Y/f$, with f the lens focal length (valid for small angles, i.e., a large lens f-number). The z location in (9.75) can be taken as fixed for the lens location and dropped from the derivation (i.e., we take $z = 0$ at the lens).

On the other hand, we have already seen that an RPP breaks up the field into elementary fields from each RPP element following Eq. (9.4), i.e.,

$$\hat{\mathcal{E}}_{\text{RPP}} = \hat{E}_0 \sum_{j=1}^{N} \text{rect}\left(\frac{k_x - k_{xj}}{\delta k}\right) \text{rect}\left(\frac{k_y - k_{yj}}{\delta k}\right) e^{i\phi_j} , \tag{9.81}$$

where N is the number of RPP elements. For simplicity, we assume that both the intensity profile at the lens and the RPP elements are square, but this analysis is easily generalizable to other shapes.

Combining the two gives the electric field at the lens with RPP and SSD (the SSD contribution is merely a space- and time-dependent phase shift and can thus be directly multiplied to the RPP field):

$$\hat{\mathcal{E}}_{\text{RPP+SSD}} = \hat{E}_0 \sum_{j=1}^{N} \text{rect}\left(\frac{k_x - k_{xj}}{\delta k}\right) \text{rect}\left(\frac{k_y - k_{yj}}{\delta k}\right) e^{-i\delta_m \sin(\omega_m t + \alpha f k_y / k_0) + i\phi_j}$$

$$(9.82)$$

(in the following, we drop the "RPP+SSD" subscript).

To derive the field at best focus, it is easier to use the Jacobi–Anger expansion (Eq. (9.70)) before performing the Fourier transform and invert it afterward. The expansion gives

$$\hat{\mathcal{E}} = \hat{E}_0 \sum_{j=1}^{N} \sum_{n=-\infty}^{\infty} \text{rect}\left(\frac{k_x - k_{xj}}{\delta k}\right) \text{rect}\left(\frac{k_y - k_{yj}}{\delta k}\right) J_n(\delta_m) e^{-in(\omega_m t + \alpha f k_y / k_0) + i\phi_j}.$$

$$(9.83)$$

We can now inverse Fourier transform and obtain

$$\mathcal{E} = \frac{E_0}{\sqrt{N}} \sum_{j=1}^{N} \sum_{n=-\infty}^{\infty} \text{sinc}\left(\frac{\delta k x}{2}\right) \text{sinc}\left(\frac{\delta k (y - n\alpha f / k_0)}{2}\right) \times$$

$$J_n(\delta_m) e^{i\mathbf{k}_{\perp j} \cdot \mathbf{r} - i n k_{yj} \alpha f / k_0 - i n \omega_m t + i\phi_j}, \qquad (9.84)$$

where we used the same normalization already introduced in Sect. 9.1, $\hat{E}_0 \delta k^2 / (2\pi) = E_0 / \sqrt{N}$ such that E_0 represents the average electric field near the center of the focal spot (averaged over many RPP realizations, or in space near the center of the spot provided that the number of RPP elements—i.e., the number of speckles—is large enough).

The two sinc functions define the large-scale envelope of the focal spot without the speckles, as we already discussed in Sect. 9.1. However, we now see that the envelope is spread along y (the direction of the SSD grating dispersion); it is the effect we mentioned earlier in our introduction to SSD, where the angular dispersion of the grating translates into a spatial shift at best focus. The spatial extent of the spread can easily be estimated: we know that only harmonics up to $\sim \pm \delta_m$ contribute to the SSD spectrum, so the spatial spread of the focal spot due to SSD is simply

$$\boxed{\Delta y_{\text{SSD}} = 2\delta_m \alpha f / k_0 = 2\lambda_0 \delta_m N_c f_\#}. \qquad (9.85)$$

This quantity represents the full extent of the focal spot motion along y due to SSD. It provides some very simple and practical information:

- For SSD to smooth out the speckle structure when averaging over the modulator period $1/\nu_m$, we need the motion to extend over a distance larger than the transverse speckle size $\rho_{sp} \approx \lambda_0 f_\#$, i.e., we need $2\delta_m N_c \geq 1$;

- To avoid SSD to spread out the beam too much with respect to Δy_{env}, the size of the beam envelope along y as designed by the RPP or CPP, we also want to keep $\Delta y_{SSD} \ll \Delta y_{env}$.

Assuming that the focal spot spread along y is negligible, $\Delta y_{SSD} \ll \Delta y_{env}$, we can simplify the large-scale envelope of the electric field as $\propto \mathrm{sinc}(\delta ky/2)$, so the electric field becomes

$$
\mathcal{E} = \frac{E_0}{\sqrt{N}} \mathrm{sinc}\left(\frac{\delta kx}{2}\right) \mathrm{sinc}\left(\frac{\delta ky}{2}\right) \sum_{j=1}^{N} \sum_{n=-\infty}^{\infty} J_n(\delta_m) e^{i\mathbf{k}_{\perp j}\cdot\mathbf{r} - ink_{yj}\alpha f/k_0 - in\omega_m t + i\phi_j} .
$$

(9.86)

We can now revert the Jacobi–Anger expansion, Eq. (9.70), and finally obtain

$$
\mathcal{E} = \frac{E_0}{\sqrt{N}} \mathrm{sinc}\left(\frac{\delta kx}{2}\right) \mathrm{sinc}\left(\frac{\delta ky}{2}\right) \sum_{j=1}^{N} e^{i(\mathbf{k}_{\perp j}\cdot\mathbf{r} + \phi_j) - i\delta_m \sin(\omega_m t + \alpha f k_{yj}/k_0)} . \quad (9.87)
$$

This expression is very similar to the one obtained in Eq. (9.11) for the field envelope of an RPP beam at best focus. The large-scale envelope, given by the product of sinc functions, is the same (except for the slight spread along y discussed above, which is neglected in this expression); the only effect of SSD is to modulate the phase of the field. At any given time, e.g., arbitrarily setting $t = 0$, SSD just adds a fixed dephasing to each RPP element, $\exp[-i\delta_m \sin(\alpha f k_{yj}/k_0)]$: because the phases are randomized by the RPP via the ϕ_j terms, this will not change the speckle statistics, meaning that at any given time, the speckle statistics described earlier for RPP is still valid. SSD merely evolves the speckle pattern as a function of time; just like we studied the speckle size for an RPP beam, the relevant quantity to study at this point is the speckle lifetime, which we describe in the next section.

9.3.3   Speckle Lifetime

Just like for the speckle size, the lifetime can be approximated via the temporal autocorrelation function, which is related to the spectral density via the Wiener–Khinchin theorem:

$$
C_{\mathcal{E}\mathcal{E}^*}(t) = \mathcal{F}^{-1}\left[|\mathcal{E}(\omega)|^2\right], \tag{9.88}
$$

where the Fourier transform is taken with respect to time.

We saw above that the power spectrum in the presence of SSD is composed of series of Dirac delta functions at the harmonics of ω_m,

$$
\mathcal{E}(\omega) \propto \sum_{n=-\infty}^{\infty} J_n(\delta_m)\delta(\omega - n\omega_m), \tag{9.89}
$$

covering a bandwidth of approximately $\Delta\omega_{SSD} \approx 2\delta_m\omega_m$ (cf. Fig. 9.12).

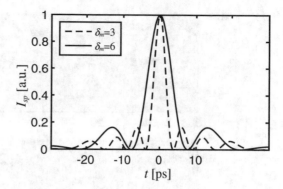

Fig. 9.13 Characteristic temporal intensity profile of speckles, for $\nu_m = 17\,\text{GHz}$, $\lambda_0 = 1.054\,\mu\text{m}$, and $\delta_m = 3$ and 6. The pattern is periodic and repeats itself every $T_m = 1/\nu_m \approx 59\,\text{ps}$

Just like we saw for the speckle size, which can be interpreted as the coherence length, the typical electric field temporal evolution is characterized by $C_{\mathcal{E}\mathcal{E}^*}(t)$, meaning that the typical temporal intensity profile $I_{sp}(t) \propto |C_{\mathcal{E}\mathcal{E}^*}(t)|^2$.

We can already infer that the typical coherence time (or speckle lifetime) τ_{sp} will be comparable to the inverse of the SSD bandwidth, $\tau_{sp} \approx 1/\Delta\nu_{\text{SSD}}$. The characteristic temporal intensity profile $I_{sp}(t)$ cannot be easily expressed analytically but can readily be plotted numerically after a pair of forward and inverse Fourier transforms, via

$$I_{sp}(t) \propto \left| \mathcal{F}^{-1}\left\{ \left|\mathcal{F}\left[e^{i\delta_m \sin(\omega_m t)}\right]\right|^2 \right\} \right|^2. \tag{9.90}$$

The corresponding curve is shown in Fig. 9.13, for $\nu_m = 17\,\text{GHz}$ (i.e., a modulation period $T_m \approx 59\,\text{ps}$), $\lambda_0 = 1.054\,\mu\text{m}$, and $\delta_m = 3$ and 6.

The full width at half maximum of the coherence time for a sinusoidal modulation as provided by SSD is given by

$$\boxed{\tau_{sp} \approx 0.7\frac{1}{\Delta\nu_{\text{SSD}}} = \frac{0.35}{\delta_m\nu_m}}. \tag{9.91}$$

For example, for a modulation at $\nu_m = 17\,\text{GHz}$ and $\delta_m = 4$, which is the modulation frequency and a typical modulation amplitude on the NIF laser,[8] we obtain $\tau_{sp} \approx 5.3\,\text{ps}$.

Figure 9.14 represents simulations of an RPP field showing the intensity vs. x (at a fixed y) or y (at a fixed x) and as a function of time with y the direction of the

---

[8] The modulation is applied at the fundamental laser wavelength of 1.054 μm, so it needs to be tripled to account for the frequency tripling before focusing: in other words, the actual modulation applied by the EOM is 1.3, which becomes 4 after the frequency conversion crystals where the wavelength becomes 351 nm.

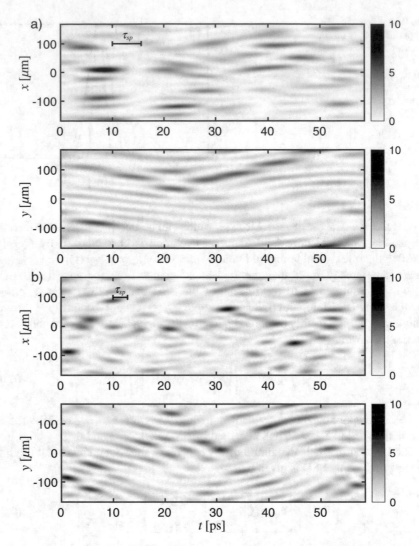

Fig. 9.14 Motion of speckles from an RPP beam with SSD over one modulation period ($T_m = 1/\nu_m = 59$ ps for $\nu_m = 17$ GHz) along x and y (the SSD grating dispersion is along y), for an SSD modulation depth of (**a**) $\delta_m = 3$ and (**b**) $\delta_m = 6$. The laser wavelength is 1.054 μm. The corresponding speckle lifetimes τ_{sp} of 7 and 3.5 ps for $\delta_m = 3$ and 6 are indicated in the figure for reference

SSD grating dispersion, for modulation depths $\delta_m = 3$ (a) and 6 (b), corresponding to lifetimes of ≈ 7 and 3.5 ps, respectively.

Because of the sinusoidal modulation imparted by the EOM, we clearly see some oscillation of the speckles along y. The lifetime at a fixed location in space is consistent with 7 and 3.5 ps as given by Eq. (9.91) (as is more easily seen on the plots of $I(x, t)$), and however for LPI considerations it will be useful to consider the speckle motion along y and not only the random "blinking" along x, as the lifetime in the frame moving with the speckles can be longer than τ_{sp} [25, 26].

9.4 Induced Spatial Incoherence (ISI)

Induced spatial incoherence (ISI), invented by Lehmberg and Obenschain [27, 28], combines both the spatial and temporal decoherence ideas from RPPs and SSD.[9] It is more efficient than SSD in terms of randomizing the temporal evolution of speckles in the focal spot; however, it relies on the finite bandwidth of the laser source, as we will see below, which makes it inconvenient to implement on frequency-tripled glass lasers[10] but particularly well suited for excimer lasers (like KrF or ArF) such as the Nike laser at the Naval Research Laboratory [29, 30]. The short wavelengths of these lasers constitute another beneficial factor to mitigate laser–plasma instabilities, which generally scale like $I\lambda_0^2$ as we have seen throughout this book.

The general setup is illustrated in Fig. 9.15. A pair of echelons, sitting orthogonal to each other in the beam path in the near-field (just before the lens), breaks up the beam into small beamlets with random dephasings, similar to an RPP. Here the dephasings are random simply because the size δl of each echelon step along z (the propagation direction) is only known up to an uncertainty greater than the laser wavelength, i.e., for each echelon step j, the longitudinal size δl_j is (typically) a random variable with mean δl and standard deviation $\sigma_l \gg \lambda_0$.

Overall, the beam is then divided into $N = N_s^2$ beamlets, where N_s is the number of steps in each echelon. Following the same steps we used to derive the expression for the field of an RPP beam at best focus in Sect. 9.1.1, we start with the expression for the electric field in the near-field, after propagation through both echelons:

$$\hat{E}(k_x, k_y, z, t)$$

$$= \frac{1}{2} \sum_{j=1}^{N} \hat{E}_0(t - \tau_j) \, \text{rect}\left(\frac{k_x - k_{xj}}{\delta k}\right) \text{rect}\left(\frac{k_y - k_{yj}}{\delta k}\right) e^{-i\omega_0(t-\tau_j)+ik_0 z} + c.c.,$$

$$(9.92)$$

where δk is related to the size of an echelon step in the transverse direction δ_{ech} (assumed to be the same for both echelons) via $\delta k = k_0 \delta_{ech}/f$, with f the lens' focal length, similar to Eq. (9.3). τ_j represents the delay of the beamlet j after

---

[9] ISI was invented prior to SSD.

[10] While Nd-based glass lasers can maintain some amount of bandwidth up to the frequency-doubling process, the frequency-tripling is intrinsically narrowband with an efficiency that typically drops for bandwidths larger than $\Delta\nu/\nu_0 \sim 5 \times 10^{-4}$.

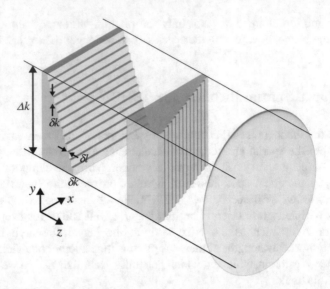

Fig. 9.15 General setup for ISI: two orthogonal echelons are placed in the near-field before the focusing lens, producing both spatial incoherence (similar to an RPP) and temporal decoherence— as long as the induced path delay between two consecutive echelon steps is larger than the laser coherence time

propagating through the echelons, i.e., $\tau_j = \delta l_j n / c$, where δl_j is the propagation distance through the two combined echelons and n is the echelons refractive index. Note that in this expression $\hat{E}_0(t)$ represents the slowly varying temporal profile of the laser pulse.

An inverse Fourier transform along the transverse spatial coordinates gives the field at best focus:

$$E = \frac{1}{2}\mathcal{E}(\mathbf{r}, t)e^{i\psi_0} + c.c., \tag{9.93}$$

$$\mathcal{E} = \frac{\delta k^2}{2\pi}\text{sinc}\left(\frac{x\delta k}{2}\right)\text{sinc}\left(\frac{y\delta k}{2}\right)\sum_{j=1}^{N}\hat{E}_0(t - \tau_j)e^{i(\phi_j + \mathbf{k}_{\perp j} \cdot \mathbf{r})}, \tag{9.94}$$

where $\phi_j = \omega_0 \tau_j = \delta l_j k_0$ is equivalent to a random variable with uniform distribution in $[0, 2\pi]$, given the uncertainty on δl_j mentioned earlier.

We recover an expression very similar to the one from an RPP beam; in particular, if we express the ensemble-averaged intensity (meaning averaged over different statistical realizations of the echelons with a small random component in the longitudinal step size), we obtain

$$\left\langle |\mathcal{E}|^2 \right\rangle = \frac{\delta k^4}{(2\pi)^2}\text{sinc}^2\left(\frac{x\delta k}{2}\right)\text{sinc}^2\left(\frac{y\delta k}{2}\right)\sum_{j=1}^{N}\left\langle |\hat{E}_0(t - \tau_j)|^2 \right\rangle, \tag{9.95}$$

where we used the fact that $\langle \exp i(\phi_j - \phi_k) \rangle = \delta_{jk}$, where δ is a Kronecker delta function. If the delays imposed by the echelons remain small compared to the temporal variations of the pulse envelope, such that at any time t, $|t_N - t_1| d_t \hat{E}_0(t) \ll \hat{E}_0(t)$ (from a Taylor expansion), where $t_N - t_1$ represents the maximum delay between two beamlets due to the propagation in the echelons, then we have

$$\sum_{j=1}^{N} \left\langle |\hat{E}_0(t - \tau_j)|^2 \right\rangle \approx N |\hat{E}_0(t)|^2 , \qquad (9.96)$$

and we recover the same expression for the ensemble-averaged intensity as the one from an RPP, Eq. (9.8):

$$\langle I(\mathbf{r}, t) \rangle = \frac{c\varepsilon_0 \langle |\mathcal{E}|^2 \rangle}{2} = \frac{c\varepsilon_0}{2} \frac{\hat{E}_0^2(t) \Delta k^2 \delta k^2}{(2\pi)^2} \mathrm{sinc}^2 \left(\frac{\delta k x}{2} \right) \mathrm{sinc}^2 \left(\frac{\delta k y}{2} \right) , \quad (9.97)$$

where Δk is related to the beam's f-number $f_\#$ via $\Delta k = k_0 / f_\#$, similar to Eq. (9.2) for an RPP.

The key difference from RPP lies in the variable delays imposed by the echelons. When the source laser is broadband, the resulting coherence time will be relatively short, $\tau_c \sim 1/\Delta\nu$, where $\Delta\nu$ is the laser bandwidth. If the coherence time becomes shorter than the minimum delay between two consecutive echelon elements, then all the beamlets will be incoherent, and the speckle pattern will be smoothed out when averaging over a time larger than τ_c.

To quantify this, let us consider the intensity averaged over a duration T, which is proportional to $\langle |\mathcal{E}|^2 \rangle_T$, where $\langle f \rangle_T = \int_{-T/2}^{T/2} f(t) dt / T$. We have

$$\left\langle |\mathcal{E}(t)|^2 \right\rangle_T \propto \sum_{j=1}^{N} \sum_{l=1}^{N} \left\langle \hat{E}_0(t - \tau_j) \hat{E}_0^*(t - \tau_l) \right\rangle_T e^{i(\phi_j - \phi_l + (\mathbf{k}_{\perp j} - \mathbf{k}_{\perp l}) \cdot \mathbf{r})} . \quad (9.98)$$

We now introduce the coherence function $\gamma(t)$ of the laser, defined as

$$\left\langle \hat{E}_0(t) \hat{E}_0^*(t + \tau) \right\rangle_T = |\hat{E}_0(t)|^2 \gamma(T) , \qquad (9.99)$$

where $\gamma(T)$ is a function that equals 1 at $T = 0$ and goes to 0 for $T \gg \tau_c$; typically, γ will resemble a Gaussian function of width τ_c like the one illustrated in Fig. 9.16.

If we have $\tau > \tau_e$, where $\tau_e = n\delta l / c$ is the minimum delay between two echelon steps (with δl the echelon step size along z), and $\tau_e \gg \tau_c$, then we can approximate $\gamma(\tau_j - \tau_l) \approx \delta_{jl}$, and

$$\left\langle \hat{E}_0(t - \tau_j) \hat{E}_0^*(t - \tau_l) \right\rangle_T \approx |\hat{E}_0(t)|^2 \delta_{jl} . \qquad (9.100)$$

Fig. 9.16 Illustration of the laser coherence function $\gamma(t)$

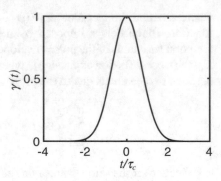

The derivation of the time-average then follows the one for the ensemble-average, leading to

$$\left\langle |\mathcal{E}(t)|^2 \right\rangle_T \approx \left\langle |\mathcal{E}(t)|^2 \right\rangle \ , \ \ T \geq \tau_c , \tag{9.101}$$

meaning that when averaging over a duration longer than τ_c, the speckle pattern gets "smoothed out" and we recover the large-scale envelope without the speckles.

A key design criterion for ISI is thus

$$\boxed{\tau_c \leq \tau_e \ll N\tau_e \leq \frac{\hat{E}_0(t)}{d\hat{E}_0(t)/dt}} , \tag{9.102}$$

where, as we saw earlier, the last inequality means that the maximum delay between echelon elements should remain small compared to the slowly varying temporal variations of the laser pulse (i.e., the echelons will smooth out time variations in the laser pulse shorter than $N\tau_e$). This last constraint can be eliminated by using the "echelon-free" ISI method, which consists in using an incoherent laser source and imaging an apertured plane (via a pinhole) in the front end directly at the target plane [31, 32].

Like for SSD, speckles are going to "blink" on time scales on the order of $\sim 1/\delta\nu$, the inverse of the bandwidth (for SSD, it was $1/\Delta\nu_{SSD}$ per Eq. (9.91)). But unlike SSD, with ISI, the speckle pattern does not repeat itself every cycle; also, the speckle does not have a preferred motion like with SSD along the direction of angular dispersion (Fig. 9.14, motion along y), which effectively extends the speckle lifetime (for SSD) in the frame of the moving speckle.

To give typical practical numbers, we take the example of the application of ISI in KrF lasers at the Naval Research Laboratory [33, 34]. The KrF laser operates at 248 nm and has a coherence time τ_c as short as 1 or 2 ps. The echelons have minimum delays of τ_e of 2 ps also, which means that the longitudinal step size of the echelons is approximately $\delta l = \tau_e c/n \approx 0.4$ mm (for $n = 1.5$). The combined number of echelon steps in both transverse directions is $N = 350$, meaning that the laser pulse is temporally smoothed by the echelons by a filter window of approximately 0.7 ns.

9.5 Laser–Plasma Instabilities from Optically Smoothed Laser Beams

Optical smoothing can impact laser–plasma instabilities in many different ways. One of the most important aspects of LPI in optically smoothed laser beams is the presence of speckles from RPPs or CPPs, whose intensity can be many times higher than the average beam intensity. While intense speckles can trigger instabilities, the other smoothing techniques we presented in the previous sections, such as PS, SSD, or ISI, can help mitigate the nefarious effects from these intense speckles.

Here we consider the behavior of backscatter and filamentation for smoothed laser beams. Note that the two processes can be strongly correlated, as filamentation tends to create local intensity increases in speckles which can then trigger backscatter. Conversely, smoothing techniques such as PS, SSD and ISI, which all reduce the contrast and lifetime of speckles, can reduce backscatter *indirectly*, via a reduction in filamentation. This has been observed in many experiments [35–40] as well as in simulations [41, 42].

In the following we study the impact of optical smoothing on filamentation and backscatter separately; while the problem is generally difficult, some simple estimates and derivations can provide a useful physics understanding of these processes. The analyses below provide some simple metrics that can help assess the risk of instability and have been successfully validated against simulations and experiments.

9.5.1 Filamentation of Optically Smoothed Laser Beams

9.5.1.1 RPP Beams

The use of optical smoothing via random or continuous phase plates can efficiently reduce the filamentation of high-power lasers as they propagate in plasma. Laser beams used in HED or ICF experiments are typically characterized by very large spot sizes and very high powers (e.g., the NIF laser typically operates near 2.5 TW per beam, with typical spot sizes of several hundred microns), which usually greatly exceed the critical power for self-focusing: for example, using typical values of $T_e = 3$ keV and $n_e/n_c = 0.1$, the critical power for ponderomotive self-focusing is approximately 1 GW according to Eq. (5.73), meaning that the power in a beam can easily be thousands of times above the critical power.

Using RPPs breaks up the beam into wavelength-scale speckles that are independent from each other as their phases are uncorrelated. The criterion for filamentation or self-focusing then takes a very different form, as the critical parameter is the power contained in a speckle, rather than the whole beam. The power contained in a speckle of peak intensity I_{sp} is approximately $P_{sp} \approx I_{sp}\pi\rho_{sp}^2$, with $\rho_{sp} = \lambda_0 f_\#$ (Sect. 9.1). Self-focusing occurs if $P_{sp} \geq P_c$, or, using Eq. (5.73),

$$I_{sp}[\text{W/cm}^2] \geq \frac{3.4}{\pi} 10^{15} n_0 \frac{T_{e,keV}}{n_e/n_c} \frac{1}{f_{\#}^2 \lambda_{\mu}^2}, \tag{9.103}$$

where $n_0 = (1 - n_e/n_c)^{1/2}$ is the plasma refractive index and $T_{e,keV}$ and λ_{μ} the temperature in keV and wavelength in μm, respectively.

The focal spot of an RPP beam contains many speckles whose intensities follow a probability distribution, as we discussed in Sect. 9.1. Therefore, for a given average intensity in the focal spot, we can estimate the fraction of speckles whose intensity is beyond the self-focusing threshold. A useful filamentation figure of merit (FFOM), inspired by Refs. [43, 44], is

$$\boxed{\text{FFOM} = \frac{\langle I \rangle\,[\text{W/cm}^2]}{10^{13}} \lambda_{\mu}^2 \frac{n_e}{n_c} \frac{3}{T_{e,keV}} \left(\frac{f_{\#}}{8}\right)^2,} \tag{9.104}$$

such that FFOM $= 1$ marks the onset of "noticeable" filamentation. Here $\langle I \rangle$ is the average beam intensity. Comparing this expression to the critical intensity for self-focusing in a speckle above, Eq. (9.103), shows that FFOM $= 1$ means that the self-focusing threshold is at approximately $5 \langle I \rangle$, i.e., speckles at five times the average intensity and above will self-focus when FFOM=1. Note that the threshold for filamentation of an RPP beam is now in intensity, not in power; the $\propto \lambda_0^2 f_{\#}^2$ scaling expresses the fact that the relevant spot size for self-focusing is not the overall beam envelope but the speckle size (with $\langle I \rangle \lambda_0^2 f_{\#}^2 \propto P_{sp}$).

To get an idea of what this represents, it is useful to apply the formula from Eq. (9.39) giving the fraction of power above the average intensity (FOPAI). The definition of the FFOM above, such that speckles above $5\langle I \rangle$ exceed the self-focusing threshold, means that approximately 4% of the beam power (FOPAI(5) $= 6e^{-5} \approx 0.04$) will self-focus and spray outside the initial beam aperture for FFOM=1.

Reduction of filamentation in laser–plasma experiments using an RPP and filamentation of an RPP beam were measured experimentally [35, 45–47]. In particular, the filamentation figure of merit has been shown to accurately predict the onset of measurable filamentation in simulations as well as in experiments, as evidenced by an angular spread of the laser beam as measured in the transmitted near-field [48].

We shall keep in mind that while the literature usually uses the term filamentation for laser beams equipped with RPP or CPP, the key process at play is really self-focusing of individual speckles; filamentation is rather supposed to describe the break-up of a plane wave-like beam into small-scale filaments [43]. With an RPP, the beam is already broken up, and per our discussion in Sect. 7.3.6 the size δx of the filaments that a speckle would break up into (given by Eq. (7.126)) can only be much smaller than the speckle size (a necessary condition for filamentation to occur and for the filamentation model to be valid) if the speckle is much beyond its self-focusing threshold.

Finally, we note that the condition that the power per speckle exceeds the critical power for self-focusing, $P_{sp}/P_c \geq 1$, which constitutes the basis for the figure of merit, is exactly equivalent to saying that the spatial amplification of the filamentation instability derived in Sect. 7.3.2, Eq. (7.123) over a speckle length is larger than unity (cf. Problem 9.3; this metric was initially proposed by Berger in Ref. [49]).

9.5.1.2   Mitigation of Filamentation Using PS, SSD, and ISI

Polarization smoothing (PS) further helps mitigate filamentation. We saw that the fraction of power above the average intensity (FOPAI) with PS is $(2n^2+2n+1)e^{-2n}$ (cf. Eq. (9.67)). The fraction of power above $5\langle I \rangle$ drops from 4% without PS to $\approx 0.28\%$ with PS [50]. Inferring a new FFOM with PS can only be approximate since the statistical distribution of intensity and FOPAI are different; we can still take a rough estimate by noting that $\text{FOPAI}_{\text{RPP}}(5) \approx 4\% \approx \text{FOPAI}_{\text{RPP+PS}}(3.3)$: the amount of beam spread marking the onset of filamentation without PS is recovered with PS if the average intensity is increased by $\approx 50\%$. In other words, PS raises the FFOM by approximately 50%,

$$\boxed{\text{FFOM}_{\text{PS}} \approx \frac{\langle I \rangle \, [\text{W/cm}^2]}{1.5 \times 10^{13}} \lambda_\mu^2 \frac{n_e}{n_c} \frac{3}{T_{e,keV}} \left(\frac{f_\#}{8} \right)^2 .} \tag{9.105}$$

Note that this is a very approximate metric, since the contribution of PS in reducing the fraction of intense speckles is not constant but intensity-dependent and becomes more important for intense speckles, as we discussed in Sect. 9.2.

Smoothing by spectral dispersion (SSD) also contributes to mitigating filamentation of RPP beams. The physics picture and key criteria can easily be obtained by going back to our description of the ponderomotive "digging" of a density perturbation in Sect. 5.2. There we saw that the minimum time it takes to dig the density depression is on the order of the acoustic transit time through a speckle, $\tau_w = \rho_{sp}/c_s$ [cf. Eq. (5.39) and Problem 5.1], which is the time it takes for the excess density to move out of the speckle. We also saw in Sect. 9.3 that the typical lifetime of a speckle when SSD is present (Eq. (9.91)) is $\tau_{sp} \approx 0.35/(\nu_m \delta_m)$, where ν_m is the SSD modulation frequency and δ_m is the modulation depth (possibly tripled if the laser is frequency tripled).

Therefore, if the speckle lifetime is shorter than the time it takes to dig the density depressions, the depth of the density channel and thus the speckle self-focusing will not be as strong. In other words, the criterion for efficient mitigation of filamentation of an RPP beam from SSD is

$$\boxed{\tau_{sp} = 0.35 \frac{1}{\delta_m \nu_m} \leq \tau_w = \frac{\lambda_0 f_\#}{c_s},} \tag{9.106}$$

where the speckle size was simply estimated as $\rho_{sp} = \lambda_0 f_\#$ (cf. Sect. 9.1).

As an example, consider the NIF laser, which typically operates at $\delta_m = 1.3 \times 3$ (after frequency tripling) with $\nu_m = 17\,\text{GHz}$: assuming a plasma with $ZT_e \ll 3T_i$ and $Z/A = 1/2$, we can easily show that $\tau_{sp} = \tau_w$ is reached for $T_e \approx 6\,\text{keV}$, i.e., SSD will have a mitigating effect for plasma temperatures below $6\,\text{keV}$ (because the resulting sound speed is low enough to allow the speckles to "blink" before the density depressions are fully developed).

Experiments have confirmed that SSD can effectively mitigate filamentation; this can be seen either "directly," by measuring the divergence of the transmitted beam (i.e., its angular spray in the near-field [36]), or indirectly, by observing a reduction in backscatter attributed to the mitigation of local intensity increase in the speckles due to filamentation.

ISI plays a similar role, by limiting the speckle lifetime to the laser coherence time τ_c.

9.5.2   Backscatter from Smoothed Beams

In this section we investigate the nonlinear behavior of backscatter due to the presence of speckles in an RPP beam and its mitigation via PS and SSD. As mentioned in the previous section, backscatter can be indirectly reduced by optical smoothing (PS or SSD), due to the mitigation of filamentation which can otherwise exacerbate the intensity of the most intense speckles and hence trigger backscatter [35–42]; however, as we will see PS or SSD can mitigate backscatter directly even in the absence of filamentation.

9.5.2.1   Backscatter from RPP Beams: Onset of Nonlinear Behavior due to Speckles

Since RPP beams comprise speckles whose intensity can largely exceed the average beam intensity and that the amplification of backscatter is an exponential function of intensity, we are going to encounter nonlinear effects due to speckles; this is simply due to the fact that $\langle e^I \rangle \neq e^{\langle I \rangle}$, where the brackets represent an ensemble-average over RPP realizations, i.e., the average amplification (which is the macroscopic quantity related to the total reflectivity and which we are concerned about) is not equal to the amplification of the average. In fact, as we will see, these nonlinear effects will lead to a divergence of the backscatter amplification as soon as the "average gain per speckle," i.e., the amplification gain exponent over one speckle length at the average laser intensity, will exceed one.

We can easily derive a simple model that captures the onset of nonlinearity of backscatter due to speckles. Let us consider a volume of uniform plasma containing many speckles, in all three dimensions (i.e., the length of the volume is assumed much longer than a speckle length and its transverse sizes much larger than a speckle radius). We assume a uniform background of noise acting as a seed for backscatter. We have seen in previous chapters that the convective amplification of the backscatter seed by a laser intensity I_0 over a length L can simply be expressed as

$$A(I_0, L) = \exp[g I_0 L], \tag{9.107}$$

where g is a gain coefficient depending on the plasma conditions and on the process considered (SRS or SBS).

To model the contribution of speckles, we can average the amplification over the intensity distribution function of an RPP beam, given by $p(I_0) = \exp[-I_0/\langle I_0 \rangle]/\langle I_0 \rangle$ (cf. Sect. 9.1). However, we also need to account for the correlation length of the field along the propagation axis z, which is on the order of a speckle length: in other words, if a speckle is located at (x, y, z), we can assume that there will not be another speckle at x, y, $z + L_{sp}$. This means that the amplification $\langle A(\langle I_0 \rangle, L) \rangle = \int_0^\infty A(I_0) p(I_0) dI_0$ is valid over a distance $L < L_{sp}$ only. For distances longer than a speckle length, we can simply take the product of the amplification over each "row" of speckles, which leads to

$$\langle A(\langle I_0 \rangle, L) \rangle = \langle A(\langle I_0 \rangle, L_{sp}) \rangle^{L/L_{sp}}, \tag{9.108}$$

as illustrated in Fig. 9.17.

Performing the integration over intensities with $p(I_0) = \exp[-I_0/\langle I_0 \rangle]/\langle I_0 \rangle$ and defining $u = I_0/\langle I_0 \rangle$ lead to the following expression for the average amplification over one speckle length:

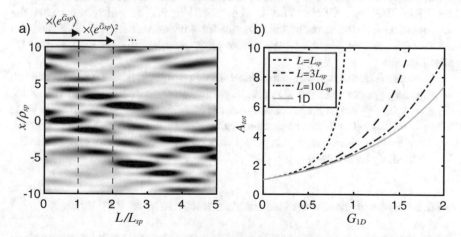

Fig. 9.17 (a) Laser intensity I_0 (with color scale saturated at $4\langle I_0 \rangle$), illustrating the average amplification over many speckle length: the background seed for backscatter gets amplified by a factor $\langle A(\langle I_0 \rangle, L_{sp}) \rangle = \langle \exp[\bar{G}_{sp}] \rangle$ every speckle length, and the total amplification is the product of the amplification from each "row" of speckles. (b) Total backscatter amplification as a function of the 1D gain for $L/L_{sp} = 1, 3$, and 10 and 1D amplification $A_{1D} = \exp[G_{1D}]$

$$\langle A(\langle I_0 \rangle, L_{sp}) \rangle = \int_0^\infty \exp[u(\bar{G}_{sp} - 1)]du$$

$$= \frac{1}{1 - \bar{G}_{sp}}, \quad \bar{G}_{sp} \leq 1, \tag{9.109}$$

where we defined $\bar{G}_{sp}$, the gain over one speckle length at average intensity (often simply referred to as the average gain per speckle) as

$$\bar{G}_{sp} = g \langle I_0 \rangle L_{sp}. \tag{9.110}$$

The integral is only convergent for $\bar{G}_{sp} < 1$: this means that as soon as the average gain per speckle exceeds one, the average amplification diverges, since the strong amplification due to the most intense speckles "overwhelms" the rarefaction of intense speckles following their probability distribution.

Inserting the expression for the average amplification over one speckle length into Eq. (9.108) gives

$$\boxed{\langle A(\langle I_0 \rangle, L) \rangle = \exp\left[-\frac{L}{L_{sp}} \ln(1 - G_{1D} L_{sp}/L)\right],} \tag{9.111}$$

where $G_{1D} = gL \langle I_0 \rangle = \bar{G}_{sp} L / L_{sp}$ is the 1D gain over the amplification distance L. As expected, in the limit of $L_{sp}/L \ll 1$ (i.e., amplification over many speckle lengths, with the average amplification per speckle $\bar{G}_{sp} \ll G_{1D}$), we recover $\langle A(\langle I_0 \rangle, L) \rangle \approx \exp[G_{1D}]$—meaning that the amplification is not influenced by speckles anymore and follows the simple 1D expressions derived in earlier chapters for SRS, SBS, and CBET. This is shown in Fig. 9.17b, which represents the amplification as a function of the 1D gain for various values of L/L_{sp} (i.e., the number of rows of speckles over the amplification length), as well as the 1D amplification.

This also shows that the amplification when speckles are present can be significantly greater than the 1D case; in particular, the amplification diverges for $G_{1D} = L/L_{sp}$—and as mentioned above, the amplification diverges for $\bar{G}_{sp} = 1$, when $L = L_{sp}$. Significant backscatter can occur in ICF or HED experiments in conditions where the average gain per speckle remains smaller than one, due to the long scale-lengths involved.

9.5.2.2   A Simple Model for Reflectivity Saturation of an RPP Beam

When the average gain per speckle exceeds one, the behavior becomes strongly nonlinear and an ensemble-average of the amplification cannot be used anymore. Here we present a simple model for this nonlinear situation, first introduced by Rose and Dubois [51].

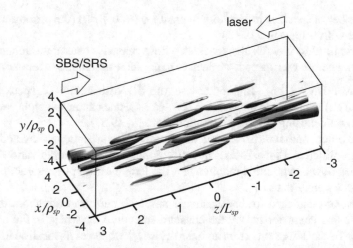

Fig. 9.18 Speckle intensity iso-contour, illustrating the Rose & Dubois nonlinear reflectivity model: the total reflectivity is calculated by summing up the power contained in each speckle whose intensity exceeds the threshold value I_{th} such that the individual speckle reflectivity should be ≥ 1 (i.e., we assume 100% reflectivity in each of these speckles, sum up the power, and ignore the rest)

The main idea of this model consists in finding the speckles in the plasma volume whose intensity is such that the local reflectivity should be larger than 100%, assume that the reflectivity is instead saturated at 100% (i.e., full pump depletion[11]) and sum up the reflected power from these speckles only while ignoring the rest. This is illustrated in Fig. 9.18.

To that end, we need to use the speckle statistics introduced in Sect. 9.1.4; we denote by $M(I_0)$ the number of speckles in the volume V whose normalized intensity is larger than I_0; per Eq. (9.58), defining the intensity normalized to the average $u = I_0/\langle I_0 \rangle$, we have[12]

$$M(u) = \frac{\pi^{3/2}\sqrt{5}V}{27\rho_{sp}^2 L_{sp}} \left(u^{3/2} - \frac{3}{10}u^{1/2} \right) e^{-u} . \tag{9.112}$$

The probability distribution function (pdf) $p(u)$ is such that

$$p(u)V du = -M'(u)du \tag{9.113}$$

---

[11] The model was initially developed for SBS; for SRS, one might want to account for the energy fraction deposited in the EPW per the Manley–Rowe relations, cf. Sect. 6.4.6.

[12] Here we use the hot spot statistical model revisited by Garnier in Ref. [8] instead of the one initially used by Rose in Ref. [51]: this leads to a correction in the final expression of the average reflectivity compared to [51], as we will see below.

is the number of speckles in V whose intensity $\in [u, u + du]$ (the prime denotes the derivative with respect to u).

We first need to identify the speckles whose intensity causes the reflectivity to exceed one; before we proceed, we need to make some assumptions for our model:

- We must have $L > L_{sp}$, otherwise the 3D distribution of speckles from Eq. (9.112) would be inappropriate; if $L \ll L_{sp}$, one can in principle construct a similar model using the 2D distribution from Eq. (9.57).
- On the other hand, we also assume that the speckles are independent and that the light amplified in a speckle does not get significant further amplification by other speckles downstream: this requires that the length of the plasma volume be *not too much longer* than L_{sp}.
- The background seed for backscatter is assumed uniform throughout the entire volume and characterized by the constant fraction ε anywhere in the box, such that the backscatter starts from an intensity $\varepsilon \langle I_0 \rangle$ whatever its location in the box.
- We ignore *macroscopic* pump depletion, consistent with the previous assumption of a uniform background seed (i.e., $\langle I_0 \rangle$ does not have a z-dependence from pump depletion).

Since we assume that the background seed for backscatter is $\varepsilon \langle I_0 \rangle$ everywhere in the volume, the reflectivity of a single speckle is estimated as the amplified background seed divided by the speckle intensity, i.e.,

$$R_{sp} = \frac{\varepsilon \langle I_0 \rangle A}{I_0} = \varepsilon A / u, \qquad (9.114)$$

where $A = \exp[\bar{G}_{sp} u]$ is the speckle amplification factor defined in the previous section (Eq. (9.107)), with $\bar{G}_{sp}$ the average gain per speckle (cf. Eq. (9.110)). We can thus extract a threshold intensity u_{th} at which the reflectivity equals 1, defined as

$$\varepsilon \frac{\exp[\bar{G}_{sp} u_{th}]}{u_{th}} = 1. \qquad (9.115)$$

This leads to $\bar{G}_{sp} u_{th} = \ln(u_{th}) - \ln(\varepsilon) \approx -\ln(\varepsilon)$, where we assumed that $u_{th} = O(1)$, whereas $\varepsilon \ll 1$. Therefore, the threshold intensity in a speckle at which the reflectivity equals one may be approximated as

$$u_{th} \approx -\frac{\ln(\varepsilon)}{\bar{G}_{sp}}. \qquad (9.116)$$

The average reflectivity from a distribution of speckles in V can then be estimated by summing up the power in each speckle, $P_{sp} = I_0 \sigma = u \langle I_0 \rangle \sigma$ (where $\sigma \approx \pi \rho_{sp}^2$ is the speckle cross-sectional area), over all speckles above u_{th}, and normalizing by the total laser power $P_{tot} = \langle I_0 \rangle S$ (with S the cross-sectional area of the full plasma volume):

$$\langle R \rangle = \frac{1}{P_{tot}} \int_{u_{th}}^{\infty} p(u) V P_{sp}(u) du = -\frac{\sigma}{S} \int_{u_{th}}^{\infty} M'(u) u \, du \,. \tag{9.117}$$

Using the expression for $M(u)$ above, we have

$$\langle R \rangle = \frac{\sqrt{5}\pi^{5/2}}{27} \frac{L}{L_{sp}} \int_{u_{th}}^{\infty} \left(u^{5/2} - \frac{9}{5} u^{3/2} + \frac{3}{20} u^{1/2} \right) e^{-u} du \,. \tag{9.118}$$

Here we simply give the following integration formula:

$$\int_{a}^{\infty} \left(u^{5/2} - \frac{9}{5} u^{3/2} + \frac{3}{20} u^{1/2} \right) e^{-u} du \tag{9.119}$$

$$= \frac{1}{10} \left[\sqrt{a} e^{-a} (10a^2 + 7a + 12) + 6\sqrt{\pi} \, \mathrm{erfc}(\sqrt{a}) \right] \,. \tag{9.120}$$

Since u_{th} is expected to be significantly above 1 (to satisfy the weak nonlinear approximation from this model), the dominant term is $\approx a^{5/2} e^{-a}$, and we arrive at

$$\langle R \rangle = \frac{\sqrt{5}\pi^{5/2}}{27} \frac{L}{L_{sp}} u_{th}^{5/2} e^{-u_{th}} \,. \tag{9.121}$$

Inserting the expression for u_{th} above leads to the final expression for the reflectivity:

$$\boxed{\langle R \rangle \approx 1.4 \frac{L}{L_{sp}} \left(\frac{\ln(1/\varepsilon)}{\bar{G}_{sp}} \right)^{5/2} \varepsilon^{1/\bar{G}_{sp}} \,,} \tag{9.122}$$

where we used $\sqrt{5}\pi^{5/2}/27 \approx 1.4$.

This expression shows that the reflectivity is linearly dependent on the propagation length L, as expected from the assumption of full local pump depletion in the speckles (e.g., doubling the length L doubles the number of speckles and thus doubles the reflectivity, given our model assumptions). It is similar to the original expression by Rose & Dubois, except for the 5/2 exponent (vs. 2 in Rose and Dubois), which comes from the slightly different model used for the speckles statistics, as explained in footnote 12. The behavior of $\langle R \rangle$ is illustrated in Fig. 9.19.

Similar independent speckle models were subsequently used in other publications with various levels of refinement and improvements, in particular regarding the reflectivity of a single speckle [52, 53] or the use of optical smoothing techniques [54, 55].

Fig. 9.19 Average reflectivity per speckle length as a function of the average gain per speckle $\bar{G}_{sp}$, for different values of the background seed level ε

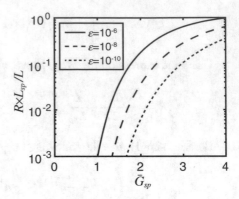

9.5.2.3   Mitigation of Backscatter Using PS

To quantify the effects of PS on backscatter, we wish to calculate the amplification of a background seed by a pump laser equipped with RPP and PS. We now need to keep track of the polarization of the pump; its electric field consists of two independent RPP patterns for each of the two polarization directions orthogonal to the beam propagation direction, a_{0x} and a_{0y}. We neglect pump depletion and want to model the amplification of a backscatter seed a_1. This model was introduced by Divol in Ref. [56].

The beginning of the derivation is exactly the same as for the description of CBET with arbitrary polarization, which was presented in Sect. 7.2.5 (the following is therefore a description of SBS but can easily be generalized to SRS). For counter-propagating waves in 1D, we obtain the same system of coupled equations except for a sign change on the right-hand side in the equation for the pump (since it is counter-propagating) and $\psi = \pi$. Substituting a_{0p}, a_{0s} with a_{0x}, a_{0y}, we arrive at Eqs. (7.100), (7.106) (we ignore pump depletion and are merely concerned with the amplification of the seed). The rest of the derivation in Sect. 7.2.5 assumed a linearly polarized pump; here we cannot make that assumption and need to keep the two fields a_{0x} and a_{0y}. To summarize, we have the same derivation up to the following solution for the amplification of the seed over a speckle length L_{sp}:

$$|a_1(z + L_{sp})\rangle = PGP^{-1} |a_1(z)\rangle , \qquad (9.123)$$

with

$$G = \begin{pmatrix} \exp[\frac{1}{2}\bar{G}_{sp}u] & 0 \\ 0 & 1 \end{pmatrix} , \quad P = \begin{pmatrix} \tilde{a}_{0x} & -\tilde{a}_{0y}^* \\ \tilde{a}_{0y} & \tilde{a}_{0x}^* \end{pmatrix} , \quad P^{-1} = \frac{1}{J_0} \begin{pmatrix} \tilde{a}_{0x}^* & \tilde{a}_{0y}^* \\ -\tilde{a}_{0y} & \tilde{a}_{0x} \end{pmatrix} ,$$

$$(9.124)$$

and where $J_0 = J_{0x} + J_{0y} = |\tilde{a}_{0x}|^2 + |\tilde{a}_{0y}|^2$. Like in the previous section, $u = I_0/\langle I_0 \rangle$ is the pump intensity normalized to its ensemble-average, with $I_0 \propto J_0$, and $\bar{G}_{sp}$ is the amplification gain exponent over one speckle length at the average intensity (the factor 1/2 denotes the fact that the amplification gain in amplitude is half the one in intensity). Note that the inverse matrix of P is also proportional to its conjugate transpose (as indicated by the dagger symbol), $P^{-1} = (1/I_0)P^{\dagger}$.

We want to calculate the seed amplification (in intensity) over one speckle length, given by

$$A_{\mathrm{PS}}(L_{sp}) = \frac{I_1(z + L_{sp})}{I_1(z)} = \frac{\langle a_1(z + L_{sp})|a_1(z + L_{sp})\rangle}{\langle a_1(z)|a_1(z)\rangle}. \tag{9.125}$$

Taking the conjugate transpose of $|a_1(z + L_{sp})\rangle$ from Eq. (9.123) above gives $\langle a_1(z + L_{sp})| = \langle a_1(z)|\, PGP^{-1}$, where we used $G^{\dagger} = G$ and $P^{-1} = (1/I_0)P^{\dagger}$, and therefore

$$\langle a_1(z + L_{sp})|a_1(z + L_{sp})\rangle = \langle a_1(z)|\, PG^2P^{-1}\, |a_1(z)\rangle. \tag{9.126}$$

Developing that expression via the matrix products eventually gives

$$I_1(z + L_{sp}) = I_{1x}(z)\frac{I_{0x}\exp[\bar{G}_{sp}u] + I_{0y}}{I_0} + I_{1y}(z)\frac{I_{0x} + I_{0y}\exp[\bar{G}_{sp}u]}{I_0}$$

$$+2\mathrm{Re}\left[\tilde{a}_{0x}\tilde{a}_{0y}^*\tilde{a}_{1x}^*\tilde{a}_{1y}\left(\exp[\bar{G}_{sp}u] - 1\right)\right], \tag{9.127}$$

where $I_{jx} \propto |\tilde{a}_{jx}|^2$, $I_{jy} \propto |\tilde{a}_{jy}|^2$ for $j \in \{0, 1\}$ and $I_j = I_{jx} + I_{jy}$.

To calculate the average amplification, we take the ensemble-average of that expression. We assume that the background seed and the laser are independent, i.e., $\langle I_0 I_1 \rangle = \langle I_0 \rangle \langle I_1 \rangle$, that the x and y components of the laser polarization are also independent—as is always the case with PS (thus the last term on the right-hand side averages to zero), and assume that the background noise contains the same power in both polarization directions, such that $\langle I_{1x} \rangle = \langle I_{1y} \rangle = \langle I_1 \rangle /2$. The average amplification over one speckle length becomes

$$\langle A_{\mathrm{PS}}(L_{sp})\rangle = \frac{\langle I_1(z + L_{sp})\rangle}{\langle I_1(z)\rangle} = \frac{\langle \exp[\bar{G}_{sp}u]\rangle + 1}{2}. \tag{9.128}$$

Finally, we need to calculate the average of the exponential term. Using the statistical distribution of the intensity with PS, as given by Eq. (9.64), gives

$$\langle \exp[\bar{G}_{sp}u]\rangle = \int_0^{\infty} p_{\mathrm{PS}}(I_0)\exp[\bar{G}_{sp}u]dI_0, \tag{9.129}$$

where $u = I_0/\langle I_0 \rangle$ and $p_{\mathrm{PS}}(I_0)dI_0 = 4u\exp[-2u]du$ is the intensity probability distribution with PS, from Eq. (9.64).

Proceeding with the integration yields

$$\langle \exp[\bar{G}_{sp}u]\rangle = \int_0^\infty 4u e^{u(\bar{G}_{sp}-2)}du \qquad (9.130)$$

$$= \left(1 - \frac{\bar{G}_{sp}}{2}\right)^{-2}, \quad \bar{G}_{sp} \le 2. \qquad (9.131)$$

The first important result is that the integral, and therefore the average amplification, is only defined for $\bar{G}_{sp} < 2$. While with RPP the divergence of the ensemble-averaged amplification of the backscatter seed starts for an average gain per speckle of 1 (Eq. (9.109)), when PS is used, the divergence threshold is raised to an average gain per speckle larger than 2. This is an obvious consequence of the $\propto \exp[-2I/\langle I\rangle]$ statistical distribution for a beam with PS vs. the $\propto \exp[-I/\langle I\rangle]$ for a beam with RPP only, which means that the fraction of intense speckles decays twice as fast vs. intensity with PS, thus providing a factor of two margin in intensity for the onset of divergent amplification compared to RPP.

The average amplification over one speckle length with PS is then

$$\langle A_{PS}(L_{sp})\rangle = \frac{1}{2}\left[1 + (1 - \bar{G}_{sp}/2)^{-2}\right]. \qquad (9.132)$$

The average amplification over a length L is then obtained like we did for RPP beams, taking the product of the amplification per speckle length over each row of speckle, $\langle A(L)\rangle = \langle A(L_{sp})\rangle^{L/L_{sp}}$, leading to

$$\boxed{\langle A_{PS}(L)\rangle = \exp\left[\frac{L}{L_{sp}}\ln\left(\frac{1 - \bar{G}_{sp}/2 + \bar{G}_{sp}^2/8}{(1 - \bar{G}_{sp}/2)^2}\right)\right].} \qquad (9.133)$$

When $\bar{G}_{sp} \ll 1$, a first-order expansion gives

$$\langle A_{PS}(L)\rangle \approx \exp\left[\frac{G_{1D}}{2}\right], \qquad (9.134)$$

where, like earlier, $G_{1D} = \bar{G}_{sp}L/L_{sp}$ is the 1D gain at the average intensity over L; in other words, PS reduces the 1D gain by a factor of two compared to RPP (cf. Eq. (9.111)). This is the same situation we already discussed for CBET, which also assumed a 1D model (which we now know is justified as long as the average gain per speckle is $\ll 1$, which is often the case for CBET in ICF conditions as the overall gains are typically of order $O(1)$ or less), and where PS also gave a factor of two reduction in gain as seen in Eq. (7.117) with $\psi = \pi$.

To finish this discussion, it is also interesting to quantify how much intensity margin is provided by PS for the onset of backscatter compared to using RPP only. Following Divol [56], we look for the coefficient α such that

Fig. 9.20 Backscatter amplification vs. 1D gain for RPP and PS, with $L/L_{sp} = 5$, 10, and ∞ (in which case $\langle A \rangle = \exp[G_{1D}]$ for RPP and $\langle A \rangle = \exp[G_{1D}/2]$ for PS)

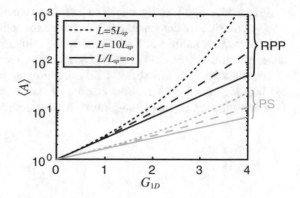

$$\langle A_{\text{RPP}}(\bar{G}_{sp}) \rangle = \langle A_{\text{PS}}(\alpha \bar{G}_{sp}) \rangle, \tag{9.135}$$

i.e., the backscatter amplification with RPP is the same as the amplification with PS with the average intensity multiplied by α.

After substituting in the expressions of $\langle A_{\text{RPP}} \rangle$ and $\langle A_{\text{PS}} \rangle$, we obtain

$$\alpha = 2 - \frac{\bar{G}_{sp}}{2} \frac{1 - \bar{G}_{sp}/2}{1 - \bar{G}_{sp}/4}. \tag{9.136}$$

The margin factor α varies slowly from 2 when $\bar{G}_{sp} \ll 1$ to 5/3 when $\bar{G}_{sp} \to 1$. In other words, using PS gives an intensity margin of about a factor of 2 compared to using RPP only, i.e., we can expect a similar level of backscatter with twice the laser intensity when using PS compared to RPP only. This has been confirmed in experiments and simulations [56]. The reduction in backscatter amplification for PS vs. RPP depending on L/L_{sp} is shown in Fig. 9.20.

9.5.2.4 Backscatter Mitigation via SSD or ISI

SSD mitigation of backscatter usually occurs indirectly, via the reduction of filamentation, which prevents the intensity increase in the most intense speckles as mentioned earlier. The typical bandwidth associated with SSD, at least in glass-based lasers, is typically too small to change the nature of the backscatter instabilities and impact their growth rates (cf. Ref. [57] and the references therein).

However, it has been estimated that SSD could in principle directly help mitigate SBS in certain regimes (SRS usually cannot be affected, as the associated time scales are too short). The direct mitigation processes identified for SBS via SSD occurs when the speckle pattern moves faster than the convective saturation time for SBS (i.e., the time it takes for SBS to reach steady state), $\tau_{sat} \approx G/\nu$, where G is the convective gain and ν is the IAW damping. This can be the case in weakly damped plasmas and for sufficiently high SSD bandwidths [39, 54, 55].

ISI can also significantly reduce backscatter. Experiments using ISI implemented on glass lasers have demonstrated SRS mitigation, although like for SSD, the effect was most likely indirect and attributed to the mitigation of filamentation (since the coherence time of the laser remained too long to have a direct effect on SRS) [34]. ISI on glass lasers demonstrated SBS mitigation as well [58]. The combination of ISI with high-bandwidth excimer lasers such as ArF or KrF is estimated to provide further mitigation against laser–plasma instabilities [30, 59, 60].

Problems

9.1 Aberration vs. RPPs

* Describe the similarities (and differences) between a beam whose phase front is strongly distorted by aberrations and a beam with an RPP.
* Describe the consequences of having a laser with phase front aberrations whose typical spatial scales are smaller than the RPP elements size. What will be the consequences in terms of focal spot characteristics?

9.2 Effect of Finite Pulse Duration on the Spectrum of an SSD Beam

* Consider a laser beam modulated by SSD with a square intensity profile in time of duration Δt, i.e., $I(t) = \text{rect}(t/\Delta t)$. Derive the expression of the beam power spectrum, and show that the finite duration of the pulse ensures that the "Bessel peaks" in the temporal spectrum of the beam (Fig. 9.12) are not delta functions but maintain a finite width $\delta \omega = 2\pi/\Delta t$ (in FWHM). Pay attention to the phase of the SSD modulation relative to the beam envelope: introduce a variable φ_0 in the SSD modulation term in the electric field expression from Eq. (9.68), i.e., $\hat{E} \propto \exp[-i\delta_m \sin(\omega_m t + \varphi_0)]$. The SSD phase at the beginning of the pulse ($t_0 = -\Delta t/2$) is then $\varphi_0 - \omega_m \Delta t/2$.
* Notice that when $\Delta t = 1/\nu_m$, significant broadening of the peaks occurs. Describe the evolution of the SSD spectrum when the pulse duration becomes shorter than $1/\nu_m$ and the effect of the SSD modulation phasing (i.e., φ_0) relative to the beginning of the pulse.

9.3 Self-focusing Threshold on a Speckle vs. Gain per Speckle Length

Show that the ratio P_{sp}/P_c of the power contained in a speckle to the critical power for self-focusing (cf. Sect. 5.6) is equal to the spatial amplification gain exponent for the filamentation instability (Sect. 7.3.2, Eq. (7.123)) over one speckle length. In other words, the power in a speckle exceeds the self-focusing threshold when the filamentation spatial gain exponent over a speckle length exceeds unity.

References

1. J. Goodman, *Statistical Optics*. Wiley Series in Pure and Applied Optics (Wiley, 2015)
2. J. Goodman, *Speckle Phenomena in Optics: Theory and Applications*. SPIE Press monograph (SPIE Press, 2020)
3. J.W. Goodman, J. Opt. Soc. Am. **66**, 1145 (1976)
4. Y. Kato, K. Mima, N. Miyanaga, S. Arinaga, Y. Kitagawa, M. Nakatsuka, C. Yamanaka, Phys. Rev. Lett. **53**, 1057 (1984)
5. S.N. Dixit, I.M. Thomas, B.W. Woods, A.J. Morgan, M.A. Henesian, P.J. Wegner, H.T. Powell, Appl. Opt. **32**, 2543 (1993)
6. S.N. Dixit, J.K. Lawson, K.R. Manes, H.T. Powell, K.A. Nugent, Opt. Lett. **19**, 417 (1994)
7. S.N. Dixit, M.D. Feit, M.D. Perry, H.T. Powell, Opt. Lett. **21**, 1715 (1996)
8. J. Garnier, Phys. Plasmas **6**, 1601 (1999)
9. E. Hecht, *Optics*. Pearson Education (Addison-Wesley, 2002)
10. H.A. Rose, D.F. DuBois, Phys. Fluids B: Plasma Phys. **5**, 590 (1993)
11. A. Gut, *An Intermediate Course in Probability*, 2nd edn. (Springer, Berlin, 2009)
12. J. Garnier, C. Gouédard, A. Migus, J. Mod. Opt. **46**, 1213 (1999)
13. R.J. Adler, *The Geometry of Random Fields* (Society for Industrial and Applied Mathematics, 2010)
14. C. Delmas, Comptes Rendus de l'Académie des Sciences - Series I - Mathematics **327**, 393 (1998)
15. K. Tsubakimoto, M. Nakatsuka, H. Nakano, T. Kanabe, T. Jitsuno, S. Nakai, Opt. Commun. **91**, 9 (1992)
16. T.R. Boehly, V.A. Smalyuk, D.D. Meyerhofer, J.P. Knauer, D.K. Bradley, R.S. Craxton, M.J. Guardalben, S. Skupsky, T.J. Kessler, J. Appl. Phys. **85**, 3444 (1999)
17. D.H. Munro, S.N. Dixit, A.B. Langdon, J.R. Murray, Appl. Opt. **43**, 6639 (2004)
18. S. Dixit, D. Munro, J. Murray, M. Nostrand, P. Wegner, D. Froula, C. Haynam, B. MacGowan, in *Journal de Physique IV (Proceedings)*, vol. 133 (EDP Sciences, 2006), pp. 717–720
19. S. Skupsky, R.S. Craxton, Phys. Plasmas **6**, 2157 (1999)
20. T.R. Boehly, V.N. Goncharov, O. Gotchev, J.P. Knauer, D.D. Meyerhofer, D. Oron, S.P. Regan, Y. Srebro, W. Seka, D. Shvarts, S. Skupsky, V.A. Smalyuk, Phys. Plasmas **8**, 2331 (2001)
21. S. Skupsky, R.W. Short, T. Kessler, R.S. Craxton, S. Letzring, J.M. Soures, J. Appl. Phys. **66**, 3456 (1989)
22. D.H. Kalantar, M.H. Key, L.B. DaSilva., S.G. Glendinning, J.P. Knauer, B.A. Remington, F. Weber, S.V. Weber, Phys. Rev. Lett. **76**, 3574 (1996)
23. V.N. Goncharov, S. Skupsky, T.R. Boehly, J.P. Knauer, P. McKenty, V.A. Smalyuk, R.P.J. Town, O.V. Gotchev, R. Betti, D.D. Meyerhofer, Phys. Plasmas **7**, 2062 (2000)
24. M.L. Spaeth, K.R. Manes, D.H. Kalantar, P.E. Miller, J.E. Heebner, E.S. Bliss, D.R. Spec, T.G. Parham, P.K. Whitman, P.J. Wegner, P.A. Baisden, J.A. Menapace, M.W. Bowers, S.J. Cohen, T.I. Suratwala, J.M.D. Nicola, M.A. Newton, J.J. Adams, J.B. Trenholme, R.G. Finucane, R.E. Bonanno, D.C. Rardin, P.A. Arnold, S.N. Dixit, G.V. Erbert, A.C. Erlandson, J.E. Fair, E. Feigenbaum, W.H. Gourdin, R.A. Hawley, J. Honig, R.K. House, K.S. Jancaitis, K.N. LaFortune, D.W. Larson, B.J.L. Galloudec, J.D. Lindl, B.J. MacGowan, C.D. Marshall, K.P. McCandless, R.W. McCracken, R.C. Montesanti, E.I. Moses, M.C. Nostrand, J.A. Pryatel, V.S. Roberts, S.B. Rodriguez, A.W. Rowe, R.A. Sacks, J.T. Salmon, M.J. Shaw, S. Sommer, C.J. Stolz, G.L. Tietbohl, C.C. Widmayer, R. Zacharias, Fusion Sci. Technol. **69**, 25 (2016)
25. L. Videau, C. Rouyer, J. Garnier, A. Migus, J. Opt. Soc. Am. A **16**, 1672 (1999)
26. J. Garnier, L. Videau, Phys. Plasmas **8**, 4914 (2001)
27. R. Lehmberg, S. Obenschain, Opt. Commun. **46**, 27 (1983)
28. R.H. Lehmberg, A.J. Schmitt, S.E. Bodner, J. Appl. Phys. **62**, 2680 (1987)

29. S.P. Obenschain, S.E. Bodner, D. Colombant, K. Gerber, R.H. Lehmberg, E.A. McLean, A.N. Mostovych, M.S. Pronko, C.J. Pawley, A.J. Schmitt, J.D. Sethian, V. Serlin, J.A. Stamper, C.A. Sullivan, J.P. Dahlburg, J.H. Gardner, Y. Chan, A.V. Deniz, J. Hardgrove, T. Lehecka, M. Klapisch, Phys. Plasmas **3**, 2098 (1996)
30. S.P. Obenschain, A.J. Schmitt, J.W. Bates, M.F. Wolford, M.C. Myers, M.W. McGeoch, M. Karasik, J.L. Weaver, Philoso. Trans. R. Soc. A: Math. Phys. Eng. Sci. **378**, 20200031 (2020)
31. R.H. Lehmberg, J. Goldhar, Fusion Technol. **11**, 532 (1987)
32. S. Obenschain, R. Lehmberg, D. Kehne, F. Hegeler, M. Wolford, J. Sethian, J. Weaver, M. Karasik, Appl. Opt. **54**, F103 (2015)
33. S.P. Obenschain, J. Grun, M.J. Herbst, K.J. Kearney, C.K. Manka, E.A. McLean, A.N. Mostovych, J.A. Stamper, R.R. Whitlock, S.E. Bodner, J.H. Gardner, R.H. Lehmberg, Phys. Rev. Lett. **56**, 2807 (1986)
34. S.P. Obenschain, C.J. Pawley, A.N. Mostovych, J.A. Stamper, J.H. Gardner, A.J. Schmitt, S.E. Bodner, Phys. Rev. Lett. **62**, 768 (1989)
35. O. Willi, T. Afshar-rad, S. Coe, A. Giulietti, Phys. Fluids B: Plasma Phys. **2**, 1318 (1990)
36. B.J. MacGowan, B.B. Afeyan, C.A. Back, R.L. Berger, G. Bonnaud, M. Casanova, B.I. Cohen, D.E. Desenne, D.F. DuBois, A.G. Dulieu, K.G. Estabrook, J.C. Fernandez, S.H. Glenzer, D.E. Hinkel, T.B. Kaiser, D.H. Kalantar, R.L. Kauffman, R.K. Kirkwood, W.L. Kruer, A.B. Langdon, B.F. Lasinski, D.S. Montgomery, J.D. Moody, D.H. Munro, L.V. Powers, H.A. Rose, C. Rousseaux, R.E. Turner, B.H. Wilde, S.C. Wilks, E.A. Williams, Phys. Plasmas **3**, 2029 (1996)
37. D.S. Montgomery, J.D. Moody, H.A. Baldis, B.B. Afeyan, R.L. Berger, K.G. Estabrook, B.F. Lasinski, E.A. Williams, C. Labaune, Phys. Plasmas **3**, 1728 (1996)
38. J. Fuchs, C. Labaune, S. Depierreux, H.A. Baldis, A. Michard, Phys. Rev. Lett. **84**, 3089 (2000)
39. J.D. Moody, B.J. MacGowan, J.E. Rothenberg, R.L. Berger, L. Divol, S.H. Glenzer, R.K. Kirkwood, E.A. Williams, P.E. Young, Phys. Rev. Lett. **86**, 2810 (2001)
40. S.H. Glenzer, D.H. Froula, L. Divol, M. Dorr, R.L. Berger, S. Dixit, B.A. Hammel, C. Haynam, J.A. Hittinger, J.P. Holder, O.S. Jones, D.H. Kalantar, O.L. Landen, A.B. Langdon, S. Langer, B.J. MacGowan, A.J. Mackinnon, N. Meezan, E.I. Moses, C. Niemann, C.H. Still, L.J. Suter, R.J. Wallace, E.A. Williams, B.K.F. Young, Nat. Phys. **3**, 716 (2007)
41. S. Hüller, P. Mounaix, V.T. Tikhonchuk, Phys. Plasmas **5**, 2706 (1998)
42. R.L. Berger, E. Lefebvre, A.B. Langdon, J.E. Rothenberg, C.H. Still, E.A. Williams, Phys. Plasmas **6**, 1043 (1999)
43. H.A. Rose, D.F. DuBois, Phys. Fluids B: Plasma Phys. **5**, 3337 (1993)
44. E.A. Williams, Phys. Plasmas **13**, 056310 (2006)
45. C. Labaune, S. Baton, T. Jalinaud, H.A. Baldis, D. Pesme, Phys. Fluids B: Plasma Phys. **4**, 2224 (1992)
46. G. Sarri, C.A. Cecchetti, R. Jung, P. Hobbs, S. James, J. Lockyear, R.M. Stevenson, D. Doria, D.J. Hoarty, O. Willi, M. Borghesi, Phys. Rev. Lett. **106**, 095001 (2011)
47. J.L. Kline, D.S. Montgomery, K.A. Flippo, R.P. Johnson, H.A. Rose, T. Shimada, E.A. Williams, Rev. Sci. Instrum. **79**, 10F551 (2008)
48. D.H. Froula, L. Divol, N.B. Meezan, S. Dixit, J.D. Moody, P. Neumayer, B.B. Pollock, J.S. Ross, S.H. Glenzer, Phys. Rev. Lett. **98**, 085001 (2007)
49. R.L. Berger, B.F. Lasinski, T.B. Kaiser, E.A. Williams, A.B. Langdon, B.I. Cohen, Phys. Fluids B: Plasma Phys. **5**, 2243 (1993)
50. E. Lefebvre, R.L. Berger, A.B. Langdon, B.J. MacGowan, J.E. Rothenberg, E.A. Williams, Phys. Plasmas **5**, 2701 (1998)
51. H.A. Rose, D.F. DuBois, Phys. Rev. Lett. **72**, 2883 (1994)
52. H.A. Rose, Phys. Plasmas **2**, 2216 (1995)
53. V.T. Tikhonchuk, P. Mounaix, D. Pesme, Phys. Plasmas **4**, 2658 (1997)
54. P. Mounaix, L. Divol, S. Hüller, V.T. Tikhonchuk, Phys. Rev. Lett. **85**, 4526 (2000)
55. L. Divol, Phys. Rev. Lett. **99**, 155003 (2007)

56. D.H. Froula, L. Divol, R.L. Berger, R.A. London, N.B. Meezan, D.J. Strozzi, P. Neumayer, J.S. Ross, S. Stagnitto, L.J. Suter, S.H. Glenzer, Phys. Rev. Lett. **101**, 115002 (2008)
57. D. Pesme, R.L. Berger, E.A. Williams, A. Bourdier, A. Bortuzzo-Lesne, A statistical description of parametric instabilities with an incoherent pump (2007). https://arxiv.org/abs/0710.2195
58. A.N. Mostovych, S.P. Obenschain, J.H. Gardner, J. Grun, K.J. Kearney, C.K. Manka, E.A. McLean, C.J. Pawley, Phys. Rev. Lett. **59**, 1193 (1987)
59. R.K. Follett, J.G. Shaw, J.F. Myatt, C. Dorrer, D.H. Froula, J.P. Palastro, Phys. Plasmas **26**, 062111 (2019)
60. J.W. Bates, J.F. Myatt, J.G. Shaw, R.K. Follett, J.L. Weaver, R.H. Lehmberg, S.P. Obenschain, Phys. Rev. E **97**, 061202 (2018)

Chapter 10
Saturation of Laser–Plasma Instabilities and Other Nonlinear Effects

In this chapter we investigate some of the most well-known saturation and nonlinear processes occurring in laser–plasma interactions. This is a vast area of research, and our list is clearly incomplete; our goal here is merely to provide an introduction to some selected physics processes with simple and intuitive scalings that can be used for the design or analysis of experiments.

We are going to describe three types of saturation processes: (i) pump depletion, i.e., saturation by depletion of the pump energy as it is transferred to scattered waves, (ii) fluid nonlinearities, more specifically secondary instabilities (i.e., when a scattered wave grows enough to start a decay process of its own) and the generation of harmonics, and (iii) kinetic effects, related to particle trapping and potentially affecting the growth of instabilities via the formation of non Maxwellian velocity distributions.

10.1 Saturation by Pump Depletion in Uniform Plasma

In this section we discuss the saturation of three-wave coupling instabilities involving two light waves and a plasma wave (SRS, SBS, or CBET) by depletion of the pump energy. Simple analytic solutions exist under certain assumptions: steady-state, weakly coupled regime, strongly damped plasma waves (in the sense of strong damping compared to spatial amplification rate, as explained in Sect. 6.4.4), and simple geometries. More specifically, geometries that do admit analytic solutions include 1D counter-propagating geometry, 1D co-propagating (which is only valid in some restricted region of space in the case of real, finite-envelope beams, as we will discuss later), and 2D. We analyze these situations in more detail below.

P. Michel, *Introduction to Laser-Plasma Interactions*, Graduate Texts in Physics, https://doi.org/10.1007/978-3-031-23424-8_10

10.1.1   Pump Depletion for 1D Backscatter: The "Tang formula"

Let us begin with the simple situation of pump depletion in a 1D backscatter geometry. The following analysis is valid for SBS or SRS (the coupling constants will be specified below, in their fluid or kinetic form). We write the coupled mode equations for the slowly varying field envelopes, Eqs. (6.70)–(6.72), assuming steady-state ($\partial_t = 0$), neglecting absorption of the two light waves ($\nu_0 = \nu_1 = 0$), and taking a strongly damped plasma wave assumption (cf. Sect. 6.4.4). We obtain, with the subscripts 0, 1, and 2 assigned to the pump laser wave, the scattered EMW and the plasma wave (EPW or IAW), respectively:

$$v_{g0}\tilde{\alpha}_0' = -i K \tilde{\alpha}_1 \tilde{\alpha}_2 , \tag{10.1}$$

$$-v_{g1}\tilde{\alpha}_1' = -i K \tilde{\alpha}_0 \tilde{\alpha}_2^* , \tag{10.2}$$

$$\nu\tilde{\alpha}_2 = -i K \tilde{\alpha}_0 \tilde{\alpha}_1^* , \tag{10.3}$$

where ν is the damping of the plasma wave, the prime denotes the derivative with respect to z, and we assumed that the pump propagates toward z and the backscattered wave (wave #1) toward $-z$ (the group velocities are all assumed positive). Like in Sect. 6.4.6, the variable $K = \gamma / |\tilde{\alpha}_0(0)|$ represents the temporal growth rate normalized to the initial value of the pump amplitude, and the quantities $\tilde{\alpha}_j$ for the three waves ($j \in \{0, 1, 2\}$) are the action amplitude envelopes[1] as defined in Sect. 1.3.5, Eqs. (1.214)–(1.216).

Inserting Eq. (10.3) into Eqs. (10.1) and (10.2) gives the following coupled equations for the two light waves:

$$\tilde{\alpha}_0' = -\Gamma \frac{v_{g1}}{v_{g0}} \frac{\mathcal{A}_1}{\mathcal{A}_0(0)} \tilde{\alpha}_0 , \tag{10.4}$$

$$\tilde{\alpha}_1' = -\Gamma \frac{\mathcal{A}_0}{\mathcal{A}_0(0)} \tilde{\alpha}_1 , \tag{10.5}$$

where $\mathcal{A}_j = |\tilde{\alpha}_j|^2$ is the wave action (cf. Sect. 1.3.5) and we recognize the spatial amplification rate $\Gamma = \gamma^2/(v_{g1}\nu)$ first introduced in Sect. 6.4.4, Eq. (6.52). Γ can take the following form depending on the process under consideration (cf. Eqs. (7.70), (8.18), (7.13)):

---

[1] The tildes denote slowly varying envelopes, such that for each of the three waves, $\alpha_j = \frac{1}{2}\tilde{\alpha}_j e^{i\psi_j} + c.c.$, with $\psi_j = \mathbf{k}_j \cdot \mathbf{r} - \omega_j t$.

$$\Gamma = \begin{cases} \dfrac{k_2^2}{8k_1}\mathrm{Im}(F_\chi)\,J_0(0) & \text{(kinetic)}, \\[2ex] \dfrac{k_2^2\,J_0(0)}{16k_1\nu/\omega_{pe}} & \text{(fluid, SRS)}, \\[2ex] \dfrac{J_0(0)}{16k_0\lambda_{De}^2\nu/\omega_{pe}} & \text{(fluid, SBS)}, \end{cases} \tag{10.6}$$

where $J_j = |\tilde{a}_j|^2$ with $a_j = eE_j/(mc\omega_j) = \frac{1}{2}\tilde{a}_j e^{i\psi_j} + c.c.$ the normalized vector potential for the light waves and $\psi_j = \mathbf{k}_j \cdot \mathbf{r} - \omega_j t$. Switching back to these more convenient variables throughout (cf. Eq. (1.214)), we arrive at

$$J_0' = -\frac{k_1}{k_0}\frac{2\Gamma}{J_0(0)}J_0 J_1, \tag{10.7}$$

$$J_1' = -\frac{2\Gamma}{J_0(0)}J_0 J_1. \tag{10.8}$$

When pump depletion is negligible, then $J_0 \approx J_0(0)$ in the second equation and we simply recover the spatial amplification obtained earlier for SRS, SBS, or CBET, i.e., $J_1' = -2\Gamma J_1$ (the minus sign comes in as the scattered wave propagates toward $z < 0$).

The two coupled equations (10.7) and (10.8) admit the following conserved quantity (from the conservation of flux according to Manley–Rowe relations, the "missing" energy being transferred to the plasma wave—cf. Sect. 6.4.6):

$$C = k_0 J_0(z) - k_1 J_1(z), \tag{10.9}$$

such that $C' = 0$.

We consider a bounded region of plasma from $z=0$ to L. The boundary conditions for the laser and scattered waves are on the opposite sides of the domain: $J_0(0)$ is the incoming pump laser (normalized) intensity, and $J_1(L)$ is the background ("seed") level initially present for the scattered light wave.

Eliminating J_1 from Eq. (10.7) using Eq. (10.9) leads to

$$J_0' = -\frac{2\Gamma}{J_0(0)}(J_0^2 - C J_0/k_0), \tag{10.10}$$

or

$$\frac{dJ_0}{J_0^2 - (C/k_0)J_0} = -\frac{2\Gamma}{J_0(0)}dz. \tag{10.11}$$

Integrating using the formula

$$\int \frac{dx}{x^2 - ax} = \frac{1}{a}\ln\left(1 - \frac{a}{x}\right) \tag{10.12}$$

eventually leads to

$$1 - \frac{C}{k_0 J_0(z)} = C_1 \exp\left[-2C\frac{\Gamma}{J_0(0)k_0}z\right],\tag{10.13}$$

where the constant of integration is evaluated at $z = 0$ as $C_1 = 1 - C/[k_0 J_0(0)]$.

Noting from the definition of C that $1 - C/[k_0 J_0(z)] = \rho J_1(z)/J_0(z)$ with $\rho = k_1/k_0$ and introducing the reflectivity

$$R = \frac{J_1(0)}{J_0(0)},\tag{10.14}$$

we obtain

$$J_1(z) = R J_0(z) \exp[-G_1(1 - \rho R)z/L],\tag{10.15}$$

where $G_1 = 2\Gamma L$ is the amplification gain exponent in intensity over the interaction length L for the initial (un-depleted) pump intensity.

Eliminating $J_0(z)$ via the expression of C from Eq. (10.9) gives

$$J_1(z) = J_1(0)\frac{1 - \rho R}{\exp[G_1(1 - \rho R)z/L] - \rho R}.\tag{10.16}$$

In the limit of small reflectivity, $R \ll 1$, with consequently $J_0(z) \approx J_0(0)$ (negligible pump depletion), we recover the result of the previous chapters on spatial amplification without pump depletion, i.e., $J_1(0) \approx J_1(L)e^{G_1}$ (remember that the scattered light wave goes from L to 0).

However, when R is not negligible, the equation becomes transcendental. It is then convenient to introduce

$$\varepsilon = \frac{J_1(L)}{J_0(0)},\tag{10.17}$$

which represents the level of background fluctuations seeding the backscatter relative to the laser. After a few manipulations, we can recast Eq. (10.16) into

$$\boxed{R(1 - \rho R) = \varepsilon\left(e^{G_1(1-\rho R)} - \rho R\right),}\tag{10.18}$$

where $\rho = k_1/k_0$.

This equation, often referred to as the Tang formula after [1] (where it was derived for SBS, i.e., with $\omega_0 \approx \omega_1$ and therefore $\rho \approx 1$), describes the general behavior of nonlinear pump depletion for the reflectivity as a function of the "un-depleted gain" G_1 for a given seed level ε. It is transcendental, meaning that the reflectivity with pump depletion cannot be expressed analytically for the

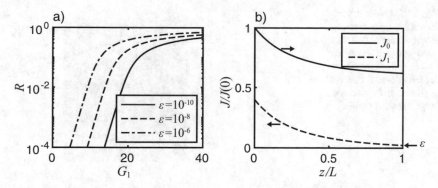

Fig. 10.1 Nonlinear pump depletion for pure backscatter geometry in a uniform plasma (assuming SBS, i.e., $\omega_0 \approx \omega_1$): (**a**) "Tang formula" (Eq. 10.18), showing the reflectivity R as a function of the linear gain exponent G_1 for several seed levels $\varepsilon = J_1(L)/J_0(0)$. (**b**) Example of intensity profiles for the pump J_0 and the seed J_1, for a gain of 4 and a seed level $\varepsilon \approx 2.3\%$ (the arrows indicate the propagation directions)

counter-propagating geometry; however, it is easily plotted numerically as shown in Fig. 10.1a. Figure 10.1b represents a typical intensity profile along z for the pump and the scattered light wave.

The similar shape between the pump and the backscatter wave's spatial intensity profiles is simply interpreted via the conservation of flux, Eq. (10.9); the drop in pump intensity is most severe near the entrance of the plasma where the backscatter is reaching its highest intensity and is thus most prone to deplete the pump.

10.1.2   Pump Depletion for 1D Forward Scatter and Co-Propagating Geometry

Analytic solutions for pump depletion can also be derived for a co-propagating geometry, as can occur for forward scatter or CBET (i.e., scattering at a fixed angle) as long as the geometry remains 1D. This is illustrated in Fig. 10.2: for two waves crossing at some angle θ and a plasma region limited to $0 \leq z \leq L$, the 1D assumption may only be valid in the shaded region in the figure. This would typically correspond to a rather thin region where resonance conditions are met and the plasma can be assumed uniform [2]. The treatment of regions of overlap outside the shaded region can still be performed analytically, as described in the next section.

The derivation is essentially the same as for the counter-propagating geometry except for a change of sign. Applying projections onto an axis z such that $\theta_j = \mathbf{k}_j \cdot \mathbf{e}_z / k_j$ (with $j \in \{0, 1\}$) as illustrated in Fig. 10.2, with $k_{jz} = \mathbf{k}_j \cdot \mathbf{e}_z$, we obtain the two coupled equations:

Fig. 10.2 Geometry of forward or near-forward propagation for two waves overlapping at an angle $\theta = \theta_0 + \theta_1$. The 1D pump depletion model may only be applied to the darker area (with a dashed outline)

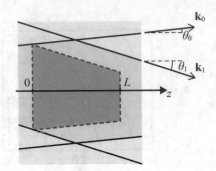

$$J_0' = -\frac{k_{1z}}{k_{0z}} \frac{2\Gamma}{J_0(0)} J_0 J_1 , \tag{10.19}$$

$$J_1' = \frac{2\Gamma}{J_0(0)} J_0 J_1 . \tag{10.20}$$

The conserved quantity is now

$$K = k_{0z} J_0(z) + k_{1z} J_1(z) , \tag{10.21}$$

with $K' = 0$. Proceeding exactly like for the counter-propagating geometry eventually leads to

$$\boxed{J_1(z) = J_1(0) \frac{1 + \rho R}{\exp[-(1 + \rho R)G_1 z/L] + \rho R}} , \tag{10.22}$$

where similarly to the previous section, $\rho = k_{1z}/k_{0z}$, $G_1 = 2\Gamma L$, and $R = J_1(0)/J_0(0)$. Note that since both waves propagate toward $z > 0$, R is not a reflectivity anymore but simply the ratio of the (presumably known) boundary conditions for the intensities at $z = 0$. $J_0(z)$ is simply obtained from the definition of K:

$$\boxed{J_0(z) = J_0(0) \frac{1 + \rho R}{1 + \rho R \exp[(1 + \rho R)G_1 z/L]}} . \tag{10.23}$$

Unlike for the counter-propagating case, the expressions for the intensities of both light waves are now perfectly well defined in terms of boundary conditions at $z = 0$ by Eqs. (10.22)–(10.23). These solutions can describe a forward scattering process, in which case $J_1(0)$ is the background seed level for the scattered light wave or CBET. Typical intensity profiles are shown in Fig. 10.3, for $G_1 = 4$ and $R = 0.2$.

Fig. 10.3 Intensities for
co-propagating propagation
with 1D pump depletion, for
$R = 0.2$ and $G_1 = 4$

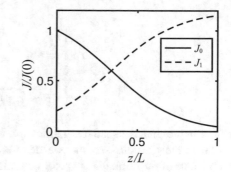

Fig. 10.4 Geometry of
CBET in two dimensions. P_0
represents the flux per unit
length of beam 0 across the
dashed arrow

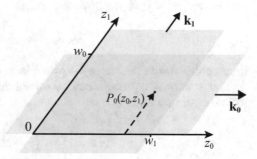

10.1.3 Pump Depletion in 2D for CBET

Unlike backscatter instabilities like SBS and SRS, where the nonlinear pump
depletion can often be treated as a 1D problem since the scattered light wave tends
to propagate exactly backward from the incoming laser, CBET always involves a
finite crossing angle between the interacting beams which makes the pump depletion
process intrinsically 2D. Here the two relevant dimensions are along the propagation
directions of the two beams, z_0 and z_1 (cf. Figs. 7.5, 10.4); the following results are
valid for any "slice" at a fixed location along the third dimension, normal to the
plane of incidence (z_0, z_1). The problem was analyzed in [3], and its solution gives
some interesting physical insight into the nonlinear pump depletion problem in 2D.

Consider the geometry from Fig. 10.4; we will work in the 2D coordinate system
(z_0, z_1). The widths of beams 0 and 1 are w_0 (along z_1) and w_1 (along z_0),
respectively. We follow the same procedure as in the previous section, i.e., start with
Eqs. (6.70)–(6.72) with the same assumptions (steady-state, no EMW absorption,
strong damping of the plasma wave) and eventually obtain

$$\partial_{z_0} J_0(z_0, z_1) = -2\hat{\Gamma} J_0 J_1 , \tag{10.24}$$

$$\partial_{z_1} J_1(z_0, z_1) = 2\hat{\Gamma} J_0 J_1 , \tag{10.25}$$

where

$$
\hat{\Gamma} =
\begin{cases}
\dfrac{k_2^2}{8k_1}\mathrm{Im}(F_\chi) & \text{(kinetic)}, \\[2ex]
\dfrac{1}{16k_0\lambda_{De}^2\,\nu/\omega_{pe}} & \text{(fluid)}.
\end{cases}
\tag{10.26}
$$

$\hat{\Gamma}$ represents the spatial amplification rate without the intensity dependence, and like in the previous section, we denote $J_j = |\tilde{a}_j|^2$ with $\tilde{a}_j$ the slowly varying envelope of the normalized vector potential for wave j ($j \in \{0, 1\}$). Since $\omega_0 \approx \omega_1$, we assumed $k_0 \approx k_1$ as well. We notice that

$$
\partial_{z_0} J_0 + \partial_{z_1} J_1 = 0
\tag{10.27}
$$

(in accordance with the Manley–Rowe relations).

This system of equations can be solved analytically [3, 4], by introducing the following quantities:

$$
P_0(z_0, z_1) = \int_0^{z_1} J_0(z_0, z_1')dz_1',
\tag{10.28}
$$

$$
P_1(z_0, z_1) = \int_0^{z_0} J_1(z_0', z_1)dz_0',
\tag{10.29}
$$

where z_0' and z_1' are dummy integration variables.

The quantities P_0 and P_1 represent the flux per unit length of beams 0 and 1, integrated up to the point of coordinates (z_0, z_1), as illustrated in Fig. 10.4. Defining input boundary conditions for the beams' intensities at the entrance of the interaction region, $J_{0i} = J_0(0, z_1)$ and $J_{1i} = J_1(z_0, 0)$, we find from the definition of P_0 and using Eq. (10.27)

$$
\partial_{z_0} P_0 = J_{1i} - J_1.
\tag{10.30}
$$

Inserting the solution of Eq. (10.25), $J_1 = J_{1i}\exp[2\hat{\Gamma}P_0]$ gives the equation:

$$
\frac{\partial_{z_0} P_0}{1 - e^{2\hat{\Gamma}P_0}} = J_{1i}.
\tag{10.31}
$$

This equation can be integrated using the formula:

$$
\int \frac{dx}{1 - e^x} = x - \ln\left(1 - e^x\right),
\tag{10.32}
$$

yielding the following solution for P_0:

$$P_0 = -\frac{1}{2\hat{\Gamma}} \ln \left(1 + C \exp \left[-2\hat{\Gamma} J_{1i} z_0 \right] \right) , \tag{10.33}$$

where C is an integration constant. The boundary condition at $z_0 = 0$ gives $P_0(0, z_1) = J_{0i} z_1$, which is then used to eliminate the integration constant, giving the final expression for P_0:

$$P_0(z_0, z_1) = \frac{-1}{2\hat{\Gamma}} \ln \left(1 + e^{-2\hat{\Gamma} J_{1i} z_0} \left[-1 + e^{-2\hat{\Gamma} J_{0i} z_1} \right] \right) . \tag{10.34}$$

The intensities can then be derived via the relationships $J_0 = \partial_{z_1} P_0$ and $J_1 = J_{1i} \exp[2\hat{\Gamma} P_0]$, leading to

$$J_0(z_0, z_1) = J_{0i} \frac{e^{-2\hat{\Gamma} J_{0i} z_1}}{e^{2\hat{\Gamma} J_{1i} z_0} - 1 + e^{-2\hat{\Gamma} J_{0i} z_1}} , \tag{10.35}$$

$$J_1(z_0, z_1) = J_{1i} \frac{e^{2\hat{\Gamma} J_{1i} z_0}}{e^{2\hat{\Gamma} J_{1i} z_0} - 1 + e^{-2\hat{\Gamma} J_{0i} z_1}} . \tag{10.36}$$

These solutions are illustrated in Fig. 10.5, representing the intensities in the 2D region of overlap (cf. Fig. 10.4). The coupling term $\hat{\Gamma}$ was chosen positive in these simulations, with $2\hat{\Gamma} J_{0i} w_0 = 2\hat{\Gamma} J_{1i} w_1 = 1$ (the energy transfer goes from beam 0 to beam 1).

Beam 0 experiences the linear interaction (i.e., with beam 1 assumed constant) along the $z_1 = 0$ boundary, since this is where beam 1 enters the interaction region; it decays by $e^{-1} \approx 0.37$ as it reaches $(z_0 = w_1, z_1 = 0)$. Likewise, for beam 1, the amplification is also linear (i.e., without pump depletion from beam 0) along the $z_0 = 0$ boundary, where beam 0 comes in. It is amplified by $e^1 \approx 2.7$ as it reaches $(z_0 = 0, z_1 = w_0)$.

On the other side of the overlap region, along $z_1 = w_0$, beam 0 interacts with a "stronger" beam 1 (as beam 1 has been amplified from $z_1 = 0$ to $z_1 = w_0$) and experiences more depletion as it propagates to $(z_0 = w_1, z_1 = w_0)$. Likewise, along $z_0 = w_1$ (i.e., downstream from beam 0 propagation), beam 1 interacts with a significantly depleted beam 0 and experiences a much weaker amplification as it reaches $(z_0 = w_1, z_1 = w_0)$.

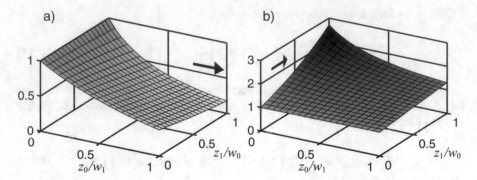

Fig. 10.5 2D pump depletion in CBET. (**a**) Normalized intensity of beam 0, J_0/J_{0i}, and (**b**) normalized intensity of beam 1, J_1/J_{1i}. Beam 0 and beam 1 are propagating toward z_0 and z_1, respectively. Distances are normalized to the length of the overlap region (cf. Fig. 10.4); the maximum amplification gain exponent is $2\hat{\Gamma} J_{0i} w_0 = 2\hat{\Gamma} J_{1i} w_1 = 1$

10.2   Fluid Nonlinearities: Secondary Instabilities and Harmonics Generation

All the three-wave instabilities we discussed in the previous chapters (SRS, SBS, CBET, TPD) can drive high-amplitude plasma waves (EPWs or IAWs). When the driven plasma wave becomes large enough, it can drive its own three-wave instability as a pump and decay into daughter waves; the two main instabilities seen in experiments and simulations are the Langmuir decay instability (LDI), whereby an EPW decays into another EPW and an IAW, and the two-ion decay instability (TID), where an IAW decays into two daughter IAWs (cf. Sect. 6.1 and Table 6.1). These processes can saturate the growth of the primary three-wave instability (involving a high-intensity laser), by limiting the growth of the daughter plasma wave. We will also describe the electromagnetic decay instability (EDI), where an EPW decays into an EMW and an IAW.

Another potential saturation mechanism is the generation of harmonics of the driven plasma wave. As we will see, harmonic generation is not an instability and follows a linear (vs. exponential) growth vs. propagation distance. It can, however, limit the growth of the plasma wave from the primary instability and act as a saturation mechanism.

10.2.1   Langmuir Decay Instability (LDI)

The Langmuir decay instability (LDI) is the process whereby a high-amplitude EPW decays into another EPW and an IAW (cf. Table 6.1). LDI can be an efficient saturation process for the three-wave instabilities that involve an EPW as a secondary wave, like SRS and TPD: if the primary instability (SRS/TPD) develops to the point where the excited EPW reaches a high enough amplitude, then this EPW will decay via LDI, which will saturate the growth of the primary instability. LDI was investigated theoretically by Dubois in [5] in the context of laser–plasma experiments, only a few years after the laser was invented. It was (much) later suggested as a saturation mechanism for SRS and TPD in laser–plasma experiments [6, 7]. Experimental evidence for the process in laser–plasma experiments started to emerge a few years later, first via the observation of increased SRS saturation in plasmas with low IAW damping (attributed to the facilitated growth of IAWs from LDI) [8], then via the direct measurement of Thomson scattering off the secondary EPW [9, 10], and finally via the simultaneous measurement (also from Thomson scattering) of both secondary waves (EPW and IAW) [11]. LDI was also measured by radar observations in the ionosphere, as a secondary process driven by the decay of a high-power radio wave (cf. [12] and the references therein). In the following we derive the main characteristics of LDI, such as its geometry, growth rate, and threshold, and will briefly describe the so-called LDI cascade and evolution toward turbulence.

We identify the three waves via their indices 0 (primary EPW), 1 (daughter EPW), and 2 (daughter IAW) as in the earlier chapters, being careful to remember that wave 0, the "pump," is not a light wave anymore but an EPW. The phase-matching conditions are as usual for any three-wave coupling process,

$$\omega_0 = \omega_1 + \omega_2 \,, \tag{10.37}$$

$$\mathbf{k}_0 = \mathbf{k}_1 + \mathbf{k}_2 \,, \tag{10.38}$$

where the three waves satisfy the dispersion relations:

$$\omega_{0,1}^2 = \omega_{pe}^2 + 3k_{0,1}^2 v_{Te}^2 \,, \tag{10.39}$$

$$\omega_2 \approx k_2 c_s \tag{10.40}$$

(we assumed quasi-neutrality, i.e., $ZT_e \gg T_i$, and $k_2 \lambda_{De} \ll 1$ for the IAW for simplicity).

Since $\omega_2 \ll \omega_0, \omega_1$, we have $k_0 \approx k_1$, with a wave-vector geometry reminiscent of SBS, as illustrated in Fig. 10.6. However, we should keep in mind that all waves here are longitudinal, so the electric field will be aligned with the wave-vector for each of the three waves.

We proceed in the same way as with the other three-wave instabilities investigated in Chaps. 7 and 8. The quantities δn_j and $\delta \mathbf{v}_j$ for $j = 0, 1, 2$ denote the electron density and velocity perturbations associated with each wave; we introduce the following enveloped quantities, which we will then use in the slowly varying envelope approximation:

$$\delta n_j = \frac{1}{2} \delta \tilde{n}_j e^{i\psi_j} + c.c. \,, \tag{10.41}$$

Fig. 10.6 LDI geometry; for a given EPW pump wave-vector $\mathbf{k}_0$, the possible secondary EPW wave-vectors k_1 are on a circle of radius $\approx k_0$, while the secondary IAW wave-vector satisfies $\mathbf{k}_2 = \mathbf{k}_0 - \mathbf{k}_1$

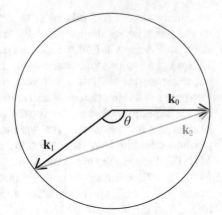

$$\delta \mathbf{v}_j = \frac{1}{2} \delta \tilde{\mathbf{v}}_j e^{i\psi_j} + c.c., \tag{10.42}$$

with $\psi_j = \mathbf{k}_j \cdot \mathbf{r} - \omega_j t$. For a given wave j, the two quantities are related by the linearized continuity equation, Eq. (1.92), i.e.,

$$\frac{\delta \tilde{n}_j}{n_{e0}} = \frac{1}{\omega_j} \mathbf{k}_j \cdot \delta \tilde{\mathbf{v}}_j = \frac{k_j \delta \tilde{v}_j}{\omega_j} = \frac{\delta \tilde{v}_j}{v_{\phi j}}, \tag{10.43}$$

where $v_{\phi j} = \omega_j / k_j$ is the phase velocity of wave j (since all waves are longitudinal, we have $\delta \mathbf{v}_j /\!/ \mathbf{k}_j$).

In the following we neglect pump depletion. Starting from the second-order coupled wave equations for EPWs and IAWs for the two daughter waves, Eqs. (6.24) and (6.29) from Chap. 6, where we assume $v_{\phi 2} = c_s \ll v_{Te} \ll v_{\phi 0}, v_{\phi 1}$, we obtain (cf. Problem 10.2)

$$\left(\partial_t^2 + \omega_{pe0}^2 - 3v_{Te}^2 \nabla^2 \right) \frac{\delta n_1}{n_{e0}} = -\partial_t \nabla \left(\frac{\delta n_2}{n_{e0}} \delta \mathbf{v}_0 \right), \tag{10.44}$$

$$\left(\partial_t^2 - c_s^2 \nabla^2 \right) \frac{\delta n_2}{n_{e0}} = \frac{Zm}{M_i} \nabla^2 (\delta \mathbf{v}_0 \cdot \delta \mathbf{v}_1). \tag{10.45}$$

Introducing the enveloped quantities defined above, eliminating $\delta \tilde{v}_j$ to keep only $\delta \tilde{n}_j$ as the main field variable for each wave, and applying the slowly varying envelope approximation (cf. Eq. (6.35) in Sect. 6.4) to these two coupled equations lead to the following after collecting terms oscillating at the phases ψ_1 and ψ_2:

$$(\partial_t + \mathbf{v}_{g1} \cdot \nabla) \frac{\delta \tilde{n}_1}{n_{e0}} = -\frac{i}{4} \omega_{pe} \cos(\theta) \frac{\delta \tilde{n}_0}{n_{e0}} \frac{\delta \tilde{n}_2^*}{n_{e0}}, \tag{10.46}$$

Fig. 10.7 LDI growth rate
(normalized to the growth
rate for the backscatter
geometry γ_0) vs. angle θ
between $\mathbf{k}_0$ and $\mathbf{k}_1$

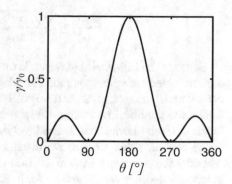

$$(\partial_t + \mathbf{v}_{g2} \cdot \nabla) \frac{\delta \tilde{n}_2}{n_{e0}} = -\frac{i}{2} \sqrt{\frac{Zm}{M_i}} \frac{\omega_{pe}}{k_0 \lambda_{De}} \sin(\theta/2) \cos(\theta) \frac{\delta \tilde{n}_0}{n_{e0}} \frac{\delta \tilde{n}_1^*}{n_{e0}}, \quad (10.47)$$

where θ is the angle between $\mathbf{k}_0$ and $\mathbf{k}_1$, such that $k_2 \approx 2k_0 \sin(\theta/2)$ (cf. Fig. 10.6),
and we used $\omega_0 \approx \omega_1 \approx \omega_{pe}$ for the coupling terms on the right-hand sides of the
equations and $Zm/M_i \approx c_s^2/v_{Te}^2$.

The temporal growth rate γ is obtained by following the same procedure as in
Sect. 6.4.3 (cf. Eq. (6.44)), leading to

$$\gamma^2 = \frac{\omega_{pe}^2}{8k_0 \lambda_{De}} \sqrt{\frac{Zm}{M_i}} \sin(\theta/2) \cos^2(\theta) \left| \frac{\delta \tilde{n}_0}{n_{e0}} \right|^2. \quad (10.48)$$

The growth rate strongly peaks for the backscatter geometry, as shown in
Fig. 10.7. This is not only due to the dependence of the growth rate on k_2 like for
SRS and SBS but also due to the fact that the backscatter geometry maximizes the
alignment of the waves' electric fields (and thus the ponderomotive force from their
beat waves), which are all parallel to the wave-vectors since all the waves involved
are longitudinal.

The peak LDI growth rate for the backscatter geometry ($\theta = \pi$) is then

$$\gamma_0 = \frac{\omega_{pe}}{\sqrt{8k_0 \lambda_{De}}} \left(\frac{Zm}{M_i} \right)^{1/4} \left| \frac{\delta \tilde{n}_0}{n_{e0}} \right|. \quad (10.49)$$

It can also be useful to express the instability threshold in terms of the density
modulation of the primary EPW. We saw in Sect. 6.4, Eq. (6.48) that the threshold
condition is $\gamma_0^2 \geq \nu_1 \nu_2$, where ν_1 and ν_2 are the damping rates of the two daughter
waves (here the secondary EPW and IAW). Inserting the LDI growth rate from
Eq. (10.49) gives the threshold condition:

$$\left|\frac{\delta\tilde{n}_0}{n_{e0}}\right| \geq \frac{\sqrt{8\nu_1\nu_2 k_0\lambda_{De}}}{\omega_{pe}}\left(\frac{Zm}{M_i}\right)^{1/4}. \tag{10.50}$$

Because the LPI threshold depends on the damping of the daughter IAW, using a plasma medium with strong ion damping like a light and heavy ion mixture (e.g., carbon and hydrogen, cf. Sect. 1.3.6) will raise the threshold of LDI. In turn, this can inhibit the nonlinear saturation of high-amplitude EPWs, which can lead to higher levels of laser–plasma instabilities involving EPWs such as SRS or TPD.

To conclude this section, we discuss the LDI cascade, which is the process whereby the secondary EPW from LDI is strong enough to generate its own LDI decay into tertiary waves, and so forth. Evidence of LDI cascade was observed in radar measurements after excitation of plasma modes with intense radio waves in the ionosphere, as first noted by Dubois who matched the observations with Zakharov simulations [12].

The LDI cascade process is illustrated in Fig. 10.8. For better clarity, we slightly change our notations compared to the discussion above and use $(\omega_{e,0}, \mathbf{k}_{e,0})$ for the frequency and wave-vector of the primary EPW, $(\omega_{e,1}, \mathbf{k}_{e,1})$ and $(\omega_{a,1}, \mathbf{k}_{a,1})$ for the secondary EPW (subscript e) and IAW (subscript a), then $(\omega_{e,2}, \mathbf{k}_{e,2})$ and $(\omega_{a,2}, \mathbf{k}_{a,2})$ for the tertiary EPW and IAW (driven by EPW #1), etc.

For the first step of the cascade, writing $\omega_{e,0} \approx \omega_{pe}(1 + \frac{3}{2}k_{e,0}^2\lambda_{De}^2)$ and $\omega_{e,1} \approx \omega_{pe}(1 + \frac{3}{2}k_{e,1}^2\lambda_{De}^2)$ (assuming $k\lambda_{De} \ll 1$ for both waves) and substituting in $\omega_{a,1} = \omega_{e,0} - \omega_{e,1} = (k_{e,0} + k_{e,1})c_s$ (the dispersion relation of IAW #1) leads to

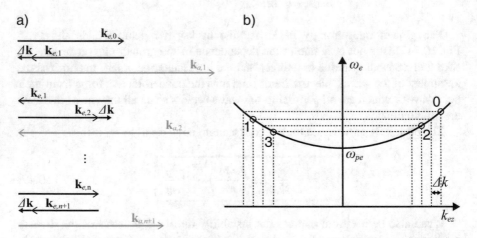

Fig. 10.8 (**a**) LDI cascade: the primary EPW (indexed 0) first decays into secondary EPW and IAW (index 1); EPW1 can subsequently decay into tertiary EPW and IAW (index 2), etc. (**b**) Illustration of the cascade on the curve of the EPW dispersion relation ω_e vs. k_{ez} (z being the common axis of the waves) (not represented at scale). The circles on the curve mark the different EPWs of the cascade, and the numbers (0,1,2,3) are their order in the cascade

$$k_{e,0} - k_{e,1} \equiv \Delta k = \frac{2}{3} \frac{1}{\lambda_{De}} \frac{c_s}{v_{Te}} = \frac{2}{3} \frac{1}{\lambda_{De}} \sqrt{\frac{Zm}{M_i}} \,. \tag{10.51}$$

In other words, Δk represents the small difference between the two EPWs' wave-vector amplitudes, which we had ignored until now. For typical plasmas with $Z \approx A/2$, we have $\Delta k \lambda_{De} = (2/3)\sqrt{m/2M_p}$, where M_p is the proton mass, i.e., $\Delta k \lambda_{De} \approx 1\%$, and thus $\Delta k / k_{e,0} \approx 0.01/(k_{e,0}\lambda_{De})$. For example, for $k_{e,0}\lambda_{De} = 0.2$, we have $\Delta k / k_{e,0} \approx 5\%$.

The wave-numbers of the daughter EPW and IAW in the first step are then

$$k_{e,1} = k_{e,0} - \Delta k \tag{10.52}$$

$$k_{a,1} = k_{e,0} + k_{e,1} = 2k_{e,0} - \Delta k \,. \tag{10.53}$$

Repeating the iteration gives the wave-vector amplitudes for the nth step:

$$k_{e,n} = k_{e,0} - n\Delta k \,, \tag{10.54}$$

$$k_{a,n} = 2k_{e,0} - (2n - 1)\Delta k \,. \tag{10.55}$$

This shows that at each new step of the LDI cascade, the daughter waves swap directions and have a wave-vector and frequency that keep getting a little bit smaller at every step. This constitutes a clear experimental signature of the LDI cascade, measurable with Thomson scattering. The first experimental indication of LDI cascade in laser-plasma experiments was made by Depierreux et al. using Thomson scattering [11, 13], which allowed to observe both backward and forward IAW ($k_{a,2}$ in Fig. 10.8a) as well as a broadened k-vector spectrum for the EPW on the longer wavelength side only (i.e., shorter k's). Montgomery et al. [14] and Kline et al. [15, 16] made the first definitive measure of LDI cascade, resolved in both frequency and wave-vector, eliminating any potential ambiguity about the cascade process (as opposed to scattering off a broad turbulent spectrum of waves [17]).

10.2.2 Two-Ion Decay (TID)

Two-ion decay is the coupling of a high-amplitude IAW to other two IAWs. Because IAWs have low frequencies compared to the other two types of waves, it is the only possible decay process for an IAW (cf. Sect. 6.2). TID was investigated theoretically and experimentally in the early 1970s, not in the context of laser–plasma; three-wave analyses similar to the one presented below are described in Refs. [18, 19], where the authors also investigated wave-mixing situations similar to TID by exciting two IAWs at different frequencies. Later on, TID was proposed and investigated as a saturation process for SBS by Karttunen [20, 21] and first measured in laser–plasma experiments using Thomson scattering by Bandulet [22] and Niemann [23].

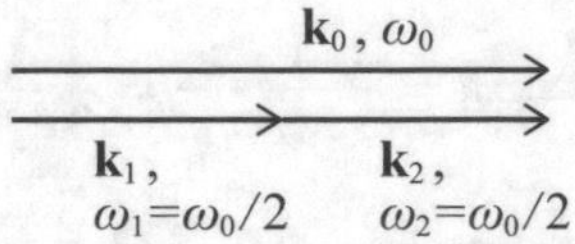

$$\mathbf{k}_0, \omega_0$$

$$\mathbf{k}_1, \qquad \mathbf{k}_2,$$
$$\omega_1 = \omega_0/2 \qquad \omega_2 = \omega_0/2$$

Fig. 10.9 Two-ion decay (TID) geometry: the process is primarily 1D, with the high-amplitude IAW decaying into two co-propagating daughter IAWs at half the frequency and wave-vector of the pump

It was also observed and analyzed for a wide range of conditions in numerical simulations [24–26]. A more general theoretical investigation including both Stokes and anti-Stokes modes was published by Pesme [27]; here we restrict ourselves to the simplest (and most unstable, in most cases) case of three-wave coupling (i.e., Stokes only) between a pump IAW and two daughter waves.

Continuing with the same notations as before, we identify the three waves by the indices 0, 1, and 2, with 0 referring to the high-amplitude "pump." We also have the usual phase matching relations $\omega_0 = \omega_1 + \omega_2$, $\mathbf{k}_0 = \mathbf{k}_1 + \mathbf{k}_2$.

In the limit of $ZT_e \gg T_i$, and $k\lambda_{De} \ll 1$, the dispersion relation for each wave is simply $\omega_j = k_j c_s$, with $j = 0$, 1 or 2. In this case the frequency-matching relation implies that $k_0 = k_1 + k_2$; since we also have the same expression in vector form, $\mathbf{k}_0 = \mathbf{k}_1 + \mathbf{k}_2$, the only geometry that allows these two relations to be simultaneously valid is the one illustrated in Fig. 10.9, with all three wave-vectors parallel and pointed in the same direction; the particular case shown in the figure where both daughter waves are at half the frequency of the pump happens to maximize the instability, as we will see below. Note that *exact* phase matching is never possible for finite $k\lambda_{De}$ because the three waves follow the same dispersion relation; it is only approximately satisfied in the $k\lambda_{De} \ll 1$ limit [20, 28].

To describe the key features of the TID instability, we start from the driven IAW equation, Eq. (6.29):

$$\left(\partial_t^2 - c_s^2 \nabla^2\right)\frac{\delta n_e}{n_{e0}} = \frac{Zm}{2M_i}\nabla^2 \delta \mathbf{v}_e^2 + \frac{1}{2}\nabla^2 \delta \mathbf{v}_i^2 - \partial_t \nabla \left(\frac{\delta n_i}{n_{i0}}\delta \mathbf{v}_i\right)$$

$$- \frac{1}{2}\left[\gamma_i v_{Ti}^2 \nabla^2 \left(\frac{\delta n_i}{n_{i0}}\right)^2 + \frac{Zm}{M_i}\gamma_e v_{Te}^2 \nabla^2 \left(\frac{\delta n_e}{n_{e0}}\right)^2\right]. \quad (10.56)$$

Since all the waves involved are IAWs, we have $\gamma_e = 1$ and $\gamma_i = 3$. We also know that for IAWs, $\delta v_e \approx \delta v_i \approx c_s \delta n_e/n_{e0}$ (cf. Sect. 1.3.1 Eq. (1.139)): therefore, the first term on the right-hand side is negligible compared to all the other terms. The three waves being identified by their density or velocity perturbation $\delta n_e = \sum_j \delta n_j$, $\delta \mathbf{v}_i = \sum_j \mathbf{v}_j$ $(j=0,1,2)$, we start by considering the wave equation Eq. (10.56) for wave 1 (i.e., δn_1 on the left-hand side), coupled to waves 0 and 2 (i.e., we only consider products between 0 and 2 on the right-hand side): we obtain, in 1D along z,

$$\left(\partial_t^2 - c_s^2\partial_z^2\right)\frac{\delta n_1}{n_{e0}} = \partial_z^2(v_0 v_2) - \partial_t\partial_z\left(\frac{\delta n_0}{n_{e0}}v_2 + \frac{\delta n_2}{n_{e0}}v_0\right) - c_s^2\partial_z^2\left(\frac{\delta n_0}{n_{e0}}\frac{\delta n_2}{n_{e0}}\right),$$

$$(10.57)$$

where we used $c_s^2 = 3v_{Ti}^2 + (Zm/M_i)v_{Te}^2$.

Next we envelope the density and velocity perturbations along their main frequency and wave-vector,

$$\frac{\delta n_j}{n_{e0}} = \frac{1}{2}\frac{\delta\tilde{n}_j}{n_{e0}}e^{i\psi_j} + c.c.,$$

$$(10.58)$$

$$v_j = \frac{1}{2}\tilde{v}_j e^{i\psi_j} + c.c.$$

$$(10.59)$$

with $\psi_j = k_j z - \omega_j t$. We can then apply the slowly varying envelope approximation (Eq. 6.35) to the left-hand side of Eq. (10.57) and collect terms $\propto e^{i\psi_1}$ on both sides (with $\psi_1 = \psi_0 - \psi_2$ for the terms on the right-hand side); after eliminating the velocity via $\tilde{v}_j = c_s\delta\tilde{n}_j/n_{e0}$, we obtain

$$(\partial_t + c_s\partial_z)\frac{\delta\tilde{n}_1}{n_{e0}} = -\frac{i}{2}\omega_1\frac{\delta\tilde{n}_0}{n_{e0}}\frac{\delta\tilde{n}_2^*}{n_{e0}}.$$

$$(10.60)$$

Proceeding similarly for wave 2, we have

$$(\partial_t + c_s\partial_z)\frac{\delta\tilde{n}_2}{n_{e0}} = -\frac{i}{2}\omega_2\frac{\delta\tilde{n}_0}{n_{e0}}\frac{\delta\tilde{n}_1^*}{n_{e0}}.$$

$$(10.61)$$

Equations (10.60) and (10.61) are the coupled equations describing TID in the $k\lambda_{De} \ll 1$ limit. The temporal growth rate is immediately obtained by combining the terms on the right-hand side following Sect. 6.4.3, Eq. (6.44) (i.e., by taking $\partial_z = 0$ in the coupled equations and solving vs. time), which leads to

$$\gamma = \frac{1}{2}\sqrt{\omega_1\omega_2}\left|\frac{\delta\tilde{n}_0}{n_{e0}}\right|.$$

$$(10.62)$$

Since $\omega_2 = \omega_0 - \omega_1$, we can recast this expression as $\gamma = \sqrt{\omega_1(\omega_0 - \omega_1)}|\delta\tilde{n}_0/n_{e0}|/2$. We can easily verify that the value of ω_1 maximizing the growth rate is $\omega_0/2$, corresponding to the half-harmonic decay of the primary IAW following

$$\omega_1 = \omega_2 = \frac{\omega_0}{2},$$

$$(10.63)$$

$$\mathbf{k}_1 = \mathbf{k}_2 = \frac{\mathbf{k}_0}{2},$$

$$(10.64)$$

like we had illustrated in Fig. 10.9. The corresponding maximum growth rate for half-harmonic decay is

$$\gamma_0 = \frac{\omega_0}{4} \left| \frac{\delta \tilde{n}_0}{n_{e0}} \right|. \tag{10.65}$$

It is also interesting to express the instability threshold; from Eq. (6.48), the threshold is $\gamma_0^2 \geq \nu_1 \nu_2$, with ν_1 and ν_2 the damping rates of the two daughter waves. We also know that for any of the three IAWs involved, ν_j / ω_j is the same, as it is a function of $Z T_e / T_i$ only per Eq. (1.189): therefore, since $\omega_1 = \omega_2 = \omega_0/2$, we have $\nu_1 = \nu_2 = \nu_0/2$ and the threshold can be expressed as

$$\left| \frac{\delta \tilde{n}_0}{n_{e0}} \right| \geq 2 \frac{\nu_0}{\omega_0}. \tag{10.66}$$

10.2.3   Electromagnetic Decay Instability (EDI)

The last secondary process is the electromagnetic decay instability (EDI), whereby an EPW (indexed "0") decays into an EMW (indexed "1") and an IAW (indexed "2"; cf. Table 6.1). The phase-matching conditions are, as usual, $\omega_0 = \omega_1 + \omega_2$, $\mathbf{k}_0 = \mathbf{k}_1 + \mathbf{k}_2$ with $\omega_0 \approx \omega_{pe}$ (since wave 0, the pump, is an EPW) and $\omega_2 = k_2 c_s \ll \omega_0$ which implies $\omega_0 \approx \omega_1$. In turn, using the dispersion relations of the EMW and EPW, $\omega_0^2 = \omega_{pe}^2 + 3 k_0^2 v_{Te}^2$ and $\omega_1^2 = \omega_{pe}^2 + k_1^2 c^2$, we have $k_1 = \sqrt{3}(v_{Te}/c) k_0 \ll k_0$, and therefore $k_0 \approx k_2$. Finally, we must note that since $\mathbf{E}_0 /\!/ \mathbf{k}_0$ whereas $\mathbf{E}_1 \perp \mathbf{k}_1$, the pump EPW and daughter EMW cannot have co-aligned wave-vectors, otherwise their electric fields would not interfere and drive the IAW. The geometry of EDI is illustrated in Fig. 10.10.

Note that since $\omega_1 \approx \omega_0 \approx \omega_{pe}$, the daughter EMW is created near the critical density for its wavelength and will propagate in the direction opposite to the density gradient, toward lower densities. When the pump EPW is the product of a primary instability from a laser–plasma instability like SRS or TPD, the wavelength λ_1 of the EDI-generated EMW can be related to the wavelength of the laser λ_{laser} via $\lambda_1 / \lambda_{laser} = 1/\sqrt{n_e/n_c}$, where n_c is the critical density for the laser wavelength. Figure 10.11 shows the ratio of the EDI-generated EMW wavelength to the laser wavelength over the range of densities where the laser can generate EPWs for the

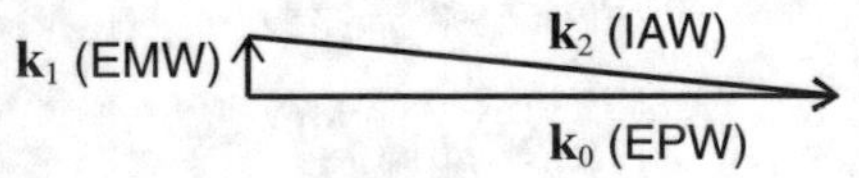

Fig. 10.10 Electromagnetic decay instability (EDI) geometry: growth is maximum when the daughter EMW is orthogonal to the primary EPW

Fig. 10.11 Wavelength λ_1 of EDI-generated EMW, normalized to the laser wavelength λ_{laser}, as a function of n_e/n_c, where n_c is the critical density at the laser wavelength

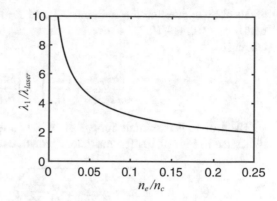

EDI instabilities, i.e., from 0 to $n_c/4$. For EPWs generated near $n_c/4$, the EDI-EMW has a wavelength of $\approx 2\lambda_{laser}$; λ_1 increases at lower densities due to the decrease in $\omega_0 \approx \omega_1 \approx \omega_{pe}$.

To derive the coupled equations in the slowly varying envelope approximation and the growth rate, we proceed as usual and start from the wave equations with quadratic coupling terms for an EPW, EPW, and IAW from Eqs. (6.25), (6.21), and (6.30). We describe the EPW, EMW, and IAW by their electron density modulation, normalized vector potential, and electron density modulation, respectively:

$$\frac{\delta n_0}{n_{e0}} = \frac{1}{2}\frac{\delta \tilde{n}_0}{n_{e0}}e^{i\psi_0} + c.c.,\tag{10.67}$$

$$\mathbf{a}_1 = \frac{1}{2}\tilde{a}_1\mathbf{e}_1e^{i\psi_1} + c.c.,\tag{10.68}$$

$$\frac{\delta n_2}{n_{e0}} = \frac{1}{2}\frac{\delta \tilde{n}_2}{n_{e0}}e^{i\psi_2} + c.c.,\tag{10.69}$$

where $\mathbf{e}_1$ is a unit vector along the direction of $\mathbf{a}_1$. Under the slowly varying envelope approximation and assuming perfect phase matching in a uniform plasma ($\psi_0 = \psi_1 + \psi_2$), the three coupled equations can be expressed as (the derivation is left as an exercise, cf. Problem 10.4)

$$(\partial_t + v_{g0}\partial_{z_0})\frac{\delta \tilde{n}_0}{n_{e0}} = -i\frac{k_0c\omega_{pe}}{4\omega_0}\frac{\delta \tilde{n}_2}{n_{e0}}\tilde{a}_1(\mathbf{e}_0 \cdot \mathbf{e}_1),\tag{10.70}$$

$$(\partial_t + v_{g1}\partial_{z_1})\tilde{a}_1 = -i\frac{\omega_{pe}^2}{4k_0c}\frac{\delta \tilde{n}_0}{n_{e0}}\frac{\delta \tilde{n}_2^*}{n_{e0}}(\mathbf{e}_0 \cdot \mathbf{e}_1),\tag{10.71}$$

$$(\partial_t + v_{g2}\partial_{z_2})\frac{\delta \tilde{n}_2}{n_{e0}} = -i\frac{c\omega_{pe}}{4c_s}\frac{Zm}{M_i}\frac{\delta \tilde{n}_0}{n_{e0}}\tilde{a}_1^*(\mathbf{e}_0 \cdot \mathbf{e}_1),\tag{10.72}$$

where $\mathbf{e}_0$ is a unit vector along $\mathbf{k}_0$ and we used $k_0 \approx k_2$. The instability growth rate, first derived in [29] in the context of electron beam–plasma interaction, can be obtained by expressing the constants K_1 and K_2 from Eqs. (6.40) by identification with Eqs. (10.71)–(10.72) above and using Eq. (6.44) to get the growth rate γ, leading to

$$\gamma = \frac{\omega_{pe}}{4\sqrt{k_0\lambda_{De}}} \left(\frac{Zm}{M_i}\right)^{1/4} \left|\frac{\delta\tilde{n}_0}{n_{e0}}\right| (\mathbf{e}_0 \cdot \mathbf{e}_1). \qquad (10.73)$$

The growth is maximum for $\mathbf{e}_0 \cdot \mathbf{e}_1 = 1$, i.e., when $\mathbf{k}_0$ and $\mathbf{k}_1$ are orthogonal as illustrated in Fig. 10.10. The maximum growth rate γ_0 then takes the form:

$$\boxed{\gamma_0 = \frac{\omega_{pe}}{4\sqrt{k_0\lambda_{De}}} \left(\frac{Zm}{M_i}\right)^{1/4} \left|\frac{\delta\tilde{n}_0}{n_{e0}}\right|.} \qquad (10.74)$$

We notice that the growth rate is exactly the same as for LDI, up to a factor $\sqrt{2}$— cf. Eq. (10.49). EDI and LDI are the two primary mechanisms for the parametric decay of high-amplitude EPWs; in fact, EDI often has a lower threshold, since the damping of the daughter EMW for EDI will typically be smaller than the one of the daughter EPW for LDI [30].

However, spatial amplification in non-uniform plasmas will be much stronger for LDI than EDI, due to the slower propagation of the daughter wave #1 for LDI (an EPW) vs. EDI (an EMW) through the amplification region (since the second daughter wave is an IAW in both cases, the convective velocity is dominated by wave #1, cf. Sect. 6.6). Indeed, in both cases, the expression of the exponential amplification factor is given by the Rosenbluth gain formula Eq. (6.105), $G_R = 2\pi\gamma_0^2/|\kappa' v_{g1}v_{g2}|$; here γ_0^2 is the same except for a factor 2 and v_{g2} is also the same ($= c_s$), so for a fixed spatial gradient κ' the only difference is v_{g1}.

For LDI, we have $v_{g1} = 3v_{Te}^2 k_1/\omega_1 \approx 3v_{Te}k_0\lambda_{De}$ since $k_1 \approx k_0$ and $\omega_1 \approx \omega_{pe}$ (cf. Sect. 10.2.1). For EDI, we have $v_{g1} = c^2 k_1/\omega_1 \approx \sqrt{3}ck_0\lambda_{De}$ since $k_1 \approx \sqrt{3}k_0v_{Te}/c$ and $\omega_1 \approx \omega_{pe}$ (cf. above). Therefore, we obtain

$$G_{R,EDI} = \frac{\sqrt{3}}{2}\frac{v_{Te}}{c} G_{R,LDI} \ll G_{R,LDI}, \qquad (10.75)$$

confirming that the spatial amplification for EDI is much smaller than for LDI. The dominance of LDI vs. EDI as a decay mechanism for EPWs has been confirmed experimentally using optical Thomson scattering [31].

Note that EDI has been proposed as a possible origin of the half-harmonic emission of the laser frequency observed in experiments and associated with high-amplitude EPWs generated near the quarter-critical density (cf. Fig. 10.11, and Sect. 8.2.4) [32].

10.2.4 Harmonics Generation

When the amplitude of a plasma wave at frequency and wave-number (ω_p, k_p) is not purely sinusoidal (e.g., when $\delta n_e / n_{e0}$ is not a very small quantity anymore), its spectrum consists of harmonics at $(n\omega_p, nk_p)$ with $n \in \mathbb{N}$. The fundamental mode can then scatter into higher-order harmonics, which can saturate its growth via energy transfer to the harmonics.

Unlike for the secondary instabilities described in the previous sections, harmonics generation is not an instability. Practically, this means that while the coupling between the fundamental and its harmonics occurs via nonlinear terms in the fluid equations, the growth of the higher-order harmonics is not exponential, and there is no feedback loop like in three-wave couplings.

To quantify the effect, we take the example of IAWs; like for TID, we start from the second-order IAW equation,

$$
\left(\partial_t^2 - c_s^2 \nabla^2\right) \frac{\delta n_e}{n_{e0}} = \frac{1}{2} \nabla^2 \left[\frac{Zm}{M_i} \delta \mathbf{v}_e^2 + \delta \mathbf{v}_i^2 \right] - \partial_t \nabla \left(\frac{\delta n_i}{n_{i0}} \delta \mathbf{v}_i \right)
$$
$$
- \frac{1}{2} \left[\gamma_i v_{Ti}^2 \nabla^2 \left(\frac{\delta n_i}{n_{i0}} \right)^2 + \frac{Zm}{M_i} \gamma_e v_{Te}^2 \nabla^2 \left(\frac{\delta n_e}{n_{e0}} \right)^2 \right]. \tag{10.76}
$$

As we will see, it is usually sufficient to consider only the second harmonic. We decompose the electron density perturbation into the fundamental IAW and its second harmonic, i.e., $\delta n_e = \delta n_1 + \delta n_2$ with

$$
\delta n_{1,2} = \frac{1}{2} \delta \tilde{n}_{1,2} e^{i\psi_{1,2}} + c.c., \tag{10.77}
$$

with $\psi_1 = kz - \omega t$ and $\psi_2 = 2\psi_1$. We can see from Eq. (10.76) that the primary wave $\propto e^{i\psi_1}$ can couple to terms $\propto \delta n_2 \delta n_1^* \propto e^{i\psi_1}$ and that the second harmonic $\propto e^{i\psi_2} = e^{2i\psi_1}$ can couple to terms $\propto (\delta n_1)^2 \propto e^{2i\psi_1}$. Taking the limit of $T_i \ll T_e$, with $\delta n_e/n_e \approx \delta n_i/n_i \approx \delta v_e/c_s \approx \delta v_i/c_s$ (cf. Sect. 1.3.1.5 and Problem 1.3) and collecting terms $\propto e^{i\psi_1}$ and $\propto e^{i\psi_2}$, respectively, give the following coupled equations for the amplitudes of the primary IAW and its second harmonic:

$$
(\partial_t + c_s \partial_z) \frac{\delta \tilde{n}_1}{n_{e0}} = -\frac{i}{2} \omega \frac{\delta \tilde{n}_2}{n_{e0}} \frac{\delta \tilde{n}_1^*}{n_{e0}}, \tag{10.78}
$$

$$
(\partial_t + c_s \partial_z) \frac{\delta \tilde{n}_2}{n_{e0}} = -\frac{i}{4} \omega \left(\frac{\delta \tilde{n}_1}{n_{e0}} \right)^2. \tag{10.79}
$$

Similar coupled equations were derived in [19] and later in Refs. [33, 34] while also coupled to 1D equations for the propagation of two counter-propagating light waves in order to study the saturation of SBS due to harmonics generation. The important point is that in steady-state, the spatial growth of the second harmonic is given by the simple expression:

$$\partial_z \frac{\delta \tilde{n}_2}{n_{e0}} = -\frac{i}{4} k \left(\frac{\delta \tilde{n}_1}{n_{e0}} \right)^2.$$

(10.80)

This shows that as mentioned above, the growth of the second harmonic for a fixed amplitude of the primary IAW is linear in space (and not exponential via a feedback loop like TID or the other three-wave instabilities) and that the growth of the second harmonic requires the primary wave amplitude to be significant ($\delta \tilde{n}_1 / n_{e0} \lesssim 1$). The system of coupled equations above can be completed with higher-order harmonics, but we can easily show that the growth of any harmonics requires all the lower-order harmonics to be significant, which in turn requires $\delta n_1 / n_{e0} \approx 1$ for the primary.

As shown in [34], SBS saturation via harmonics generation requires high-amplitude IAW. Harmonics generation is accompanied by a nonlinear frequency shift which can detune the primary IAW (cf. [35] for a similar treatment for EPWs); however, while harmonics generation scales like $\propto (\delta n_1 / n_{e0})^2$, kinetic effects such as those we will describe in the following section scale like $\sqrt{\delta n_1 / n_{e0}}$ and can kick in and saturate the IAW growth before the harmonics generation does.

Second harmonic generation is however routinely observed in simulations (e.g., [24, 25, 36]) and is a clear signature of high-amplitude IAWs. The process was measured experimentally, and Eq. (10.80) was validated to extremely good accuracy in [19], using a discharge plasma column where IAWs were excited using an applied field and directly measured (i.e., $\delta n_e / n_{e0}$) using a collecting grid. Second harmonic generation from high-amplitude, SBS-driven IAWs was also measured in laser–plasma experiments using CO_2 lasers in [37], using Thomson scattering as a diagnostic.

10.3 Kinetic Effects

Particle trapping, which we introduced in Sect. 2.2.3, can impact the damping of plasma waves driven by laser–plasma instabilities by reducing Landau damping. Such kinetic effects are routinely observed in simulations; this area of research spans more than five decades, starting with the early "classic" works from Refs. [38, 39] for un-driven waves and Refs. [25, 40] for driven waves. In this section, we discuss the basic physics ideas for this process, known as kinetic inflation, and derive its threshold due to competition with de-trapping mechanisms and its saturation by the nonlinear frequency shift that accompanies the reduction in damping.

10.3.1 Reduction of Landau Damping and "kinetic inflation"; Threshold from De-Trapping Processes

10.3.1.1 Kinetic Inflation Process

We saw in Sect. 2.2.3 that a plasma wave (EPW or IAW) can exchange energy with the plasma particles even in the absence of collisions. This occurs when the wave's phase velocity is not too high compared to the thermal velocity, such that particles in the tail of the velocity distribution can have a velocity close to the wave's and resonantly interact with it.

For a particle velocity distribution initially Maxwellian, this wave–particle exchange process leads to an energy increase of the particles whose velocity is near the wave's phase velocity v_ϕ, as they tend to uniformly fill the velocity distribution for a width v_{trap} around v_ϕ as they follow the phase-space trajectories derived in Sect. 2.2. The energy gain by the particles is compensated by an energy loss from the wave via the Landau damping.

As the particles start to fill the phase-space in that region, the distribution function flattens at v_ϕ, leading to a reduction of the Landau damping, since $v_L \propto f'(v_\phi)$ [38]. As a result, the process saturates: the particle energy gain and wave energy loss (Landau damping) both stop. The characteristic time it takes to reach this saturation state is on the order of a bounce period τ_B given by Eq. (2.35), i.e., the time it takes a deeply trapped particle to execute a full "bounce" along its trapped orbit. In reality, $f(v)$ never gets truly flat at v_ϕ, as this is an asymptotic state which is never *exactly* reached because $\tau_B \to \infty$ for particles close to the phase-space separatrix as explained in Sect. 2.2. However, the Landau damping can still be significantly reduced.

If the plasma wave is driven by a ponderomotive source, like in three-wave instabilities, the reduction of damping can lead to an increase of the wave amplitude, since the spatial amplification rate Γ is inversely proportional to the plasma wave damping (Eq. 6.52). Practically, this means that simple estimates based on convective gain calculations can underestimate the growth of the instability. This process, often referred to as "kinetic inflation," was invoked to help explain experimental results on SRS where the reflectivity levels exceeded predictions from linear gains [14, 15, 41], and was observed in numerical simulations [42, 43]. This was also found to be a plausible contributor to the high levels of SRS in early ICF experiments on the NIF [44, 45].

Kinetic inflation is a process that must compete against several mechanisms that tend to "de-trap" particles. These include collisions and convective effects, i.e., particles escaping a plasma wave with finite spatial extent (typically, localized in a laser speckle) due to their transverse momentum. If de-trapping occurs on time scales that are faster than the bounce period, then the flattening of the distribution at v_ϕ, and thus the inflation process, will not be significant. Since the bounce frequency $\omega_B \approx \omega_{pe}\sqrt{\delta n_e/n_{e0}}$, this will set a threshold for inflation with a minimum wave amplitude below which inflation remains negligible. In the next subsection,

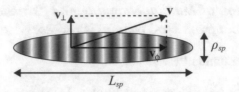

Fig. 10.12 Illustration of de-trapping by side- or end-loss: a trapped particle with longitudinal velocity $\approx v_\phi$ and transverse velocity $v_\perp \approx v_T$ will escape a plasma wave localized in a laser speckle in a time equal to the minimum of $\tau_{d,SL} \approx \rho_{sp}/v_T$ and $\tau_{d,EL} \approx L_{sp}/v_\phi$

we derive threshold amplitudes for kinetic inflation from collisional effects and convective effects (often referred to as "side-loss").

10.3.1.2   Inflation Threshold from Convective De-Trapping

We first consider the convective de-trapping process, which is often the most effective in realistic ICF or HED plasma experimental conditions. In these experiments, laser beams are typically smoothed by phase plates (Sect. 9.1), with focal volumes characterized by high-intensity speckles of radius $\rho_{sp} = \lambda_0 f_\#$ and length $\pi f_\#^2 \lambda_0$. Driven plasma waves (EPWs or IAWs) from any of the three-wave instabilities discussed in the previous chapters will typically be localized in these speckles, with a spatial extent comparable to the speckle's. Convective de-trapping occurs as electrons trapped in a wave (localized in a speckle) leave the speckle due to their finite transverse velocity and will prevent inflation if the typical time to escape the wave is shorter than the bounce period along the trapped orbit. The process was first suggested by Rose et al. [46, 47].

Considering a plasma wave directed along z, a trapped particle will have a longitudinal velocity $\approx v_\phi$, the wave's phase velocity, and a transverse velocity that is typically on the order of the thermal velocity v_T (cf. Fig. 10.12). Therefore, a trapped particle will cross the speckle along the side and escape the plasma wave's potential after a "side-loss" de-trapping time

$$\tau_{d,SL} \approx \frac{\rho_{sp}}{v_T} . \tag{10.81}$$

Inflation can only happen if $\tau_B < \tau_{d,SL}$, i.e., if the particles have time to execute a full bounce orbit (and thus flatten the distribution function near v_ϕ) before they get de-trapped via side-loss. Since $\omega_B = 2\pi/\tau_B = \omega_{pe}\sqrt{\delta n_e/n_{e0}}$ for an EPW (cf. Eq. (2.36)), the condition for inflation $\tau_B < \tau_{d,SL}$ can be expressed as a minimum threshold on the plasma wave amplitude:

$$\boxed{\frac{\delta n_e}{n_{e0}} > \frac{\delta n_e}{n_{e0}}\bigg|_{th,SL} = \frac{v_{Te}^2}{c^2}\frac{1}{f_\#^2 n_{e0}/n_c}}. \tag{10.82}$$

A similar expression was derived in [14]. For example, for typical ICF conditions with $T_e = 2$ keV, $n_{e0}/n_c = 0.1$, and $f_\# = 8$, the inflation threshold from side-loss is $\delta n_e/n_{e0} \approx 6\times10^{-4}$. Estimates where the transverse velocity is integrated over the full distribution function (as opposed to our simple assumption of $v_\perp \approx v_T$) lead to a similar scaling except for a numerical pre-factor $\sim O(1)$ [45].

Note that in principle, de-trapping can also occur if the particle leaves the wave from the end, which is sometimes referred to as "end-loss" (and is more easily accessible to 1D PIC simulations). In this case, the end-loss de-trapping time (with the trapped particle now moving approximately at the wave's phase velocity along the wave direction) is estimated as

$$\tau_{d,EL} \approx \frac{L_{sp}}{v_\phi}. \tag{10.83}$$

Therefore, we have $\tau_{d,SL}/\tau_{d,EL} = (v_\phi/v_T)/(L_{sp}/\rho_{sp})$; for typical NIF parameters, $f_\# = 8$, and assuming that v_ϕ/v_T is at most ≈ 3 (otherwise kinetic effects do not play a role as the phase velocity is too fast for the particles), we have $\tau_{d,SL}/\tau_{d,EL} \ll 1$, i.e., de-trapping typically occurs much faster via side-loss than end-loss.

10.3.1.3 Inflation Threshold from Collisional De-Trapping

Collisions can also lead to particle de-trapping. In this case, de-trapping happens in velocity space (vs. real space for convective de-trapping), as the changes in a trapped particle's velocity imparted by collisions will eventually move it out of its trapped orbit.

Let us consider the simpler case of electron de-trapping from an EPW due to electron–ion (e–i) collisions. The wave is assumed to move along z; a trapped electron has its velocity z-component $v_z \in [v_\phi - v_{trap}, v_\phi + v_{trap}]$, where $v_\phi = \omega/k$ is the EPW phase velocity and v_{trap} the half-width of the trapping region in phase-space (cf. Sect. 2.2, and Fig. 2.3).

We also know from Sect. 1.4 that a single electron moving at velocity $\mathbf{v}$ will experience a loss of momentum along its propagation direction at a rate $\nu_{/\!/}$ such that $dv_{/\!/}/dt = -\nu_{/\!/}v_{/\!/}$, where $v_{/\!/}$ is the parallel velocity component (along the initial direction of the electron) and $\nu_{/\!/}$ is given by Eq. (1.250).

Therefore, an electron will be kicked out of its trapped orbit due to collisions once its longitudinal velocity has been reduced by approximately v_{trap}, as illustrated in Fig. 10.13 (note that the initial transverse velocity components of the electron should typically be on the order of $v_{Te} \ll v_\phi$ and can be ignored). The time it takes for the electron to have its longitudinal velocity reduced by v_{trap} is approximately

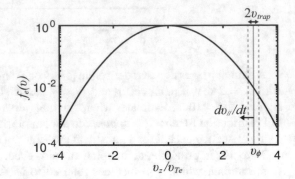

Fig. 10.13 Illustration of de-trapping by collisions: electrons trapped in an EPW, i.e., whose velocity is $\in [v_\phi - v_{trap}, v_\phi + v_{trap}]$, experience a loss of momentum due to collisions with ions which will eventually move them out of the trapped region

$$\tau_{d,\parallel} \approx \frac{v_{trap}}{v_\parallel v_\phi}, \qquad (10.84)$$

where we assumed that the initial electron velocity was approximately equal to the wave's phase velocity, $v_z \approx v_\phi$, which is appropriate for small amplitude waves where $v_{trap} \ll v_\phi$.

Kinetic inflation can only exist if a trapped electron has time to execute a full bounce in its trapped orbit before being de-trapped by collisions, i.e., if

$$\tau_B \leq \tau_{d,\parallel}, \qquad (10.85)$$

where $\tau_B = 2\pi/\omega_B$ is the bounce period defined by Eq. (2.35), with $\omega_B = k v_{trap}$ and $v_{trap} = 2\omega_B/k_p \approx 2v_\phi \sqrt{\delta n_e/n_{e0}}$ per Eq. (2.39). Substituting these expressions in Eq. (10.85) above gives the following threshold for kinetic inflation for the EPW amplitude:

$$\boxed{\frac{\delta n_e}{n_{e0}} \geq \pi \frac{v_\parallel(v_\phi)}{\omega_{pe}}}, \qquad (10.86)$$

where we took $\omega \approx \omega_{pe}$ and $v_\parallel(v_\phi)$ is the loss of momentum rate for an electron propagating with a velocity $\approx v_\phi = \omega/k$ from Eq. (1.250):

$$v_\parallel(v_\phi) = \frac{4\pi n_i}{m^2 v_\phi^3} \left(\frac{Ze^2}{4\pi\varepsilon_0}\right)^2 \ln(\Lambda). \qquad (10.87)$$

Our simple analysis captures the essential physics ideas behind the threshold for kinetic inflation due to electron–ion collisions. More sophisticated analyses for trapping in EPWs and IAWs are presented in Refs. [45] and [48], respectively, where all types of collisional process are taken into consideration (electron–electron, ion–ion, as well as all the other collisional relaxation processes in plasmas such as transverse or longitudinal diffusion or loss of kinetic energy).

10.3.2 Saturation of Inflation by Nonlinear Frequency Shifts

If the inflation threshold described in the previous section is exceeded, then a plasma wave driven by an "external" ponderomotive driver (such as the beat wave between other two waves in a three-wave instability, as discussed in earlier chapters) will experience rapid growth as the reduction of the Landau damping from inflation increases the spatial amplification.

However, as the plasma wave amplitude varies in time, so does its frequency. The analysis of the frequency shift, which is in a way complementary to O'Neil's classic paper on nonlinear Landau damping [38], was derived by Morales and O'Neil [39]. The wave's electric field is then represented by the WKB method under the approximation of a slowly varying envelope:

$$E = \frac{1}{2}\tilde{E}\exp\left[ikz - i\int_0^t \omega(t')dt'\right] + c.c., \tag{10.88}$$

where $\omega(t)$ is the instantaneous frequency (cf. Sect. 3.4). For simplicity, we will only consider EPWs in the following, but a similar treatment can be applied to IAWs. In addition to the linear perturbation of the distribution function associated with the wave, which satisfies $\varepsilon[k, \omega(t = 0)] = 0$ (the linear dispersion relation at early times, before the wave amplitude starts to evolve due to Landau damping), we now include another contribution to the distribution function, assumed small, to represent the plasma particles' response to the wave evolution (the two being self-consistently connected, as the shape of the distribution function modifies the damping and thus the wave amplitude). We write

$$f(\mathbf{r}, \mathbf{v}, t) = f_0(\mathbf{v}) + \delta f_L(\mathbf{r}, \mathbf{v}, t) + \delta f_{NL}(\mathbf{v}, t), \tag{10.89}$$

where f_0 is the background distribution (typically Maxwellian), δf_L the linear perturbation associated with the wave, and f_{NL} the small and slowly varying nonlinear perturbation. Inserting into the Vlasov equation like we did for the kinetic treatment of plasma waves in Sect. 1.3.3 leads to

$$i\left[kv_z - \omega(t)\right]\delta f_L = -\frac{qE}{m}\partial_{v_z}\left(f_0(\mathbf{v}) + \delta f_{NL}(\mathbf{v}, t)\right), \tag{10.90}$$

where z is the direction of the wave (i.e., $\mathbf{k} = k\mathbf{e}_z$, where $\mathbf{e}_z$ is a unit vector). Inserting into Poisson's equation, $\partial_z E = (q/\varepsilon_0)\int d^3v[\delta f_L(\mathbf{r}, \mathbf{v}, t) + \delta f_{NL}(\mathbf{r}, \mathbf{v}, t)]$, and eliminating E on both sides (we follow the same steps as in Sect. 1.3.3) leads to the dispersion relation:

$$\varepsilon = 1 + \chi_e = 0, \tag{10.91}$$

with the electric susceptibility:

$$\chi_e = \chi_{eL} + \chi_{eNL}, \tag{10.92}$$

$$\chi_{eL} = -\frac{e^2}{k^2 m \varepsilon_0} \int d^3 v \frac{\partial_{v_z} f_0(\mathbf{v})}{v_z - \omega/k}, \tag{10.93}$$

$$\chi_{eNL} = -\frac{e^2}{k^2 m \varepsilon_0} \int d^3 v \frac{\partial_{v_z} \delta f_{NL}(\mathbf{v}, t)}{v_z - \omega/k} \tag{10.94}$$

(again, the integrals are over the three velocity dimensions, with $d^3 v = dv_x dv_y dv_z$).

The nonlinear frequency shift of the wave is obtained by a perturbative expansion around the initial frequency $\omega_0 = \omega(t = 0)$:[2]

$$\varepsilon(\omega) \approx \varepsilon(\omega_0) + \delta\omega \left. \frac{\partial \varepsilon}{\partial \omega} \right|_{\omega_0}, \tag{10.95}$$

where $\delta\omega(t) = \omega(t) - \omega_0$. Since the wave satisfies the linear dispersion relation at early times, we have $\varepsilon_L(\omega_0) = 1 + \chi_{eL}(\omega_0) = 0$; inserting the expression for the dielectric constant above, with f_{NL} considered a small perturbation on top of f_0, we obtain the expression for the red-shift:

$$\delta\omega(t) = -\frac{\chi_{eNL}(t)}{\partial \varepsilon_L / \partial \omega |_{\omega_0}}. \tag{10.96}$$

For the denominator, we can usually use the simple fluid approximation, $\chi_{eL} \approx -\omega_{pe}^2/\omega^2$, leading to $\partial_\omega \varepsilon_L|_{\omega_0} \approx 2\omega_{pe}^2/\omega_0^3$. The difficult part is the evaluation of the nonlinear susceptibility, via the modified distribution function $f_{NL}(\mathbf{v}, t)$.

The derivation by Morales and O'Neil assumes that the wave turns on instantaneously and then calculates the evolution of the particles in phase-space on the orbits derived in Sect. 2.2.3 (e.g., Fig. 2.3b). The mathematical derivation is nontrivial.[3] Its key result is an expression for the asymptotic solution at times greater than the bounce period, when the nonlinear Landau damping has saturated due to the flattening of the distribution function. The asymptotic frequency shift can be expressed as

$$\boxed{\frac{\delta\omega}{\omega_0} = -\alpha v_{trap} v_\phi^2 \left. \frac{d^2 f_0}{dv^2} \right|_{v_\phi}}, \tag{10.97}$$

---

[2] We take the frequency at $t = 0$ with the implicit assumption that the wave is turned on immediately at $t = 0$, but the true meaning here is that the frequency should be the frequency at early times, while the wave amplitude still has not significantly decreased due to Landau damping, i.e., for times $\ll \tau_B$.

[3] The details of the calculation, not reported in the original paper, can be found in [49].

where $v_\phi = \omega_0/k$ is the (linear) phase velocity of the wave, $v_{trap} = 2\sqrt{e\Phi_0/m}$ is the trapping width defined previously in Eq. (2.39), and $\alpha \approx 0.41$ is a numerical factor coming from the numerical integration of the polynomial of complete elliptic integrals. Note that a similar derivation for an adiabatic rise of the wave was derived by Dewar [50], leading to a similar result except for a different numerical factor $\alpha \approx 0.27$ (cf. also [48]).[4]

While the full derivation of this result goes beyond the scope of this book, here we will show a different approach proposed by Williams and Divol [51, 52], which while being more heuristic also remains much simpler mathematically and provides some useful physical insight into the process.

The method consists in taking an educated guess for the final shape of the distribution function. We know from the physical arguments discussed in Sect. 2.2.3 that in our simple 1D picture, the distribution function should evolve asymptotically over long time scales ($t \gg \tau_B$) toward a modified shape with a flattening around the phase velocity, with a width approximately equal to v_{trap} (for example, see Fig. 2.6a). This behavior is routinely observed in kinetic simulations.

Therefore, we can try to construct such a distribution; Williams and Divol propose the following:

$$F_\infty(v) = F_0(v) + \beta F_1(v - \omega/k) + \gamma F_2(v - \omega/k), \qquad (10.98)$$

where the distribution functions are one-dimensional with v representing v_z and normalized to the background density (i.e., $F(v) = \int d^2v_\perp f(\mathbf{v})/n_{e0}$ and $\int_{-\infty}^{\infty} F(v)dv = 1$). Here F_0 is the Maxwellian background distribution and F_1 and F_2 are small perturbations to F_0 chosen to create a plateau of width v_{trap} at the phase velocity ω/k:

$$F_0(v) = \frac{1}{v_{Te}\sqrt{2\pi}} \exp\left[-\frac{v^2}{2v_{Te}^2}\right], \qquad (10.99)$$

$$F_1(w) = w \exp\left[-\frac{w^2}{2v_{trap}^2}\right], \qquad (10.100)$$

$$F_2(w) = \frac{v_{trap}^2}{3}\left(\frac{w^2}{v_{trap}^2} - 1\right) \exp\left[-\frac{w^2}{2v_{trap}^2}\right]. \qquad (10.101)$$

By construction, F_1 and F_2 satisfy $\int_{-\infty}^{\infty} F_{1,2}(v)dv = 0$, which ensures that the number of particles is conserved. From the expressions above, it is also easy to show that $F_\infty'(v_\phi) = F_0'(v_\phi) + \beta$ and $F_\infty''(v_\phi) = F_0''(v_\phi) + \gamma$, where the prime denotes the derivative with respect to v; therefore, the first and second derivatives of F_∞ can be set to zero at v_ϕ with

---

[4] The numerical factor α differs from [48] by a factor 2 due to the different definition of v_{trap}.

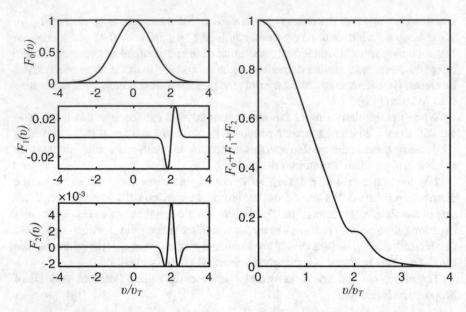

Fig. 10.14 Ad hoc distribution function after nonlinear Landau damping equilibrium, for times $\gg$ τ_B, constructed to obtain a plateau of width v_{trap} at the phase velocity v_ϕ according to Eqs. (10.99)–(10.101). For this illustration, we used $v_\phi = 2v_{Te}$ and $v_{trap} = 0.2v_{Te}$

$$\beta = -F_0'(v_\phi), \tag{10.102}$$

$$\gamma = -F_0''(v_\phi). \tag{10.103}$$

The resulting function has the desired shape, with a plateau at v_ϕ of width v_{trap}, as shown in Fig. 10.14.

We can now proceed with the calculation of the red-shift, by inserting this ad hoc distribution function in Eq. (10.96) with the expression for the nonlinear susceptibility:

$$\chi_{eNL} = -\frac{e^2}{k^2 m \varepsilon_0} \int_{-\infty}^{\infty} dv \frac{\partial_v [\beta F_1(v - v_\phi) + \gamma F_2(v - v_\phi)]}{v - v_\phi}. \tag{10.104}$$

Since $F_1'(x)/x$ is odd, as can be seen from deriving its expression from the definition of F_1 above, its integral is zero and it does not contribute to the red-shift. On the other hand, we have

$$\frac{F_2'(x)}{x} = \left[1 - \frac{x^2}{3v_{trap}^2}\right] e^{-x^2/2v_{trap}^2}, \tag{10.105}$$

which once inserted back into Eq. (10.96) gives the following expression for the asymptotic red-shift:

$$\frac{\delta\omega}{\omega} = \frac{\sqrt{2\pi}}{3} F_0''(v_\phi) v_{trap} v_\phi^2 . \qquad (10.106)$$

Since $\sqrt{2\pi}/3 \approx 0.836$, we see that this expression recovers that of Morales and O'Neil to within a factor of 2.[5]

While the change in the first-order derivative of the distribution function at v_ϕ does not contribute to the red-shift, it is responsible for the energy loss from the wave. In fact, we can show (see Problem 10.3) that the energy gained by the plasma electrons (via the distortion introduced by F_1) is approximately equal (up to a numerical factor of order one) to the energy lost by the wave via Landau damping during a bounce period, i.e.,

$$\int_{-\infty}^{\infty} \frac{1}{2} mv^2 \left[F_\infty(v) - F_0(v)\right] dv \propto -\nu_L \tau_B . \qquad (10.107)$$

The physical interpretation is clear: since τ_B is approximately the time it takes for the distribution to flatten and the damping to saturate, this relation is simply the consequence of energy conservation between the wave and the plasma particles.

To conclude on kinetic frequency shifts, we note that a similar analysis can be carried out for IAWs; here, a noticeable feature is that both electron and ion trapping can contribute to the frequency shift, with a scaling similar to Eq. (10.97). However, because for electrons, $v_\phi \ll v_{Te}$, trapping occurs in the bulk of the distribution, where the curvature of the distribution (i.e., its second derivative) is positive. This means that electron trapping in an IAW tends to blue-shift the wave, whereas ion trapping introduces a red-shift (because $f''(v) < 0$ for $v > v_T$ in a Maxwellian distribution). The relative contribution of these two effects depends on ZT_e/T_i, with the electron trapping-induced blue-shift dominating for $ZT_e/T_i > 10$ and the ion trapping-induced red-shift dominating for $ZT_e/T_i < 10$ [48]. The two contributions can also nearly cancel each other, in which case the kinetic inflation leads to rapid growth and saturation via other processes such as TID [53].

Nonlinear frequency shifts will typically lead to a bursty behavior for instabilities involving a driven plasma wave, as the wave keeps moving in and out of resonance based on its dephasing with respect to the ponderomotive driver [42].

Saturation levels of SBS consistent with saturation of the IAW growth by kinetic effects have been reported in several experiments [54–57]. Later on, a sharp transition from EPW saturation via LDI to saturation via kinetic inflation was observed by Kline et al. [15] following an increase in $k\lambda_{De}$. The (k, ω) spectrum of EPWs measured via Thomson scattering transitioned from discrete peaks corresponding to the satellites of the LDI cascade at a lower $k\lambda_{De}$ to a

---

[5] The definition of v_{trap} in Refs. [51, 52] differs from ours by a factor of 2.

single peak with frequency broadening and down-shift (and narrow k) at higher $k\lambda_{De}$, consistent with the nonlinear frequency shift from kinetic inflation. The same nonlinear frequency downshift from particle trapping was also observed for driven IAWs by Froula et al. [58], with a measured scaling of frequency shift vs. ion wave amplitude in agreement with Eq. (10.97).

10.3.3   Plasma Heating

We saw in the previous sections that particle trapping can lead to a flattening of the distribution function near a plasma wave's phase velocity and thus to kinetic inflation (i.e., enhanced growth of the plasma wave amplitude via a three-wave instability due to the reduction of Landau damping) but must compete with de-trapping processes such as side-loss and collisions.

Landau damping and wave–particle energy transfer can still occur even if de-trapping processes prevent significant flattening of the distribution function near the wave's phase velocity. The transfer of energy from the plasma wave to particles at velocities near the wave's phase velocity will typically result in the formation of a suprathermal distribution of particles in addition to the bulk distribution at the background temperature; this distribution is usually quasi-Maxwellian due to turbulence effects, with an effective temperature on the order of $\frac{1}{2}mv_\phi^2$ (cf. Sect. 8.4). The presence of this hot particle distribution will typically increase Landau damping and act to saturate the growth of the plasma wave.

Saturation of IAW growth from CBET was directly observed in a microwave experiment [59]; two microwave sources were used in a discharge plasma column in a counter-propagating geometry, with a tunable frequency shift between the two sources and a direct measurement of the IAW amplitude in the plasma. The IAW amplitude was observed to peak when the frequency shift matches the IAW frequency, and saturation of the IAW amplitude (at a few percent level) was directly measured and correlated with a measurement of the tail of the ion distribution, confirming the creation of a hot ion distribution with temperature on the order of $\frac{1}{2}M_i c_s^2$. Hot ion tail formation is also observed in kinetic simulation of driven IAWs [53].

On longer time scales, collisions will act to thermalize the suprathermal distribution; equilibration will restore a Maxwellian and increase its temperature, which can also increase the Landau damping and saturate the plasma wave growth. This process was observed in simulations of SBS [60] and CBET [61, 62] and later confirmed in CBET laser–plasma experiments where the ion temperature was measured with Thomson scattering [63].

Problems

10.1 Energy Conservation in 2D Nonlinear CBET
Derive the expression for $P_1(z_0, z_1)$. Show that $P_0(z_0, z_1) + P_1(z_0, z_1) = P_0(0, z_1) + P_1(z_0, 0)$. Interpret this result in terms of energy conservation in a limited sub-domain of the whole overlap region.

10.2 Simplification of the Wave Equations for LDI
Justify neglecting the nonlinear right-hand side terms from Eqs. (6.24) and (6.29) when establishing Eqs. (10.44)–(10.45). Hint: use dimensional analysis; e.g., since both sides of Eq. (10.44) have fast oscillations at $(\omega_1, \mathbf{k}_1)$, we have $\nabla^2 \sim k_1^2, \partial_t \nabla \sim \omega_1 \mathbf{k}_1$ for the right-hand side terms. Eliminate the δv_j variables to keep only δn_j; use $\delta v_j / v_{\phi j} = \delta n_j / n_{e0}$ with $v_{\phi 2} = c_s \ll v_{Te} \ll v_{\phi 0}, v_{\phi 1}$.

10.3 Wave–Particle Energy Conservation for a Flattened Distribution Function

- Using the ad hoc expression for the flattened distribution function from Eqs. (10.98)–(10.101), calculate the total energy gained by the plasma electrons while establishing the flattened distribution compared to the original Maxwellian distribution,

$$\Delta W_p = \int_{-\infty}^{\infty} \frac{1}{2} m v^2 \left[F_\infty(v) - F_0(v) \right] dv, \qquad (10.108)$$

and show that

$$\Delta W_p = -\sqrt{2\pi} m v_{trap}^3 \left[v_\phi F_0'(v_\phi) + \frac{v_{trap}^2}{3} F_0''(v_\phi) \right], \qquad (10.109)$$

where $v_\phi = \omega/k$ and $v_{trap} = 2\pi/k\tau_B = \sqrt{eE_0/km}$ (cf. Eq. (2.39)), with E_0 the peak amplitude of the wave electric field at $t = 0$.
- Show that the second term in the brackets in the expression above (proportional to $F_0''(v_\phi)$) is smaller than the first by a factor $\propto (v_\phi/v_{trap})$, assumed $\ll 1$ for the linear waves considered here.
- Now using the expression for Landau damping,

$$\nu_L = \frac{\pi}{2} \frac{\omega^3}{k^2} F_0'(v_\phi), \qquad (10.110)$$

where $\omega \approx \omega_{pe}$ is the wave frequency, and the wave energy density $\varepsilon_0 E_0^2/4$, calculate the energy lost by the wave during a bounce period τ_B and show that it is equal, up to a numerical factor $\sim O(1)$, to ΔW_p, the energy gained by the plasma electrons while going from the initial Maxwellian distribution $F_0(v)$ to the flattened distribution $F_\infty(v)$.

10.4 Coupled Mode Equations for EDI

Following the same procedure as with the other three-wave instabilities, derive the coupled mode equations for EDI, Eqs. (10.70)–(10.72), starting from the wave equations with second-order coupling terms, Eqs. (6.25), (6.21), and (6.29).

References

1. C.L. Tang, J. Appl. Phys. **37**, 2945 (1966)
2. E. Kur, M. Lazarow, J.S. Wurtele, P. Michel, Opt. Express **29**, 1162 (2021)
3. C.J. McKinstrie, J.S. Liu, R.E. Giacone, H.X. Vu, Phys. Plasmas **7**, 3 (1996)
4. T. Speziale, Phys. Fluids **27**, 2583 (1984)
5. D.F. DuBois, M.V. Goldman, Phys. Rev. Lett. **14**, 544 (1965)
6. S.J. Karttunen, Phys. Rev. A **23**, 2006 (1981)
7. J.A. Heikkinen, S.J. Karttunen, Phys. Fluids **29**, 1291 (1986)
8. R.P. Drake, S.H. Batha, Phys. Fluids B Plasma Phys. **3**, 2936 (1991)
9. K.L. Baker, R.P. Drake, B.S. Bauer, K.G. Estabrook, A.M. Rubenchik, C. Labaune, H.A. Baldis, N. Renard, S.D. Baton, E. Schifano, A. Michard, W. Seka, R.E. Bahr, Phys. Rev. Lett. **77**, 67 (1996)
10. C. Labaune, H.A. Baldis, B.S. Bauer, V.T. Tikhonchuk, G. Laval, Phys. Plasmas **5**, 234 (1998)
11. S. Depierreux, J. Fuchs, C. Labaune, A. Michard, H.A. Baldis, D. Pesme, S. Hüller, G. Laval, Phys. Rev. Lett. **84**, 2869 (2000)
12. D.F. DuBois, H.A. Rose, D. Russell, Phys. Rev. Lett. **66**, 1970 (1991)
13. S. Depierreux, C. Labaune, J. Fuchs, D. Pesme, V.T. Tikhonchuk, H.A. Baldis, Phys. Rev. Lett. **89**, 045001 (2002)
14. D.S. Montgomery, J.A. Cobble, J.C. Fernández, R.J. Focia, R.P. Johnson, N. Renard-LeGalloudec, H.A. Rose, D.A. Russell, Phys. Plasmas **9**, 2311 (2002)
15. J.L. Kline, D.S. Montgomery, B. Bezzerides, J.A. Cobble, D.F. DuBois, R.P. Johnson, H.A. Rose, L. Yin, H.X. Vu, Phys. Rev. Lett. **94**, 175003 (2005)
16. J.L. Kline, D.S. Montgomery, L. Yin, D.F. DuBois, B.J. Albright, B. Bezzerides, J.A. Cobble, E.S. Dodd, D.F. DuBois, J.C. FernÃndez, R.P. Johnson, J.M. Kindel, H.A. Rose, H.X. Vu, W. Daughton, Phys. Plasmas **13**, 055906 (2006)
17. D.S. Montgomery, Phys. Rev. Lett. **86**, 3686 (2001)
18. T. Ohnuma, Y. Hatta, J. Phys. Soc. Jpn. **29**, 1597 (1970)
19. L.P. Mix, L.N. Litzenberger, G. Bekefi, Phys. Fluids **15**, 2020 (1972)
20. S.J. Karttunen, J.N. McMullin, A.A. Offenberger, Phys. Fluids **24**, 447 (1981)
21. S. Karttunen, R. Salomaa, Phys. Lett. A **72**, 336 (1979)
22. H.C. Bandulet, C. Labaune, K. Lewis, S. Depierreux, Phys. Rev. Lett. **93**, 035002 (2004)
23. C. Niemann, S.H. Glenzer, J. Knight, L. Divol, E.A. Williams, G. Gregori, B.I. Cohen, C. Constantin, D.H. Froula, D.S. Montgomery, R.P. Johnson, Phys. Rev. Lett. **93**, 045004 (2004)
24. T. Chapman, S. Brunner, J.W. Banks, R.L. Berger, B.I. Cohen, E.A. Williams, Phys. Plasmas **21**, 042107 (2014)
25. B.I. Cohen, B.F. Lasinski, A.B. Langdon, E.A. Williams, Phys. Plasmas **4**, 956 (1997)
26. C. Riconda, S. Hüller, J. Myatt, D. Pesme, Physica Scripta **T84**, 217 (2000)
27. D. Pesme, C. Riconda, V.T. Tikhonchuk, Phys. Plasmas **12**, 092101 (2005)
28. R.N. Franklin, Rep. Progress Phys. **40**, 1369 (1977)
29. P.K. Shukla, M.Y. Yu, M. Mohan, R.K. Varma, K.H. Spatschek, Phys. Rev. A **27**, 552 (1983)
30. J.C. Fernández, J.A. Cobble, B.H. Failor, D.F. DuBois, D.S. Montgomery, H.A. Rose, H.X. Vu, B.H. Wilde, M.D. Wilke, R.E. Chrien, Phys. Rev. Lett. **77**, 2702 (1996)
31. S. Depierreux, C. Labaune, J. Fuchs, H.A. Baldis, Rev. Sci. Instrum. **71**, 3391 (2000)
32. K.L. Baker, K.G. Estabrook, R.P. Drake, B.B. Afeyan, Phys. Rev. Lett. **86**, 3787 (2001)

33. V. Silin, V. Tikhonchuk, ZhETF Pisma Redaktsiiu **34**, 385 (1981)
34. S. Karttunen, R. Salomaa, Phys. Lett. A **88**, 350 (1982)
35. B.J. Winjum, J. Fahlen, W.B. Mori, Phys. Plasmas **14**, 102104 (2007)
36. S. Hüller, P.E. Masson-Laborde, D. Pesme, M. Casanova, F. Detering, A. Maximov, Phys. Plasmas **13**, 022703 (2006)
37. C.J. Walsh, H.A. Baldis, Phys. Rev. Lett. **48**, 1483 (1982)
38. T. O'Neil, Phys. Fluids **8**, 2255 (1965)
39. G.J. Morales, T.M. O'Neil, Phys. Rev. Lett. **28**, 417 (1972)
40. B.I. Cohen, A.N. Kaufman, Phys. Fluids **20**, 1113 (1977)
41. J.C. Fernández, J.A. Cobble, D.S. Montgomery, M.D. Wilke, B.B. Afeyan, Phys. Plasmas **7**, 3743 (2000)
42. H.X. Vu, D.F. DuBois, B. Bezzerides, Phys. Rev. Lett. **86**, 4306 (2001)
43. H.X. Vu, D.F. DuBois, B. Bezzerides, Phys. Plasmas **14**, 012702 (2007)
44. N.B. Meezan, L.J. Atherton, D.A. Callahan, E.L. Dewald, S. Dixit, E.G. Dzenitis, M.J. Edwards, C.A. Haynam, D.E. Hinkel, O.S. Jones, O. Landen, R.A. London, P.A. Michel, J.D. Moody, J.L. Milovich, M.B. Schneider, C.A. Thomas, R.P.J. Town, A.L. Warrick, S.V. Weber, K. Widmann, S.H. Glenzer, L.J. Suter, B.J. MacGowan, J.L. Kline, G.A. Kyrala, A. Nikroo, Phys. Plasmas **17**, 056304 (2010)
45. D.J. Strozzi, E.A. Williams, H.A. Rose, D.E. Hinkel, A.B. Langdon, J.W. Banks, Phys. Plasmas **19**, 112306 (2012)
46. D.S. Montgomery, R.J. Focia, H.A. Rose, D.A. Russell, J.A. Cobble, J.C. Fernández, R.P. Johnson, Phys. Rev. Lett. **87**, 155001 (2001)
47. H.A. Rose, D.A. Russell, Phys. Plasmas **8**, 4784 (2001)
48. R.L. Berger, S. Brunner, T. Chapman, L. Divol, C.H. Still, E.J. Valeo, Phys. Plasmas **20**, 032107 (2013)
49. S. Brunner, "Advanced theory of plasmas," Lecture Notes, École Polytechnique fédérale de Lausanne, https://crppwww.epfl.ch/brunner/NonLinear.pdf
50. R.L. Dewar, Phys. Fluids **15**, 712 (1972)
51. E.A. Williams, B.I. Cohen, L. Divol, M.R. Dorr, J.A. Hittinger, D.E. Hinkel, A.B. Langdon, R.K. Kirkwood, D.H. Froula, S.H. Glenzer, Phys. Plasmas **11**, 231 (2004)
52. L. Divol, E.A. Williams, B.I. Cohen, A.B. Langdon, B.F. Lasinski, in *Proceedings of the 3rd International Conference on Inertial Fusion Sciences Applications* (2003) p. TU07.2, https://www.osti.gov/servlets/purl/15005092
53. C. Riconda, A. Heron, D. Pesme, S. Huller, V.T. Tikhonchuk, F. Detering, Phys. Plasmas **12**, 112308 (2005)
54. M.J. Herbst, C.E. Clayton, F.F. Chen, Phys. Rev. Lett. **43**, 1591 (1979)
55. C.E. Clayton, C. Joshi, F.F. Chen, Phys. Rev. Lett. **51**, 1656 (1983)
56. J.E. Bernard, J. Meyer, Phys. Rev. Lett. **55**, 79 (1985)
57. S.H. Glenzer, L.M. Divol, R.L. Berger, C. Geddes, R.K. Kirkwood, J.D. Moody, E.A. Williams, P.E. Young, Phys. Rev. Lett. **86**, 2565 (2001)
58. D.H. Froula, L. Divol, A.A. Offenberger, N. Meezan, T. Ao, G. Gregori, C. Niemann, D. Price, C.A. Smith, S.H. Glenzer, Phys. Rev. Lett. **93**, 035001 (2004)
59. C.J. Pawley, H.E. Huey, N.C. Luhmann, Phys. Rev. Lett. **49**, 877 (1982)
60. P.W. Rambo, S.C. Wilks, W.L. Kruer, Phys. Rev. Lett. **79**, 83 (1997)
61. P. Michel, W. Rozmus, E.A. Williams, L. Divol, R.L. Berger, R.P.J. Town, S.H. Glenzer, D.A. Callahan, Phys. Rev. Lett. **109**, 195004 (2012)
62. P. Michel, W. Rozmus, E.A. Williams, L. Divol, R.L. Berger, S.H. Glenzer, D.A. Callahan, Phys. Plasmas **20**, 056308 (2013)
63. A.M. Hansen, K.L. Nguyen, D. Turnbull, B.J. Albright, R.K. Follett, R. Huff, J. Katz, D. Mastrosimone, A.L. Milder, L. Yin, J.P. Palastro, D.H. Froula, Phys. Rev. Lett. **126**, 075002 (2021)

Appendix A
Formulary

A.1 Plasma Parameters

$$\omega_{pe} = \sqrt{\frac{n_e e^2}{m \varepsilon_0}} \approx 5.641 \times 10^4 \sqrt{n_e [\text{cm}^{-3}]} \ [\text{s}^{-1}] \tag{A.1}$$

$$\omega_{pi} = \sqrt{\frac{n_i Z^2 e^2}{M_i \varepsilon_0}} \approx 1.316 \times 10^3 \sqrt{\frac{Z^2 n_i [\text{cm}^{-3}]}{A}} \ [\text{s}^{-1}] \tag{A.2}$$

$$\text{Note: if } n_e = Z n_i, \ \omega_{pi} = \omega_{pe} \sqrt{\frac{Zm}{M_i}} \tag{A.3}$$

$$v_{Te} = \sqrt{\frac{T_e}{m}} \approx 1.326 \times 10^7 \sqrt{T_e [\text{keV}]} \ [\text{m/s}] \tag{A.4}$$

$$v_{Ti} = \sqrt{\frac{T_i}{M_i}} \approx 3.1 \times 10^5 \sqrt{\frac{T_i [\text{keV}]}{A}} \ [\text{m/s}] \tag{A.5}$$

$$\lambda_{De} = \frac{v_{Te}}{\omega_{pe}} \approx 235 \sqrt{\frac{T_e [\text{keV}]}{n_e [\text{cm}^{-3}]}} \ [\text{m}] \tag{A.6}$$

$$\lambda_{Di} = \frac{v_{Ti}}{\omega_{pi}} = \sqrt{\frac{T_i}{Z T_e}} \lambda_{De} \tag{A.7}$$

© The Author(s), under exclusive license to Springer Nature Switzerland AG 2023
P. Michel, *Introduction to Laser-Plasma Interactions*, Graduate Texts in Physics,
https://doi.org/10.1007/978-3-031-23424-8

$$N_{De} = n_e \lambda_{De}^3 \approx 1.3 \times 10^{13} \sqrt{\frac{T_e^3[\text{keV}]}{n_e[\text{cm}^{-3}]}} \tag{A.8}$$

$$\nu_{ei} = \frac{Z}{3(2\pi)^{3/2}} \frac{\omega_{pe}}{N_{De}} \approx 9.2 \frac{Z \ln(\Lambda) n_e [10^{20} \text{cm}^{-3}]}{T_e^{3/2}[\text{keV}]} \; [\text{ns}^{-1}]. \tag{A.9}$$

A.2   Waves in Plasmas

A.2.1   EMWs

$$\mathbf{E} = -\partial_t \mathbf{A} = -\frac{mc}{e} \partial_t \mathbf{a} \tag{A.10}$$

$$I = \frac{\varepsilon_0 c E_0^2}{2} \tag{A.11}$$

$$a_0 = \frac{eE_0}{mc\omega_0} \approx 0.855 \times 10^{-9} \sqrt{I[\text{W/cm}^2] \lambda_0^2[\mu\text{m}]} \tag{A.12}$$

$$n_c = \frac{m\varepsilon_0 \omega_0^2}{e^2} \approx \frac{1.115 \times 10^{21}}{\lambda_0^2[\mu\text{m}]} \; [\text{cm}^{-3}] \tag{A.13}$$

$$\text{Note: } \frac{\lambda_0}{\lambda_{De}} = 2\pi \sqrt{\frac{mc^2 n_e/n_c}{T_e}} \approx 142 \sqrt{\frac{n_e/n_c}{T_e[\text{keV}]}} \tag{A.14}$$

$$v_\phi = \frac{c}{n} \tag{A.15}$$

$$v_g = cn \tag{A.16}$$

$$n = \sqrt{1 - \frac{n_e}{n_c}}. \tag{A.17}$$

A.2.2   EPWs

$$\Phi = -\frac{m}{e} \frac{\omega_{pe}^2}{k^2} \frac{\delta n_e}{n_e} = -\frac{m}{e} \frac{v_\phi^2}{1 + 3k^2 \lambda_{De}^2} \frac{\delta n_e}{n_e} \tag{A.18}$$

$$\delta \mathbf{v}_e = -i \frac{e \omega \mathbf{E}}{m \omega_{pe}^2} = (1 + 3k^2 \lambda_{De}^2) \mathbf{v}_{os,e} \ , \ \mathbf{v}_{os,e} = -ie\mathbf{E}/m\omega \qquad (A.19)$$

$$\delta \mathbf{v}_e = -\frac{e \mathbf{v}_\phi}{m v_\phi^2} (1 + 3k^2 \lambda_{De}^2) \Phi \qquad (A.20)$$

$$\delta \mathbf{v}_e = \mathbf{v}_\phi \frac{\delta n_e}{n_e} \qquad (A.21)$$

$$v_\phi = v_{Te} \frac{\sqrt{1 + 3k^2 \lambda_{De}^2}}{k \lambda_{De}} \qquad (A.22)$$

$$v_g = 3 v_{Te} \frac{k \lambda_{De}}{\sqrt{1 + 3k^2 \lambda_{De}^2}}. \qquad (A.23)$$

A.2.3   IAWs

$$\frac{\delta n_i}{n_{i0}} = \frac{\delta n_e}{n_{e0}} (1 + k^2 \lambda_{De}^2) \qquad (A.24)$$

$$\delta \mathbf{v}_i = (1 + k^2 \lambda_{De}^2) \delta \mathbf{v}_e \qquad (A.25)$$

$$c_s = \sqrt{\frac{Z T_e + 3 T_i}{M_i}} \approx 3.1 \times 10^5 \sqrt{\frac{Z T_e + 3 T_i}{A}} \text{[keV] [m/s]} \qquad (A.26)$$

$$\text{Note: } \omega_{pi} = \frac{c_s}{\lambda_{De}} \left(1 + \frac{3 T_i}{Z T_e} \right)^{-1/2} \qquad (A.27)$$

$$\Phi = \frac{m}{e} v_{Te}^2 \frac{\delta n_e}{n_e} = \frac{T_e}{e} \frac{\delta n_e}{n_e} = \frac{m v_{Te}^2}{e} \frac{\delta v_e}{v_\phi} \qquad (A.28)$$

$$\frac{\delta n_e}{n_{e0}} = \frac{\delta v_e}{v_\phi} = \frac{\delta v_e}{c_s} \sqrt{1 + k^2 \lambda_{De}^2} \qquad (A.29)$$

$$v_\phi = \frac{c_s}{\sqrt{1 + k^2 \lambda_{De}^2}} \qquad (A.30)$$

$$v_g = \frac{c_s}{(1 + k^2 \lambda_{De}^2)^{3/2}}. \qquad (A.31)$$

Multi-ion species plasmas:

- Fast mode phase velocity:

$$v_{fast}^2 = \frac{Z^* T_e}{M_i(1 + k^2\lambda_{De}^2)} \ , \ Z^* = \frac{\langle Z^2 \rangle}{\langle Z \rangle}.$$ (A.32)

- Slow mode: (if present)

$$v_{slow}^2 = 3v_{Th}^2 + f_h \frac{Z_h^2}{M_h \langle Z \rangle} \frac{T_e}{1 + k^2\lambda_{De}^2 + f_l Z_l^2 T_e/(\langle Z \rangle T_i)}$$ (A.33)

with the subscripts h and l referring to the heavy and light ions, respectively, and f_h and f_l the atomic fractions of heavy and light ions.

A.3   Plasma Dispersion Function

$$Z(\zeta) = \frac{1}{\sqrt{\pi}} \int_{-\infty}^{\infty} \frac{e^{-t^2}}{t - \zeta} dt.$$ (A.34)

- $\zeta \ll 1$:

$$Z(\zeta) \approx i\sqrt{\pi}e^{-\zeta^2} - 2\zeta + O(\zeta^3)$$ (A.35)

$$Z'(\zeta) \approx -2 - 2i\sqrt{\pi}\zeta + 4\zeta^2 + O(\zeta^3).$$ (A.36)

- $\zeta \gg 1 \ (\zeta \in \mathbb{R})$:

$$Z(\zeta) \approx i\sqrt{\pi}e^{-\zeta^2} - \frac{1}{\zeta} - \frac{1}{2\zeta^3} + O\left(\frac{1}{\zeta^3}\right)$$ (A.37)

$$Z'(\zeta) \approx \frac{1}{\zeta^2} + \frac{3}{2}\frac{1}{\zeta^4} + O\left(\frac{1}{\zeta^6}\right).$$ (A.38)

A.4   Landau Damping

- EPWs:

$$\frac{v_{eL}}{\omega_{pe}} = \sqrt{\frac{\pi}{8}} (k\lambda_{De})^{-3} \exp\left[-\frac{1}{2k^2\lambda_{De}^2} - \frac{3}{2}\right]. \tag{A.39}$$

- IAWs:

$$\frac{v_{iL}}{\omega} = \sqrt{\frac{\pi}{8}} \frac{\omega_R^3}{k^3 v_{Ti}^3} \exp\left[-\frac{\omega^2}{2k^2 v_{Ti}^2}\right] \tag{A.40}$$

$$\approx \sqrt{\frac{\pi}{8}} \left(\frac{ZT_e}{T_i}\right)^{3/2} \exp\left[-\frac{ZT_e}{2T_i}\right] , \quad ZT_e \gg T_i. \tag{A.41}$$

Index

Printed in the United States
by Baker & Taylor Publisher Services